Experiments in Modern Physics

EXPERIMENTS IN MODERN PHYSICS

BY

Adrian C. Melissinos

ASSOCIATE PROFESSOR OF PHYSICS
UNIVERSITY OF ROCHESTER

ACADEMIC PRESS · New York and London

ACADEMIC PRESS, INC.
111 Fifth Avenue, New York, New York 10003

United Kingdom Edition published by
ACADEMIC PRESS, INC. (LONDON) LTD.
Berkeley Square House, London W1X 6BA

LIBRARY OF CONGRESS CATALOG CARD NUMBER: 65-26404

Seventh Printing, 1972

PRINTED IN THE UNITED STATES OF AMERICA

Εἰς μνήμην τοῦ πατρός μου

Κωνσταντίνου Ι. Μελισσινοῦ

PREFACE

It is generally accepted that training in the sciences, especially at the undergraduate level, is not complete without a fair amount of laboratory experience. This is particularly true in physics where the basic freshman and sophomore courses are supplemented by concurrent laboratory exercises.

At the junior and senior level, however, laboratory training becomes more important and forms the subject of an independent course. Rather than simple laboratory exercises, the students now perform complete experiments and one could list the aims of the course as follows:

(a) To teach the student the methods and procedures of experimental physics at an advanced level; and to give him confidence in his own ability to measure physical entities and relationships between them.

(b) To familiarize the student with modern research equipment and its use; also to make him aware of the most basic techniques presently used in widely varying fields of physics.

(c) To convince the student that the material he studied and covered in his lecture courses can indeed be tested experimentally; and to give him the satisfaction of doing so himself.

On the other hand the real professional training for students who will become experimental physicists takes place in graduate school during their thesis work; this is a period of intensive involvement in research but within a highly specialized field. It therefore appears that the best opportunity for a broad look at the general experimental methods of physics still remains in the junior and senior laboratory courses.

The present text is an outgrowth of such a laboratory course given by the author at the University of Rochester between 1959 and 1963. It consisted of a one-year course with two 3-hour meetings in the laboratory and two 1-hour lecture meetings weekly; the students had access to the laboratory at all times and, in general, worked during hours of their own choice well in excess of the scheduled periods. The students worked in pairs, which in most cases provides a highly motivating and successful relationship.

The material included in this course was selected from those experiments in atomic and nuclear physics that have laid the foundation and provided the evidence for modern quantum theory. The experiments were set up in such a fashion that they could be completed in a two- to four-week period of normal work taking into account the other demands on the student's time. A frequent tendency of students (especially the more enthusiastic ones) is to become involved in experiments that are "almost original" or in setting up new experiments; this, however, requires construction of their own equipment and can result in considerable "gadgeteering" as well as leading to extended involvement, which a senior cannot afford. We found this to be a common trap eventually leading to frustration and discouragement with a student having only a "progress report" or a marginal result to show for one term of work.

For these reasons we used, whenever possible, commercial equipment, and all experiments were carefully tested before being handed over to the student. The emphasis was on the "physics" of the experiment and the interpretation of the results obtained; clearly, to obtain correct results the student had to properly adjust, use, and understand his equipment. Furthermore, a time limit could be set so that eight to ten different experiments could be completed in one academic year. This variety not only brings the student in contact with a broader segment of physics and of techniques, it also gives him the opportunity of a "fresh start" several times throughout the course; and, most important, it keeps the student continuously interested in spite of any setback or difficulty he may encounter in one or more experiments.

The experiments described in the first four chapters of this text are, in general, easier than the ones discussed later; each can usually be completed in a one-week period, and at the University of Rochester are performed in the second term of the junior year. This leaves then the two terms of the senior year for the more advanced experiments described in the later chapters. The various experiments have been grouped according to the basic physical principle rather than the special technique. For each experiment the underlying theoretical ideas are first introduced, then the experimental apparatus is described in considerable detail and, finally, the results obtained by the students are given and discussed. In this respect we believe that this text is not a "laboratory manual"; instead we have aimed at a fairly coherent presentation of experimental physics in spite of the limited and occasionally random selection of the experiments. We feel that our approach is similar to that of G. P. Harnwell and J. J. Livingood in their classic text "Experimental Atomic Physics" which appeared originally in 1933.

The reader may occasionally be surprised by the great detail with which we describe apparatus or special procedures for analysis of data. We have done so to assist those who may wish to set up a similar laboratory and because these are the details the student has usually to find out by himself; but also we believe that only through such detail can one acquire the real flavor of experimental physics. We have placed special emphasis on numerical results and on simple calculations emphasizing the use of the correct units.

Contrary to accepted practice we have included only a minimum number of references; instead, we have given a selected bibliography to each subject through which the interested reader may find all pertinent information. It is, however, expected that the student is familiar or is concurrently taking a course on modern physics. The usual mathematical level of calculus is considered as a prerequisite and is freely used throughout.

As mentioned before, modern commercial equipment is used whenever practicable; this is the same type of equipment as used in present-day research and frequently is the basis for a successful teaching laboratory. It is true, however, that similar equipment can be obtained from several manufacturers and that special apparatus is preferably built in one's own shop. We do have on file the prints of all such special equipment and we will be glad to supply them on request.

The list of experiments in this text is not complete. For example, we have not included a discussion of "coherent scattering" (diffraction) experiments, of "electromagnetic spectrometers," and of "visual techniques" (bubble chamber, spark chamber, and nuclear emulsion) in spite of their successful performance by several students. We hope to be able to remedy these omissions in a future edition. We also realize that in some cases a better, or more educational, technique might be available for the experiments presented here. We would be grateful to our readers if they wish to indicate to us these alternatives.

In line with our original intention, all the data and results presented in this book were obtained by students of the "Senior Laboratory" of the University of Rochester and the appropriate credit is given in the text. Exceptions to this rule are the data of Figs. 5.43, 5.44, and 9.28–9.31 that were obtained at Brookhaven National Laboratory; also the data of Figs. 7.27, 7.30, 7.35, 7.36, and 8.35 were obtained at Massachusetts Institute of Technology. The results presented here could not have been achieved without the support of the Physics Department of the University of Rochester; also major equipment was purchased through a grant from the United States Atomic Energy Commission and a matching funds grant from the National Science Foundation. As is always the case, whatever success this laboratory did enjoy is due to the combined efforts of many individuals, a large part of which was supplied by the participating students. It is a special pleasure to thank from here the graduate assistants during the 1959–1963 period, Drs. E. Griffin, J. Robbins, J. Mochel, and J. Reed for their contributions to the laboratory. More than to anyone else the laboratory is indebted to Mr. F. L. Reynolds who has been in charge of all technical matters and has kept the equipment in operating condition; I wish to express to him my personal gratitude for his friendship and for many helpful suggestions connected with this text. I also wish to acknowledge discussions with many of my colleagues in Rochester and, in particular, Dr. W. P. Alford, Dr. M. F. Kaplon, and Dr. R. E. Marshak.

In the preparation of the manuscript I benefited from the art work of Messrs. Yu-Chang Lee, W. Stinson, and J. Pinero; most of the manuscript was typed by Mrs. B. M. Marsh, and to all of them I express my appreciation for their excellent work. I am also indebted to the following of my colleagues for reading early parts of the manuscript and making many valuable suggestions and corrections: Dr. P. Baumeister on Chapter 2; Dr. T. Castner on Chapter 3; Dr. D. Cline on Chapter 5; Dr. R. Ellsworth on Chapter 6; Dr. L. Bradley on Chapter 7; Mr. C. Cook on Chapter 8; and Dr. J. Reed on Chapter 9. Still, however, the responsibility for all errors is mine and I would appreciate it if the readers could indicate them to me. Finally, I would like to thank my wife, Joyce, for her encouragement and assistance during the course of this work.

A. C. M.

Rochester, New York

ACKNOWLEDGMENTS

We wish to thank the following organizations for their permission to use the respective material: E. Leybold's Nachfolger for use of Fig. 1.14; General Electric Company for use of Fig. 3.11 and Table 3.1; Computer Measurements Corporation for use of Fig. 4.6; Tektronix Inc. for use of Fig. 4.11; Keithley Instruments Inc. for use of Figs. 4.12, 4.13, and 4.14; Consolidated Vacuum Corporation for use of Figs. 4.16 and 4.17; the Lawrence Radiation Laboratory for use of Tables 4.5, 4.6, and 4.7; John Wiley and Sons, Inc. for use of Table 5.3; Radiation Counter Laboratories for use of Fig. 5.19; Harshaw Chemical Corporation for use of Fig. 6.12; Laboratory for Electronics Inc. for use of Fig. 8.13; Brookhaven National Laboratory for use of Figs. 9.5 and 9.16; Drs. R. Ellsworth and M. Tannenbaum for Figs. 9.28 and 9.29. Figures 8.22, 8.23, and 8.24 are based on drawings from the U. S. Army Manual for TS/13-AP.

CONTENTS

1

EXPERIMENTS ON QUANTIZATION

1. Introduction

A prominent characteristic of present-day physics is that many of the quantities used to describe physical phenomena are quantized. That is, such quantities cannot take any one of a continuum of values, but are restricted to a set (perhaps an infinite set) of discrete values. Common examples are the intensity of radiation of the electromagnetic field, the energy of atomic systems, and the electric charge. Strong evidence for such quantization is obtained from the three experiments that will be described in this chapter:

(a) Millikan's experiment by which the charge on individual oil droplets is measured. The experiment shows that the charge is always an integer multiple of the smallest charge observed; this is identified with the charge of the electron.

(b) The Frank–Hertz experiment on the excitation by electron bombardment of atomic vapors. It is found that only for discrete bombarding energies is such excitation possible, and the first excited state of the mercury (Hg) atom is thus measured.

(c) The photoelectric effect, in which the emission of electrons from a cesium (Cs) surface under the influence of electromagnetic radiation in the visible spectrum is studied. It is then seen that in this process each individual electron absorbs from the radiation field (which has a frequency ν) energy $h\nu$, where h is Planck's constant.

2. The Millikan Oil Drop Experiment

2.1 GENERAL

In 1909, R. Millikan reported a reliable method for measuring ionic charges. It consists of observing the motion of small oil droplets under the influence of an electric field. Usually the drops acquire a few electron charges and thus conventional fields impart to them velocities that permit isolation of a drop and continuous observation for a considerable length of time; further, the mass of the oil droplet remains almost constant (there is very slight evaporation) during these long observation times.

In principle, if we measure the force due to the electric field E,

$$F_e = qE = neE \tag{2.1}$$

we can obtain ne; repeating this measurement for several (or the same) drops but with different values of the integer n, we can extract the charge of the electron e.

The electric force can be measured either by a null method—that is, by balancing the drop against the gravitational force—or as will be described here, by observing the motion of the drop under the influence of both forces. Oil droplets in air, acted on by a constant force, soon reach a terminal velocity given by Stokes' law,

$$F = 6\pi a\eta v \tag{2.2}$$

where a is the radius of the (assumed spherical) droplet, η the viscosity of the oil, and v the terminal velocity. To obtain the radius of the drop (needed in Eq. 2.2) we observe the free fall of the drop; the gravitational force is

$$F_g = \tfrac{4}{3}\pi a^3(\rho - \sigma)g \tag{2.3}$$

with ρ and σ the density of air and oil and g the acceleration of gravity.

Schematically, as shown in Fig. 1.1, the apparatus consists of two parallel plates which can be alternatively charged to a constant potential $+V$, $-V$, or 0. The drop is then observed (with a telescope) and the time t it takes to travel through a distance d is measured. Let F_+ denote the force on a negatively charged drop with electric field up (time t_+; electric force aiding gravity) and F_- the force with electric field down (time t_-; electric

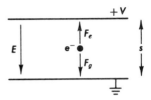

F_{IG.} 1.1   Forces on a charged oil drop between the plates of the Millikan apparatus.

force opposing gravity). Then

$$F_{\pm} = \pm ne(V/s) - \tfrac{4}{3}\pi a^3(\rho - \sigma)g = 6\pi a\eta d(1/t_{\pm}^{(n)})$$

$$F_0 = -\tfrac{4}{3}\pi a^3(\rho - \sigma)g = 6\pi a\eta d(t_0)^{-1}$$

(2.4)

where the sign conventions hold if t is considered >0 when the drop moves up, $t < 0$ when it is moving down. A convenient method of analysis is to write Eq. 2.4 as

$$\frac{1}{t_{\pm}^{(n)}} = \pm An - B \qquad A = \frac{Ve}{6s\pi a\eta d}$$

(2.5)

$$\frac{1}{t_0} = -B \qquad B = \frac{2}{9}\frac{a^2(\rho - \sigma)g}{\eta d}$$

so that A and B can be easily determined.†

Indeed, a plot of $1/t_{\pm}^{(n)}$ against n reveals the linear relationship and the fact that only *integer* values of n appear, proving that the drop has acquired one, two, three, or more electric charges of value e, and never a fraction of that value. Thus we have clear evidence that the ionic charge picked up by the oil drops is *quantized*. Furthermore, the absolute value of this minimal electric charge is in good agreement with inferred measurements of the charge carried by the atomic electrons,‡ and therefore is accepted as the most accurate value of the charge of the electron.

2.2 T_{HE} E_{XPERIMENT}

The apparatus used in this laboratory (Fig. 1.2) consists of two parallel brass plates $\tfrac{1}{4}$ in. thick and approximately 2 in. in diameter, placed in a lucite cylinder held apart by three ceramic spacers 4.7 mm long. This assembly is in turn enclosed in a cylindrical brass housing with provisions for electrical connections and containing two windows, one for illumination of the drops and one for observation. The top plate has a small hole in its

---

† These expressions are in cgs units so that V must be expressed in statvolts if e is to be obtained in esu.

‡ As in e/m experiments, shot noise measurements, etc.

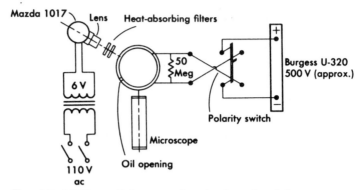

FIG. 1.2   Millikan oil drop experiment schematic of the apparatus.

center for the admission of the oil drops, which are produced by spraying oil with a regular atomizer.

To charge the plates, a 500-V battery and a reversing switch are used; the plates are shunted by a 50-megohm resistor to prevent them from remaining charged when the switch is open. For observations a 10-cm focal length microscope is used (Cenco 72925), while illumination is provided by a Mazda 1017-watt lamp and condensing lens. To avoid convection currents inside the apparatus, a heat-absorbing filter (Corning infrared-absorbing) is placed in the illuminating beam.

The plates should be made perpendicular to the gravitational field by means of the three leveling screws at the base of the apparatus and a level placed on the top plate. Being a cosine error, the deviation introduced by an angular displacement of the gravitational component from perpendicular by 8 degrees is 1 percent. A value for the plate spacing s may be obtained by using the stage micrometer. The micrometer should be focused on a wire inserted in the oil hole in the center of the top plate, and the crosshair of the micrometer should be moved along the length of the wire. Several measurements should be taken and their results averaged.

The velocities are determined by measuring with a stop watch the time required for the droplet to cover a specified number of divisions of the microscope scale. Care must be taken to avoid drafts and vibrations in the vicinity of the apparatus; for that reason and because of Brownian motion, the drop may wander or be displaced out of the field of the microscope. It may then be necessary to *reposition* the microscope between measurements on a single drop. Moreover, the drop should be kept in focus to avoid parallax errors.

Both the microscope and the light source may be adjusted by viewing a small wire inserted in the oil hole. The light should be adjusted so that the focal point is somewhat ahead or behind the wire and the wire is more or less evenly illuminated. To light the scale, a small light is placed next to the slit just ahead of the eyepiece of the microscope. The actual distance to which a scale division corresponds may be found by using a microscope slide on which a subdivided millimeter scale has been scratched.† The eyepiece focus of the microscope should not be changed during a run, since moving the eyepiece changes the effective distance of the scale. (To bring the drop back into focus the entire microscope should be moved.)

It is important to be *sparing* in the amount of oil sprayed into the chamber. In addition to gumming up the interior more quickly, large quantities create so many particles in the microscope field that without excessive eyestrain it is virtually impossible to single out and follow a single droplet.

Under the influence of gravity, droplets will fall at various limiting speeds. If the plates are charged, some of the drops will move down more rapidly, whereas others will reverse their direction of motion since in the process of spraying some drops become positively charged and others negatively charged. By concentrating on one drop which can be controlled by the field, and manipulating the sign of the electric field so that this particular drop is retained, it is possible to remove all other drops from the field. The limiting velocity is reached very quickly and the measurement should be started near the top or bottom of the plate. Measurement should be completed before the drop has reached a point in its travel where application of the reverse potential is insufficient to save the drop from being "gobbled up."

The density in air of the oil used was‡ 0.883 ± 0.003 gm/cm^3. It is desirable to take measurements in the shortest possible time since, as previously mentioned, the mass of the drop changes through evaporation.

It is obviously desirable to make measurements on as many different charges on the same or different drops as possible. Thus after four or five measurements of $t_+^{(n)}$, $t_-^{(n)}$, and t_0 have been taken, the charge on the drop must be changed; this is accomplished by bringing close to one of the windows a radioactive source (10 to 100 μCi of Co^{60} will do).§ The droplet should be brought close to the top plate and allowed to fall with the *field off*; on its way down it will sweep up a few ions created by the source. This can be checked by occasionally turning the field on to see if the charge has changed; rarely will a drop pick up any charge when the field is on.

The battery voltage should be checked with a 1-percent resistor divider and potentiometer; microscope calibration should be checked before and

† Note that the focal length of the microscope must not be changed, but instead the slide should be brought into the focal plane.

‡ This may be found by a simple measurement.

§ Ci ≡ curie.

after the measurements; the same holds true for air temperature and pressure, which are needed for a correction to Stokes' law.

Indeed, when the diameter of the drop is comparable to the mean free path in air, the viscosity η in Eq. (2.2) should be replaced by

$$\eta(T) = \eta_0(T)\left[1 + \frac{b}{aP}\right]^{-1} \tag{2.6}$$

where $\eta_0(T)$ is the viscosity of air as a function of T (Fig. 1.3), $b = 6.17 \times 10^{-4}$, P is the air pressure in centimeters of mercury, and a is the radius of the drop in centimeters (of the order of 10^{-4} cm). In analyzing the data it is convenient to calculate a_0 by letting $\eta = \eta_0(T)$ in Eq. 2.5; a_0 is then inserted in Eq. 2.6 to obtain $\eta(T)$ and thus a more accurate value for a.

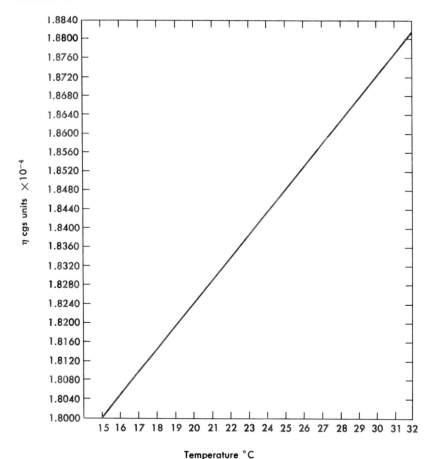

FIG. 1.3   Viscosity of dry air as a function of temperature.

TABLE 1.1
DATA ON MILLIKAN OIL DROP EXPERIMENT

	$\bar{t}_0$	$\bar{t}_-^{(n)}$	$\bar{t}_+^{(n)}$	$(1/t_+)-(1/t_0)$	$(1/t_-)-(1/t_0)$	n	$\dfrac{(1/t_+)-(1/t_-)}{2n}$
Drop	−27.9	+8.69	−5.65	−0.142	+0.150	1	−0.146
1	−29.6	+1.36	−1.18	−0.813	+0.770	5	−0.158
	−28.2	+3.66	−3.00	−0.299	+0.308	2	−0.152
	−29.3	+0.75	−0.716	−1.362	+1.368	9	−0.152
	−29.4	+2.35	−1.97	−0.473	+0.460	3	−0.155

$1/t_0 = -0.0346$

 0.1535

$$a = 4.91 \times 10^{-5} \text{ cm} \qquad \eta = 1.60 \times 10^{-4}$$

	$\bar{t}_0$	$\bar{t}_-^{(n)}$	$\bar{t}_+^{(n)}$	$(1/t_+)-(1/t_0)$	$(1/t_-)-(1/t_0)$	n	$\dfrac{(1/t_+)-(1/t_-)}{2n}$
Drop	−24.22	+3.98	−3.071	−0.285	+0.291	2	−0.144
2	−25.75	+9.73	−5.65	−0.137	+0.143	1	−0.140
	−25.4	+2.5	−2.12	−0.432	+0.440	3	−0.145
	−25.22	+9.67	−5.42	−0.145	+0.143	1	−0.144
	−25.22	+4.1	−3.07	−0.286	+0.288	2	−0.143
	−24.4	+1.73	−1.73	−0.538	+0.618	4	−0.144
	−24.4	+9.95	−6.02	−0.126	+0.141	1	−0.133

$1/t_0 = -0.0400$

 0.1433

$$a = 5.06 \times 10^{-5} \text{ cm} \qquad \eta = 1.59 \times 10^{-4}$$

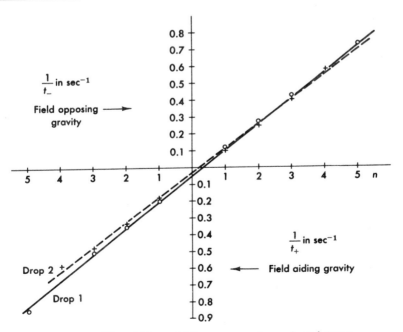

FIG. 1.4 Plot of $1/t_+$ and $1/t_-$ versus n where n is an integer.

2.3 ANALYSIS OF THE DATA

Table 1.1 is a sample of data obtained by a student.† Two drops were used and several charges were measured; for each charge six measurements were performed and averaged as shown in Table 1.1.

The pertinent parameters for these data were

Distance of fall	$d = 7.63 \times 10^{-2}$ cm
Temperature	$T = 25°$ C
Pressure	$P = 76.01$ cm Hg
Density	$\rho' = (\rho - \sigma) = 8.82 \times 10^{-1}$ gm/cm³
Potential	$V = 500$ V $= 1.666...$ statvolts
Plate separation	$s = 4.71 \times 10^{-1}$ cm

A plot of the data and the linear least squares fit are shown in Fig. 1.4. From the least squares fit we obtain (see Eq. 2.5)

$$A_1 = 0.1533 \pm 0.0011 \qquad B_1 = -0.0380 \pm 0.0053$$

$$e = (4.90 \pm 0.1) \times 10^{-10} \text{ esu}$$

$$A_2 = 0.1439 \pm 0.0017 \qquad B_2 = -0.0402 \pm 0.0038$$

$$e = (4.69 \pm 0.1) \times 10^{-10} \text{ esu}$$

where the values of e are calculated‡ from A; they are in good agreement with the accepted value

$$e = 4.803 \times 10^{-10} \text{ esu}$$

3. The Frank-Hertz Experiment

3.1 GENERAL

From the early spectroscopic work it was clear that atoms emitted radiation at discrete frequencies; from Bohr's model the frequency of the radiation ν is related to the change in energy levels through $\Delta E = h\nu$. Further experiments demonstrated that the absorption of radiation by atomic vapors also occurred only for discrete frequencies.

It is then to be expected that transfer of energy to atomic electrons by any mechanism should always be in discrete amounts§ and related to the atomic spectrum through the equation given above. One such mechanism of energy transfer is through the inelastic scattering of electrons from the

---

† D. Peters, class of 1962.

‡ It is seen that in this special case (partly because of the low voltage) the diameter of the drops is so small that the correction to the Stokes equation is considerable (15 percent).

§ They are still bound after the process.

entire atom. If the atom that is bombarded does not become ionized, and since little energy is needed for momentum balance, almost the entire kinetic energy of the bombarding electron can be transferred to the atomic system.

Frank and Hertz in 1914 set out to verify these considerations—namely, that (a) it is possible to excite atoms by low-energy electron bombardment, (b) that the energy transferred from the electrons to the atoms always had discrete values, and (c) that the values so obtained for the energy levels were in agreement with the spectroscopic results.

The necessary apparatus consists of an electron-emitting filament and an adequate structure for accelerating these electrons to a desired (variable) potential. The accelerated electrons are allowed to bombard the atomic vapor under investigation and the excitation of the atoms is studied as a function of accelerating potential.

For detecting the excitation of the atoms in the vapor it is possible to observe, for example, the radiation emitted when the atoms return to the ground state, or the change in absorption of a given spectral line, or some other related phenomenon; however, a much more sensitive technique consists in observing the electron beam itself. Indeed, if the electrons have been accelerated to a potential just *equal* to the energy of the first excited level, some of them will excite atoms of the vapor and as a consequence will lose almost all their energy; clearly, if a small retarding potential exists before the collector region, electrons that have scattered inelastically will be unable to overcome it and thus will not reach the anode.

These conditions are created in the experimental arrangement by using two grids between the cathode and collector. When the potentials are distributed as in Fig. 1.5a, the beam is accelerated between the cathode and grid 1; then it is allowed to drift in the interaction region between the two grids and has to overcome the retarding potential between grid 2 and the anode. When the threshold for exciting the first level is reached, a sharp decrease in electron current is observed, proportional to the number of collisions that have occurred (product of atomic-density and cross section). It is clear that when the threshold of the next level is reached, a further dip in the collector current will be observed. These current decreases (dips) are superimposed on a monotonically rising curve; indeed the number of electrons reaching the anode depends on V_{acc}, inasmuch as it reduces space charge effects and elastic scattering in the dense vapor. In addition, the dips are not perfectly sharp because of the distribution of velocities of the thermionically emitted electrons, and the rise of the excitation cross section.

An alternate distribution of potentials is shown in Fig. 1.5b, where V_{acc} is applied at grid 2 so that an electron can gain further energy after a col-

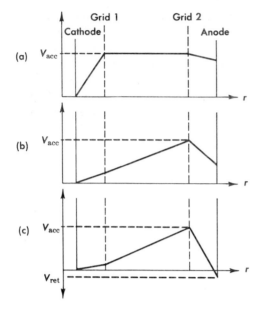

FIG. 1.5 Different configurations of the potential in a Frank-Hertz arrangement. (a) For observation of single excitation. (b) For observation of multiple excitation. (c) For measuring the ionization potential.

lision in the space between the two grids. In this case when V_{acc} reaches the first excitation potential, inelastic collisions are again possible and the decrease in electron current is observed at the anode; when, however, V_{acc} reaches a value twice that of the first excitation potential, it is possible for an electron to excite an atom halfway between the grids, lose all its energy, and then gain anew enough energy to excite a second atom and end with practically zero energy at grid 2; thus it is not able to overcome the retarding potential and reach the anode, giving rise to a second dip in the current.

The advantage of this setup is that the current dips are much more pronounced, and it is easy to obtain fivefold or even larger multiplicity in the excitation of the first level. However, it is practically impossible to observe the excitation of higher levels. As before, a slight retarding potential is applied between grid 2 and the anode, and an accelerating potential between the cathode and grid 1, sufficient to overcome space charge effects and to provide adequate electron current. It is evident that the density of the atomic vapor through which the electron beam passes greatly affects the observed results. Low densities result in large electron currents but very small dips; on the contrary, high density has as a consequence weaker currents but proportionally larger dips. When mercury vapor is used, obviously the adjustment of the tube temperature provides control of the density.

Another important point is that in principle the experiment must be performed with a monatomic gas; since if a molecular vapor is bombarded, it is possible for the electrons to transfer energy to the molecular energy levels which form almost a continuum. Some of the preferred elements for the Frank-Hertz experiment are mercury, neon, and argon.

The same apparatus can be used for the measurement of the ionization potential—that is, the energy required to remove an electron completely from the atom. In this case, instead of observing the bombarding electron beam, it is easier to detect the ions that are formed. The distribution of potentials is as shown in Fig. 1.5c, where the anode is made slightly negative with respect to the cathode; no electrons can then reach the anode, which becomes an ion collector. The accelerating potential is increased until a sharp rise in the ion current measured at the anode is observed.

In both types of measurements the values obtained for the accelerating potential have to be corrected for the contact potential difference (cpd) between cathode and anode.† If in the excitation experiment the same level has been observed two or more times, however, the potential difference between adjacent peaks is an exact measure of the excitation energy, since the contact potential difference shifts the whole voltage scale. Once the excitation energy has been found the contact potential difference is given by the difference between this true value and the first peak; in turn the contact potential difference so found can be used to correct the ionization potential measurement.

3.2 THE EXPERIMENT

In this laboratory a mercury-filled tube made by the Leybold Company (55580) is used; the electrode configuration is shown in Fig. 1.6; the circuit diagrams for the measurement of excitation and of ionization potential are given in Figs. 1.7a and 1.7b respectively.

As can be seen from the circuit diagram, grid 1 is operated in the neighborhood of 1.5 V, and the retarding potential is of the same order. The anode currents are of the order of 10^{-9} amp and are measured either with a sensitive galvanometer (for example Leeds and Northrup No. 2500) or with a Keithley 600A electrometer (see Chapter 4); adequate shielding of the leads is required to eliminate a-c pickup and induced voltages. The diagram of Fig. 1.7a uses the distribution of potentials as shown in Fig. 1.5b and the accelerating voltage can be measured with an ordinary voltmeter (for example, Triplet 625) in steps of 0.1 V, or with a vacuum tube voltmeter.

---

† See Chapter 3. Briefly this is because the "work function" for the metal of which the anode is made is usually higher than that of the cathode. The work function is a measure of the "ionization potential" of the metal; that is, of the energy needed to extract an electron from it.

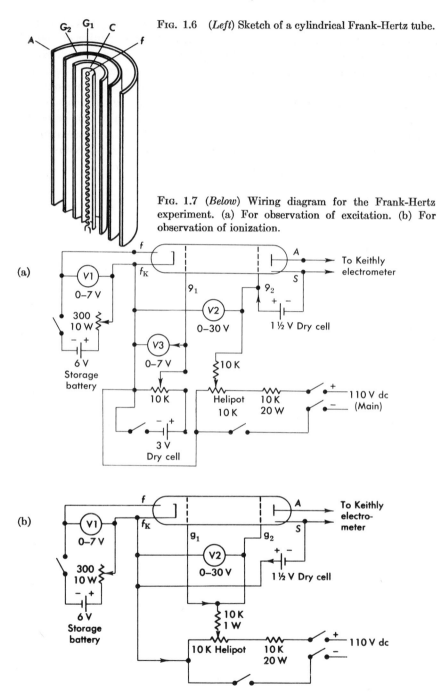

FIG. 1.6 (*Left*) Sketch of a cylindrical Frank-Hertz tube.

FIG. 1.7 (*Below*) Wiring diagram for the Frank-Hertz experiment. (a) For observation of excitation. (b) For observation of ionization.

The Frank-Hertz tube is placed in a small oven which is heated by line voltage through a variac; it should be operated in the vicinity of 200° C for the excitation curve and between 100° to 150° C for the ionization curve. To measure the temperature a copper-constantan thermocouple should be inserted through the small hole of the furnace. The junction should be positioned on the side of the tube near the electrodes. The other junction is immersed in a thermos of ice and water bath. The potential developed across the thermocouple is measured with a potentiometer (usually set on its lowest scale); Fig. 1.8 gives a calibration curve for a copper-constantan thermocouple.

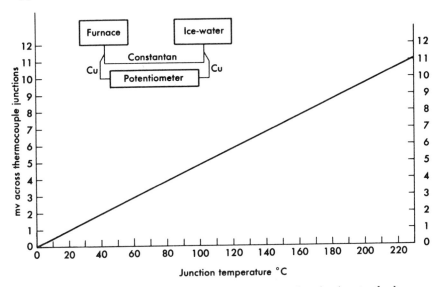

FIG. 1.8 Calibration of copper constant thermocouple using ice standard.

The resolution and definition of both the excitation and ionization curves is a function of atom density (temperature) and electron beam density (filament and grid 1 voltage) and the experimenter has to find the optimum conditions. However, for large beam densities a discharge occurs, which obviously is to be avoided.

A suggested adjustment procedure is to set grid 2 at 30 V and then advance grid 1 until the discharge sets in, as evidenced by the immediate build-up of the anode current. Grid 2 should then be quickly returned to 0 V and grid 1 set slightly below the breakdown voltage; a reasonable filament voltage is between 4 and 6 V. To determine whether the tube is overheated it can be taken out of the oven for about 30 sec; the collector current will then increase and maxima may appear if such is the case. If

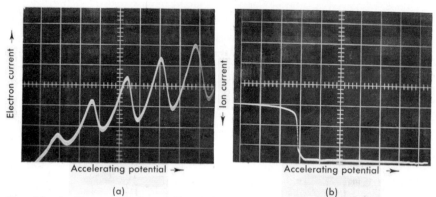

Fig. 1.9 Oscilloscope display of Frank-Hertz experiment. (a) Beam current vs. accelerating potential. (b) Ion current vs. accelerating potential.

the tube is too cool, the emission current will be large, and the maxima, particularly those of higher order, will be washed out.

With present-day techniques it is possible to use an oscilloscope for a simultaneous display of the electron or ion current against accelerating potential. In this laboratory a Tektronix 545 oscilloscope was used; its sweep generator (sawtooth) output is fed to the accelerating grid, while synchronously drives the horizontal sweep; the output of the Keithley is fed to the vertical input. An excitation curve as well as an ionization curve obtained by a student† in this fashion are shown in Fig. 1.9. The oscillocope method can be very useful in finding optimum operating conditions for mercury vapor pressure and electron beam density.

3.3 ANALYSIS OF THE DATA

Two sets of data obtained by a student‡ for the excitation potential point by point are shown in Fig. 1.10; both curves are obtained at a temperature of 195° C and with +1 V on grid 1. The filament voltage is 2.5 V for curve C and 1.85 V for curve D with the consequent decrease of the electron current by a whole decade.

Readings are taken for 1-V changes on grid 2 with smaller steps in the vicinity of the peak. A significant decrease in electron (collector) current is noticed every time the potential on grid 2 is increased by approximately 5 V, thereby indicating that energy is transferred from the beam in (bundles) "quanta" of 5 eV only. Indeed, a prominent line in the spectrum of mercury exists at 2537 Å, corresponding to 12378/2537 = 4.86 eV, arising from the

† D. Statt, class of 1963.
‡ D. Owen, class of 1963.

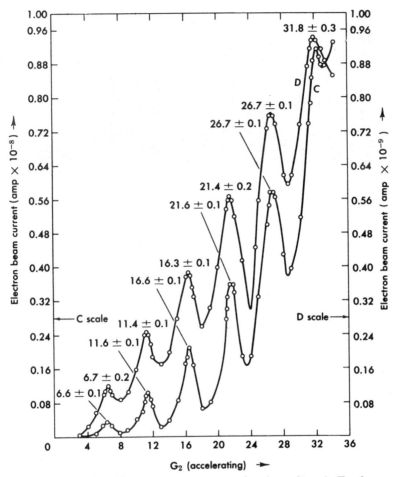

Fɪɢ. 1.10   Plot of beam current versus accelerating voltage in Frank-Hertz experiment. Curve C (left-hand scale) is obtained with the filament set at 2.5 volts while Curve D (right-hand scale) with filament at 1.85 volts.

transition of the $6s6p$ $^3P_1$ excited state to the $6s6s$ $^1S_0$ ground state.† Thus our interpretation is that the electrons in the beam excite the mercury atom from the ground state to the $^3P_1$ state, thereby losing 4.86 eV in the process.

The location of the peaks is indicated in Fig. 1.10 and was measured in this case with a vacuum tube voltmeter (VTVM). The average value ob-

† See Chapter 2.

tained for the spacing between peaks is

$$5.02 \pm 0.1 \text{ V}$$

to be compared with the accepted value of

$$4.90 \text{ V}$$

while the spectroscopic value for the energy level difference (as mentioned before) is 4.86 eV.

Using the value found for the spacing between peaks and the location of the first peak, we obtain the contact potential:

$$(6.65 \pm 0.15) - (5.02 \pm 0.1) = 1.63 \pm 0.18 \text{ V}$$

As mentioned in Section 3.1, with the configuration of potentials used (Fig. 1.5b) it is more probable that the same energy level will be excited twice rather than that several different levels will be excited; indeed this is the way in which the data of Fig. 1.10 have been interpreted. This is not surprising if one considers the excitation probabilities for the energy levels lying closest to the ground state of mercury. It is possible, however, by using different grid and voltage configurations (for example, Fig. 1.5a) and improved resolution, to observe the excitations to other levels, namely, the $6 \, ^3P_2$, $6 \, ^3P_0$ and $6 \, ^1P_1$.

For the ionization potential, data obtained by a student[†] are shown in Fig. 1.11. A word of caution is to be added to the interpretation of such ionization curves, which seem strongly dependent on filament voltage and vapor pressure: indeed the very sharp increase observed in ion current is due to an avalanche (regenerative effect) of the ejected electrons ionizing more atoms, the thus-ejected electrons ionizing still more atoms and so on; this avalanche does not necessarily occur as soon as the ionization threshold is crossed. If the vapor is too dense, the ions recombine before reaching the anode, thus masking the effect until complete breakdown sets in.

The curve shown was taken at a temperature of 155° C with a filament voltage of 2.6 V. If, then, the onset of ion current is taken to be at 11.4 $\pm$ 0.2 V, and using the value for the contact potential previously determined (from the excitation curve), 1.63 $\pm$ .18 V, the ionization potential is obtained as

$$(11.4 \pm 0.2) - (1.63 \pm 0.18) = 9.77 \pm 0.25 \text{ eV}$$

only in fair agreement with the accepted value of 10.39 eV.

An additional feature of the curve of Fig. 1.11 is a "knee" in the ion current, setting in at approximately 8 V; the observation of this "knee" as well is strongly dependent on the temperature and current density, but

---

[†] J. Reed, class of 1961.

can be consistently reproduced over a considerable range of these parameters. In order to understand this behavior we remember that the arrival of ions at the anode is equivalent to the departure of electrons; indeed the observed behavior is due to a photoelectric effect produced at the anode, by short-wavelength light quanta (the electrons are further accelerated by grid 2). When the electron beam reaches 8 V, it can excite the 6 $^1P_1$ level (lying at 6.7 eV above the ground state, plus 1.63 V for contact potential difference), so the mercury atoms radiate the 1849 Å ultraviolet line when returning to the ground state. These quanta are very efficient in ejecting photoelectrons, and the cylindrical geometry of the anode is most favorable for this process.

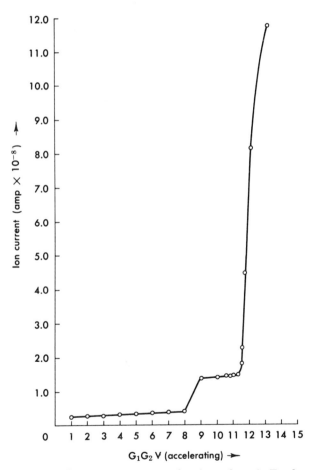

FIG. 1.11   Ion current versus accelerating voltage in Frank-Hertz experiment. Knee at 8 V is due to photoeffect.

4. The Photoelectric Effect

4.1 GENERAL

It was observed as early as 75 years ago that most metals under the influence of radiation (light), especially ultraviolet radiation, emit electrons. This phenomenon was termed photoelectric emission, and detailed study of it has shown:

(a) That the emission process depends strongly on the frequency of the light, and that for each metal there exists a critical frequency such that light of lower frequency is absolutely unable to liberate electrons while light of higher frequency always does. Indeed, for a given surface, if the frequency of the incident radiation is increased, the energy of the emitted electrons increases in some linear relation.

(b) The emission of the electrons occurs within a very short time interval after the arrival of the radiation, and the number of electrons emitted is strictly proportional to the intensity of the radiation.

The experimental facts given above are among the strongest evidence for our present-day belief that the electromagnetic field is *quantized*. They cannot be explained in terms of a continuous energy distribution in the radiation field, but it must be assumed that the field consists of "quanta" of energy

$$E = h\nu$$

where ν is the frequency of the radiation and h is Planck's constant (an expression we have already used in Section 3). These quanta are called *photons*.

Further it is assumed† that the electrons are bound inside the metal surface with an energy $e\phi$, where ϕ is called† the "work function," and that all such electrons have equal probability of absorbing a photon. It then follows that if the frequency of the light ν is such that

$$h\nu > e\phi$$

it will be possible to eject photoelectrons, while if

$$h\nu < e\phi$$

this is impossible, since the probability that an electron will absorb two photons simultaneously is minimal. In the former case, the excess energy of the quantum appears as kinetic energy of the electron, so that

$$h\nu = \tfrac{1}{2}mv^2 + e\phi \qquad (4.1)$$

---

† See Chapter 3.

which is the famous photoelectric equation formulated by Einstein in 1905. In writing the equation in this form we express the fact that the energy is shared between the electron and the photon only; however, to balance momentum† a third body is needed, which in this case is the crystal lattice, which recoils with negligible energy.

Thus in the photoelectric effect we observe the transfer of the total energy of a photon to an electron bound in a lattice. We will see later another transfer mechanism, prevalent at higher photon energies, whereby only part of the energy of the photon is transferred to a free electron: this is the Compton scattering.

Equation 4.1 has been extensively verified for many materials and over a broad range of frequencies. What is experimentally measured is the energy of the emitted photoelectrons against frequency, either with a magnetic field or in a simpler way by a retarding potential technique, as is done in this laboratory exercise. Since the "work function" ϕ is usually not known beforehand, the kinetic energy of the photoelectrons $E_e = \frac{1}{2}mv^2$ is obtained as a function of ν so that the slope of the straight line

$$E_e = h\nu - e\phi$$

yields h, and the intercept at the extrapolated point $\nu = 0$, can give $e\phi$. When a retarding potential V is used to measure E_e, we have $E_e = eV_0$, so that really it is the ratio h/e that is determined:

$$V_0 = (h/e)\nu - \phi \tag{4.2}$$

The arrangement generally used consists of a clean surface of the metal to be investigated, and an anode facing or surrounding the cathode, both sealed in vacuum. When radiation is incident on the cathode, electrons are emitted which reach the anode giving rise to a detectable current if the circuit between anode and cathode is completed through a sensitive current meter as shown in Fig. 1.12. If a negative potential V is applied to the anode, only electrons with $E_e > eV$ can reach the anode, and for some potential V_0 no electrons at all arrive at the anode; this retarding potential multiplied by e is equal to the energy of the fastest electrons emitted. In practice all electrons are not emitted with the same energy, and therefore the threshold at V_0, is not very sharp; space charge effects further reduce the definition.

An additional consideration, already encountered in the Frank-Hertz experiment is the contact potential difference; namely, the fact that the potential applied and measured across the anode and cathode leads does

---

† Note that in any event the electrons are emitted in a direction opposite to that of the incoming radiation.

not equal the potential that the electron traveling from the cathode to the anode has to overcome. To see this, consider Fig. 1.13, where ϕ_C represents the work function of the cathode and $\phi_A \neq \phi_C$ is the work function of the anode. The external voltage V' is applied between metallic junctions and we may neglect the ohmic voltage drop in the leads; thus the electrons *inside* the anode are at potential V' higher than the electrons *inside* the cathode. The energy losses around the loop of Fig. 1.13 must, however, be zero, and the arrows indicate the direction for which an electron *loses* energy in the field (namely, negative potential); if V is the potential seen by the free electron, we obtain

$$-e\phi_C + eV + e\phi_A - eV' = 0$$

or

$$V = V' - (\phi_A - \phi_C) \qquad (4.3)$$

The term $(\phi_A - \phi_C)$ is the contact potential difference (cpd) and usually $\phi_A > \phi_C$. Therefore the measured potentials V' must be corrected according to Eq. 4.3 in order to be used in Eq. 4.2. One way of finding the contact potential difference is to normalize all curves to the same saturation current and observe for what (common) value of V' saturation sets in; this must correspond to the point where V changes over from retarding to accelerating—namely, from Eq. 4.3, $V = 0$ or $V' = \phi_A - \phi_C = \text{cpd}$.

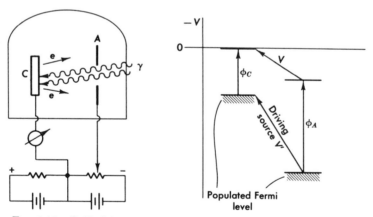

FIG. 1.12   (*Left*) Schematic of a setup for detecting the photoelectric effect; the anode can be made either negative or positive with respect to the cathode.

FIG. 1.13   (*Right*) Potential at anode (−) and cathode (+) of photoelectric cell; w and w' are the work functions of anode and cathode, respectively. Note that $\phi - V - \phi' + eV' = 0$ so that the potential seen by the free electron is $V = V' - (\phi - \phi')$.

By combining Eqs. 4.1 and 4.3 we note that if emission stops for an applied *retarding* potential V_0', then

$$V_0 = V_0' + (\phi_A - \phi_C) \tag{4.3b}$$

further

$$e\,|\,V_0\,| = \tfrac{1}{2}mv^2 = h\nu - e\phi_C \tag{4.1b}$$

so that

$$|\,V_0'\,| = (h/e)\nu - \phi_A \tag{4.2b}$$

namely, a plot of the *applied* stopping potential (without cpd corrections) vs. ν yields a line of slope h/e but with an intercept at $\nu = 0$ equal to the work function of the *anode* rather than to the work function of the cathode predicted by Eq. 4.2.

4.2 THE EXPERIMENT

To perform the experiment we need a source of monochromatic light, at several frequencies, the photoelectric cell, and a sensitive current detecting device.

In this laboratory the photocell is of special construction (Leybold catalogue 55877) with a potassium (coated) cathode and an anode which consists of a platinum ring (Fig. 1.14). A special casing is available for pro-

FIG. 1.14 A photocell made by the Leybold Company (Catalog No. 55877).

tection, electrostatic shielding, and adequate insulation of the anode contact.

The optical system is mounted on an optical bench as shown in Fig. 1.15. The light source is a high-pressure mercury discharge (Will Corp. 17391); it is focused onto slit 1 and then dispersed in the direct vision prism (Leybold 46604). Slit 2 selects the desired wavelength. The direct vision prism consists of a combination of two crown glass and one flint glass prisms and passes medium wavelengths without refraction.

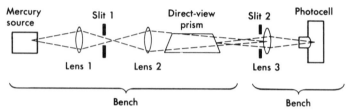

FIG. 1.15   Optical system for photoelectric effect measurement.

When the system is set up, the source must be focused on slit 1, and the image of the slit focused on lens 3; a sharp image of the spectral line can then be obtained on the photocathode by adjusting the prism and photocell positions. Since the system is chromatic, each line must be refocused independently. Obviously it is easier to make these adjustments in a darkened room or with the help of black cloth.

Care must be taken that the incident radiation contains only the line chosen for the investigation, and that it does not hit the platinum anode.† This can be achieved by placing an appropriate mask in front of the photocell.

The mercury lines most readily available are

Yellow	at	5770 Å and 5790 Å
Green	at	5461 Å
Blue-green	at	4916 Å (weak)
Blue	at	4358 Å mainly, and at 4343, 4339 Å
Violet	at	4047 Å

The electrical connections are shown in Fig. 1.16. The photocurrent is measured with a Keithley electrometer connected to the photocathode while the anode is returned to ground through a variable d-c voltage which provides the desired retarding or accelerating field. Since the voltages applied between anode and cathode are very small, appropriate corrections must be made for the voltage drop produced in the meter, due to the photocurrent flowing in the large input impedance.

---

† Due to scattering a small fraction of the incident radiation reaches the anode.

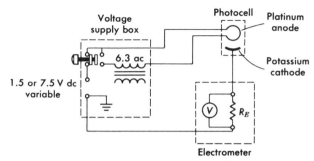

Fɪɢ. 1.16   Wiring diagram for photoelectric effect measurement.

Provisions exist for connecting the anode ring to a 6.3-V a-c supply; thus the anode may be heated in order to evaporate any potassium traces that have deposited on it. It is usually adequate to close the circuit momentarily, since long heating will destroy the anode. The retarding (or accelerating) potential is obtained from a fixed source and a calibrated resistive network, which must, however, be checked against a potentiometer.

In performing the experiment, after the desired line has been focused on the photocathode, the current is measured as a function of the voltage applied between anode and cathode. In principle the accelerating potential should be increased (in appropriate steps) until saturation is reached; this is, however, difficult to achieve with the photocell described here.† The decelerating potential should also be decreased until zero current is observed and beyond that point, to assure that the current remains zero.

Frequently, as is the case with this particular apparatus, a reverse current is observed leveling off at approximately 10^{-12} amp; this is attributed to photoemission from the anode and introduces difficulties in the exact determination of the stopping potential V_0'. Nevertheless, it is possible to obtain significant results if the same consistent criterion is used at all wavelengths for obtaining V_0'.

Since the current variations in the vicinity of the stopping potential are of the order of 10^{-13} amp, their measurement becomes difficult and special care must be exercised. Leakage current across the glass face of the tube must be minimized, as by surrounding the bulb base with a moisture absorber (silica gel), especially on humid days. The cathode contact must be kept very clean (with alcohol and hot air), all leads on the high-impedance side must be coaxially shielded, and the appropriate connectors must be used. The electrometer and apparatus should be protected from vibrations and stray field pickup. Dark current must be monitored and appropriate corrections applied.

---

† Mainly due to geometrical considerations.

Finally, caution must be exercised in using the mercury source, since its envelope transmits ultraviolet light, which can cause serious damage to the eyes and sunburn to the skin.

4.3 ANALYSIS OF THE DATA

The data presented below were obtained by students.† The five lines of mercury mentioned in the previous section were used, and the photo-currents near saturation due to these lines were in the following proportion:

Yellow	1.00
Green	1.50
Blue-green	0.44
Blue	1.70
Violet	0.55

These yields are a combination of the intensity of the spectral lines, their attenuation in the optical system, and the photosensitivity of the cathode, which is not the same at all wavelengths. In analyzing the data as mentioned before it is useful to normalize all photocurrents to the same satura-

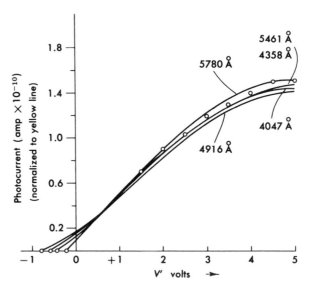

FIG. 1.17 Photocurrent as a function of anode voltage. The currents have been normalized to the yield from the yellow lines (λ = 5780 Å).

---

† D. Owen and D. Sawyer, class of 1963.

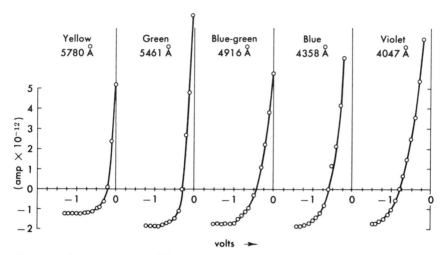

FIG. 1.18 Photocurrent for different wavelengths as a function of stopping potential; the same normalization as in Fig. 1.17 has been used. Note the reverse current due to photoemission from the anode.

tion value; the results of such normalization for the five wavelengths are shown in Fig. 1.17. We note that in the accelerating region the curves are quite similar, and the small differences can be attributed to observational errors.

On the other hand, Fig. 1.18 represents the region close to the stopping point, but separately for each wavelength; the normalized photocurrents are shown. In spite of the reverse current it is possible to read off the stopping potential for each line; the difficulty arises rather from the apparent zero slope of the curves.† From these curves values of V_0' have been obtained (a) by forming the intersection of the tangents to the limiting branches of the curves, and (b) by estimating the voltage at which the current curve begins to rise. These values are given in Table 1.2.

A plot of these stopping voltages and the least-squares fit are shown in Fig. 1.19. We see that in both cases a correct order of magnitude of h/e is obtained namely:

Method (a) $h/e = (3.84 \pm 0.55) \times 10^{-15}$ V-sec
  intercept at $\nu = 0$ $V' = +1.2$ V

Method (b) $h/e = (3.84 \pm 0.4) \times 10^{-15}$ V-sec
  intercept at $\nu = 0$ $V' = 1.6$ V

---

† It should be a finite slope, but this is not observed in the present arrangement because of the reverse current.

TABLE 1.2

RETARDING POTENTIALS REQUIRED TO STOP PHOTOEMISSION AS OBTAINED FROM
FIG. 1.18

Line	Retarding potentials	
	(a)	(b)
Yellow	-0.25 V	-0.7 V
Green	-0.40	-0.8
Blue-green	-0.70	-1.0
Blue	-0.82	-1.3
Violet	-1.15	-1.5

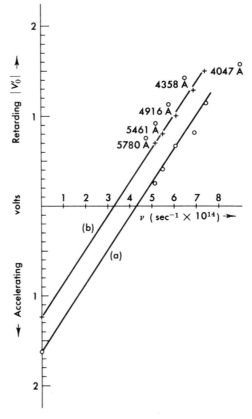

FIG. 1.19 Plot of applied stopping potential versus frequency. Lines (a) and (b) refer to the data of Table 1.2.

Multiplying by the charge of the electron $e = 1.6 \times 10^{-19}$ coulombs we obtain

$$h = (6.14 \pm 0.8) \times 10^{-34} \text{ joules-sec}$$

to be compared with the accepted value of

$$h = 6.61 \times 10^{-34} \text{ joules-sec}$$

While the value obtained for h/e is quite satisfactory it is not possible to draw any conclusions with regards to the cpd. The known values for the work functions are

$$\text{anode, platinum } \phi_A = 5.29 \text{ V}$$

$$\text{cathode, potassium } \phi_C = 2.15 \text{ V}$$

And thus cpd = 3.14 V. This value is in qualitative agreement with the saturation data of Fig. 1.17. However, from Fig. 1.18 we are inclined to deduce $\phi_A \simeq 1.5$ V rather than ≈ 5 V; this fact, is a further indication that cathode material had deposited on the anode while these data were obtained.

Thus we have seen in three basic experiments that fundamental quantities of nature, such as the electric charge, the energy of electrons bound in an atom, and the energy of the electromagnetic field are quantized: that is, they cannot take any of a continuous set of values but only discrete ones. This fundamental characteristic of our world was first formulated in 1901, when Max Planck introduced it as the basic hypothesis for his theoretical interpretation of the spectrum (continuous in frequency) of a heated black body. It has led to a serious revision of both the method of thought and the mathematical tools of physics.

A slightly more detailed description of these experiments, including references to the original literature can be found in G. P. Harnwell and J. J. Livingood, *Experimental Atomic Physics*. New York: McGraw-Hill (1961).

2

SIMPLE QUANTUM-
MECHANICAL SYSTEMS

1. Introduction

The title of this chapter requires explanation. A quantum-mechanical system is the aggregate of two or more interacting particles usually bound together; it is "simple" when the number of independent particles in the system is small, or when the system can be adequately described in terms of a few-particle model. It is characteristic of bound quantum-mechanical systems that their total energy cannot have any value, but that the system is found in one of its energy levels, or states. Transitions of the system between various of its states may occur, provided all the basic conservation laws of electric charge, energy, momentum, angular momentum, and so on, are satisfied.

Transitions from a higher energy state to a state with less energy can occur for an isolated system, and the larger the probability for this transition, the shorter the "lifetime" of that excited state. During such spontaneous transitions of a quantum-mechanical system to a lower energy state, a quantum of radiation, or one or more particles, can be emitted, which will carry away the energy lost by the system (after recoil effects have been taken into account). In the presence of a radiation field the quantum-mechanical system can either gain energy from the field and change into a state with higher energy, or lose energy to the field and revert to a lower energy state. For all quantum-mechanical systems there exists a lowest energy state, called the *ground state*.

By observing the quanta of radiation, or the particles emitted during such transitions, we gain information on the energy levels involved. The typical example is optical spectroscopy, which consists in the accurate determination of the energy of the light quanta emitted by atoms. Infrared spectroscopy deals mainly with the quanta emitted by molecules, nuclear spectroscopy with the quanta emitted in nuclear transitions, and so on. In nuclei, however, the separation between energy levels is much larger, so that the emitted quanta of electromagnetic radiations lie in the gamma-ray region; thus different techniques are employed for detection and measurement of their energy. It is also very common for nuclei to transit from one energy state to another by the emission of an electron and neutrino (beta decay) and for certain heavier nuclei by the emission of a helium nucleus (alpha particle).

By far the most common simple quantum-mechanical system is the hydrogen atom. It consists of a proton and an electron bound together. The theory of quantum electrodynamics is able to predict the position (energy) of its possible states to the best experimental accuracy (of the order of $1/10^8$)† achieved up to date. The simple Bohr theory (even though incorrect) makes the same predictions as the exact solution of the Schrödinger equation for the hydrogen system, and is therefore correct if relativistic effects, fine structure, and higher order corrections are neglected. This leads to an accuracy of $1/10^5$, which is adequate when compared to the resolution of the equipment available in a teaching laboratory.

The postulates of the Bohr theory are (a) that the electron is bound in a circular orbit around the nucleus such that the angular momentum is quantized in integral units of Planck's constant (divided by 2π); namely, $pr = mvr = n(h/2\pi) = n\hbar$; and (b) that the electron in this orbit does not radiate energy, unless a transition to a different orbit occurs. We can then calculate the radii of these orbits and the total energy of the system (potential plus kinetic energy of the electron).‡ The attractive force between the electron (charge e) and the proton (charge e) or a nucleus (of charge Ze) is the Coulomb force, which is set equal to the centripetal force, so

---

† By predicting the Lamb shift of the $2\,{}^2P_{1/2}$ level of hydrogen.

‡ Neglecting the motion of the nucleus.

that

$$\frac{1}{4\pi\epsilon_0} \frac{(Ze)e}{r^2} = \frac{mv^2}{r}$$

but

$$mvr = n\hbar$$

hence

$$\frac{Ze^2}{4\pi\epsilon_0} \frac{1}{n\hbar} = v$$

and

$$r = n^2 \frac{4\pi\epsilon_0\hbar^2}{Ze^2 m} \tag{1.1}$$

As to the energy,

$$T = \frac{1}{2}mv^2 = \frac{1}{2}\frac{Z^2e^4}{(4\pi\epsilon_0)^2}\frac{m}{\hbar^2 n^2}$$

$$V = -\frac{(Ze)e}{4\pi\epsilon_0 r} = \frac{Z^2e^4}{(4\pi\epsilon_0)^2}\frac{m^2}{\hbar^2 n^2} = -2T$$

so that

$$E = T + V = -\frac{1}{2}V = -\frac{1}{2}\frac{Z^2e^4m}{(4\pi\epsilon_0)^2\hbar^2}\frac{1}{n^2} \tag{1.2}$$

For the hydrogen atom $Z = 1$ and Eqs. 1.1 and 1.2 can be written as

$$r = n^2 a_0 \tag{1.1a}$$

where $a_0 = 0.527 \times 10^{-8}$ cm $= 0.527$ Å and is called the first Bohr radius. Similarly

$$E = -hc \, \mathrm{R}_\infty \frac{1}{n^2}$$

where

$$\mathrm{R}_\infty \, hc = 13.605 \text{ eV}; \qquad \mathrm{R}_\infty = 1.0974 \times 10^5 \text{ cm}^{-1}$$

and is called Rydberg's wave number.

The energy levels of the hydrogen atom can then be represented by Fig. 2.1. However, the lines observed in the spectrum correspond to transitions between these levels; this is shown in Fig. 2.2, where arrows have been drawn for all possible transitions. The energy of a line is given by

$$\Delta E_{if} = hc\, \mathrm{R}_\infty \left(\frac{1}{n_f^2} - \frac{1}{n_i^2} \right) \tag{1.3}$$

where the subscripts i and f stand for initial and final state respectively.

Since the frequency of the radiation is connected to the energy of each quantum through

$$E = h\nu \tag{1.4}$$

one finds that

$$\frac{1}{\lambda} = \frac{\nu}{c} = \frac{E}{hc}$$

and

$$\frac{1}{\lambda_{if}} = \mathrm{R}_\infty \left(\frac{1}{n_f^2} - \frac{1}{n_i^2} \right) \tag{1.3a}$$

Indeed the simple expression (Eq. 7.3a) is verified by experiment to a high degree of accuracy.

From Eq. 1.3 (or from Fig. 2.2) we note that the spectral lines of hydrogen will form groups depending on the final state of the transition, and that within these groups many common regularities will exist; for example in the notation of Fig. 2.2

$$\nu(L_\beta) - \nu(L_\alpha) = \nu(B_\alpha)$$

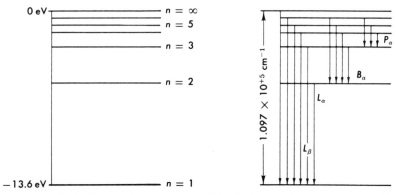

FIG. 2.1 (*Left*) Energy-level diagram of the hydrogen atom according to the simple Bohr theory.

FIG. 2.2 (*Right*) Transitions between the energy levels of a hydrogen atom. The lines L_α, L_β, etc., belong to the Lyman series, B_α, B_β, etc., to the Balmer series, and P_α, P_β, etc., to the Paschen series, and so forth.

If $n_f = 1$, then

$$\lambda_{i1} = 911 \left(\frac{n_i^2}{n_i^2 - 1} \right) \text{Å}; \qquad n_i \geq 2$$

and all lines fall in the far ultraviolet; they form the so-called Lyman series. Correspondingly if $n_f = 2$,

$$\lambda_{i2} = 3644 \left(\frac{n_i^2}{n_i^2 - 4} \right) \text{Å}; \qquad n_i \geq 3$$

and all lines fall in the visible part of the spectrum, forming the so-called Balmer series. Further, for $n_f = 3$ the series named after Paschen falls in the infrared, and so on.

2. Experiment on the Hydrogen Spectrum

2.1 GENERAL

Measurements of the frequency of the radiation emitted by an excited hydrogen atom are based on either interference, as when a plane grating is used, or variation with wavelength of the refractive index of certain media, as when prism spectrometers are used. Prism spectrometers are

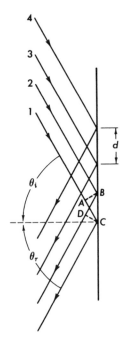

FIG. 2.3 Schematic diagram of reflection grating. A parallel beam of radiation is incident along the rays 1 through 4 at an angle θ_i with respect to the normal; the reflected radiation is observed at an angle θ_r. The spacing between the grooves of the grating is d.

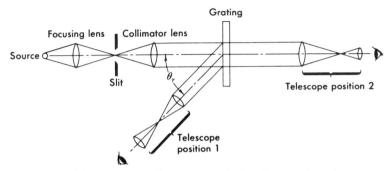

FIG. 2.4   Diagrammatic arrangement of grating spectrometer.

obviously limited to wavelength regions for which they are able to transmit the radiation; for example, in the infrared, special fluoride or sodium chloride prisms and lenses are used; in the ultraviolet, the optical elements are made of quartz. Also, the sensitivity of the detectors varies with wavelength, so that different types are used in each case (thermopile, photographic emulsion, phototube, etc.).

In this laboratory a small constant-deviation spectrograph (Bausch and Lomb type 33-82-09) and a 2-in. reflection grating spectrometer were used. We will first consider a measurement of the hydrogen spectrum with the grating, since an absolute value for the wavelengths can be obtained and visual detection is used.

From Fig. 2.3, it is evident that the path difference between rays 1 and 2 after reflection is

$$BD - AC = CB \sin \theta_r - CB \sin \theta_i$$

where CB is the grating spacing d. The angles θ_i and θ_r are both taken as positive when they lie on opposite sides of the normal. Since for constructive interference the path difference must be a multiple of the wavelength, we obtain the condition

$$n\lambda = d \ (\sin \theta_r - \sin \theta_i) \tag{2.1}$$

It can be shown that the resolution of the grating is given by

$$\frac{\lambda}{\Delta\lambda} = nN$$

where n is the order of diffraction and N the total number of rulings.

The grating is mounted on a goniometer table in the general arrangement shown in Fig. 2.4. A slit and collimating lens are used to form a beam of parallel light from the source, and a telescope mounted on a rotating arm is used for viewing the diffracted lines. It is obviously necessary to insure parallelism of the incident and reflected beam, normality of the grating, and so on. A suggested alignment procedure is as follows:

(a) The viewing telescope is focused for parallel rays (on some distant object).

(b) Then with the grating removed, the slit is viewed with the telescope (in position 2) to ascertain that the slit is aligned and in focus; in this way the collimator lens is adjusted.

(c) The source and source lens are placed in position and the alignment and focusing are again checked. The crosshairs are aligned with the slit.

(d) This position of the telescope is carefully noted since it represents the 0-degree position. The readings on the scale should be made to one minute of a degree by using the vernier and a flashlight.

(e) From now on one may have to work in dark, or by draping the apparatus with a black cloth.

(f) The grating is placed in position and aligned for normal incidence ($\theta_i = 0$). This can be done by "auto-collimation"; a strong light is focused onto the slit and a cardboard mask with a narrow slit is placed on the collimator lens. The grating is then adjusted until the reflected image of the cardboard slit coincides with the slit itself.

(g) Finally, the lines of the grating should be made parallel to the slit (hence the crosshairs); this can be done by viewing one edge of the grating with the telescope in position 1.

With any reasonable grating it is possible to observe the visible lines of the spectrum in several orders; thus we expect the measurements for λ/d to be self-consistent, since

$$\sin \theta_{m+1} - \sin \theta_m = (m + 1) \frac{\lambda}{d} - m \frac{\lambda}{d} = \frac{\lambda}{d} \qquad (2.2)$$

independently of angle of incidence θ_i, or order.† The grating spacing d is usually stated by the manufacturer; for example, a grating in this laboratory has rulings of the order of 7000 to the inch ($d = 3.692 \times 10^{-4}$ cm). However, d can be obtained by using one or more standard lines of known wavelength.

The following data were obtained by a student ‡ using the grating spectrometer. The source was a low-pressure hydrogen discharge tube (Cenco

---

† Provided certainly that both θ_m and θ_{m+1} are taken on the same side of the normal.
‡ R. S. Stevens, class of 1963.

type 87210) operated at a few thousand volts; a 5 kV transformer and variac were used to provide the variable voltage. The useful life of these discharge tubes is limited because of the appearance of strong molecular bands after some hours of operation.

2.2 DETERMINATION OF d

To obtain the grating spacing d, sodium (Na) was used as a standard, and measurement on three lines (for the shorter wavelength of the doublet) gave the following results:

TABLE 2.1
DIFFRACTION ANGLES FOR A SODIUM SOURCE

λ in Å	Order n	θ_n	$\theta_i = 19°12'$
6154.3	1	29°42′	
	2	41°27′	
	3	55°58′	
5890.0	1	29°14′	
	2	40°21′	
	3	53°49′	
	4	75°15′	
5682.7	2	39°32′	
	3	52°12′	
	4	70°48′	

Since for all the above measurements θ_i is the same, it follows that

$$d^{-1}(n_k\lambda_k) + \sin \theta_i = \sin \theta_k$$

and a least squares fit to the linear relation $\beta x + \alpha = y$ can be made; we have

$$\frac{1}{d} = \frac{N \sum (n_k\lambda_k \sin \theta_k) - \sum (\sin \theta_k) \sum (n_k\lambda_k)}{N \sum (n_k\lambda_k)^2 - [\sum (n_k\lambda_k)]^2} \tag{2.3}$$

where the sums are over k, $k = 1, 2, \cdots, N$ and N is the total number of measurements. From the data of Table 2.1 we obtain†

$$\frac{1}{d} = 2.7085 \pm 0.009 \times 10^3 \text{ cm}^{-1} \tag{2.4}$$

It may be noted that if the same orders are measured on both sides of the

† In reaching this result we have constrained $\theta_i = 19°12'$.

incident beam, the terms $\sum (n_k \lambda_k)$ become zero, and Eq. 2.3 is greatly simplified.

Some care must be exercised when comparing wavelengths, since they do depend on the refractive index, n, of the medium in which they are measured.

$$c' = \frac{c(\text{vacuum})}{n}$$

hence

$$\lambda' = \frac{\lambda(\text{vacuum})}{n}$$

The wavelengths listed in most tables are given for dry air at a pressure of 760 mm mercury. However, any theoretical calculation as Eq. 1.3a predicts the vacuum wavelengths. The refractive index of air at stp is

$$n(\text{air}) = 1.00029 \tag{2.5}$$

2.3 THE BALMER SERIES

Measurements on the first four members of the Balmer series, which lie in the visible region, can be made with the spectrometer described above. The data obtained by a student† and their reduction are given in Table 2.2.

TABLE 2.2

DATA ON THE BALMER SERIES OF HYDROGEN AS OBTAINED WITH THE SMALL GRATING SPECTROMETER. ALL WAVELENGTHS ARE IN Å.

Color	θ_n	$\sin \theta_n - \sin \theta_i$	Order	Calculated λ	Accepted λ	Balmer series identification
Violet	33°12′	0.22199	2	4107.5 ± 6	4101.7 H$_\delta$	$n_i = 6$
	41°15′	0.33378	3			
Blue	26°16′	0.11698	1			
	34°06′	0.23483	2	4338.2 ± 8	4340.5 H$_\gamma$	$n_i = 5$
	42°42′	0.35259	3			
Green	27°10′	0.13001	1			
	36°04′	0.26316	2	4857.5 ± 10	4861.3 H$_\beta$	$n_i = 4$
	46°09′	0.39559	3			
Red	30°11′	0.17720	1			
	42°57′	0.35579	2	6579.4 ± 14	6562.8 H$_\alpha$	$n_i = 3$
	59°29′	0.53532	3			

[1] $d = 36{,}921 \pm 30$ Å as determined by the previous measurements on the sodium standard lines.

[2] $\sin \theta_i = 0.32557$.

† R. S. Stevens, class of 1963.

We observe that the obtained values for the wavelengths of the Balmer series are in agreement with the accepted values at the level of 1/1000. We can now test Eq. 1.3a and obtain the Rydberg wave number. We note that

$$\frac{1}{\lambda} = R_H \left[\frac{1}{4} - \frac{1}{n^2} \right]$$

So that from least squares,

$$R_H = \frac{\sum \rho_i^2}{\sum \lambda_i \rho_i}$$

where

$$\rho_i = \frac{4 n_i^2}{n_i^2 - 4}$$

giving

$$R_H = (1.09601 \pm 0.003) \times 10^5 \text{ cm}^{-1}$$

in good agreement with the accepted value†

$$R_H = 1.096778 \times 10^5 \text{ cm}^{-1}$$

3. The Prism Spectrograph

3.1 General

Some of the limitations of prism instruments have been mentioned earlier; however, since all the light is diffracted into a single direction, they have the advantage of yielding higher intensities. The dispersion of a prism is a function of the refractive index; thus it cannot be used for absolute measurements without careful calibration. In the case of a simple

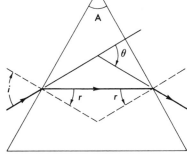

Fig. 2.5 Diffraction of a ray at minimum deviation through a prism of apex angle A.

† The difference between R_H and R_∞, on page 30 is due to the motion of the electron about the center of mass rather than about the proton.

prism at minimum deviation (see Fig. 2.5) the diffraction angle θ is given by

$$\frac{\sin \theta_i}{\sin \theta_r} = n \qquad 2\theta_r = A \qquad \theta_i = \theta_r + \tfrac{1}{2}\theta$$

thus

$$\sin\left(\frac{A + \theta}{2}\right) = n \sin \frac{A}{2} \tag{3.1}$$

where θ_i and θ_r are the angles of incidence and refraction respectively; A is the apex of the prism. In Fig. 2.6 the refractive index of flint glass as a function of wavelength is given. We note that in the determination of wavelength from the diffraction angle the relation is by no means linear and is in general of serious complexity. Further, most modern prism spectrographs do not consist of a single dispersive element, but of some combination of prisms. The instrument used in this laboratory is of the "constant-deviation" type and Fig. 2.7 gives the optical paths for an incident ray. It may be seen that the angle of incidence and the angle of exit can remain fixed for all wavelengths by an appropriate rotation of the prism; this has obvious advantages for positioning and alignment of source and detector.

The rotation of the prism is calibrated, to give rough wavelength indications, but measurements are made on exposed photographic plate (or film). A known spectrum is superimposed on the spectrum that is to be

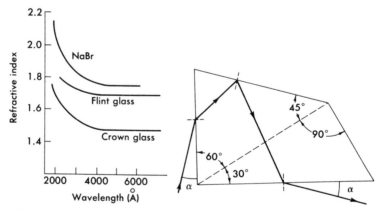

FIG. 2.6   (*Left*) Refractive index of various materials as a function of wavelength.

FIG. 2.7   (*Right*) A constant-deviation prism and the diffraction of a ray passing through it.

investigated, and an interpolation between the known lines is used. One such interpolation technique is the Hartmann method, in which it is assumed that wavelengths can be expressed as

$$\lambda_i = A + \frac{B}{d_i - C} \qquad (3.2)$$

where d_i is the distance of the line from some fixed origin and A, B, and C are constants. If three lines of wavelength,[†] λ_1, λ_2, λ_3, are known, the constants can be found:

$$A = \lambda_i + \frac{B}{C - d_i} \qquad i = 1, 2, 3$$

$$B = (\lambda_1 - \lambda_2)\left[\frac{(C - d_1)(C - d_2)}{d_1 - d_2} \right] \qquad (3.3)$$

$$C = \frac{(Rd_1) - d_3}{R - 1}$$

where

$$R = \frac{\lambda_1 - \lambda_2}{\lambda_2 - \lambda_3} \times \frac{d_2 - d_3}{d_1 - d_2}$$

This method requires a desk calculator and is obviously most accurate for wavelengths contained between the known ones.

As a standard for the interpolation, the well-analyzed spectra of mercury, sodium, helium, or other elements can be used. A brief discussion of the first two spectra and of their most intense lines is given in Section 4 below.

Since the data in this experiment will be recorded on photographic emulsion, we should keep in mind that different types of film are sensitive to different parts of the spectrum. Also, the speed of an emulsion (usually designated by $\log \gamma$) can vary by factors as large as 1000 from one type to another; pertinent information can be found in the Kodak manuals. In this laboratory the following fast emulsions are used:

Plates 103 − F   Spectral region 6900 − 4500 Å

103 − 0   5000 − 2500 Å

and for preliminary testing,

Royal Pan film sheets   6200 − 4200 Å

The actual time of the exposure depends on the source intensity, slit width

---

[†] We choose widely separated wavelengths λ_1, λ_2, and λ_3.

(resolution desired), and type of film; it will have to be determined experimentally for each exposure but is of the order of a fraction of a minute to a few minutes. Development is of the order of 4 to 6 min, followed by 10 to 20 min fixing.

3.2 EXPERIMENTAL TECHNIQUE

The general arrangement of the spectrograph is shown in Fig. 2.8. Source, lens, and slit should be aligned and the source focused on the slit. By viewing through the eyepiece and varying the prism position, one can get a feeling for the dispersion and the range of the instrument.

To obtain photographs of a spectrum, the telescope should be replaced by the camera assembly. The focal plane of the spectrum should be adjusted, however, so as to coincide with the photographic plate; this can be achieved by using a frosted glass in the plate holder and adjusting (1) the camera lens and (2) the orientation of the plate-holder frame. More elaborate techniques using a traveling microscope or other viewing device can be used, when such accuracy is warranted. It should be noted that it is possible to photograph almost the entire visible region from 6500 to 4000 Å in one exposure; however, the prism should be rotated in each case so as to provide maximum linear dispersion at the desired wavelength. Obviously several exposures can be had on the same plate by moving the plate holder across the camera area; to distinguish different spectra superimposed at the same location on the plate, the "fishtail," which controls the length of the slit, can be used.

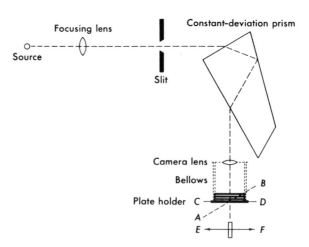

FIG. 2.8   Schematic arrangement of the constant-deviation spectrograph.

Usually several trial exposures are taken before the final adjustment of the instrument and the exact determination of exposure and development parameters. The operating conditions of the source should not be overlooked either; hard-driven sources produce broad lines (unimportant in low-resolution work) and may produce molecular bands, or other undesired spectra. When it is desired to observe fine structure, the lines should not be overexposed, as this would broaden the image on the emulsion; since, however, the intensity of different spectroscopic lines varies greatly, the use of optical filters may be useful.

Once the plates have been obtained, they can be read on a comparator or with the help of a traveling microscope. In this procedure the plate is placed on a diffusely illuminated surface and securely fastened to it (an inverted Pyrex tray with a strong light source underneath it can be very useful). During the measurement, the microscope should always be moved in the same direction to avoid backlash, and an effort should be made to locate the center of the lines; several complete readings of the plate should be obtained and their average taken.

3.3 Analysis of Data

The data that are analyzed here were obtained by a student.† Helium was used as a standard and the following measurements obtained for the positions of the lines are given in Table 2.3.

TABLE 2.3

DATA ON THE HELIUM AND HYDROGEN SPECTRUM OBTAINED WITH THE CONSTANT DERIVATION SPECTROGRAPH

Helium

$d_1 = 0$	$\lambda = 6678.1$ Å
$d_2 = 0.8232$ cm	$\lambda = 5875.6$
$d_3 = 2.2330$	$\lambda = 5047.9$
$d_4 = 0$	$\lambda = 4921.9$
$d_5 = 1.3639$	$\lambda = 4471.6$
$d_6 = 2.8172$	$\lambda = 4143.7$

Hydrogen

$d_1 = 0.0966$ cm	referred to the same origin as lines d_1, d_2, d_3 of helium
$d_2 = 0.1541$ cm	referred to the same origin as lines d_4, d_5, d_6 of helium
$d_3 = 1.8864$	
$d_4 = 2.9383$	

---

† D. Peters, class of 1962.

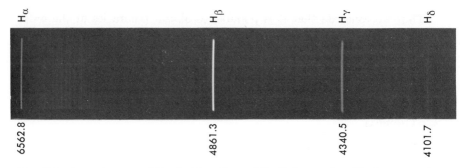

H$_\alpha$ H$_\beta$ H$_\gamma$ H$_\delta$

6562.8   4861.3   4340.5   4101.7

FIG. 2.9   A spectrogram of the first four lines of the Balmer series of hydrogen as obtained with the constant-deviation spectrograph.

Now Eqs. 3.3 may be applied to the above data, and from the first group of helium lines, we obtain

$$A = 2.5799 \times 10^{-5}$$
$$B = 1.38655 \times 10^{-4}$$
$$C = -3.3833$$

so that Eq. 3.2 yields

$$\lambda_1 = 6564.3 \text{ Å}$$

From the second group of helium lines, we obtain

$$A = 2.7104 \times 10^{-5}$$
$$B = 1.1798 \times 10^{-4}$$
$$C = -5.3347$$

and thus

$$\lambda_2 = 4859.8 \text{ Å} \qquad \lambda_3 = 4344.2 \text{ Å} \qquad \lambda_4 = 4136.4 \text{ Å}$$

where it should be noted that λ_4 cannot be expected to be very accurate, since it lies outside the three standard helium lines d_4, d_5, d_6.

As was done in section 2.3 a least-squares fit is made to obtain the best value of the Rydberg, through Eq. 1.3a

$$\frac{1}{\lambda_i} = R_H \left(\frac{1}{4} - \frac{1}{n_i^2} \right) \tag{1.3a}$$

by using only λ_1, λ_2, and λ_3 we obtain†

$$R_H = (1.0974 \pm 0.0014) \times 10^5 \text{ cm}^{-1}$$

in good agreement with the accepted value of

$$R_H = 1.096778 \times 10^5 \text{ cm}^{-1}$$

---

† Note that this value needs no correction for the refractive index of air since the helium standard lines were already corrected.

Thus we conclude that it is possible to obtain accuracies of the order of a few parts in 10^4. Figure 2.9 is a reproduction of the hydrogen spectrum (as obtained by a student†) showing the first four lines of the Balmer series.

4. The Sodium and Mercury Spectrum

4.1 GENERAL

Mention has been made in the previous section of the spectrum of sodium (Na) and mercury (Hg); a brief analysis will be given here, since both elements have been investigated in detail and are representative of the one-electron and two-electron spectra. Sodium has 11 electrons, so that the $n = 1$ and $n = 2$ shells are completely filled and one electron ($n = 3$) exists outside closed shells. In this respect the sodium spectrum should be equivalent to that of hydrogen except for the central charge that the free electron sees. Indeed, since the nucleus with $Z = 11$ is "screened" by 10 negative charges (the $n = 1$ and $n = 2$ electrons) the free electron sees a potential $-e/r$ when far from the nucleus and a potential $(-Ze)/r + C$ when close to it, where C is the potential generated at the nucleus by the other electrons. However, whereas in hydrogen only one energy level was found for each value of n, a more complex situation arises in sodium, with several levels corresponding to the same n. This splitting is to be attributed to the fact that the time-independent Schrödinger equation for the hydrogen-like atom,

$$\nabla^2 \psi + \frac{2m}{\hbar^2}(E - V)\psi = 0$$

admits solutions with a principal quantum number n, and angular momentum quantum number l, such that $n \geq l + 1$; when the potential that the electron sees is exactly of the Coulomb type (as in the case of hydrogen, $V = (-Ze^2)/r$) the energy eigenvalues

$$E_n = \frac{1}{2} \frac{Z^2 e^4}{(4\pi\epsilon_0)^2} \frac{m}{\hbar^2} \frac{1}{n^2} \tag{1.2}$$

are independent‡ of l, and agree with the Bohr theory. However, the screened potential that the free electron sees is no longer of the simple Coulomb type, and the energy of the level depends on both n and l. Orbits with smaller values of l are expected to come closer to the nucleus and thus be bound with greater strength; as a consequence their energy will be

---

† Ch. Georgalas, class of 1963 G.

‡ This is the so-called Coulomb degeneracy: a peculiar coincidence for the Coulomb field when used in the Schrödinger equation.

depressed. The energy level diagram of sodium is shown in Fig. 2.10, where
the levels have been grouped according to their l value. The customary
notation is used, namely:

$l = 0$ S state
$l = 1$ P state
$l = 2$ D state
$l = 3$ F state
$l = 4$ G state, and so on, alphabetically.

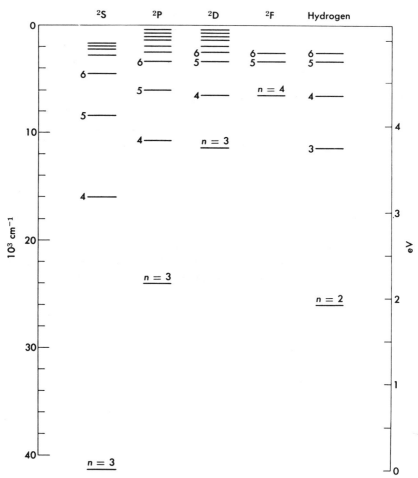

FIG. 2.10   The energy-level diagram of sodium, grouped according to the orbital angular
momentum. The last column gives the corresponding position of the levels of hydrogen.
The left-hand scale is in 10^3 cm^{-1} referred to 0 for the singly ionized sodium atom; the
right-hand scale is in electron volts referred to 0 at the ground state of the sodium atom.

The last column in Fig. 2.10 gives the position of the levels of a hydrogen-like atom.

We note that the higher the value of l, the smaller the departures from the hydrogen-like levels (as suggested qualitatively previously) and that for *given* l the energy levels for different n's follow the same ordering as the hydrogen-like atom, but with an effective charge Z^*, which for sodium is as follows:

s states	$Z^* \sim 11/9.6$
p states	$Z^* \sim 11/10.1$
d states	$Z^* \sim 1$
f states	$Z^* \sim 1$

4.2 SELECTION RULES

As we know, the spectral lines that we observe are due to transitions from one energy state to a lower one; however, in analyzing the spectrum of sodium, it becomes immediately evident that not *all possible* transitions occur. Thus. certain "selection rules" for atomic transitions must be operative and it is found that for all spectral lines†

$$\Delta l = \pm 1 \qquad (4.1)$$

This selection rule is readily explained by the quantum-mechanical theory of radiation; it then means that only "electric dipole" transitions occur. Indeed, the transition probability for "electric dipole" is larger by a factor of $(c/v)^2$ (c, velocity of light) from the next order, while under no conditions a transition can be had in which the angular momentum does not change at all ($\Delta l = 0$).

By applying the selection rule of Eq. 4.1 to the energy-level diagram of Fig. 2.10, we obtain Fig. 2.11, which gives the principal lines of the sodium spectrum; since l must change by one unit, transitions will always occur between adjacent columns and never within the same one.

Figure 2.12 is a reproduction of the visible part of the above spectrum obtained by a student‡ with the constant deviation spectrograph.

Beginning from the left (long wavelengths) we recognize the following lines:

(a)	Red	6154.3–6160.7 Å
(b)	Yellow	5890.0–5895.9 Å
		(famous Na D lines)

---

† Exceptions (as quadrupole transitions) are found in stellar spectra.
‡ Ch. Georgalas, Class of 1963G.

(c) Green   5682.7–5688.2 Å
(d)   5149.1–5153.6 Å
(e)   4978.6–4982.9 Å
(f) Blue   4748.0–4751.9 Å
(g)   4664.9–4668.6 Å
(h) Blue-Violet 4494.3–4497.7 Å

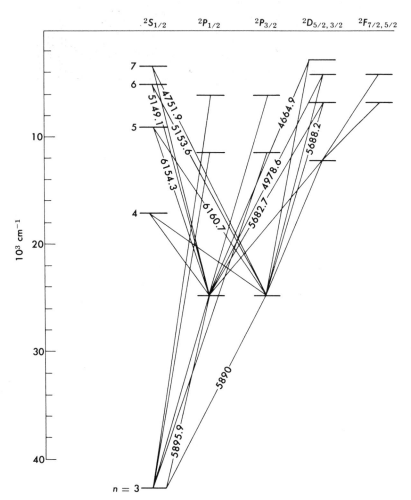

FIG. 2.11 The "allowed" transitions between the energy levels of sodium. The wavelengths in Angstroms of some of the principal lines are indicated. Note that the P states have now been shown in two columns, one referred to as $P_{1/2}$ the other as $P_{3/2}$; the small difference between their energy levels is the "fine structure."

4.3 FINE STRUCTURE

In the data above two wavelengths were given for each sodium line. Indeed, by viewing through the constant deviation or the grating spectrometer it is easy to resolve into a doublet each of the lines that appear in Fig. 2.12; the spacing is of the order of a few angstroms.

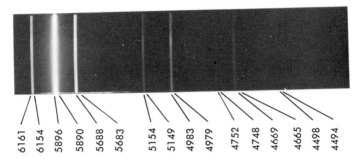

FIG. 2.12 Photograph of the visible spectrum of sodium as obtained with the constant-deviation spectrograph.

The data in Table 2.4 on the red, yellow, and green lines of sodium, viewed with the grating, were obtained by a student† simultaneously with the data used for the determination of d (Section 2.2 above):

To reduce the data we note from Eq. 2.1 that

$$n_k\lambda = d(\sin \theta_k - \sin \theta_i)$$

TABLE 2.4

DATA ON THE FINE STRUCTURE OF SODIUM AS OBTAINED WITH THE SMALL GRATING SPECTROMETER

Line	Order	θ_1	θ_2	$\Delta\theta$ (in radians)
Red	2	41°27′	41°29′	5.8×10^{-4}
	3	55°58′	56°00′	5.8
Yellow	2	40°21′	40°23′	5.8
	3	53°49′	53°52′	8.7
	4	75°15′	75°23′	23.2
Green	2	39°32′	39°33′	2.9
	4	70°48′	70°56′	23.2

† R. S. Stevens, class of 1963.

where θ_i is the angle of incidence. Also

$$\theta_2 = \theta_1 + \Delta\theta$$

By letting $\sin \Delta\theta_k \simeq \Delta\theta_k$; $\cos \Delta\theta_k \simeq 1$,

$$n_k(\Delta\lambda) = d(\cos \theta_k)\Delta\theta_k \qquad (4.2)$$

Using $d = 36{,}921$ Å and averaging over orders within each line

$$\Delta\lambda = d \frac{\sum \cos \theta_k \, \Delta\theta_k}{\sum n_k} \qquad (4.3)$$

we obtain for $\Delta\lambda$:

Line	$\Delta\lambda$ (Experiment)	$\Delta\lambda$ (Exact value)
Red	5.7 Å	6.51 Å
Yellow	6.3 Å	5.97 Å
Green	5.9 Å	5.55 Å

The experimental data are thus in fair agreement with the exact values.

This splitting of spectral lines was named "fine structure" and must reflect a splitting of the energy levels of sodium; if we express the wavelengths of sodium given in Section 4.2 in wave numbers, ($\bar{\nu} = 1/\lambda = \nu/c$) (in a scale proportional to energy), it becomes evident that the spacing in all doublets is exactly the same and equal to 17.3 cm⁻¹. Indeed, the doublet structure of all the above lines is due to the splitting of only the $3P$ ($n = 3$, $l = 1$) level as can be seen by referring back to Fig. 2.11. As we now know, the splitting of the $3P$ level is due to the effect of the electron "spin" and its coupling to the orbital angular momentum (designated by l). According to the Dirac theory, the electron possesses an additional degree of freedom, called the "spin," which has the properties of angular momentum of magnitude $s = \hbar/2$ (and therefore two possible orientations with respect to any axis, $m_s = +\frac{1}{2}$ or $m_s = -\frac{1}{2}$). The spin s can then be coupled to l according to the quantum-mechanical rules of addition for angular momenta; this will result in a total angular momentum of magnitude $J = l + \frac{1}{2}$ or $J = l - \frac{1}{2}$ and the energy of the state will depend on J. In the case of sodium, the $3P$ level splits into two levels, with $J = \frac{1}{2}$ and $J = \frac{3}{2}$ designated as $3P_{1/2}$ and $3P_{3/2}$ and separated by 17.3 cm⁻¹.

4.4 ELECTRON-ELECTRON COUPLING; THE MERCURY SPECTRUM

The mercury atom has $Z = 80$, hence 80 electrons; these fill shells $n = 1$, $n = 2$, $n = 3$, and $n = 4$ completely—(60 electrons) and in addition, from the $n = 5$ shell, the $l = 0, 1, 2$ subshells account for another 18 electrons. The remaining two electrons instead of occupying the $l = 3$

and $l = 4$ subshells are in the $n = 6$ shell with $l = 0$, giving rise to a configuration exactly equivalent to that of the helium atom.

We thus have an atom with two electrons outside closed shells in contrast to the one-electron systems of the hydrogen and sodium type. In the two-electron system, we can hardly speak of the n number of the system (atom), since each electron may be in a different shell; however we can still assign a total angular momentum $\mathbf{J}$ to the system, which will be the resultant of the values of each of the two electrons, and (as we saw in the previous section) of their additional degree of freedom, the "spin." The addition of these four angular momenta, $\mathbf{l}_1$, $\mathbf{l}_2$, $\mathbf{s}_1$, $\mathbf{s}_2$ to obtain the resultant $\mathbf{J}$ can be done in several ways; for the helium or mercury atom, the Russell-Saunders coupling scheme holds, in which $\mathbf{l}_1$ and $\mathbf{l}_2$ are coupled into a resultant orbital angular momentum $\mathbf{L}$ and $\mathbf{s}_1$ and $\mathbf{s}_2$ into a resultant spin $\mathbf{S}$; finally $\mathbf{L}$ and $\mathbf{S}$ are coupled to give the total angular momentum of the system $\mathbf{J}$†. Since $\mathbf{s}_1$ and $\mathbf{s}_2$ have necessarily magnitude $\frac{1}{2}$, the resultant $\mathbf{S}$ has magnitude $S = 0$ or $S = 1$. It is customary to call the states with $S = 0$ singlets, those with $S = 1$ triplets, since when $S = 0$ for any value of L, only a single state can result, with $J = L + S = L$; when $S = 1$, however, three states can result with $J = L + S, L, L - S$, namely, $J = L + 1, L, L - 1$ (provided $L \neq 0$). In systems where energy states have total angular momentum J, the selection rules for optical transitions are different—namely,

$$\Delta L = \pm 1$$

$$\Delta J = 0, \pm 1 \quad \text{but not} \quad J = 0 \rightarrow J = 0,$$

(4.4)

and in principle no transitions between triplet and singlet states.

With these remarks in mind we consider the energy-level diagram of mercury. Since there are two electrons outside a closed shell, in the ground state they will both be in the $n = 6$, $l = 0$ orbit, and hence (due to the Pauli principle) must have opposite orientations of their spin, leading to $S = 0$; the spectroscopic notation is $^1S_0$. For the excited states one should expect both a family of singlet states and a family of triplet states; the singlets $S = 0$, will be

$^1S_0$ for $L = 0$, and necessarily $J = 0$

$^1P_1$ for $L = 1$, and necessarily $J = 1$

$^1D_2$ for $L = 2$, and necessarily $J = 2$ etc.

---

† In the ensuing discussion the quantum-mechanical rules of addition of angular momentum are considered known. However, even if the student is not familiar with them, he can infer them from following carefully the development of the argument.

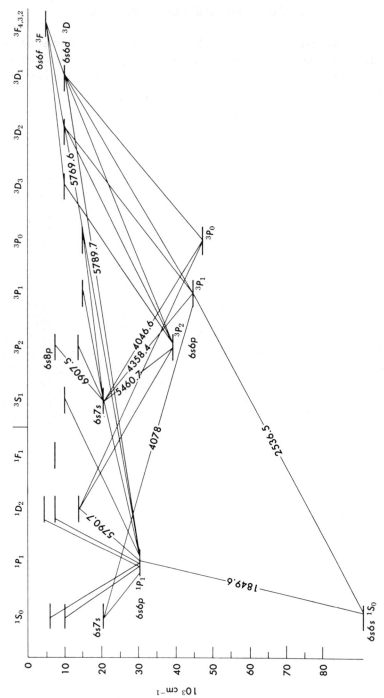

FIG. 2.13 Energy-level diagram and the principal lines in the spectrum of the hydrogen atom.

Note the spectroscopic notation, where the upper left index is $2S + 1$ indicating the singlet nature of the state; the capital letter indicates the total L of the system (according to the convention); and the lower right index stands for J. For the triplets, $S = 1$, and the states are

$$^3S_1 \qquad \text{for } L = 0 \quad J = 1$$
$$^3P_{0,1,2} \qquad \text{for } L = 1 \quad J = 0, 1, 2$$
$$^3D_{1,2,3} \qquad \text{for } L = 2 \quad J = 1, 2, 3, \text{ etc.}$$

The energy levels for mercury are shown in Fig. 2.13 with some of the strongest lines of the spectrum. It is seen that the selection rules on ΔL, and ΔJ always hold, but that transitions with $\Delta S \neq 0$ do occur. It is also to be noted that the "fine structure," that is, the splitting of the $6s6p \, ^3P$ level, is of considerable magnitude: $^3P_0 - {}^3P_1 = 190$ cm^{-1}; $^3P_1 - {}^3P_2 = 460$ cm^{-1}.

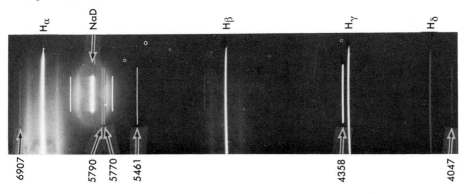

F$_{IG}$. 2.14 Photograph of the superimposed spectra of hydrogen (long slit), mercury (medium slit length), and sodium (short slit).

Figure 2.14 is a reproduction of a superimposed spectra of hydrogen (longest lines), mercury (medium length), and sodium (shortest lines) obtained by a student with the prism spectrograph. Beginning with long wavelengths (from the left) one identifies the following lines of mercury:

(a)	Red	6907.5 Å
(b)	Yellow doublet	5789.7–5769.6
(c)	Green	5460.7
(d)	Blue }triplet	4358.4
(e)	Violet	4046.6

This concludes the discussion of the spectra and energy levels of the sodium and mercury atoms. The same treatment applies to all other one-

or two-electron atoms, as well as to those with a one- or two-electron deficiency (hole) from a closed shell.

Systems with more electrons outside closed shells are treated on analogous lines, but the coupling schemes become more complicated, giving rise, as in the case of the rare earths, to extremely complex spectra.

5. Line Spectra of Nuclei

If atoms are quantum-mechanical systems and a typical manifestation of this fact is the emission of spectral lines of light, it should be expected that nuclei, when excited, would emit similar line spectra.

Since the nuclear radius is three to five orders of magnitude smaller than that of atoms, the forces that bind the nucleus (against the repulsion of the positive charges confined in its volume) must be correspondingly stronger than the forces that bind the atomic electrons to the nucleus. As a consequence, the energy levels and the quanta of energy emitted in a nuclear transition are also orders of magnitudes larger than those of atomic transitions. Indeed, the quanta of electromagnetic radiation emitted in a nuclear transition fall in the gamma-ray region, and new techniques are needed for their detection and for the measurement of their (wavelength) energy.

Further, because of the larger spacing between energy levels, it is not easy to excite a nucleus from its ground state by the simple means of electric discharges or arc sources such as are used for the atoms; instead, beams of neutrons or high-energy gamma rays, or high-energy charged particles, are required. However, in distinction from atomic transitions (probability of the order of 10^8/sec) some nuclear transitions have a very small probability (such as 10^{-7}/sec, corresponding to a lifetime of 100 days). Thus it is possible to excite a sample of nuclei inside a nuclear reactor, or by subjecting them to cyclotron bombardment, or by other means, and subsequently bring them to the laboratory for measuring their spectrum or for other uses. As a matter of fact, some of the nuclei that have very long lifetimes can be found in nature in their excited state; these are the naturally radioactive elements.

Another important difference is that in atomic transitions the only quanta of energy we considered were the quanta of electromagnetic radiation, the photons, while in a nuclear transition both alpha, beta, and gamma rays are emitted. A detailed discussion of how these quanta are detected and how their energy is measured is given in Chapter 5, but some results obtained in this laboratory are shown in the Figs. 2.15, 2.16, and 2.17.

Figure 2.15 gives the gamma-ray (photon) spectrum of the Co^{60} nucleus, and Fig. 2.16 that of the Cs^{137} nucleus. The detector used is a NaI crystal viewed with a photomultiplier tube, which gives electrical pulses propor-

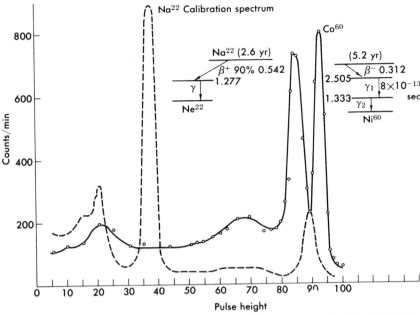

FIG. 2.15   The γ-ray spectrum of Co^{60} obtained with a NaI crystal. The dotted line gives the spectrum of Na^{22} which is used for calibration purposes. The decay schemes of these nuclei are also indicated.

tional to the number of light quanta *in the visible region* that strike its cathode. The NaI crystal acts essentially as a converter, absorbing the high-energy gamma-ray quantum and emitting in a fast sequence (10^{-7} sec) many quanta of lower energy (to which the photomultiplier is sensitive), so that the total energy equals the energy of the gamma ray. In Figs. 2.15 and 2.16 a plot of the number of electrical pulses (ordinate) per unit time with a given amplitude† (abscissa) is shown; hence this is not an absolute measurement of the energy, but the abscissa is *proportional* to the energy of the gamma quanta emitted by the source.

By using some gamma line of known energy,‡ it is possible to calibrate this "pulse height" spectrometer, and we obtain

$$Co^{60} \begin{cases} \gamma_1 = 1.200 \text{ MeV} \\ \\ \gamma_2 = 1.320 \text{ MeV} \end{cases} \tag{5.1}$$

$$Cs^{137} \; \gamma \; = 0.663 \text{ MeV} \tag{5.2}$$

---

† More correctly, that is within a given amplitude window.

‡ The 0.511-MeV line of positron annihilation and the 1.28-MeV gamma ray from Na^{22} were used.

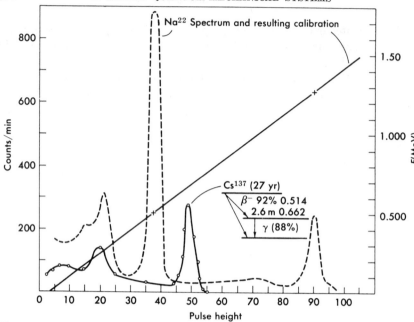

FIG. 2.16 Same as Fig. 2.15 above but for Cs¹³⁷. The calibration resulting from Na²² is indicated by the straight line which is to be referred to the right-hand energy scale.

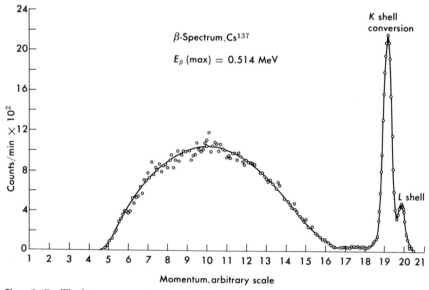

FIG. 2.17 The beta-ray spectrum of Cs¹³⁷. Note the continuous part (due to beta decay), and the monoenergetic line. The latter is due to "internal conversion" resulting in the ejection of an electron from the K shell; the subsidiary peak at the extreme right is due to the ejection of an electron from the L shell.

The accepted energy levels for these nuclei (obtained by correlating data of many types) are also shown in the figures, and the agreement with the measured values is good.

Figure 2.17, on the other hand, gives the spectrum of beta rays (electrons) emitted again from the Cs^{137} nucleus, as obtained by a student.[†] Here the electrons are detected by using a Geiger counter. Their momentum was measured by subjecting them to a path of known curvature in a magnetic field, whose strength was varied (thin-lens spectrometer). The graph gives the number of counts at the detector per unit time with given momentum p (within an interval Δp). In principle, the abscissa gives an absolute scale, but most frequently the magnet current is calibrated with a known beta line.

One notices in the spectrum a spectral line at $p = 1.012$ MeV/c and a continuous spectrum ending[‡] at $p = 0.900 \pm 20$ MeV/c. The mono-energetic line in this case is due to the ejection of an electron from the K shell of the cesium atom. This phenomenon is called "internal conversion"; the energy of the nuclear transition shown in Fig. 2.16 has been transferred (in 12 percent of all transitions) to an atomic electron.

Let us then calculate the kinetic energy E_β of the ejected electron. Since $p = 1.012$ MeV/c and $m_e = 0.511$ MeV/c^2,

$$mc^2 + E_\beta = (c^2 p^2 + m^2 c^4)^{1/2} = 1.135 \text{ MeV}$$

hence

$$E_\beta = 0.624 \text{ MeV} \tag{5.3}$$

However, E_β is related to the energy of the nuclear transition through

$$E_\beta = E_\gamma + E_K \tag{5.4}$$

where E_K is the energy of the electron before it absorbed the energy of the nuclear transition. Clearly E_K is the (negative) binding energy of the K shell of cesium; from x-ray data it is known to be

$$E_K(\text{Cs}) = -0.038 \text{ MeV} \tag{5.5}$$

Combining Eqs. 5.3, 5.4, and 5.5 we obtain

$$E_\gamma = 0.662 \text{ MeV} \tag{5.6}$$

in remarkable agreement with the value obtained from the direct measurement of the gamma-ray energy (Eq. 5.2).

[†] T. Walter, class of 1962.

[‡] The origin of the continuous spectrum cannot be detected in this setup because of the absorption of the slow electrons in the exit window of the spectrometer and the Geiger counter window.

Finally we turn to the continuous part of the beta spectrum; this no longer has the form of discrete energy lines with which we were so familiar from the study of the atomic systems, and which we also encountered in some nuclear spectra. If we wish to retain the scheme of nuclear energy levels, then the electrons must be carrying away only part of the energy of the nuclear transition, the remainder being taken away by one or more other quanta.

Indeed, we know today that an electron and a *neutrino* are simultaneously emitted and share the energy of the nuclear transition, while momentum is balanced between the nucleus itself and the electron and neutrino. The neutrino is a noncharged, weakly interacting particle with very small or zero mass. Because of this sharing process, and since the mass of the neutrino is negligible, the upper end of the continuous beta spectrum should give the energy difference between the initial and final state of the transition; this corresponds to 0.522 ± 15 MeV, and is indicated in the level diagram of Fig. 2.17. Transitions involving the emission of an electron (and neutrino) are shown by arrows slanting to the right (increase in nuclear Z); positron emission is shown by arrows slanting to the left. We usually write this as

$$A(Z) \rightarrow A(Z+1) + \beta^- + \nu \tag{5.7}$$

In Eq. 5.7 A and Z stand for the atomic number and charge of the nucleus, while β^- is the symbol for the electron and ν for the neutrino.

Thus we have seen that simple quantum-mechanical systems, such as the free hydrogen atom or other free atoms, or even nuclei, are characterized by discrete energy levels. The systems can undergo transitions between these levels with the emission or absorption of energy quanta, in the form of photons, electrons, or other particles. When only one particle is emitted, it carries away the whole energy of the transition; otherwise the energy is shared between the participating particles. The energy of these particles can be measured, as was shown, by various techniques, and thus provide information on the structure of the energy levels of these systems. For a more sophisticated treatment of atomic spectroscopy the reader should consult Chapter 7 and the references therein.

3

QUANTUM-MECHANICAL SYSTEMS WITH LARGE NUMBERS OF INTERACTING PARTICLES

1. Introduction

In this chapter systems with very large numbers of interacting particles will be considered. Actually most of matter as it can be perceived with our senses belongs in this category. In matter in the gaseous state, the distances between molecules are large and therefore the forces are weak. In solids, however, the forces are much stronger. Understanding of the thermodynamic properties of "bulk" matter, based on the microscopic behavior of the constituent molecules or atoms, was first achieved through the statistical mechanics developed by Boltzmann. Because of the immense number of interacting bodies, the statistical approach is quite valid and has proved highly successful. Classical statistical mechanics, however, was unable to explain several phenomena until quantum mechanical principles were incorporated. According to our current ideas, particles with half-integral spin—such as the electrons—obey so called "Fermi-Dirac" statistics, while particles with integral spin—such as photons and helium atoms—obey "Bose-Einstein" statistics. The fundamental distinction is that the former tpye of particles must have a completely antisymmetric

57

wave function, whereas the latter ones must have a symmetric wave function; this leads to a different distribution function for the probability that a particle will occupy a certain cell in phase space. Since the electronic properties of solids are determined by the behavior of their electrons, it is Fermi statistics that are relevant and will be applied to the description of thermionic emission and the properties of semiconductors.

Most solid-state materials, however, have a crystalline structure; that is, the atoms form a periodic lattice. Advantage can be taken of this periodicity so that the macroscopic behavior of the crystal is predicted from the general parameters of the lattice and the atoms that form it. It is found that the free electrons, instead of occupying distinct energy levels—as they do in the simpler quantum-mechanical systems—are contained in certain energy bands. Knowledge of the "band structure" is necessary in most considerations of the solid state and specifically in the understanding of the behavior of semiconductors. The motion of the free electrons or holes (contained in the valence band) through the lattice can be studied in terms of a single-particle approach. Such phenomena as scattering and the absorption or emission of vibrational quanta (phonons) are invoked and are useful in explaining further details in the macroscopic behavior of the sample.

An experiment in which the resistivity and Hall effect of germanium are measured as a function of temperature is described in Section 3. In Section 4 is given a brief sketch of junction theory and the principles involved; it is relevant to one of the most important applications of semiconductors, the transistor.

1.1 THE FERMI-DIRAC DISTRIBUTION

Let us consider a large ensemble of free Fermi particles (such as electrons); the assumption is made that in phase space† there exist many states that these electrons can occupy. Each "cell" has a phase-space volume of h^3 (where h is again Planck's constant), so that the number of available cells for a differential volume of phase space is

$$n = (h^3)^{-1} dp_x \, dp_y \, dp_z \, dx \, dy \, dz \qquad (1.1)$$

According to the exclusion principle, however, each cell can be occupied by two electrons (one with spin up and one with spin down), so that the number of available electron states is $2n$. If we integrate over the space coordinates and divide by the volume, we obtain the number of states n' per

---

† Phase space is a space spanned by the momentum and position vector of a particle. Thus a particle moving in ordinary three-dimensional space will have six components in phase space.

unit volume per differential element in momentum space:

$$n' = \frac{2}{V_0} \int_x \int_y \int_z n = \left(\frac{2}{h^3}\right) dp_x \, dp_y \, dp_z$$

Further, we can obtain the number of states per unit volume per unit energy interval

$$n_i = \frac{n'}{dw_i} = \frac{2}{h^3} 4\pi p_i^2 \, dp_i \frac{1}{dw_i}$$

and since

$$w_i = \frac{p_i^2}{2m} \qquad dw_i = \frac{2p_i \, dp_i}{2m}$$

$$\frac{dN(w_i)}{dw_i} = n_i = \frac{8\pi}{h^3} \sqrt{2m^3 w_i} \tag{1.2}$$

Equation (1.2), which was obtained from very simple considerations, represents the number of states per unit volume per unit energy interval (at a given energy) and is called the "energy density of states." We note that for a simple ensemble of free Fermi particles (a) all energies are permissible (since $dN(w)/dw$ is a continuous and not singular function); namely, the energy is *not quantized* and (b) the number of states increases with increasing energy.

Proceeding further to specify our system, we would like to know which of these infinitely many states are occupied, or in a statistical fashion, what is the probability that a state i of given energy w_i be occupied. This is the Fermi-Dirac distribution and is given by

$$\frac{N_i}{2n} = \left[\exp\left(\frac{w_F - w_i}{kT}\right) + 1 \right]^{-1} \tag{1.3}$$

where

k is the Boltzmann constant,
T is the temperature of the system, and
w_F is a characteristic energy, called the Fermi energy or Fermi-level energy.

It is interesting to note the properties of this function, graphed in Fig. 3.1:

(a) It is properly bounded, so that it can represent a probability

$$0 < N_i/2n < 1$$

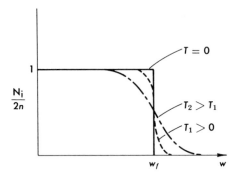

(b) For large values of w it assumes the form of the Boltzmann distribution

$$\text{Const} \times \exp\left(-w/kT\right)$$

(c) For $T = 0$ it is a step function, with

$$N_i/2n = 1 \qquad w_i < w_F$$

$$N_i/2n = 0 \qquad w_i > w_F$$

(d) For $T \neq 0$, w_F has the property that $N(w_F) = \frac{1}{2}$, and as many states above w_F are occupied, that many states below w_F are empty.

(e) In solids and for average $T \neq 0$, the distribution function is only slightly modified from its shape at $T = 0$ (for solids w_F is of the order of a few electron volts, while $1/kT = 40$ at $T = 300$).

Combining the Fermi-Dirac distribution (Eq. 1.3) with the energy density of states (Eq. 1.2) it is possible to obtain any desired distribution. For example, the number of electrons per unit volume (density) at an energy w in the interval dw is given by

$$N(w)\,dw = \frac{8\pi}{h^3} \sqrt{2m^3w} \; \{\exp\left(\frac{w_F - w}{kT}\right) + 1\}^{-1} \, dw \qquad (1.4)$$

If we express Eq. 1.4 in terms of the cartesian coordinates of the velocity, v_x, v_y, and v_z, and integrate over v_x and v_y, we obtain the number of electrons per unit volume with a given velocity in the z direction, v_z (in the interval dv_z). The result of this integration is[†]

$$N(v_z)\,dv_z = \frac{8\pi}{h^3} \frac{m^2kT}{2} \ln\left\{1 + \exp\left(\frac{w_F - mv_z^2/2}{kT}\right)\right\} dv_z \qquad (1.4a)$$

The two distributions given by Eqs. 1.4 are shown in Fig. 3.2.

[†] A. Sommerfeld, *Thermodynamics and Statistical Mechanics*. New York: Academic Press, 1956, p. 280.

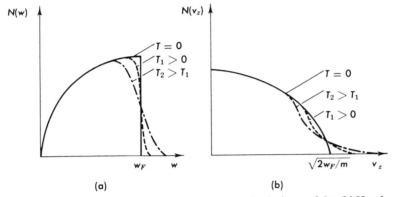

Fig. 3.2 (a) Number of electrons with an energy w in the interval dw. (b) Number of electrons with z component of velocity v_z in the interval dv_z.

Even though the majority of the electrons in a solid are not free (as we originally assumed) Fermi-Dirac statistics are applicable, especially to metals. In metals at least one electron per atom has several states available (is in the conduction band), so that it can be considered free; since there will be 6×10^{23} free electrons per gram mole, statistical methods are well justified.

1.2 ELEMENTS FROM THE BAND THEORY OF SOLIDS

Up to now, no account has been taken of the interatomic or intramolecular forces that might act on the free electrons. Indeed, we expect (from previous experience) that the consideration of some potential in the region where the electrons move will result in the appearance of energy levels; however, because of the periodic structure of this potential, instead of energy levels, *energy bands* appear, and only the states contained in these bands can be occupied (with any significant probability). In the following paragraphs we will sketch two approaches toward the understanding of the physical origin of the energy bands.

Consider first the one-dimensional problem† of an electron moving in a potential consisting of an infinite sequence of "square" wells of depth V_0, width b and spaced at a distance l from one another (Fig. 3.3). The solution of the Schördinger equation for such a potential gives for the electron a wave function:

$$\Psi_k = u_k(x)e^{ikx} \tag{1.5}$$

with $k = 2\pi/\lambda = p/\hbar$ the wave number of the electron. This wave function

† V. B. Rojansky, *Introductory Quantum Mechanics*. New York: Prentice-Hall, 1938, paragraph 49; or E. Mertzbacher, *Quantum Mechanics*. New York: Wiley, 1962.

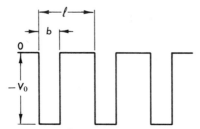

FIG. 3.3 A periodic potential that may be considered as an idealization to the actual potential of a crystal lattice.

consists of the plane wave part e^{ikx}, and $u_k(x)$, which have the must periodicity of the lattice, namely, $u_k(x \pm l) = u_k(x)$. If there are N lattice sites, the length of the crystal is Nl and we impose the following periodic boundary condition, $\Psi_k(x + N) = \Psi_k(x)$. This leads to $e^{ikNl} = 1$, or

$$kNl = n2\pi$$

$$k = n2\pi/Nl \qquad n = 0, \pm1, \pm2, \cdots \qquad (1.6)$$

Equation 1.6 determines the allowed values of k, which form almost a continuum because of the very large integer value of N. Note that for $N = 1$ one obtains the familiar "particle in a box" energy levels, with

$$E = \frac{p^2}{2m} = \frac{k^2\hbar^2}{2m} = \frac{n^2 h^2}{2ml^2}$$

Having determined the wave function, it is possible to solve the Schrödinger equation for the energy eigenvalues

$$H\Psi_k = E(k)\Psi_k \qquad \text{or} \qquad \Psi_k^*H\Psi_k = E(k) \qquad (1.7)$$

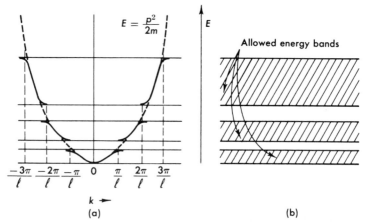

(a)

(b)

FIG. 3.4 Results of the solution of the simplified one-dimensional lattice problem. (a) Plot of energy E versus wave number $k = p\hbar$ for an electron in a crystal lattice. (b) The allowed and forbidden energy bands.

where H is the one-dimensional Hamiltonian operator.

$$H = -\frac{h^2}{2m}\frac{d^2}{dx^2} + V(x)$$

and $V(x)$ is now the potential of Fig. 3.3.

The solution of Eq. 1.7 is given in graphical form in Fig. 3.4. We note the following:

(a) Even though all values of k are allowed, discontinuities arise at $k = n\pi/l$ (note that for this particular electron wavelength, Bragg reflection from the lattice will occur with a half-angle $\theta = 90°$; $n\lambda = 2l \sin \theta$, hence $\lambda = 2l/n$, and since $\lambda = 2\pi/k$, it follows $k = n\pi/l$).

(b) Not all values of the energy are allowed, but only certain "bands"; other bands of energy are forbidden.

(c) The relation between E and p (or k) is no longer the familiar parabolic

$$E = \frac{p^2}{2m} = \frac{k^2\hbar^2}{2m} \tag{1.8}$$

We can, however, retain this relation if the mass m is assumed variable and a function of k, namely,

$$m^* = \frac{\hbar^2}{(d^2E/dk^2)} \tag{1.9}$$

In three dimensions the same formalism is carried over, but now the bands are replaced by allowed (Brillouin) surfaces and the axes of symmetry of the crystal must be taken into account.

A different approach is to start with a molecular wave function and study its behavior as the number of identical atoms is increased. In Fig. 3.5 are plotted the energy levels against interatomic distance for the $1s$ and $2s$ states of a linear array of six atoms (after Shockley).

If, then, in the limit the (almost infinite) array of the crystal is considered, the energy levels coalesce into bands. This is shown in the left-hand side of Figs. 3.6 and 3.7, where the energy bands plotted against interatomic spacing are given (after Kimball) for diamond, which is an insulator, and (after Slater) for sodium, which is a conductor.

If the lattice spacing for the particular crystal is known (as from observation), it is possible to read off from the graphs the limits of the energy bands. This is done diagrammatically on the right hand side of Figs. 3.6 and 3.7; also indicated is the position (in electron volts), of the Fermi level (as it can be calculated, for example, from Eq. 1.4 and the electron density within each band).

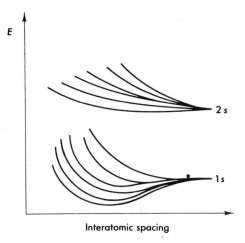

FIG. 3.5 Energy levels of a system of six similar atoms placed in a linear array.

From these considerations it is possible to understand the difference between conductors, insulators, and semiconductors. For diamond, for example, the valence band is completely filled (this fact follows also from the atomic structure of carbon and the deformation of the energy levels). The next available states are approximately 5.4 eV higher and hence cannot be reached by the electrons, with a consequent inhibition of their mobility; diamond therefore behaves as an insulator. For sodium, on the contrary, the Fermi level lies in the middle of an energy band, so that many states are available for the $(3s)$ electron, which can move in the crystal freely;

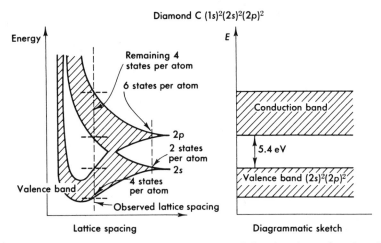

FIG. 3.6 The energy band structure of diamond (insulator) as a function of lattice spacing. The observed lattice spacing is also indicated.

sodium behaves as a conductor. Pure semiconductors, such as germanium, have a configuration such that the valence band is completely filled, but the conduction band lies fairly closely to it (0.80 eV). At high enough temperatures (that is, of the order of a few thousands of degrees), the electrons in the valence band acquire enough energy to cross the gap and occupy a state in the conduction band; when this happens the material which was previously an insulator becomes intrinsically conducting.

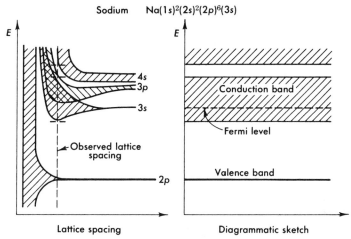

Fig. 3.7   The energy band structure of sodium (conductor) as a function of lattice spacing. The observed lattice spacing and position of the Fermi level are also indicated.

Both the electric and thermal conductivity of a solid depend on the density and mobility of the free electrons. Completely analogous to the motion of electrons is the motion of "holes"; holes can be thought of either as "vacancies" in an almost-filled band, or as electrons with negative effective mass.† Due to their thermal energy, the carriers have a random motion characterized by $(3/2)kT = E = m^*v^2/2$. When electric potential or temperature gradients are applied, a drift velocity is superimposed on the random motion of the carriers in such a direction as to establish a steady state current flow.

2. Thermionic Emission of Electrons from Metals

It is well known that when a metal is heated to high temperatures it emits electrons, as in the case of the filament of an electron tube. To ob-

† This can be seen from Eq. 1.9 and the negative curvature of some parts of the $E(k)$ curve of Fig. 3.4a.

serve thermionic current, both the emitter and detector are placed in an evacuated vessel and electric field acceleration is provided. However, the electric fields present and the geometrical configuration of filament and cathode complicate the interpretation of the emission process itself. We will proceed by first deriving the expression for the emitted current density (Richardson's equation) and the modifications needed when an accelerating or decelerating potential is applied; next we will give the expression for the space charge limited current† (Child's law) and consider the experimentally measurable quantities; finally, the equipment and experimental procedure will be discussed and analysis of specific data will follow.

2.1 DERIVATION OF RICHARDSON'S EQUATION

Let us first consider the potential in the vicinity of a metal boundary. We assume that the potential ϕ inside the metal is constant (hence $E = 0$), but at the boundary proper, there must exist strong forces if the electrons are to remain contained; Fig. 3.8a shows such a square barrier potential. However, an electron which is outside a plane conducting surface is attracted to it by the force exerted between itself and its fictitious image. (The image is introduced so that for $z \geq 0$ the potential satisfies both Laplace's equation and the boundary conditions on the metal surface, Fig. 3.8b). Since $F = e^2/(4\pi\epsilon_0 4z^2)$, it follows that

$$V = -\int E \, dz = -\frac{e}{16 \pi\epsilon_0 z} \tag{2.1}$$

Combining Eq. 2.1 with the square barrier of Fig. 3.8a, we obtain the more

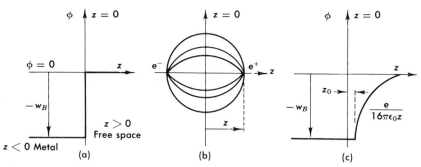

FIG. 3.8   Idealized potentials to which an electron near the surface of a metal is subjected. (a) Square barrier at metal boundary. (b) The lines of force between an electron and its image. (c) The resulting potential from combining (a) and (b).

† See, for example, N. H. Frank, *Introduction to Electricity and Optics*. New York: McGraw-Hill, 1950, p. 220.

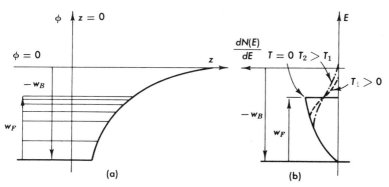

FIG. 3.9 The energy distribution of electrons inside a metal with respect to the potential barrier. (a) The ordinate gives energy, whereas the abscissa gives the distance from the boundary; the density of horizontal lines is proportional to the number of electrons that have this particular energy. (b) The density of occupied electron states versus energy; this is shown for different Temperatures, T.

realistic image field barrier shown in Fig. 3.8c. The Coulomb potential has been cut off at $z = z_0$ where $V(z_0) = -e/(16\,\pi\epsilon_0 z_0) = w_B$; w_B is the depth of the potential well in which the *free* electrons are bound; or in band structure terminology, the potential energy of the bottom of the conduction band. In Fig. 3.9a the density of states inside the potential well is sketched by using Eq. 1.4; in Fig. 3.9b a plot of the density (abscissa) for various temperatures is given. Thus, free electrons with energy $E > w_B$ can overcome the boundary and will escape if their direction of motion is towards the boundary. As is usually done in escape problems, we calculate the number of electrons within a right cylinder of unit area, and of height v_z, with $v_z^2 > 2w_B/m$ (this gives the number of electrons that reach unit area in unit time and have the appropriate velocity component). Thus

$$J = eN(\text{escape/sec cm}^2) = e \int_{v_z = \sqrt{(2w_B/m)}}^{+\infty} \frac{dN(v_z)}{dv_z} v_z\, dv_z$$

the distribution function $dN(v_z)/dv_z$ has been given in Eq. 1.4a, and for most values of T the exponential is very small (at $T = 3000°$ K the exponential is e^{-8}) so that

$$\ln(1 + x) = x - \tfrac{1}{2}x^2 + \tfrac{1}{3}x^3 - \tfrac{1}{4}x^4 + \cdots$$

and thus

$$J = e\,\frac{4\pi m^2 kT}{h^3} \int_{\sqrt{(2w_B/m)}}^{+\infty} v_z \exp\left(w_F - \tfrac{1}{2}mv_z^2\right) dv_z$$

We make the following change of variables

$$u = \frac{w_F - \frac{1}{2}mv_z^2}{kT} \qquad du = -\left(\frac{mv_z}{kT}\right) dv_z$$

and

$$v_z = \left(\frac{2w_B}{m}\right)^{1/2} \rightarrow u = \frac{w_F - w_B}{kT}$$

$$v_z = +\infty \rightarrow u = -\infty$$

so that

$$J = -\frac{4\pi mek^2 T^2}{h^3} \int_{(w_F - w_B)/kT}^{-\infty} e^u \, du$$

giving Richardson's equation,

$$J_0 = A_0 T^2 \exp\left(-e\phi/kT\right) \tag{2.2}$$

with

$$A_0 = \frac{4\pi mek^2}{h^3} = 1.2 \times 10^6 \text{ amp/m}^2\text{-deg}^2$$

and

$$\phi = \frac{w_B - w_F}{e}$$

is the work function in volts.

We have thus obtained Richardson's equation, which shows that the thermionic emission is dominated by an exponential temperature dependence. This dependence is so strong that the T^2 factor is completely masked and cannot be directly verified by experiment. Furthermore, the constant A_0 observed experimentally is seldom in agreement with the theoretical value. This is due to the simple assumptions used in the derivation and the neglect of such effects as:

(a) Temperature dependence due to variations in ϕ. Because if

$$\phi = \phi_0 + aT,$$

then

$$J = (A_0 e^{-ae/k}) T^2 \exp\left(-e\phi_0/kT\right)$$

where the exponential can easily alter the value of A_0 by factors of 2 or larger.

(b) Quantum-mechanical reflections of the electrons at the surface. Such reflections will be equivalent to a modification of the barrier potential assumed.

(c) Nonuniformity of the emitting surface, since we really measure

$$J = A_0 T^2 \sum_i \exp\left(-\phi_i e/kT\right)$$

We can now consider the modifications to the emission-current density J_0 due to the application of external fields. The two cases of a retarding field and of an accelerating field are sketched in Figs. 3.10a and 3.10b, respectively.

For a retarding field, the barrier potential that the electron has to over-come is increased by the whole amount of the retarding potential V_0, so that

$$J' = A T^2 \exp\left(-\frac{e(\phi + V_0)}{kT}\right) = J_0 \exp\left(-\frac{V_0 e}{kT}\right) \qquad (2.3a)$$

hence an exponential reduction of the current with increasing retarding potential.

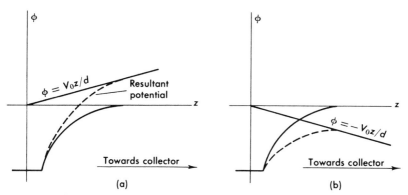

FIG. 3.10   The potential seen by an electron inside a metal when an external field is applied. (a) Retarding field. (b) Accelerating field.

An accelerating field, on the other hand, as can be seen from Fig. 3.10b, can lower the potential barrier by only a small amount. For plane geometry the resultant potential is

$$V = -\frac{e}{16\pi\epsilon_0 z} - \frac{V_0 z}{d}$$

which has a maximum at

$$z_{max} = \sqrt{\frac{ed}{16\,\pi\epsilon_0 V_0}}$$

with the value

$$V_{\max} = -2\sqrt{\frac{V_0 e}{16\,\pi\epsilon_0 d}}.$$

The barrier seen by the electrons is

$$e\phi = -V_{\max} - w_B = -w_B + 2\sqrt{\frac{V_0 e}{16\,\pi\epsilon_0 d}}$$

and

$$J_s = J_0 \exp\left[0.44(E^{1/2})/T\right], \tag{2.3b}$$

where E is the electric field in volts per meter (for a plane diode $E = V_0/d$), and J_s is the saturation current. Because of the factor $0.44/T$, the increase in emission current is very small; on a semilog plot it is proportional to the square root of the applied accelerating potential.

2.2 CHILD'S LAW AND THE MEASURABLE QUANTITIES IN THE THERMIONIC EMISSION EXPERIMENT

As mentioned before, the measured values of the thermionic current are affected by space-charge effects; that is, the emitted electrons form a sheath of negative charge in the vicinity of the cathode, inhibiting further emission. It is therefore necessary to apply a positive potential to accelerate the electron toward the anode and so reduce the space-charge layer.

By solving Poisson's equation in the space between cathode and anode and taking account of the dynamical equilibrium, we obtain expressions for the current density against voltage when accelerating fields are applied. These expressions depend on geometry and are, for a plane diode,

$$J = \frac{4\epsilon_0}{9}\sqrt{\frac{2e}{m}}\,V_0^{3/2}\,d^{-2} \tag{2.4a}$$

and for a cylindrical diode

$$J_a = \frac{4\epsilon_0}{9}\sqrt{\frac{2e}{m}}\,V_0^{3/2}r_a^{-2}\beta^{-2}, \tag{2.4b}$$

where MKS units are used and J or J_a is the current density at the *anode*. Here V_0 is the applied potential, d the anode-cathode separation for the plane diode, and r_a the anode radius for the cylindrical diode; β is a correction coefficient which is a function of (r_c/r_a) and can be found in the literature.† Equations 2.4 are known as "Child's law."

---

† Langmuir and Blodgett, "Currents Limited by Space Charge Between Coaxial Cylinders," *Phys. Rev.* **22**, 347 (1923).

As the voltage is further increased, J tends toward and finally reaches the saturation-current value given by Eq. 2.3b. We can now see what the measurable quantities are and what experimental procedure to follow.

(a) At fixed filament temperature, the anode current is measured against accelerating potential well into the saturation region. (A family of such curves is shown in a log-log plot in Fig. 3.14.) From the space-charge region we can verify Eqs. 2.4, the $V_0^{3/2}$ dependence, and obtain a value for e/m.

From the saturation region we can verify Eq. 2.3b, the $E^{1/2}$ dependence of the current density, and obtain the zero field density $J_0(T)$. It is then possible to check Eq. 2.2 (see Fig. 3.16) and find a value for the coefficient $e\phi/k$ of the exponent (hence the work function), and for the coefficient A_0.

(b) At fixed filament temperature the anode current is measured as a function of decelerating potential. One can verify the exponential dependence of J' and determine the coefficient e/kT of the exponent (Eq. 2.3a). In principle one can again determine $J_0(T)$, but values obtained from the saturation current are more accurate.

As is clear by now, a knowledge of the filament temperature is needed; the region of interest for pure tungsten (the cathode material used in this experiment) is in the region of 2000° K where thermionic emission currents can be reasonably detected. The measurement of the temperature cannot be made very accurately and is based on the change in resistivity of tungsten, or by using an optical pyrometer. The special tube used in this laboratory has a small hole in the anode so that the central part of the filament can be viewed for the optical determination of the temperature.

When the change in resistivity is used, it is important to be able to separate the effects due to lead resistance and filament resistance. While the lead resistance should remain constant, nevertheless it changes, since the leads become heated by conduction from the filament. It is also important to determine correctly the room temperature resistance, since in the measuring process the filament becomes heated. In addition, the filament itself is not at uniform temperature throughout its length, but is lower by a factor of $\frac{1}{3}$ at its terminal points. A correction for this effect can be included in the data.

2.3 Thermionic Emission Experiment and Results

In this laboratory a specially constructed cylindrical diode is used. The FP-400 tube manufactured by the General Electric Company† (Fig. 3.11) is a high-vacuum tube with a pure tungsten filament. Some relevant

† "The manufacture of these tubes has recently been discontinued; however a limited supply is available from G. J. Baberich, P. O. Box 118, Berkeley Heights, New Jersey. A possible substitute would be a diode such as 5U4."

FIG. 3.11 Photograph of the General Electric FP-400 vacuum tube appropriate for thermionic emission measurements.

data† are as follows:

Pure tungsten filament, length	3.17 cm
diameter	0.013 cm
Anode, Zr coated Ni, I.D.	1.58 cm
Average lead resistance	0.08 ohms
Maximum filament voltage	4.75 V
Maximum filament current	2.5 amp
Maximum anode voltage	125 V
Maximum anode current	0.055 amp
Maximum anode dissipation	15 W

The resistivity of tungsten is 5.64×10^{-6} ohms/cm at 27° C.

The circuit diagram is shown in Fig. 3.12. Filament heating is provided by a d-c balanced supply to minimize potential differences along the filament. The shunt and potentiometer shown in the filament circuit are used for resistivity measurements. The configuration shown in the figure is best suited for the application of accelerating potentials; for decelerating potentials a galvanometer must be introduced in the anode branch. In

† From the manufacturer's data sheet.

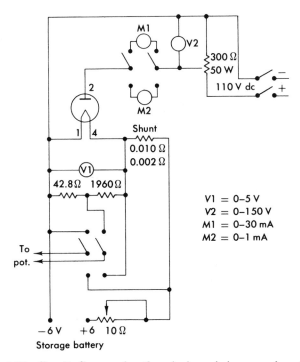

FIG. 3.12   Circuit diagram for thermionic emission experiment.

the latter instance voltage gradients along the filament become a serious consideration, and knowledge of the correct retarding potential is difficult to achieve unless a switching technique is used for filament heating.

When taking data, the maximum filament rating of 4.5 V should not be exceeded. Information necessary for the temperature determination is given in Fig. 3.13, which is a plot of the ratio of the resistivity of tungsten (W), obtained from independent calibration.†

Data obtained by a student‡ are shown in Fig. 3.14, which is a log–log plot of anode current against anode voltage. Since the geometry is cylindrical, we obtain the saturation current from Eq. 2.4b

$$I_s = l \frac{8\pi\epsilon_0}{9} \sqrt{\frac{2e}{m}} \, V_0^{3/2} r_a^{-1} \beta^{-2} \qquad (2.4c)$$

---

† Worthington and Forsythe, *Astrophys. J.*, **61**, 146–185 (1925) contains an extensive discussion of the properties of tungsten and corrections for lead losses of tungsten filaments.

‡ D. Kohler, class of 1962.

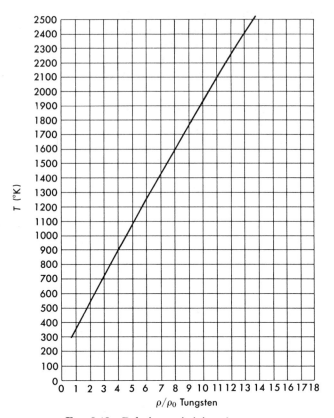

FIG. 3.13   Relative resistivity of tungsten.

where l is the length of the filament; for this particular tube the constant β^{-2} has the value 0.93. In the space-charge region, the plots are indeed straight lines and their slope is

$$\frac{\Delta \log I}{\Delta \log V} = 1.49 \pm 0.02$$

in agreement with the prediction of Eqs. 2.4. If we attempt, however, to obtain e/m using Eq. 2.4c, and accept as the current $I = 0.041 \times 10^{-3}$ amp at 1 V, the result is

$$e/m = 1.07 \times 10^{11} \text{ coul/kg}$$

while the true value is $e/m = 1.76 \times 10^{11}$ coul/kg. This discrepancy is not surprising since only a fraction of the filament length l contributes effectively to the emission process; this is due to both thermal and collection end-effects.

To obtain the filament temperature, first the resistance of room temperature was measured and extrapolated to zero current (Fig. 3.15), with the result $R' = 0.285$ ohms; next the filament resistance was calculated from its geometrical dimension and found to be $R_f = 0.135$ ohms, so that the lead resistance must be $R = R' - R_f = 0.15$ ohm. From the measurements of R' it is then possible to obtain the corresponding temperature, which has been used to label the curves of Fig. 3.14.

Once the temperature is known, the zero field current I_0 can be obtained from the saturation current I_s through Eq. 2.3b; the correction is in general small. Since the geometry is cylindrical,

$$V = \int \mathbf{E} \cdot d\mathbf{r} = (Er) \ln \frac{r_a}{r_f}$$

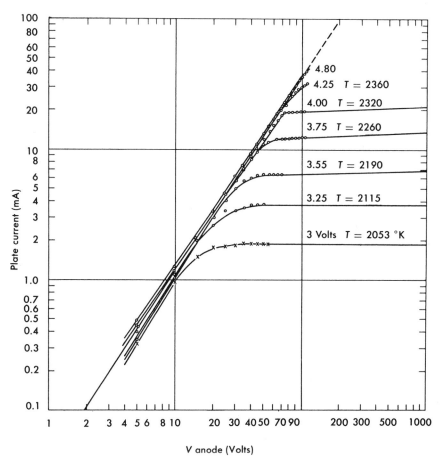

FIG. 3.14   Data on thermionic emission. Plate current versus anode voltage as a function of filament temperature.

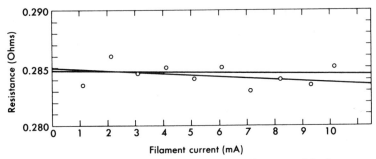

FIG. 3.15   "Cold" resistance of tungsten filament and leads.

hence

$$E = \frac{V}{r} \ln \frac{r_f}{r_a}.$$

The results are summarized in Table 3.1, where $J_0 = I_0/(2\pi r_c l)$ is the thermionic emission current density at the cathode.

TABLE 3.1
DATA OBTAINED FROM THERMIONIC EMISSION EXPERIMENT

Filament voltage	3	3.25	3.5	3.75	4.0	4.25
T (°K)	2053	2115	2190	2260	2320	2360
I_s (ma) at 110 V	2.0	3.7	6.8	12.5	20.2	33.8
I_0 (ma)	1.76	3.28	6.03	11.13	18.05	30.25
J_0/T^2 (amp deg^{-2}m^{-2})$\times 10^{-4}$	0.324	0.582	0.975	1.69	2.52	4.18

A plot of the values of J_0/T^2 against $1/T$ in a semilog plot is given in Fig. 3.16; from it a slope

$$\frac{e\phi}{k} = 37{,}500° \text{ K}$$

is obtained, hence

$$\phi = 3.22 \text{ V}$$

Similar but more precise measurements by Langmuir and Jones† yield $\phi = 4.52$ V, which is within the accepted value for the work function of tungsten. It is clear that values of A_0 obtained from these data depend

† From General Electric data sheet for FP-400.

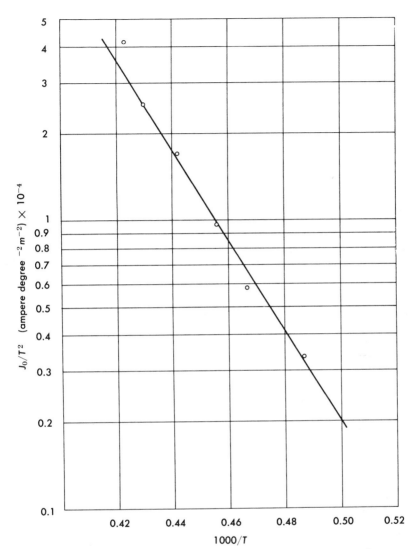

FIG. 3.16 Determination of the work function of tungsten. The saturation current density J_0/T^2 is plotted versus $1/T$ from the data of Fig. 3.14.

in a very sensitive manner on the slope of the linear fit in Fig. 3.16. For example, using these data, we obtain $A_0 = 33.5$ amp/deg^2-m^2 which is orders of magnitude off from the accepted value†

$$A_0 = 60 \times 10^4 \text{ amp/deg}^2\text{-m}^2$$

---

† This differs from the theoretical value by a factor of 2, attributed to reflections at the surface.

On the other hand, if Langmuir's value of $\phi = 4.52$ V is accepted for the slope, the data of Fig. 3.16 yield a more reasonable value:

$$A_0 = 176 \times 10^4 \text{ amp/deg}^2\text{-m}^2.$$

When a retarding potential is applied, measurements with the present setup are more difficult because of the effect of voltage gradients along the filament; also space-charge effects are still quite pronounced for anode currents of the order of 10^{-5} amp so that only data beyond this point must be considered. For such measurements see Harnwell and Livingood, *Experimental Atomic Physics*, page 210.

2.4 THE STEFAN-BOLTZMANN LAW

With the data obtained from the thermionic emission experiment, it is possible to verify the Stefan-Boltzmann law. This law, which can be derived on thermodynamic arguments, states that the total energy radiated per second by a black body of unit surface area is proportional to the fourth power of the temperature.

$$E = \sigma T^4 \tag{2.5}$$

where E is the total power radiated from unit area, T the temperature in degrees Kelvin, and σ is Stefan's constant.

By integrating Planck's radiation law over all wavelengths, Eq. 2.5 can be obtained. Since the energy density u_ν for a frequency ν is

$$u_\nu \, d\nu = \frac{8\pi h}{c^3} \frac{\nu^3}{\exp (h\nu/kT) - 1} \, d\nu \tag{2.6}$$

$$u = \int_0^\infty u_\nu \, d\nu = \frac{8\pi k^4 T^4}{h^3 c^3} \int_0^\infty \frac{z^3}{e^z - 1} \, dz = \frac{8\pi k^4}{h^3 c^3} \frac{\pi^4}{15} T^4 \tag{2.7}$$

If the total power emitted by a black body per unit area is E, the energy density when in equilibrium is

$$u = (4/c) E$$

so that

$$E = \frac{c}{4} \frac{8\pi^5 k^4}{15 h^3 c^3} T^4 \tag{2.8}$$

and substitution of the constants in Eq. 2.8 gives the experimentally observed value of $\sigma = 5.64 \times 10^{-8}$ joules m^{-2} deg^{-4} sec^{-1}.

If it is then assumed that the energy loss of the filament through conduction in the leads is small, the radiated power is given

$$P = I^2 R_f$$

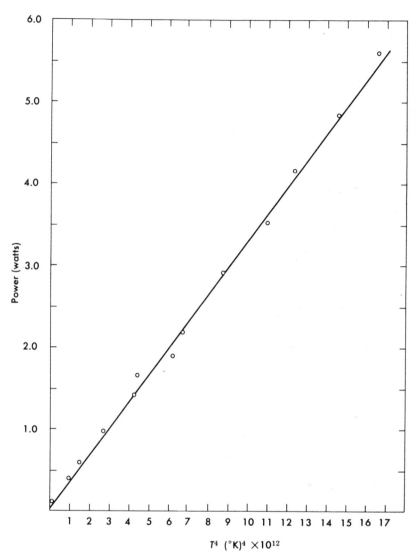

Fɪɢ. 3.17 Verification of Stefan's law. Plot of radiated power versus filament temperature.

A plot of P against T^4 as obtained by a student† is given in Fig. 3.17. The best fit to his data gives

$$P \propto T^{4.08}$$

† D. Owen, class of 1963.

and for Stefan's constant, $\sigma = 2.7 \times 10^{-8}$ joules m^{-2} deg^{-4} sec^{-1} which is of the correct order of magnitude, and smaller† than σ.

In concluding, we note that our results have confirmed the exponential dependence on temperature of the thermionic current (Richardson's equation), and the phenomena of space charge. Also, qualitative agreement has been achieved with the accepted values of the parameters involved; similarly, agreement has been achieved for Stefan's law.

3. Some Properties of Semiconductors

3.1 GENERAL

We have seen in the first section how a free-electron gas behaves, and what can be expected for the band structure of a crystalline solid. In the second section we applied the principle of free-electron gas behavior to the emission of electrons from metals, and in the present section we will apply both principles to the study of some properties of semiconductors which can be verified easily in the laboratory.

As mentioned before, a semiconductor is a crystalline solid in which the conduction band lies close to the valence band, but is not populated at low temperatures; semiconductors are unlike most metals in that both *electrons* and *holes* are responsible for the properties of the semiconductor. If the semiconductor is a pure crystal, the number of holes (positive carriers, p) is equal to the number of free electrons (negative carriers, n), since for each electron raised to the conduction band, a hole is created in the valence band; these are called the *intrinsic* carriers.

All practically important semiconductor materials, however, have in them a certain amount of impurities which are capable either of donating electrons to the conduction band (making an n-type crystal) or of accepting electrons from the valence band, thus creating holes in it (making a p-type crystal). These impurities are called *extrinsic* carriers; in such crystals $n \neq p$.

Let us then first look at the energy-band picture of a semiconductor as it is shown in Fig. 3.18; the impurities are all concentrated at a single energy level usually lying close to, but below, the conduction band. The density of states has to be different from that of a free-electron gas (Eq. 1.4 and Fig. 3.2a) since, for example, in the forbidden gaps it must be zero; close to the ends of the allowed bands it varies as $E^{1/2}$ and reduces to zero on the edge. On the other hand, the Fermi distribution function, Eq. 1.3, remains the same. The only parameter in this function is the Fermi energy, which can be found by integrating the number of *occupied* states (Fermi function

---

† Note that the tungsten filament is not a perfect black body; the emissivity of a hot tungsten filament is usually taken as one third.

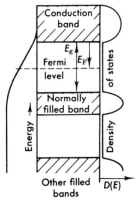

FIG. 3.18   Energy band structure of a semiconductor without impurities. On the left-hand side the Fermi distribution for a free-electron gas is shown; on the right-hand side the actual density of states $D(E)$ is shown.

times density of states) and setting it equal to the electron density. It is clear, however, that if we are to have as many empty states in the valence band as occupied ones in the conduction band, the Fermi level must lie exactly in the middle of the forbidden gap† (because of the symmetry of the trailing edge of the distribution). In Fig. 3.18, the density of states is shown to the right and the Fermi distribution function to the left. We measure the position of the Fermi level *from the conduction band* and define it by E_F; the exact value of E_F is

$$E_F = -\frac{E_g}{2} + kT \ln \left(\frac{m_h{}^*}{m_e{}^*}\right)^{3/4} \tag{3.1}$$

Since the Fermi level lies below the conduction band, E_F is a *negative* quantity; E_g is the energy gap always taken positive and $m_h{}^*$ and $m_e{}^*$ are the effective masses of holes and electrons, respectively. If w_C and w_F stand for the actual position of the conduction band and Fermi level above the zero point energy, then

$$w_F = w_C + E_F$$

To find the number of electrons in the conduction band (or holes in the valence band) we simply substitute Eq. 3.1 for w_F into Eq. 1.4, multiply by the density of states, and integrate over w from $w = w_C$ to $+\infty$. When, however, the exponent

$$-(w_F - w) \approx \frac{E_g}{2} + E \gg kT \tag{3.2}$$

the Fermi distribution degenerates to a Boltzmann distribution. (Here E is the energy of the electrons as measured from the top of the conduction band; obviously it can take either positive or negative values.) With this

---

† If the effective masses of p- and n-type carriers are the same.

assumption the integration is easy, yielding

$$n = \left(\frac{2\pi m_e k T}{h^2}\right)^{3/2} e^{E_F/kT} \simeq \left(\frac{2\pi m_e k T}{h^2}\right)^{3/2} e^{-E_g/2kT} \qquad (3.3a)$$

similarly,

$$p = \left(\frac{2\pi m_h k T}{h^2}\right)^{3/2} e^{-(E_g+E_F)/kT} \simeq \left(\frac{2\pi m_h k T}{h^2}\right)^{3/2} e^{-E_g/2kT} \qquad (3.3b)$$

It is interesting that the product np is independent of the position of the Fermi level†—especially if we take $m_e = m_h$

$$n_i^2 = np = 2.31 \times 10^{31}\ T^3\ e^{-E_g/kT}$$

Thus it should be expected that as the temperature is raised, the intrinsic carriers of a semiconductor will increase at an exponential rate characterized by $E_g/2kT$. This temperature is usually very high since $E_g \approx 0.7$ V (see Eqs. 3.5).

We have already mentioned that impurities determine the properties of a semiconductor, especially at low temperatures where very few intrinsic carriers are populating the conduction band. These impurities, when in their ground state, are usually concentrated in a single energy level lying very close to the conduction band (if they are donor impurities) or very close to the valence band (if they are acceptors). As for the intrinsic carriers, the Fermi level for the impurity carriers lies halfway between the conduction (valence) band and the impurity level; this situation is shown in Figs. 3.19a and 3.19b. If we make again the low temperature approximation of Eq. 3.2, the number of electrons in the conduction band is given by

$$n = N_d \left(\frac{2\pi m k T}{h^2}\right)^{3/2} e^{-E_d/2kT} \qquad (3.4)$$

where N_d is the number of donors and E_d the separation of the donor energy level from the conduction band. In writing Eq. 3.4, however, care must be exercised because the conditions of Eq. 3.2 are valid only for very low temperatures. Note, for example, that for germanium

$$E_g = 0.7\ \text{eV}, \qquad \text{and for } kT = 0.7\ \text{eV}, \qquad T = 8{,}000°\ \text{K}$$

whereas

$$E_d = 0.01\ \text{eV}, \qquad \text{and for } kT = 0.01\ \text{eV}, \qquad T = 120°\ \text{K} \qquad (3.5)$$

Thus at temperatures $T \gtrsim 120°$ K most of the donor impurities will be

---

† This result is very general and holds even without the approximation that led to Eqs. 3.3.

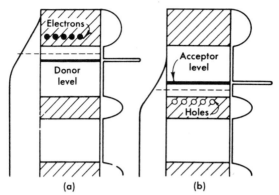

FIG. 3.19 Same as Fig. 3.18 but with the addition of impurities. (a) The impurities are of the donor type and lie at an energy slightly below the conduction band. (b) The impurities are of the acceptor type and lie slightly above the valence band. Note the shift of the Fermi level as indicated by the dotted line.

in the conduction band and instead of Eq. 3.4 we will have $n \simeq N_d$; namely, the number of impurity carriers becomes saturated. Once saturation has been reached the impurity carriers in the conduction band behave like the free electrons of a metal.

In the experiments to be discussed below, the lowest temperature achieved was $T \simeq 84°$ K; for germanium this still corresponds to almost complete saturation of the impurity carriers. We will, however, be able to study the gradual increase, as a function of the temperature, of the *intrinsic* carriers of germanium.

3.2 RESISTIVITY

We know already that conduction in solids is due to the motion of the charge carriers under the influence of an applied field. We define the following symbols:

$\mathbf{J}$ = current density
σ = conductivity, such that $\sigma\mathbf{E} = \mathbf{J}$; $\rho = 1/\sigma$ = resistivity
e = charge of the electron
n = (negative) carrier density; p = (positive) carrier density
$\bar{v}$ = drift velocity
$\mathbf{E}$ = applied electric field
μ = mobility; such that $\mu\mathbf{E} = \bar{\mathbf{v}}$
m^* = effective mass; m_e^*, m_h^* for electrons, holes
λ = mean free path between collisions

From simple reasoning, the current density is

$e \times$ (number of carriers traversing unit area in unit time)

which is equivalent to the carrier density multiplied by the drift velocity. Thus

$$\mathbf{J} = \mathbf{I}/A = e(n\bar{\mathbf{v}}_e - p\bar{\mathbf{v}}_h) \tag{3.6}$$

however, if s is the distance traveled and t the time between thermal collisions,

$$s = \frac{1}{2} \alpha t^2 = \frac{1}{2} \frac{e \, | \, \mathbf{E} \, |}{m^*} t^2$$

and

$$\bar{v} = \frac{s}{t} = \frac{1}{2} \frac{e \, | \, \mathbf{E} \, |}{m^*} t$$

in terms of the mean free path, $t = \lambda/v$, where v is now the thermal velocity,

$$\frac{m^* v^2}{2} = \frac{3}{2} kT$$

thus

$$\bar{\mathbf{v}} = \frac{e\lambda \mathbf{E}}{2\sqrt{3kTm^*}}$$

If only one type of carrier is present,

$$\mathbf{J} = en\bar{\mathbf{v}} = en\mu\mathbf{E}$$

and

$$\sigma = en\mu = \frac{ne^2\lambda}{2\sqrt{3kTm^*}}$$

Thus, if (a) the number of carriers is constant, and (b) the mean free path remains constant, the conductivity should decrease as $T^{-1/2}$. The first of these conditions holds in the extrinsic region after all impurity carriers are in the conduction band; the mean free path, however, is not constant, because higher temperatures increase lattice vibrations, which in turn affect the scattering of the carriers. A simple calculation suggests a $1/kT$ dependence for λ, so that

$$\times \qquad \sigma = ne\mu = C \frac{ne^2}{m^*} T^{-3/2} \tag{3.7}$$

where C is a constant. We will see that this dependence is not always observed experimentally.

3.3 The Hall Effect

It is clear from Eq. 3.6 that conductivity measurements cannot reveal whether one or both types of carriers are present, nor distinguish between them. However, this information can be obtained from Hall effect measurements, which are a basic tool for the determination of mobilities. The effect was discovered by E. H. Hall in 1879.

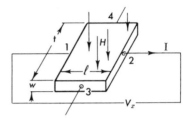

Fig. 3.20   Schematic arrangement for the measurement of the Hall effect of a crystal.

Consider a simple crystal mounted as in the Fig. 3.20, with a magnetic field H in the z direction perpendicular to contacts 1, 2 and 3, 4. If current is flowing through the crystal in the x direction (by application of a voltage V_x between contacts 1 and 2), a voltage will appear across contacts 3, 4. It is easy to calculate this (Hall) voltage if it is assumed that all carriers have the same drift velocity. We will do this in two steps: (a) by assuming that carriers of only one type are present, and (b) by assuming that carriers of both types are present.

(a) *One type of carrier.* The magnetic force on the carriers is $\mathbf{F}_m = e(\bar{\mathbf{v}} \times \mathbf{H})$ and it is compensated by the Hall field $\mathbf{F}_H = e\mathbf{E}_H = eE_y\mathbf{i}_y$ thus $\bar{v}H = E_y$, but $\bar{v} = \mu E_x$, hence $E_y = H\mu E_x$. The Hall coefficient R_H is defined as

$$| R_H | = \frac{E_y}{J_x H} = \frac{\mu E_x}{J_x} = \frac{\mu}{\sigma} = \frac{1}{ne} \tag{3.8a}$$

Hence for fixed magnetic field and fixed *input current,* the Hall voltage is proportional to $1/n$. It follows that

$$\mu_H = R_H \sigma, \tag{3.8b}$$

providing an experimental measurement of the mobility; R_H is expressed in cm³ coulomb⁻¹, and σ in ohm⁻¹ cm⁻¹; thus μ is expressed in units of cm² volt⁻¹ sec⁻¹.

In most experiments the voltage across the input is kept constant, so that it is convenient to define the Hall angle as the ratio of applied and measured voltages:

$$\phi = \frac{V_y}{V_x} = \frac{E_y t}{E_x l} = \mu \frac{t}{l} H \tag{3.8c}$$

where l is the length and t the thickness of the crystal.

The Hall angle is proportional to the mobility, and

$$\rho V_y = \left(\frac{t}{l} H\right) \frac{1}{ne} \tag{3.9}$$

is again proportional to $1/n$ and thus to $|R_H|$.

(b) *Two types of carriers.* Now it is important to recognize that for the same electric field E_x, the Hall voltage for p carriers will have opposite sign from that for n carriers. (That is, the Hall coefficient R has a different sign.) Thus, the Hall field E_y will not be able to compensate for the magnetic force on both types of carriers and there will be a transverse motion of carriers; however, the net transverse transfer of charge will remain zero since there is no current through the 3, 4 contacts; this statement is expressed as

$$e(v_y{}^+ n - v_y{}^- n) = 0$$

while

$$e(v_x{}^+ n - v_x{}^- p) = J_x \quad\text{and}\quad e(\mu^+ p + \mu^- n) = \sigma$$

where the mobility is always a positive number; however, $v_x{}^+$ has the opposite sign from $v_x{}^-$, but

$$v_y = \frac{s}{t} = \left(\frac{1}{2} \frac{F}{m^*} t^2\right) \frac{1}{t}$$

where

$$\mathbf{F}^+ = e[(\mathbf{v}_x{}^+ \times \mathbf{H}) - \mathbf{E}_y]$$

$$\mathbf{F}^- = -e[(\mathbf{v}_x{}^- \times \mathbf{H}) - \mathbf{E}_y]$$

Thus

$$v_y{}^+ = \frac{1}{2} \frac{e}{m_h{}^*} t[(\mu^+ E_x H) - E_y] = \mu^+(\mu^+ E_x H - E_y)$$

$$v_y{}^- = \frac{1}{2} \frac{e}{m_e{}^*} t[(\mu^- E_x H) - E_y] = \mu^-(\mu^- E_x H + E_y)$$

and thus

$$\mu^+ p(\mu^+ E_x H - E_y) - \mu^- n(\mu^- E_x H + E_y) = 0$$

$$E_y = E_x H \frac{(\mu_h{}^2 p - \mu_e{}^2 n)}{\mu_h p + \mu_e n}$$

and for the Hall coefficient R_H

$$R_H = \frac{E_y}{J_x H} = \frac{E_y}{\sigma E_x H} = \frac{\mu_h{}^2 p - \mu_e{}^2 n}{e(\mu_h p + \mu_e n)^2} \tag{3.10}$$

Equation 3.10 correctly reduces to Eq. 3.8 when only one type of carrier is present.[†]

Since the mobilities μ_h and μ_e are not constants but functions of T, the Hall coefficient given by Eq. 3.10 is also a function of T and it may become zero and even change sign. In general $\mu_e > \mu_h$ so that inversion may happen only if $p > n$; thus "Hall coefficient inversion" is characteristic only of "p-type" semiconductors.

At the point of zero Hall coefficient, it is possible to determine the ratio of mobilities $b = \mu_e/\mu_h$ in a simple manner. Since $R_H = 0$, we have from Eq. 3.10

$$nb^2 - p = 0 \tag{3.11}$$

Let N_a be the number of impurity carriers for this "p-type" material; then in the extrinsic region

$$p = N_a \qquad n = 0$$

whereas in the intrinsic region

$$p = N_a + N \qquad n = N$$

and Eq. 3.11 becomes

$$n = \frac{N_a}{b^2 - 1} \tag{3.12}$$

We can also express the conductivity σ at the inversion point, $T = T_0$, in terms of the mobilities

$$\sigma_0 = e[\mu_e n + \mu_h(N_a + n)] \tag{3.13}$$

and this value can be directly measured. Further, by extrapolating conductivity values *from* the extrinsic region to the point $T = T_0$ we obtain

$$\sigma_e(T = T_0) = e\mu_h N_a(T = T_0).$$

It therefore follows that

$$\frac{\sigma_0}{\sigma_e(T = T_0)} = \frac{N_a + n(1 + b)}{N_a} \tag{3.14}$$

[†] Both Eq. 3.8 and Eq. 3.10 have been derived on the assumption that all carriers have the same velocity; this is not true; but the exact calculation modifies the results obtained here by a factor of only $3\pi/8$.

Substituting the value of n from Eq. 3.12 we obtain

$$\frac{\sigma_0}{\sigma_e} = \frac{b}{b-1}$$

or

$$b = \frac{R_e(T = T_0)}{R_e(T = T_0) - R_0} \qquad (3.15)$$

where R_0 is the measured resistance of the sample at the inversion point and $R_e(T = T_0)$ is the resistance extrapolated from the extrinsic region to the value it would have at the inversion temperature (see Fig. 3.27).

We thus see that the Hall effect, in conjunction with resistivity measurements, can provide information on carrier densities, mobilities, impurity concentration, and other values. It must be noted, however, that mobilities obtained from Hall effect measurements $\mu_H = |R_H| \sigma$ do not always

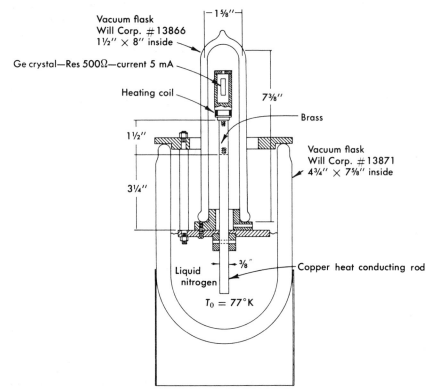

FIG. 3.21 Cryostat and assembly of equipment for Hall effect and resistivity measurements. The dewar is filled with liquid nitrogen; the crystal is mounted on top of a copper rod, which is in contact with the liquid nitrogen.

agree with directly measured values, and a distinction between the two is made; this will become clear when the experimental data are analyzed (Eq. 3.18 and Eq. 3.19).

3.4 EXPERIMENTAL ARRANGEMENT AND PROCEDURE

The sample, a small crystal of germanium, is mounted in a cryostat; a drawing of the assembly is shown in Fig. 3.21. The small dewar placed on the top permits the use of a magnet with only a 2-in. poleface separation. The heat is drawn from the sample chamber through the copper-brass rod into the liquid nitrogen heat sink. This allows the sample chamber to reach approximately 80° K. The liquid nitrogen must be kept up to level throughout the entire experiment. To raise the temperature of the sample chamber, a heating coil is placed just below it. The coil is wound non-inductively of 9 ft of No. 32 cotton-covered resistance wire; between each layer of the winding, metal foil is placed to conduct the heat quickly to the copper chamber. The maximum current is 1.5 amp and the heating coil should not be operated without liquid nitrogen in the large dewar (the maximum current with no liquid nitrogen is 35 ma.). At a current of about 1.3 amp the chamber will be at room temperature.

To measure the temperature, a copper-constantan thermocouple is fastened to the outside of the sample chamber. The standard junction is in ice-water at 0° C and Fig. 3.22 gives the appropriate calibration on this basis.

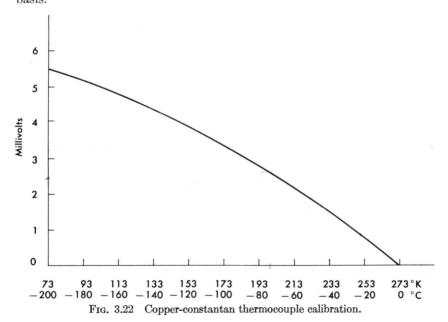

FIG. 3.22   Copper-constantan thermocouple calibration.

The equilibrium time is of the order of 15 min; however, it is not necessary to wait this long between points when taking data. Measurements should start at the lowest temperature, and during the experiment the heating coil current should be kept slightly in advance of equilibrium so as to maintain a slow but steady rise in temperature; the average millivolt reading should be recorded.

One method of mounting the crystal and making the contacts (used in this laboratory) is shown in Fig. 3.23. The copper stub used for one of the sample current contacts also provides a low-resistance heat path to keep the germanium at the sample chamber's temperature. The two wires soldered on each side of the germanium crystal allow the measurement of the Hall voltage at the top or bottom of the sample and the measurement of the conductivity on either side of the sample. Use of fine wires provides isolation from room temperature. The finite extent of the side contacts does not, to first order, affect the conductivity measurements.

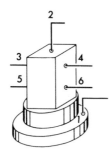

FIG. 3.23   Mounting and electrical connections to a crystal sample.

There are two problems in making the solder contacts. One is wetting the germanium with solder (that is, making the solder contact stick) and the other is to avoid making a "rectifying junction." To wet the germanium it is necessary first to etch it for about 30 sec in a solution of three parts hydrofluoric acid, three parts glacial acetic acid, and four parts nitric acid; this is called the CP4 etching solution. To avoid making a rectifying junction, the surface of the germanium where the contact is to be made must be destroyed; this is best done with a small Swiss file. A colloidal mixture of acid flux and solder is then used (it looks like a gray paste and is called "plumber's solder"). A very quick etch after the contacts have been made helps to remove any flux which would change the conductivity measurements. After etching, the germanium should be handled only with clean tweezers.

A schematic diagram of the measuring circuits is shown in Fig. 3.24. Resistivity is usually measured across contacts 1, 2 and an a-c bridge should be used; this is a more accurate measurement and also reduces the effects of rectifying contacts. There are provisions to rveerse the sample current, to measure the Hall voltage at either the top or bottom of the sample, and to balance out the zero field potential. The zero field potential comes from two sources: the first is due to the fact that the Hall contacts are not quite opposite each other; since there is a potential gradient due to the sample current, a part of this gradient will be seen between the two contacts. The second source is from the contact potential and the "rectifying action" of the contacts. To measure the Hall voltage, a Keithley electrometer is used, since its high input impedance does not affect the Hall voltage; the sample current is provided from a d-c battery, hence at fixed voltage. The Hall voltage must be measured for both directions of the magnetic field and should always be properly zeroed when the field is off; before the voltage is measured, the crystal must be rotated in the field until the position of maximum voltage is reached.

An experienced experimenter can take all the necessary data in one run. The sample is usually cooled to liquid nitrogen and then the temperature is slowly raised by control of the heater current. The following data should

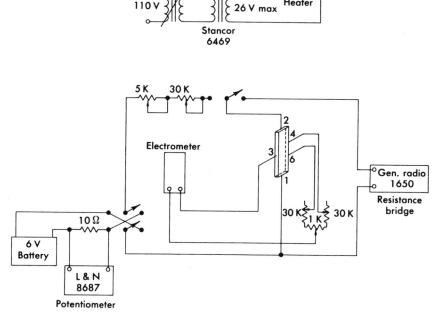

Fɪɢ. 3.24   Circuit for measuring conductivity and Hall coefficient of a crystal.

be recorded for every temperature point:

(a) Thermocouple reading
(b) Resistivity; magnetic field off
(c) Resistivity; magnetic field on
(d) Hall voltage, with field forward, off, reversed
(e) Thermocouple reading

Less experienced persons are better advised first to make a measurement of the resistivity only, over the whole temperature range from 80° K to 330° K, and then to measure the Hall voltage separately; as usual, it is advisable to plot the data as it is obtained so as to know where a greater density of measurements is desirable.

3.5 ANALYSIS OF DATA

Data on Hall effect and resistivity, obtained by students[†] using a low impurity germanium crystal, are presented and analyzed below.

Table 3.2 gives the raw data; the dimensions of the crystal were $l = 1.09$, $t = 0.183$, $w = 0.143$ cm (see Fig. 3.20). A fixed voltage of 1.32 V was applied across the long end of the sample, and the magnetic field was (800 ± 40) gauss. The Hall voltage was measured across the t dimension of the crystal.

From the data of the Table 3.2 the following plots were made:

(a) $\rho = 1/\sigma = RA/l$, hence[‡] $\rho = R \times 2.42 \times 10^{-2}$ ohms-cm; this is shown in a semilog plot against $1/T$ in Fig. 3.25. We note that for $T < 290°$ K, conduction is due mainly to the impurity carriers: this is the *extrinsic region*. For $T > 280°$ K, electrons are transferred copiously from the valence band into the conduction band and the crystal is in the *intrinsic region*. From the slope of the intrinsic region and making use of Eq. 3.3b, we have $\rho \propto 1/n \propto \exp(E_g/2kT)$ and thus $\ln \rho = E_g/2kT$. Hence

μ assumed
constant

$$\frac{E_g}{2k} = \frac{\Delta \log_{10}}{\Delta(1/T)} \frac{1}{\log_{10} e} = \frac{1.81 \times 10^3}{0.4343} \tag{3.16}$$

which leads to $E_g = 0.72 \pm 0.07$ eV, in agreement with the accepted value.

(b) A log-log plot of the resistivity in the extrinsic region against $1/T$ shown in Fig. 3.26. If a power law as in Eq. 3.7 is applicable, we would have $\rho \propto (1/T)^\alpha$ and hence

$$\alpha = \frac{\Delta \log \rho}{\Delta \log (1/T)} = -2.0 \pm 0.1 \tag{3.17}$$

[†] E. Yadlowski and P. Schreiber, class of 1962.
[‡] Note that here R is the resistance of the sample and not the Hall coefficient R_H.

TABLE 3.2

DATA ON HALL EFFECT AND RESISTIVITY MEASUREMENTS OF GERMANIUM

V_H (Volts) Hall voltage	R_m (Ohms) Magneto-resistance	R (Ohms) Resistance	T_{mv}(millivolts) Thermocouple	T (°K)
0.05 ± 0.002	157.8 ± 0.2	142.0 ± 0.2	5.37	84
0.047 ± 0.002	172.0 ± 0.2	152.0 ± 0.2	5.28	86
0.041 ± 0.002	183.9 ± 0.2	168.0 ± 0.2	5.20	93
0.038 ± 0.002	200.0 ± 0.2	184.9 ± 0.2	5.10	100
0.038 ± 0.002	207.0 ± 0.1	192.0 ± 0.2	5.036	103
0.0375 ± 0.002	219.0 ± 0.2	202.2 ± 0.1	5.00	105
0.0360 ± 0.002	229.0 ± 0.2	215.0 ± 0.4	4.94	111
0.0340 ± 0.002	238.0 ± 0.2	226.0 ± 0.4	4.86	114
0.0330 ± 0.002	261.0 ± 0.2	247.0 ± 0.1	4.78	121
0.0300 ± 0.002	290.0 ± 0.2	276.0 ± 0.2	4.65	127
0.0280 ± 0.002	321.0 ± 0.2	306.0 ± 0.5	4.53	131
0.0275 ± 0.002	335.0 ± 0.5	321.0 ± 0.5	4.45	137
0.0230 ± 0.002	371.0 ± 0.5	360.0 ± 0.5	4.30	143
0.0230 ± 0.002	407.0 ± 0.5	396.0 ± 1	4.15	150
0.0208 ± 0.002	457.5 ± 0.5	446.0 ± 1	3.95	155
0.0204 ± 0.002	495.0 ± 0.5	482.5 ± 2	3.84	164
0.0185 ± 0.001	559.0 ± 0.5	550.0 ± 1	3.60	172
0.0180 ± 0.001	610.0 ± 0.5	595.0 ± 2	3.40	182
0.0145 ± 0.0005	702.0 ± 1	700.0 ± 1	3.10	193
0.0150 ± 0.0005	827.0 ± 2	820.0 ± 1	2.72	199
0.0144 ± 0.0003	855.0 ± 1	850.0 ± 1	2.60	203
0.0131 ± 0.0003	888.0 ± 1	875.0 ± 1	2.45	212
0.0115 ± 0.0003	955.0 ± 1	950.0 ± 1	2.20	241
0.0095 ± 0.0001	1270.0 ± 5	1270.0 ± 5	1.20	253
0.0075 ± 0.0002	1440.0 ± 3	1440.0 ± 3	0.69	262
0.0059 ± 0.0002	1500.0 ± 3	1500.0 ± 3	0.45	260
0.0065 ± 0.0002	1460.0 ± 3	1450.0 ± 5	0.65	263
0.0075 ± 0.0002	1500.0 ± 3	1500.0 ± 4	0.380	266
0.0075 ± 0.0002	1570.0 ± 3	1560.0 ± 6	0.180	270
0.0067 ± 0.0002	1650.0 ± 3	1645.0 ± 5	0.000	273
0.0061 ± 0.0002	1700.0 ± 5	1700.0 ± 5	−0.400	282
0.0055 ± 0.0002	1660.0 ± 5	1665.0 ± 5	−0.630	286
0.0050 ± 0.0002	1610.0 ± 5	1590.0 ± 5	−0.860	290
0.0038 ± 0.0001	1455.0 ± 5	1440.0 ± 5	−1.10	296
0.0032 ± 0.0001	1340.0 ± 5	1310.0 ± 5	−1.22	302
0.0018 ± 0.0001	980.0 ± 5	950.0 ± 5	−1.44	305
0.0008 ± 0.0001	795.0 ± 3	785.0 ± 2	−1.67	312
0.00035 ± 0.00005	654.0 ± 2	632.0 ± 1	−1.86	315
0.00001 ± 0.00002	566.0 ± 2	545.0 ± 1	−2.00	318
0.00000 ± 0.000050	505.0 ± 1	488.0 ± 2	−2.15	323
−0.0002 ± 0.000050	395.0 ± 2	390.0 ± 2	−2.30	326
−0.0006 ± 0.00005	303.0 ± 2	293.0 ± 2	−2.65	329
−0.0008 ± 0.00005	225.0 ± 2	225.0 ± 2	−3.05	337
−0.0010 ± 0.0001	178.0 ± 1	178.0 ± 1	−3.33	346

Since in this region the carrier density is constant, this gives for the mobility a dependence

$$\mu = CT^{-2.0} \tag{3.18}$$

which is in disagreement with the prediction of Eq. 3.7. It is, however, the correct value for germanium, indicating that the simplified calculations used in deriving Eq. 3.7 are not completely adequate.

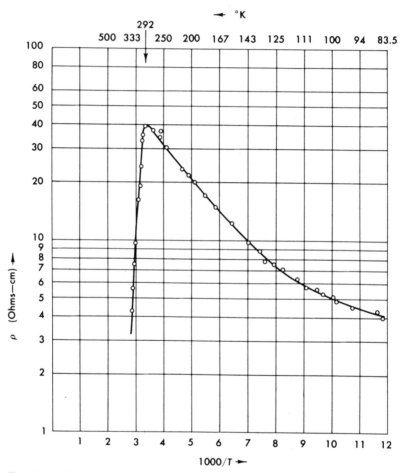

FIG. 3.25   The resistivity of a pure germanium crystal as a function of inverse temperature. For $T < 290°$ K, conduction is due mainly to the impurity carriers (extrinsic region); for $T > 290°$ K, conduction is due to electrons transferred to the conduction band (and the corresponding holes created in the valence band): this is the intrinsic region.

Turning now to the Hall-voltage measurements of Table 3.2, we can form the quantities defined by Eq. 3.8c.

$$\frac{V_H}{V_S} = \phi = \mu H \frac{t}{l}$$

hence

$$R_H = \phi R \frac{w}{H}$$

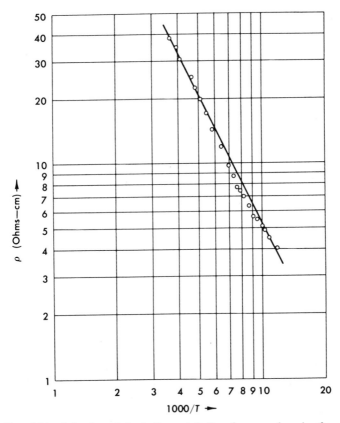

FIG. 3.26 A log-log plot of the resistivity of germanium in the extrinsic region versus $1/T$. It is assumed that the number of carriers is independent of T since saturation of the impurity carriers has already been reached.

and the Hall mobility

$$\mu_H = R_H \sigma = \phi \left(\frac{l}{t} \right) H^{-1}$$

The Hall mobility so obtained is shown in a log-log plot against T in Fig. 3.27. Since the Hall coefficient changes sign, we can immediately recognize that the crystal is of the p type; the inversion temperature of this par-

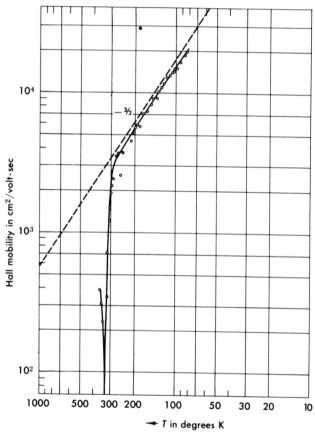

FIG. 3.27   Log-log plot of the "Hall-mobility" (Hall coefficient $\times$ conductivity) versus T. We note that the Hall coefficient becomes zero at $T = 323°$ K (inversion temperature) and changes sign beyond that point. Since negative values cannot be shown on the log plot for $T > 323°$ K, the Hall mobility of the reverse sign is again shown in the same graph. Note also the $T^{-3/2}$ dependence of the Hall mobility in the extrinsic region.

ticular sample is found to be

$$T_0 = (323 \pm 3)^\circ \text{K}$$

From the slope of the μ_H curve in the extrinsic region in Fig. 3.27, we obtain

$$\mu_H = C T^{-3/2} \tag{3.19}$$

which is different from Eq. 3.18 and is in agreement with conclusions of other observers; this is the reason why a distinction between the Hall mobility μ_H and the drift mobilities μ_D obtained from resistivity measurements is made.

By extrapolating Eq. 3.19 to the inversion temperature, we obtain the hole mobility† at $T = T_0 = 323^\circ$;

$$\mu_H(h) = 2.7 \times 10^3 \text{ cm}^2/\text{V-sec.} \tag{3.20a}$$

We can now apply the analysis indicated in Section 3.4, which led to Eq. 3.11. From Fig. 3.25 we have $R_e(T = T_0) = 2150$ ohms and $R_0 = 500$ ohms leading to $b = 1.31 \pm 0.2$. Thus we obtain for the electron mobility at $T = T_0 = 323^\circ$

$$\mu_H(e) = 3.5 \times 10^3 \text{ cm}^2/\text{V-sec,} \tag{3.20b}$$

both results being in agreement with the accepted values.

From the Hall coefficient in the extrinsic region, we can also obtain an order of magnitude for the density of impurity carriers. Since in that region only one type of carrier is present, $ne = 1/R_H$, and since

$$R_H = \frac{\phi R w}{H} \simeq 8 \times 10^5 \text{ cm}^3/\text{coulomb}$$

$$\tag{3.21}$$

$$n \simeq 8 \times 10^{12}/\text{cm}^3$$

which is reasonable for this sample, indicating an impurity concentration of the order of two parts in 10^{10}.

From the data of Table 3.2 it can be further noticed that the resistance of the sample changes when the magnetic field is turned on. This phenomenon, called magnetoresistance, is due to the fact that the drift velocity of all carriers is not the same. With the magnetic field on, the Hall voltage $V = E_y t = |\mathbf{v} \times \mathbf{H}|$ compensates exactly the Lorentz force for carriers with the *average* velocity; slower carriers will be overcompensated, and faster ones undercompensated, resulting in trajectories that are not along the applied external field. This results in an effective decrease of the mean free path and hence an increase in resistivity.

---

† Note that 1 gauss $= 10^{-4}$ weber/m^2 $= 10^{-8}$ V-sec/cm^2.

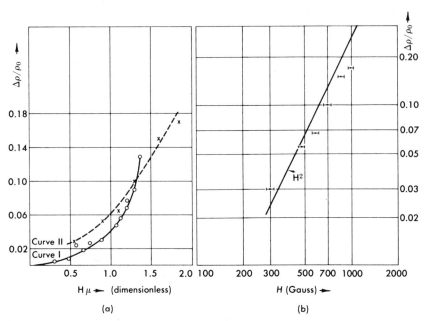

FIG. 3.28   Magnetoresistance of a germanium crystal; we plot $\Delta\rho/\rho_0 = (R_m - R)/R$ where R_m is the sample resistance with the field on, and R with field off. (a) A linear plot of $\Delta\rho/\rho_0$ versus the dimensionless parameter $H\mu_H$; curve I is obtained by varying μ at fixed H while curve II is obtained by varying H at fixed μ ($T = 84°$ K). (b) Log-log plot of $\Delta\rho/\rho_0$ versus H; note the H^2 dependence.

Several calculations of magnetoresistance have been made, but it is known that germanium in the extrinsic region exhibits many times the calculated value. One expects the magnetoresistance to increase with increased mean free path, and to reach saturation at very strong fields; for lower fields it is expected to have a quadratic dependence on the field strength. For the same reason, the Hall coefficient also has a slight dependence on magnetic field. Magnetoresistance measurements are of value in determining the exact shape of the energy surfaces.

Fig. 3.28 shows a plot of $\Delta\rho/\rho_0 = (R_m - R)/R$ (where R is the sample resistance without field, and R_m with magnetic field) obtained from the data of Table 3.2. In Fig. 3.28a, $\Delta\rho/\rho_0$ is plotted against the dimensionless parameter† $\mu H = eH\lambda(12\ m^*kT)^{-1/2}$. The points on curve I have been obtained by keeping H fixed and varying the temperature (hence μ). Curve II is obtained by varying H at a fixed $T = 84°$ K; the points from this curve are also shown on Fig. 3.28b, which is a log-log plot of $\Delta\rho/\rho_0$ against H, showing the almost quadratic dependence.

† See Eq. 3.8c.

4. Sketch of $p-n$ Semiconductor Junction Theory

As mentioned before, semiconductor materials with high impurity concentration, when properly combined, form a transistor. Transistors (of the type most used today) consist of two junctions of dissimilar-type semiconductors, one p type and one n type; the intermediate region, the base, is usually made very thin. We will briefly sketch the behavior of such a p-n junction and then see how the combination of two junctions can provide power amplification; for this we will use our knowledge of the band structure of semiconductors and the position of the Fermi level, as developed previously (Figs. 3.18 and 3.19).

When two materials with dissimilar band structure are joined, it is important to know at what relative energy level one band diagram lies with respect to the other: the answer is that *the Fermi levels of both materials must be at the same energy position* when no external fields are applied; this is shown in Fig. 3.29.

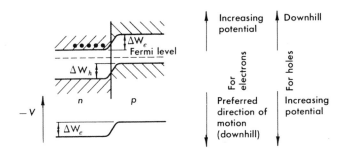

Fᴵɢ. 3.29   Structure of the energy bands at the junction of an n-type and a p-type semiconductor.

From the energy diagram of the figure, it follows that only electrons with $E_e > \Delta W_e$ will be able to cross the junction from the n material into the p region and only holes with $E_h > \Delta W_h$ from the p region to the n region. Holes in the n region or electrons in the p region are called "minority carriers." Indeed, there will be diffusion of some minority carriers across the junction, but since no electric field is present these carriers will remain in the vicinity of the junction.†

If now a *reverse bias* is applied—that is, one that opposes the further motion of the minority carriers—the Fermi levels will become displaced by the amount of the bias, as shown in Fig. 3.30a. We see that the barriers

---

† The result of such diffusion is the build-up of a local charge density, and thus potential, which prevents further diffusion. Throughout the present analysis, however, we will neglect the local effects at the junction.

ΔW_e and ΔW_h are increased by almost the full voltage, making any motion of minority carriers across the junction very improbable. Fig. 3.30b, on the other hand, shows the situation when *forward bias* is applied (favoring the motion of minority carriers). The Fermi levels are now displaced in the opposite direction so that the barriers are lowered. However, the full bias voltage does not appear as a difference between the Fermi levels because dynamic equilibrium prevails. There is a continuous flow of minority carriers in the direction of the electric field (holes obviously moving in the opposite direction from electrons) and as a result a potential gradient exists along the material; thus the entire bias voltage does not necessarily appear at the junction itself.

We will now consider two junctions put together; in Fig. 3.31a, p-type, n-type, and again p-type material are joined. When no bias is applied, we expect the Fermi levels to be at the same position, with the resulting configuration shown in the diagram; in agreement with our previous conclusions from the consideration of a simple junction, we see that barriers exist for the motion of holes from the p regions into the n region, and also for the motion of electrons from the n region into either of the p regions.

Figure 3.31b shows the double junction under operating biases; note that one junction is biased *forward*, the other is biased in the *reverse* direction. The n-type material common to both junctions is called the *base*, while the p type of the forward-biased junction is the *emitter*; the p-type material of the reverse junction is the *collector*. A completely symmetric device consisting of n-p-n materials will perform similarly when the biases are re-

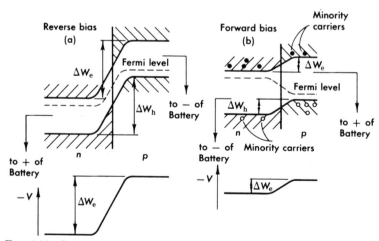

FIG. 3.30   Structure of the energy bands at a biased n-p junction. (a) Reverse bias. (b) Forward bias. The solid dots represent electrons, whereas the open dots holes.

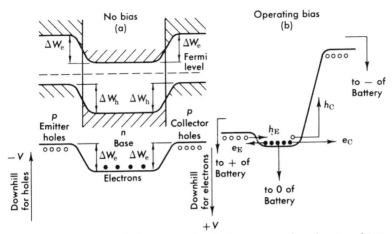

Fɪɢ. 3.31 Structure of the energy bands for a *p-n-p* junction transistor. (a) With no bias applied. (b) With operating biases. Note that the emitter is forward biased whereas the collector is reverse biased.

versed. From the energy diagram of Fig. 3.31b we can see that by varying the emitter junction bias we can control the injection of minority carriers into the base region; if the base region is made thin, it is possible for these holes to reach the collector junction, at which point they will immediately cross it, since it represents a gain in potential energy. If h_E is the minority carrier current injected into the base over a potential barrier $\Delta W_h(EB)$, the power required for injection is $P_{in} = h_E \, \Delta W_h(EB)$; similarly if h_C is the hole current into the collector down a potential drop $\Delta W_h(BC)$, the power gained is $P_{out} = h_C \, \Delta W_h(EC)$. Thus if $P_{out} > P_{in}$, the device is a power amplifier; since usually $\Delta W(CB) \gg \Delta W(EB)$, it suffices for $h_C \backsim h_E$ to give power gain.

Below we give, without proving, some quantitative formulas for the gain factors of a junction transistor. A detailed discussion can be found in Dunlap, *"An Introduction to Semiconductors."*

We introduce the following symbols and definitions:

h_E, e_E	hole current out of emitter, electron current into emitter
h_C, e_C	hole current, electron current into collector
I_E, I_B, I_C	emitter, base, collector current where hole current *leaving* a region is designated as positive
w	width of base region
D_n, D_p	diffusion coefficients for electrons, holes in base region
L_E, L_B	diffusion lengths in emitter, base
t_E, t_B	lifetime of minority carriers in emitter, base
n_E, p_B	concentration of minority carriers (electrons) in emitter, (holes) in base
ω	angular frequency

i	$\sqrt{-1}$
β	h_E/h_C diminution factor
γ	h_E/I_E injection efficiency
α	I_C/I_E current gain

If h_E holes are injected from the emitter into the base, because of recombination, only h_C reach the collector junction. Thus:

$$h_C = h_E\beta$$

where $0 < \beta < 1$, and β is given by $\beta = \operatorname{sech}(w/L_B)$. Further

$$I_E = h_E + e_E = h_E\left(1 + \frac{e_E}{h_E}\right) = \frac{h_E}{\gamma}$$

where $0 < \gamma < 1$, and is given by

$$\frac{1}{\gamma} = 1 + \frac{e_E}{h_E} = 1 + \frac{D_n}{D_p}\frac{n_E}{p_B}\frac{L_B}{L_E}\tanh\left(\frac{w}{L_B}\right)$$

Also,

$$I_C = -h_C + e_C \qquad I_B = -e_C - e_E + (h_C - h_E)$$

with the obvious conservation expression

$$I_C = -(I_E + I_B)$$

Since the collector leakage current e_C is very small, we can neglect it and obtain

$$\alpha = \frac{I_C}{I_E} \simeq \frac{h_C}{I_E} = \frac{h_E\beta}{h_E/\gamma} = \beta\gamma$$

Usually $w/L_B \ll 1$ and then $\gamma \simeq 1$, so we obtain the further approximate expressions

$$\alpha \simeq \beta = \operatorname{sech}\left(\frac{w}{L_B}\right) \simeq \left[1 + \tfrac{1}{2}\left(\frac{w}{L_B}\right)^2\right]^{-1} \tag{4.1}$$

indicating the importance of a thin base region if current gains close to unity are to be achieved. For a time-varying signal with angular frequency ω, the above expression is modified to

$$\alpha \simeq \beta \simeq \left[1 + \tfrac{1}{2}(1 + i\omega t_B)\left(\frac{w}{L_B}\right)^2\right]^{-1} \tag{4.2}$$

indicating phase shifts and reduction of gain at frequencies of the order of the reciprocal lifetime of the minority carriers in the base.

REFERENCES

For the material covered in Sections 3, 4, and 5 the reader may also consult the following texts:

W. C. Dunlap, Jr., An Introduction to Semiconductors. New York: Wiley, 1957. Brief but clear treatment.

C. Kittel, Introduction to Solid State Physics. New York: Wiley, 1953. A more general treatment of the solid state.

W. Shockley, Electrons and Holes. New York: D. Van Nostrand, 1950. A thorough presentation of the subject.

5. Contact and Thermoelectric Effects at Junctions of Metals

We now turn our attention again to metals; as mentioned in Sections 1 and 2, the free electrons can be thought of as being all in the continuum of the Fermi sea, with the density of states proportional to $\sqrt{E}$ and not restricted to allowed energy bands. If two metals with different Fermi energies $w_F(A)$ and $w_F(B)$ and different work functions $\phi(A)$ and $\phi(B)$ are joined, the energy diagram will be as shown in Fig. 3.32. Again equilibrium requirements impose the condition that the Fermi levels be at the same potential.

Let us consider, then, an electron in metal B that just overcomes the work function ϕ_B and is emitted from the metal surface. If while outside the metal the electron moves towards the dissimilar metal A (which is joined to B), it clearly sees a retarding potential $\phi_A - \phi_B$ until it enters metal B. This is called the *contact potential*; note that we had to correct for it on several occasions in discussions in Chapter 1. For an electron emitted from A and traveling towards B, the contact potential is an accelerating one ($\phi_B - \phi_A$). Thus we see that contact potential differences (cpd) arise when current flow through dissimilar metals is completed with a section in free space.

Next we will briefly mention three interrelated phenomena connecting the *reversible* flow of heat with that of current in a metal and vice versa. An application of these effects is the thermocouple of which we have repeatedly made use.

(a) *The Peltier effect* (1834): Let a circuit be completed through two dissimilar metals and a current flow through the junction. Reversible (not Joule) heating or cooling of the junction then occurs depending on the

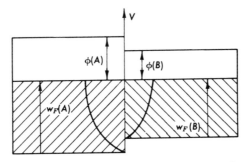

FIG. 3.32   The junction of two dissimilar metals, with different work functions and different potential barriers. Note that the Fermi levels of both metals are at the same energy. The contact potential differences equal $\phi_A - \phi_B$.

direction of current flow. Analytically we express the effect as follows

$$Q = \pi_{AB}(T)q \tag{5.1}$$

where Q is the amount of heat liberated, q is the total charge that crosses the junction, and $\pi_{AB}(T)$ is the Peltier coefficient which depends on the temperature.

(b) *The Thomson effect* (1856): This is complementary to the Peltier effect and is the reversible heating or cooling of a conductor through which current flows and along which a temperature gradient exists. Consider a conductor AB with no current flowing but with its two end points at a different temperature $T_A > T_B$; there will be heat transfer across it, effected through the motion of high-energy electrons from A towards B; to keep the charge transfer equal to zero, low-energy electrons will move from B towards A: this is shown in Fig. 3.33a. The following must hold

$$j = e(n_2 v_2 - n_1 v_1) = 0 \tag{5.2a}$$

and if S is the rate of heat transfer and $E_2 = E_A - \bar{E}$, $E_1 = \bar{E} - E_B$ are the excess or deficit from the mean thermal energy,

$$\frac{\Delta Q}{\Delta t} = (n_2 v_2 E_2 + n_1 v_1 E_1) = S \tag{5.3a}$$

If now a net current flows to the right (Fig. 3.33b),

$$j' = e(n_2' v_2' - n_1' v_1') > 0 \tag{5.2b}$$

and

$$\frac{\Delta Q}{\Delta t} = n_2' v_2' E_2 + n_1' v_1' E_1 = S' \neq S \tag{5.3b}$$

Namely, a change in $\Delta Q/\Delta t$ occurs; the balance of the heat is being supplied

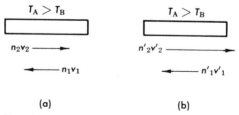

(a)   (b)

FIG. 3.33 A conductor along which a temperature gradient exists. (a) No net current flowing. (b) Current flowing to the right.

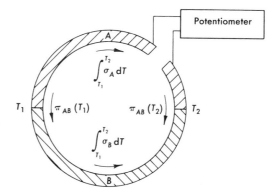

FIG. 3.34 A thermocouple made
of two dissimilar metals A and
B. Note that the two junctions
are at different temperatures
T_1 and T_2 and that there is no
net current flow. The two effects
contributing to the thermoelectric electromotive force are
shown.

or absorbed by the lattice. The situation is reversed when the current flows
to the left.

Analytically, we express the Thomson effect as follows:

$$\frac{dQ}{dx} = \sigma_A\, q \left(\frac{dT}{dx}\right) \tag{5.4}$$

where dQ/dx is the heat absorbed (or liberated) per unit length of the
metal; q is the total electric charge that has flowed through, and dT/dx the
temperature gradient; finally σ_A is the Thomson coefficient and it depends
on the metal. Conversely if a temperature gradient exists in a metal, an
electromotive force (emf) will appear at its ends. Since the product of
the charge and of the electromotive force must equal the total work done,
we find by integrating Eq. 5.4

$$\int_A^B dQ = q \int_{T_A}^{T_B} \sigma_A\, dT$$

hence

$$\text{emf (Thomson)} = \int_{T_A}^{T_B} \sigma_A\, dT \tag{5.5}$$

(c) *The Seebeck effect* (1822): This is the appearance of an electromotive force in a circuit made of two dissimilar metals when their junctions are held at different temperatures—namely, a thermocouple. It has
to be a combination of the two previously discussed effects and is shown
in Fig. 3.34. The two metals will have dissimilar Thomson coefficients σ_A
and σ_B, and the junctions (at different temperature) will have dissimilar
Peltier coefficients; the electromotive force is usually measured with

a potentiometer—namely, at zero current. Combining Eqs. 5.1 and 5.5 we obtain

$$\text{emf} = \pi_{AB}(T_2) - \pi_{AB}(T_1) + \int_{T_1}^{T_2} \sigma_A \, dT - \int_{T_1}^{T_2} \sigma_B \, dT \qquad (5.6)$$

Taking the derivative with respect to the temperature, we obtain the *thermoelectric power*, E_{AB} of two dissimilar metals

$$E_{AB} = \frac{d(\text{emf})}{dT} = \frac{d[\pi_{AB}(T)]}{dT} + (\sigma_A - \sigma_B) \qquad (5.7)$$

4

USEFUL TECHNIQUES

1. Introduction

This chapter deals with techniques that are needed repeatedly in many experiments. While the material presented is not exhaustive it is of such a nature that it is best presented by itself, detached from specific physical experiments.

The first topic is switching circuits, since they form the basic blocks of all counting and shaping circuits used with particle detectors; moreover they are the basis of all digital logic devices such as high-speed computers. Transistors, rather than vacuum tubes, are used as switching elements. This is followed by a brief mention of electronic functional assemblies to show how the basic blocks can be put together to perform logic functions.

Next, a section on means of creating and measuring vacuum is included. For most experiments in atomic or nuclear physics a good vacuum is needed, so that it is worthwhile to discuss the capabilities and limitations of modern vacuum equipment.

Finally, a discussion of radioactive safety and handling of radioactive materials is presented. This topic has been included because in many of the following experiments, especially those on nuclear physics, the student has to handle radioactive materials and prepare special sources. Knowledge of safety procedures and regulations will protect him, and others who use the same laboratory facilities, from undue or accidental exposure.

2. Transistor Switching Circuits

In this section, the following circuits will be described:

(a) Bistable multivibrator (flip-flop or Echles-Jordan circuit)
(b) Schmidt trigger
(c) Univibrator
(d) Free running multivibrator

The 2N78 transistor will be used throughout; it is an n-p-n transistor and operates with positive bias on the collector. (In this respect the n-p-n is like a vacuum tube, but the opposite is true for the p-n-p type.) In analyzing a circuit it is always wise to first measure the d-c levels at various points. Next the a-c behavior should be studied by observing wave forms on an oscilloscope; these wave forms are either generated by the circuit itself, when it is free running, or are its response to an a-c signal or pulse. For test purposes, a-c signals are obtained from signal generators and can be sinusoidal, square, or triangular waves. For testing switching circuits, it is more common to use a pulser, which generates usually square pulses of desired width, amplitude, and repetition rate. For the circuits described here, a slow oscilloscope of the 530 or 560 series of Tektronix, and a mercury switch pulser are adequate.

For each circuit, the diagram is given followed by a description of its operation, and of the wave forms that appear at various points. It is expected that students will set the circuit up on a "breadboard" and observe the change in the wave forms as they alter the values of the various components.

The voltages and wave forms shown in the figures are those measured and observed in the laboratory.

2.1 Bistable Multivibrator (Flip-Flop or Echles-Jordan Circuit)

The circuit is shown in Fig. 4.1. Because of the symmetry of the circuit, there are two stable conditions—that is, with either transistor conducting and the other cut off.

In the configuration shown, T_2 is conducting heavily with its collector at 1.3 V, and T_1 is turned off with its collector at approximately 12 V, the B^+ value. Both emitters are at common potential, which is set by the conducting transistor at approximately 1 V. The base of T_1 is held well below threshold because of the low voltage at the collector of T_2 and the voltage-

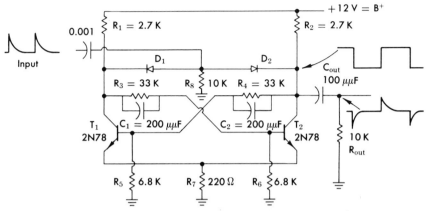

Fig. 4.1 Bistable multivibrator (flip-flop) using *n-p-n* transistors.

divider action of the $R_4 - R_5$ combination. The input steering diodes are both reverse biased, but D_1 will require a $+12$ V signal to turn on, whereas D_2 has only 1.3 V reverse bias.

If, then, a positive pulse of sufficient amplitude is applied at the input, it will pass through D_2 and be applied through C_2 at the base of T_1. This will make T_1 slightly conducting, with a consequent decrease of the T_1 collector voltage; this instantaneous decrease is transmitted through condenser C_1 and applied to the base of T_2, which is made less conducting so that its collector rises, and through C_2 raises further the base of T_1. Thus we see that a regenerative action takes place until the other equilibrium state is reached with T_1 hard on and T_2 cut off. Because of the regeneration, the rise time of the collector wave forms is very sharp.

If either of the collector outputs is differentiated, as shown in Fig. 4.1 through the $C_{out}R_{out}$ combination, a series of pulses will appear with a *positive* pulse for every *second* input pulse. Thus the circuit can be used as a *"scale of two,"* while several of them in series can be used to achieve any desirable scaling factor.

If the steering diodes are reversed and R_8 connected to the B^+ line, the circuit will respond to negative input pulses.

2.2 SCHMIDT TRIGGER

By removing one of the cross-coupling branches from a flip-flop, we obtain a Schmidt trigger. This circuit is shown in Fig. 4.2 and is used to obtain a pulse of fixed output amplitude whose duration is equal to the time that the input pulse is greater than some threshold. Its threshold stability accounts for its common use in pulse-height discriminators.

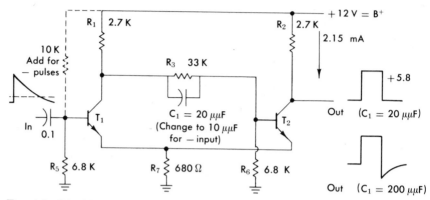

Fig. 4.2   Schmidt trigger for positive input pulses; the threshold level can be set by adjusting R_6. Note that the trigger circuit may also be used for negative input with minor changes.

With no input signal, T_1 is turned off and T_2 is conducting (not hard) because of the positive bias appearing at its base, through the $R_1R_3R_6$ divider†; the emitter of T_2 follows the base so that the common emitter point is at 1.46 V. Since the base of T_1 is connected to ground through R_5, T_1 is turned off.

A positive signal is now applied at the base of T_1, and when the amplitude exceeds the reverse bias, T_1 begins to conduct; the collector voltage drops and through the speed-up capacitor C_1 is applied to the base of T_2, resulting in regenerative action. Thus T_2 is quickly turned off and the output voltage rises to the B^+ value, which is maintained as long as T_1 is conducting.

When, however, the level of the input signal drops below the threshold level, T_1 becomes less conducting, and its collector voltage rises slightly. This is applied to the base of T_2, which is turned on, forcing T_1 to turn off because it raises the common emitter point.

The value of the speed-up capacitor determines the shape of the output pulse. For C_1 small, fast rise times are obtained, for C_1 large, the end of the output pulse overshoots because T_2 is driven hard into conduction.

The observed input threshold both for turning on and off the trigger was 1.5 V in agreement with the d-c observations. By adjusting R_6, we can vary the T_2 base voltage, thus the T_1 T_2 emitter voltage, and hence the value of the threshold.

---

† Actually the base of T_2 is at 1.52 V because of base current. The $R_1R_3R_6$ divider would keep it at 1.9 V.

$$I = (1.9 - 1.52)/6.8 \text{ K} = 56 \ \mu\text{A}.$$

For negative input pulses, the base of T_1 is connected through 10 K to the B^+ line, so that T_1 is normally conducting and T_2 off; the operation is similar and the output is a negative square pulse.

2.3 UNIVIBRATOR (MONOSTABLE FLIP-FLOP)

By a further modification of the basic circuit, we may obtain a univibrator. This circuit generates an output pulse of fixed width and amplitude for every input pulse greater than some threshold. Its threshold is not usually as stable as that of a Schmidt trigger, so we sometimes observe a Schmidt trigger (used for its threshold) driving a univibrator (used for its fixed output pulse width).

As can be seen from the circuit diagram Fig. 4.3, T_1 is biased on, at 1.45 V through the $R_4 R_5$ divider; the T_1 emitter then follows, and since the base of T_2 is isolated from the B^+ line by C_1, it is at ground potential and T_2 is off. When a negative pulse (of sufficient amplitude) is applied at the input, T_1 is being cut off and the rise of the T_1 collector charges C_1. The base of T_2 is raised, T_2 conducts, and the output voltage drops; regeneration occurs through the common emitter connection and T_1 is held off as long as T_2 is on. However, C_1 discharges through R_1, the base emitter impedance of T_2 (approximately 1 K) and R_7; the base of T_2 slowly drops until T_2 begins to cut off, by which time T_1 begins to conduct, shutting T_2 completely off. Thus, unless the input pulse is too long, the output width and amplitude are independent of the input. The width of the pulse is of the order of $(R_1 + R_7 + R_{BE}) \times C$, which for $C = 0.01$ $\mu\mu$F gives 42 μsec. The observed width was 50 μsec, and for $C = 0.1$ $\mu\mu$F 380 μsec.

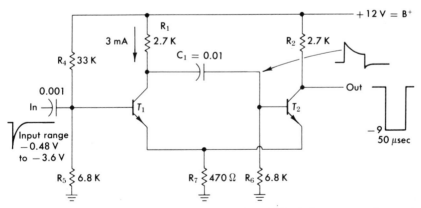

FIG. 4.3 Univibrator for negative input pulses; the width of the output pulse can be adjusted by changing condenser C_1.

2.4 MULTIVIBRATOR (FREE-RUNNING MULTIVIBRATOR)

Returning now to a symmetric circuit, a free-running multivibrator can be had by the configuration shown in Fig. 4.4.

This circuit is used for generating timing pulses or periodic gating pulses. In the vacuum tube analogue of the circuit, it is relatively easy to get very sharp rectangular output pulses whenever the plate-load resistors are much smaller than the grid-circuit impedance. The corresponding condition with transistors is not easily met, and consequently we find the recharge of the coupling capacitors spoiling the rectangular shape. We might differentiate this output and with the resultant negative spikes drive a univibrator so as to obtain a satisfactory rectangular output that repeats periodically.

In the configuration shown, the bases of both T_1 and T_2 are forward biased, through the R_3R_5 and R_4R_6 dividers. However, if one transistor (T_1) begins to conduct, the collector voltage drops, and through the coupling capacitor C_2 the drop appears at the base of T_2. Thus T_2 is being turned off, and regeneration through C_1 drives T_1 hard on, while T_2' is completely turned off. At this point, C_2, which had been charged through R_1 and T_2 to almost B^+, presents a voltage of $-B^+$ at the base of T_2. Its discharge path is through R_6 and the conducting transistor T_1 until T_2 begins to conduct again. Then regeneration occurs through C_1 and the circuit instantaneously reverses its mode of operation. It remains in this state until charge leaks off C_1 through R_5 and T_2.

If the time constants C_1R_5 and C_2R_6 are equal, the resulting output is a symmetric square wave; by making $C_1 \neq C_2$, however, it is possible to obtain an asymmetric square wave. Fig. 4.5 shows output wave forms ob-

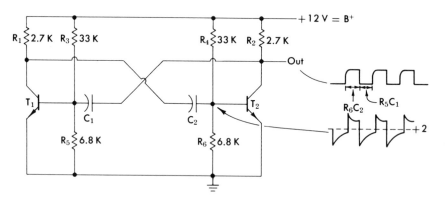

FIG. 4.4 Free-running multivibrator; the time constants of the output wave form are also indicated in the figure.

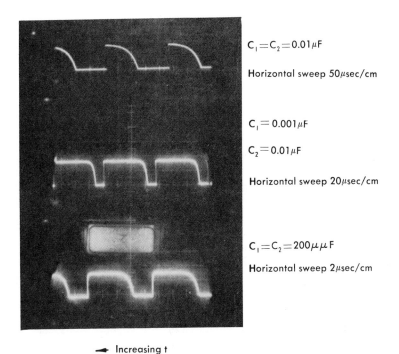

$C_1 = C_2 = 0.01\,\mu F$

Horizontal sweep 50μsec/cm

$C_1 = 0.001\,\mu F$

$C_2 = 0.01\,\mu F$

Horizontal sweep 20μsec/cm

$C_1 = C_2 = 200\,\mu\mu F$

Horizontal sweep 2μsec/cm

⬅ Increasing t

Fɪɢ. 4.5   Photograph of oscilloscope traces of the output wave forms of the free-running multivibrator of Fig. 4.4. The circuit parameters and oscilloscope scale are indicated to the right.

tained by a student†. The circuit parameters as well as the experimentally measured time constants are given below

$$C_1 = C_2 = 0.01\ \mu F \qquad RC\ \ = 80\ \mu sec,\ observed\ 110\ \mu sec$$
$$C_1 = 0.001\ C_2 = 0.01 \qquad RC_1 = 8\ \mu sec,\ observed\ 11\ \mu sec$$
$$\qquad\qquad\qquad\qquad\qquad RC_2 = 80\ \mu sec,\ observed\ 110\ \mu sec$$
$$C_1 = C_2 = 200\ \mu\mu F \qquad RC\ \ = 1.6\ \mu sec,\ observed\ 2\ \mu sec$$

3. Electronic Functional Assemblies

3.1 Sᴄᴀʟᴇ ᴏꜰ Tᴇɴ

By combining several bistable multivibrators (flip-flops) in series, it is possible to form a scaling circuit of any power of 2 (a binary). For convenience, however, scales of 10 are used, by combining four flip-flops,

† R. Kohler, class of 1962.

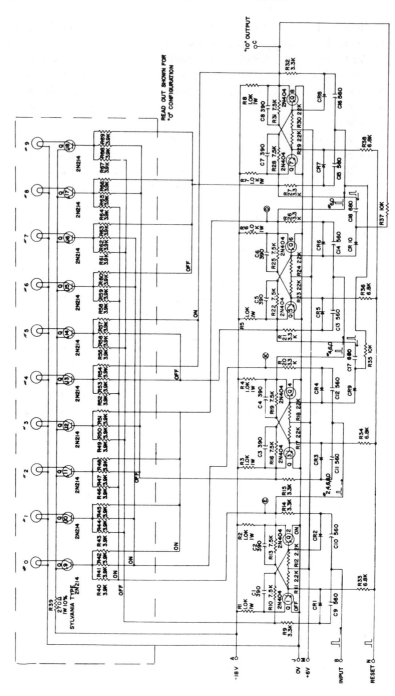

FIG. 4.6   Circuit diagram of the CMC model 100 T decade counting module.

$2^4 = 16$, and a feed-back system to eliminate the extra 6 "bits" of the scale; in reality, 6 fictitious pulses are generated by double driving the second (2 bits) and third (4 bits) flip-flop, as explained below. In this fashion an output pulse is obtained for every 10 pulses in; if, in addition, it is desired to know how many unit pulses have arrived after the last multiple of 10, a read-out system must be provided which senses the state of each of the four flip-flops.

The best way to understand the operation of a scale of 10 is to follow through its circuit diagram, such as the one shown in Fig. 4.6, adapted from the Computer Measurement Corporation's Model 100T decade-counting module. The basic structure of the flip-flops is reproduced in Fig. 4.7; state "No" is the configuration where Q_1 is off and Q_2 is on. The flip-flops are triggered by positive pulses in; indeed, through the open diode the positive pulse will be applied at the base of Q_2 and will thus turn it off, and the circuit will switch to state "Yes," Q_1 on, Q_2 off. The next flip-flop is fed from the collector of Q_2, which becomes more positive (positive pulse out) when the circuit switches from "Yes" to "No." With these points in mind, we can follow the operation of a scale of 16, as indicated in Table 4.1, where the state of each flip-flop is indicated for every pulse. It simply represents all possible combinations of the two states of the four flip-flops, in an orderly fashion, so that each combination will follow from the one before whenever an input pulse arrives.

Now, to obtain a scale of 10, configurations 4 and 5, and 8, 9, 10, and 11 of Table 4.1(a) are bypassed, as shown in Table 4.1(b). This can be easily achieved by feeding back from FF_3 to FF_2 so that when FF_2 resets to

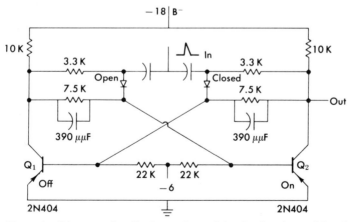

FIG. 4.7 Diagram of a flip-flop unit used in the decade module of Fig. 4.6.

TABLE 4.1

CONFIGURATION OF FLIP-FLOPS IN A SCALE OF 10

(a) No feedback (Scale of 16)

Configuration	FF$_1$ No	FF$_1$ Yes	FF$_2$ No	FF$_2$ Yes	FF$_3$ No	FF$_3$ Yes	FF$_4$ No	FF$_4$ Yes
0	X		X		X		X	
1		X	X		X		X	
2	X			X	X		X	
3		X		X	X		X	
4	X		X			X	X	
5		X	X			X	X	
6	X			X		X	X	
7		X		X		X	X	
8	X		X		X			X
9		X	X		X			X
10	X			X	X			X
11		X		X	X			X
12	X		X			X		X
13		X	X			X		X
14	X			X		X		X
15		X		X		X		X

(b) With feedback (Scale of 10)

Configuration	FF$_1$ No	FF$_1$ Yes	FF$_2$ No	FF$_2$ Yes	FF$_3$ No	FF$_3$ Yes	FF$_4$ No	FF$_4$ Yes
0	X		X		X		X	
1		X	X		X		X	
2	X			X	X		X	
3		X		X	X		X	
4	X			X		X	X	
5		X		X		X	X	
6	X		X			X		X
7		X	X			X		X
8	X			X		X		X
9		X		X		X		X

"No," advancing FF_3 to "Yes," the positive pulse from the collector of Q_5 (Fig. 4.6) is applied at the base of Q_4, turning Q_4 off and reverting FF_2 back to state "Yes"; this is configuration 6†. Similarly when FF_3 resets to "No" advancing FF_4 to "Yes," the positive pulse from Q_7 turns Q_6 off, thus returning FF_3 to a "Yes" state—namely, configuration 12. After the ninth pulse, the circuit is ready to revert back to 0, that is, all FF at "No."

In addition, in Fig. 4.6, we recognize the "reset" line, which applies a simultaneous positive pulse at Q_1, Q_3, Q_5, and Q_7, thus reverting the whole circuit to 0. The read-out consists of the n-p-n transistors Q_9 through Q_{18}, each of which has its base connected in parallel to three of the flip-flop transistors; when *all three* are simultaneously on, that is, when the collectors are positive with respect to -18, the base of Q_9 through Q_{18} is sufficiently raised, and they conduct, lighting the indicator bulbs. The reader can easily verify, using the configurations of Table 4.1 (b), that this coding is unique, and that only one light at a time will turn on, in the proper sequence.

3.2 Examples of Assembled Switching Units

To give a further example of how the basic switching units can be combined to perform certain logic operations, we will consider briefly two circuits.

The circuit shown in Fig. 4.8 is used to produce repeatedly a sequence of two square pulses spaced at a desired time interval. The width of the pulses is controlled by the time constant $R_3 C_3$, and their spacing is controlled by

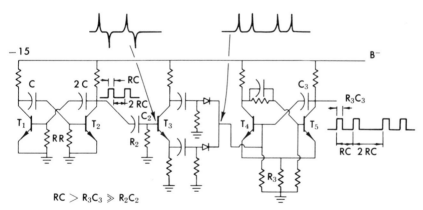

FIG. 4.8 Schematic circuit diagram of a simple double-pulse generator. Note that T_1 and T_2 form the basic multivibrator, while T_4 and T_5 are used as an output shaping univibrator.

---

† The additional loops, in Fig. 4.6, through $R35$ and $R37$ are protective and are not essential for the understanding of the principle of operation.

RC, whereas R_2C_2 is just a differentiator; thus the following conditions are set

$$RC > R_3C_3 \gg R_2C_2$$

Let us then examine the pulse shapes through the various points of the circuit. (Note that in practice this would be extremely easy to do by observing with an oscilloscope.) It is clear that the first two transistors, T_1 and T_2, form an asymmetric free-running multivibrator, with time constants $\tau_1 = RC$ and $\tau_2 = 2RC$, so that at its output the square wave indicated in the figure appears. The following R_2C_2 network differentiates this square wave, as shown in the figure, while T_3 with the two diodes is used to convert the differentiated signal into a series of positive pulses with the desired spacing. Finally, transistors T_4 and T_5 form a univibrator of time constant R_3C_3; this univibrator is triggered by the positive pulses of the previous stage and gives the desired output.

Another example of such a combination of the basic switching units is given in Fig. 4.9. This circuit is a type of time-interval counter (delayed

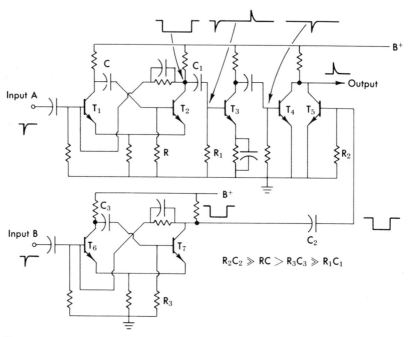

$$R_2C_2 \gg RC > R_3C_3 \gg R_1C_1$$

Fig. 4.9 Schematic circuit diagram of a delayed coincidence (time-interval counter). Note that T_1 and T_2 stretch the signal of input A so that the T_3 output is delayed with respect to A; T_6 and T_7 shape the signal of input B and T_4 and T_5 form a simple coincidence circuit.

coincidence): an output is obtained only when pulse B follows pulse A by a time interval t, within Δt. The time interval t is set by the constant RC and Δt by R_3C_3; R_1C_1 is again a differentiator, while R_2C_2 is only a coupling network; thus the following conditions hold

$$R_2C_2 \gg RC > R_3C_3 \gg R_1C_1$$

The function of the individual components is clear if we now follow the two (negative) input pulses A and B through the circuit. Transistors T_1 and T_2 form a univibrator, which is triggered by input A and has a length determined by RC; this long square pulse is differentiated in R_1C_1 and only the negative spike gets through T_3, where it might also be amplified. Thus the output of T_3 is delayed with respect to the input by a time interval t. Similarly T_6 and T_7 form a univibrator which is triggered by the input pulse B and produces a square negative output pulse of length Δt as controlled by R_3C_3; this pulse is coupled through R_2C_2 to the base of T_5 and is not distorted since R_2C_2 is a time constant much larger than R_3C_3. Finally, the two transistors T_4 and T_5, which are coupled through their collectors form a "coincidence" circuit and give an output only if both have simultaneously negative signals at their bases. This is indeed so, because if a negative signal appears at the base of T_4, T_4 will be cut off, but T_5 is still conducting heavily, so that the potential at its collector changes very slightly from ground. If, however, T_5 is at the same time turned off, there is no current flow and the common collector point has to rise towards the B^+ voltage.

3.3 Simple Pulse-Height Analyzer and Oscilloscope Assemblies

As examples of more complex assemblies, we will briefly discuss two instruments. Here we will not consider the detailed circuit diagrams, but only the assembly of the various functional blocks and the flow of pulses (information) between them.

We begin with a simplified 24-channel pulse-height analyzer; this instrument selects pulses according to their amplitude and sorts them into one of 24 channels. The basic principle is to convert the pulse height to width (time) and then measure the width of the pulse by counting the oscillations that were allowed through a gate of the same time duration as the pulse.

The block diagram and appropriate wave forms are shown in Fig. 4.10. The input pulse is first stretched and amplified. Next it is shaped and allowed through a gate, but only if no other pulse is being analyzed. The following stage converts the pulse into a negative square pulse of a width

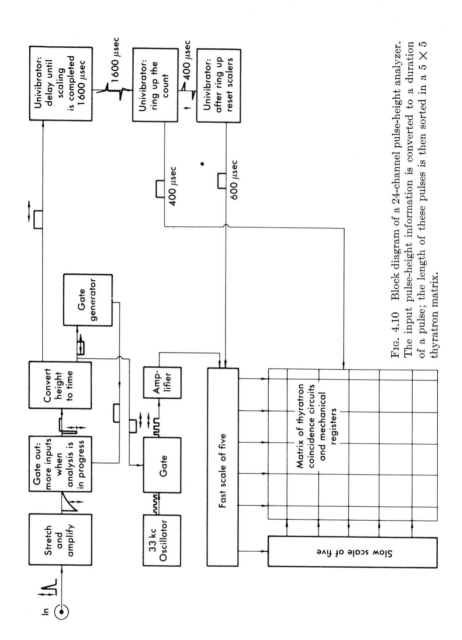

FIG. 4.10   Block diagram of a 24-channel pulse-height analyzer. The input pulse-height information is converted to a duration of a pulse; the length of these pulses is then sorted in a 5 × 5 thyratron matrix.

proportional to the height of the input. This output is used to operate a gate which allows the passage of the signal from a 33-kc oscillator. The oscillations that have passed through the gate are amplified and then scaled in two scales of 5, allowing from 0 to 25 counts; the outputs of the scalers are connected to a matrix of thyratron tubes which can drive mechanical registers when commanded to do so by the application of a positive pulse. This command pulse is obtained from the "ring up the count" univibrator, which is delayed by 1.6 msec through the preceding "delay" univibrator. In this fashion, it is ascertained that the scalers had time to register the appropriate number of oscillator pulses (those that did pass the gate), before being "read out." Finally, 400 μsec later, after the thyratrons have been activated, the scalers are reset and the instrument is ready for a new pulse.

As another example, we will consider an oscilloscope, specifically the type 514 of Tektronix. Now the block diagram is much more complicated, but the logic is again clear. Basically, what is required is a horizontal sweep for the cathode-ray tube and sufficient amplification, without distortion, of the input signal so as to drive the vertical deflection of the cathode-ray tube. Further, the sweep and signal must be properly synchronized, either by an external trigger, or with respect to one another internally.

The reader should follow the diagram in Fig. 4.11 and notice the function of each component. Starting from the top left, the first four blocks amplify and shape the trigger signal, which can be either external or internal (from the input signal), or 60 cycles/sec (from the a-c line). Next, a sweep generator provides the saw-tooth voltage, which is amplified in two circuits 180 degrees out of phase so that the cathode-ray tube is driven in a push-pull fashion, symmetric about 0 voltage. It is also necessary that during the "fly-back" when the spot returns to the origin, no trace appears on the cathode-ray tube; this is achieved through the two unblanking units, an amplifier and a cathode follower. Also, provisions are made for supplying externally a gate signal derived from the trigger, and the saw-tooth sweep signal before amplification.

As shown on the lower part of the diagram, the input signal is attenuated to the fixed input level if so required, and then delayed to provide enough time for triggering and generation of the sweep; it is amplified in two networks 180 degrees out-of phase and fed to vertical plates of the cathode-ray tube again in a push-pull arrangement. A free-running multivibrator of fixed amplitude is used to provide amplitude calibration, and a compact power supply provides the various regulated voltages required for the different units.

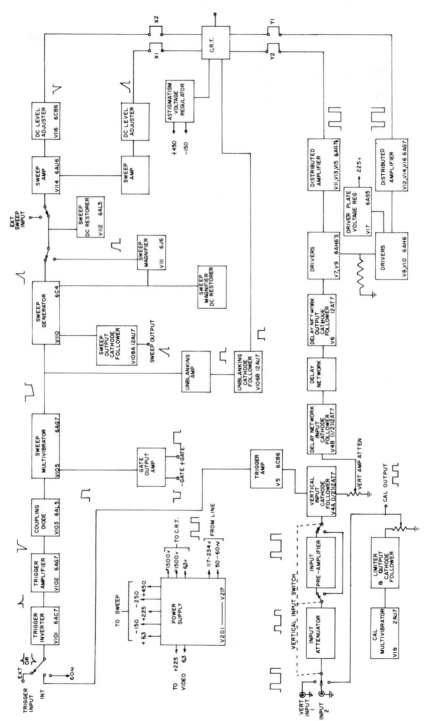

Fig. 4.11 Block diagram of the Type 514 Tektronic oscilloscope. Note the triggering network, the horizontal sweep network, and the vertical amplifier network.

122

3.4 THE KEITHLEY 600A (OR 610A) ELECTROMETER

The reader has already appreciated the difficulty of making measurements of very small currents in the micromicroampere region, and small d-c voltages in the microvolt region. Such measurements have traditionally been made in the past with sensitive galvanometers and (quadrant) electrometers. It takes a skilled experimenter, however, to use these devices in their highest sensitivity range; electronic instruments have a superior performance for most applications and are much easier to operate. Further, it is clear that a galvanometer cannot be used for a-c measurements, whereas an electronic meter can respond to a-c signals, provided the time-constant requirements are not too stringent.

In what follows we will describe the Keithley Model 600A Electrometer as it can be used for d-c or quasi-d-c measurements. This is basically an electronic (amplifier) voltmeter, of ultrahigh input impedance, fair sensitivity, and very good stability.

The input-impedance requirement is necessary so as not to load (draw current from) the voltage source that is being measured. It then also becomes possible to shunt the input with a very high resistance resistor, and thus use the electrometer as a micromicroammeter. The ultimate limit of the current that can be measured depends on the grid current of the first tube; thus, special "electrometer" vacuum tubes are used; they have very good insulation, very high vacuum, and appropriate electrode construction. The resulting characteristics for the Model 600A as given by the manufacturer are

Input impedance	$> 10^{14}$ ohms
Grid current	$< 10^{-14}$ amp
Sensitivity	from 0.01 to 10 V full scale
Drift	$< 2 \times 10^{-3}$ V/hr
Frequency response (at amplifier output)	direct current to 100 cps

The general principle of operation is shown in Fig. 4.12a, where the resistor R at the input is used only for current measurements. The balanced electrometer stage is followed by a d-c amplifier of gain 100 with phase inversion. The meter and output voltage are obtained from a cathode follower, which is driven by the amplifier.

Another consideration for all sensitive measuring devices is their response time. Since the minimum input capacitance, as determined by the connector, wiring, cabinet, and so on, is of the order of 30 $\mu\mu$F, if a 10^{13} shunt impedance is used, the time constant becomes 5 min, and any additional length of cable will hurt. To avoid this inconvenience, negative feedback from the amplifier is used to reduce the response time (and the input voltage drop) by at least a factor of 100. The principle of such an arrangement is shown in Fig. 4.12b.

To operate a Keithley successfully at high sensitivity, some precautions must be observed: since it is easy to generate electrostatic charges by friction, the signal cables must be mechanically rigid, and shielding against stray fields must be provided; radio-frequency pickup, mechanical vibrations, even walking persons can all be severe annoyances in measurements of very low currents or voltages from very high internal impedance sources.

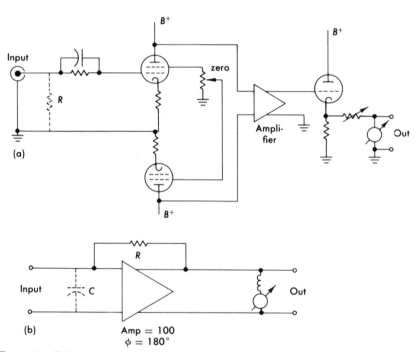

Fig. 4.12 Schematic diagram of an electrometer circuit. (a) Input stage and output cathode follower. (b) Use of feedback to speed up response.

A simplified circuit diagram of the 600A is given in Fig. 4.13 (the switching steps, heater voltages, and some decoupling links have been omitted). The three vacuum tube stages are d-c coupled through the batteries; the Model 610A is similar, but can be operated from the a-c line instead. It is evident that the amplifier ground, to which the cathodes of all four tubes return, is floating; a separate ground is established for the cathode follower (V_4) and the output: normally, the output ground is connected to the chassis and the amplifier floats. When fast response is required, the am-

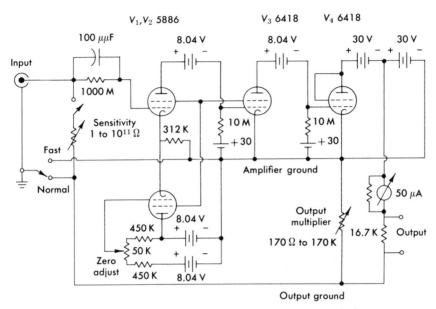

Fig. 4.13 Simplified circuit diagram of the Keithley Model 600A electrometer.

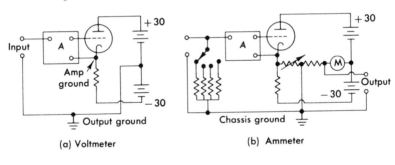

(a) Voltmeter

(b) Ammeter

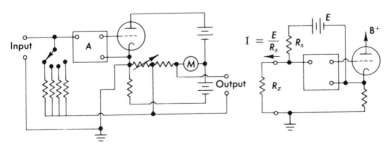

(c) Ammeter with fast response

(d) Ohmmeter

Fig. 4.14 Possible connections of a Keithley electrometer.

plifier ground is connected to the chassis and the input resistor itself becomes the feedback element from the cathode follower output to the amplifier input.

The three main controls are (a) the output multiplier which has six steps from 0.01, 0.03, etc., to 10 V, (b) the zero adjust, and (c) when used as an ammeter, the sensitivity switch; this switch selects the shunt resistor, providing a range from 3 amp to 10^{-13} amp (in conjunction with the output multiplier). The fast response should be used only in the range from 10^{-7} to 10^{-13} amp.

The functional behavior of the electrometer can be best understood from Fig. 4.14, which gives the schematics for three types of operation: (a) voltmeter, (b) ammeter, (c) ammeter with fast response, and (d) ohmmeter.

The instrument can also be used as a very high impedance ohmmeter, as a current integrator, and for various other applications; for such information the instruction manual and catalogue published by Keithley Instruments, Inc., of Cleveland, Ohio should be consulted; also a paper by J. F. Keithley† and the references given therein.

4. Vacuum Techniques

The creation and measurement of vacuum is necessary in many experimental devices and applications. Residual pressure (vacuum) is measured in millimeters of mercury, and present-day commercial devices can reach pressures as low as 10^{-10} mm. An improvement by one or two orders of magnitude has been achieved by some laboratories, and similar advances have been made in the field of vacuum-measuring devices. For comparison, the residual pressure in free space is 10^{-16} mm, corresponding to a density of one molecule per cubic centimeter.

Vacuum pumps are of three main types, each for a specific range of pressures:

(a) Mechanical pumps, which can operate against atmospheric pressure and down to a limit of 10^{-2} to 10^{-3} mm. They are used as "back-up" pumps to create the necessary rough vacuum for the operation of a high vacuum pump.

(b) Molecular diffusion pumps with capabilities down to 10^{-6} to 10^{-8} mm.

(c) Ion and sputter ion pumps with limits at $10^{-8} - 10^{-10}$ mm.

It must also be noted that the lowest pressure that can be reached is limited by the "vapor pressure" of the materials in the system. To elim-

---

† Joseph F. Keithley, "Electrometer Measurements," *Instruments and Control Systems*, January, 1962.

inate contaminations of such volatile materials in a high-vacuum system, it is standard practice to "bake out" the systems at high temperatures for several hours or days. For the same reason it is useful to use "cold traps" to condense the volatile vapors. As a matter of fact, the highest vacuums have been achieved by cryogenic techniques.

4.1 INTRODUCTORY CONCEPTS

We begin by giving some quantitative relations and by recalling some topics of gas flow. A vacuum system is characterized by its speed, S, in cm^3/sec. The speed, S, depends on the pressure, and there exists for every system a lower limit, the ultimate pressure that can be reached, P_s; the approach toward this ultimate limit is exponential, and we have

$$-\frac{dP}{dt} = \frac{S}{C}(P - P_s) \tag{4.1}$$

where C is the total volume to be evacuated, and P is the instantaneous pressure. Obviously the above expression is dimensionally consistent and can be used as a definition of S. Integration of Eq. 4.1, if S is assumed constant, yields

$$P = (P_0 - P_s) \exp\left(-\frac{tS}{C}\right) + P_s \tag{4.2}$$

where P_0 is the initial pressure at time $t = 0$.

The speed of a system is determined by the speed S_p of the vacuum pump employed, as well as by the tubing connecting the pump to the vessel to be evacuated. We define the "throughput," Q, of a system as the volume of gas entering (or leaving) it per unit time, multiplied by the pressure:

$$Q \equiv SP \tag{4.3}$$

Then, for a system of conductors (tubing) the conductance F is defined by

$$Q = F(P_1 - P_2) \tag{4.4}$$

where P_1 and P_2 are the pressures at the two ends of the system.

We consider now a volume C at a pressure P, which is evacuated at a speed S. The volume is connected to a pump of speed S_p through tubing of conductance F; let the pressure at the pump be P_p. We obtain

$$Q = SP \qquad Q_p = S_pP_p$$

where

$$Q_p = Q = F(P - P_p) \tag{4.5}$$

it follows

$$\frac{1}{S} = \frac{1}{S_p} + \frac{1}{F} \tag{4.6}$$

giving the basic relation between the speed S_p of the pump employed and the speed with which the system is evacuated.

In designing a vacuum system, if the gas flow *into* the system (through leaks, etc.) is not known and only a desired pressure is stated, a safe rule of thumb is to plan to obtain a speed $S = C/\text{sec}$; where C is as before the volume to be evacuated and S must be available at the desired pressure.

To evaluate the conductance F, we note that it depends on the type of gas flow through the system, and on geometrical factors. The gas flow can be of two types:

(a) Viscous (Poiseuille) flow, which prevails at high pressures, and is analogous to nonturbulent fluid flow;

(b) Molecular flow, which prevails at very low pressures when the mean free path of the molecules is of the same order as the diameter of the conductor (tubing), or is larger.

For viscous flow, the throughput Q depends on the pressure differential. For a long straight tubing of length l and radius a (in centimeters), if the low pressure is P_1 and the exhaust pressure P_2 (in microns $= 10^{-3}$ mm of mercury), we obtain

$$Q = \frac{10\pi}{96\eta} \left(\frac{a^4}{l}\right) (P_2{}^2 - P_1{}^2) \text{ microns-cm}^3/\text{sec} \tag{4.7}$$

where η is the coefficient of viscosity of the gas, which for air is

$$\eta = 1.845 \times 10^{-5} \text{ poise}$$

By our definition, $F = Q/(P_2 - P_1)$ and if we let $P_a = (P_2 + P_1)/2$, we obtain for viscous flow

$$F_v = \frac{10\pi}{48\eta} P_a \left(\frac{a^4}{l}\right) \text{ cm}^3/\text{sec} \tag{4.8}$$

Next, to see when molecular flow becomes important, we use the following two expressions for the mean free path:

$$\lambda = \frac{1}{\sqrt{2}} \frac{1}{\pi n \delta^2} \tag{4.9a}$$

where $n =$ number of molecules per unit volume, and $\delta =$ molecular diameter. Hence

$$\lambda = (8.59 \times 10^3)\eta \frac{1}{P} \frac{T}{M} \text{ cm} \tag{4.9b}$$

where η is again the coefficient of viscosity in poises, P the pressure in microns, T the temperature in degrees Kelvin, and M the molecular mass. Thus, for air at room temperature at a pressure of 10^{-3} mm, $\lambda = 7.3$ cm.

For molecular flow, the conductance F for a tube of length l, perimeter H, and varying cross section A is given by

$$F_m = \frac{4}{3}\bar{v}\left[\int_0^l \frac{H}{A^2} dl\right]^{-1} \tag{4.10a}$$

with $\bar{v}$ the average molecular velocity; $\bar{v} = 14.55 \times 10^3 \sqrt{T/M}$ cm/sec. Thus for a long tube of length l and radius a in centimeters,

$$F_m = \frac{2}{3}\pi \left(\frac{a^3}{l}\right) \bar{v} \text{ cm}^3/\text{sec} \tag{4.10b}$$

For a simple opening of A (in cm²)

$$F_m = \tfrac{1}{4}A\bar{v} \text{ cm}^3/\text{sec} \tag{4.10c}$$

For more details and tables of F_v and F_m the reader is referred to Dushman.†

4.2 PUMPS; OPERATION AND BRIEF DESCRIPTION

Mechanical pumps work on the usual principle of the piston pump, namely, of "intaking" a large volume at low pressure and "exhausting" it to the atmosphere by continuous reduction of this volume. However, because of the low viscosity of gases, all vacuum pumps are of rotary design and are completely immersed in oil; the immersion in oil lubricates the rotating parts but mainly acts as a seal. Some pumps have two stages connected in series; Fig. 4.15 gives the pumping speed against pressure for the Hyvac two-stage pump, and includes a sketch of one stage of this vacuum pump. Additional details can be found in Dushman.†

Diffusion or molecular pumps do not use mechanical motion. A high-velocity stream of oil (or mercury) vapor is formed and directed from the system to be evacuated toward the forepump; pumping action arises from the diffusion of the residual gas molecules into the high-velocity vapor stream, where they are carried along toward the forepump.

† S. Dushman and S. M. Lafferty, *Scientific Foundations of Vacuum Technique.* 2d ed. New York: Wiley, 1964.

The vapor stream is provided by boiling off organic oils which, however, are recovered by condensing upon collision with the cooled walls of the pump; usually two to three jets are used in series. When each jet is supplied by a separate boiler, the pumps are called "fractionating"; this procedure results in automatic selection of the lowest vapor-pressure components of the oil, for the jet closest to the high-vacuum side.

Fig. 4.16a is a sketch of the three-jet oil diffusion fractionating MCF pump manufactured by Consolidated Vacuum of Rochester, New York.

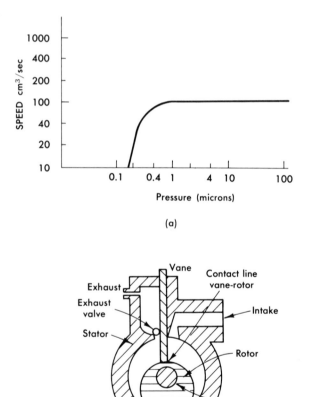

(a)

(b)

FIG. 4.15   The "Hyvac" rotary vacuum pump manufactured by Cenco Scientific Company. (a) Speed against pressure curve. (b) Cross-sectional view of one stage of this pump.

Fig. 4.16b gives the pumping speed curve for a similar two-jet glass pump, the VMF-20, manufactured by the same company. As can be seen from the curve, a diffusion pump cannot work against any forepressure, since its speed drops to zero; on the low side, however, it has an ultimate pressure which is limited only by the vapor pressure of the oil employed. The heaters of a diffusion pump should not be turned on unless a rough vacuum of 10^{-2} mm has been reached, since otherwise the oil oxidizes and is unusable for proper pumping action.

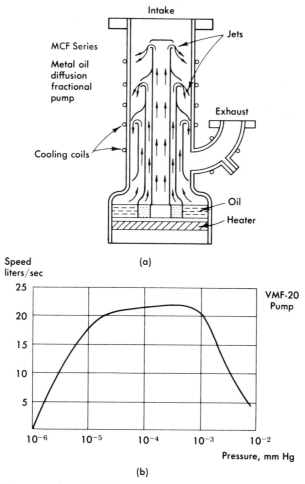

FIG. 4.16 An oil diffusion vacuum pump. (a) Cross-sectional view of the pump assembly showing the three separate jets and the flow of oil vapor. (b) Speed against pressure curve.

Ion pumps are based on the principle of ionizing the molecules of the residual gas and then accelerating them toward the walls of the system, where they get buried due to the impact. In addition, use is made of the chemical affinity of many gas molecules to combine with a "getter" material, such as titanium, so that they are removed from the residual gas atmosphere. Such getters have been used for many years in electron vacuum tubes.

Ion pumps may be of the cold cathode or hot filament type. In the former, electrons are released from the cathode by field emission toward the anode; they are then trapped in the active volume of the pump by a magnetic field. By collision, these electrons produce ions, which are accelerated back toward the cathode, as shown in Fig. 4.17a; they end up in the collector, where they get buried. Those ions that hit the titanium cathode release titanium ions, which also are attracted towards the collector; the titanium ions cover the residual gas molecules and provide a fresh titanium surface for chemical gettering.

Hot filament (evapor-ion) pumps use the electron beam from the filament to bombard a titanium wire which is continuously evaporating. The

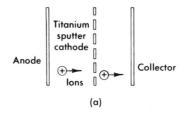

(a)

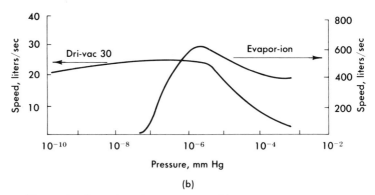

(b)

Fig. 4.17 Operation of an ion pump. (a) Schematic of the motion of ions. (b) Speed against pressure curves for a cold filament (left-hand scale) and a hot filament (right-hand scale) pump.

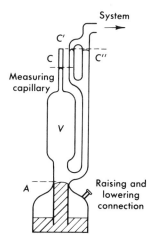

Fɪɢ. 4.18   Diagrammatic arrangement of a McLeod vacuum gauge.

titanium condenses on the cooled walls of the pump and thus provides an active getter layer which captures the molecules of the active gases that strike it; furthermore, the new layers of titanium cover the molecules that have been trapped. Obviously if air is accidentally admitted to the pump while the filament is hot, the whole system will be coated with titanium.

Figure 4.17b gives the speed curves for both the hot-filament and drivac ion pumps.

4.3 Vacuum-Measuring Devices

Measurement of residual pressure, especially at very low values, has always been a difficult problem; similarly the calibration of the various instruments in the 10^{-7}-mm and lower range is dependent on the nature of the residual gas, impurities in the system, and conditions of the surfaces of the measuring device. Present-day techniques can measure residual pressures of 10^{-11} to 10^{-12} mm and claim a sensitivity even in the 10^{-14} region. In this section, however, we will describe only three types of gauges used most commonly in laboratory and industrial applications:

(a) The MacLeod gauge, a mercury manometer with a "pressure amplifier." It is an absolute device, and covers the range from 10 to 10^{-4} mm.

(b) The thermocouple gauge, which measures the *rate of cooling*, by conduction, of a heated filament; it covers the range from 0.5 to 5×10^{-3} mm.

(c) The ionization gauge, which measures the rate of production of ions by an electron beam of standard (current) density and energy; it covers the range from 10^{-4} to 10^{-8} mm.

The principle of the MacLeod gauge is shown in Fig. 4.18; to measure the pressure, the mercury level in the gauge is raised (either mechanically or by allowing atmospheric pressure into the mercury reservoir). As the mercury rises above point A it seals off the residual gas now trapped in volume V; when the level is further raised, this gas is compressed (thus the pressure is amplified according to $PV = $ constant). Let the mercury level reach C in the measuring capillary, and $C' = C''$ in the reference capillary and connecting tube respectively; the pressure in the measuring capillary P_c must then equal the difference in levels between $C' - C = h$ plus the pressure of the system, P_s

$$P_c = h + P_s \tag{4.11}$$

If the cross section of the capillary is b, the trapped volume is $v = hb$, and since the original volume was V,

$$vP_c = VP_s = hbP_c = hb(h + P_s) \tag{4.12}$$

it follows that

$$P_s = \frac{h^2 b}{V - bh} \tag{4.13}$$

If several such amplification stages are combined in series, it becomes possible to read pressures over a wide range. There are limitations of the MacLeod gauge, however: (a) surface tension effects in the capillaries, (b) the amount of mercury (weight) that has to be lifted, (c) the fact that varpors soluble in mercury cannot be measured, and (d) the fact that since

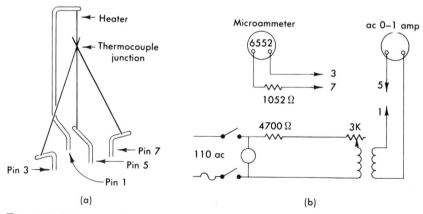

(a)   (b)

FIG. 4.19   Diagrammatic arrangement of a thermocouple vacuum gauge. The numbers in (a) refer to the NRC 501 gauge, and correspond to the connecting points shown in (b).

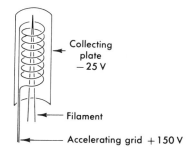

Collecting
plate
− 25 V

Filament

Accelerating grid + 150 V

Fig. 4.20   Diagrammatic arrangement of vacuum ion gauge.

the vapor pressure of mercury at room temperature is 1.8×10^{-3} mm, any high-vacuum system must be isolated from the gauge by a cold trap.

The thermocouple gauge consists of a junction which is heated by a filament through which a standard current (of the order of 0.6 amp) is flowing. The electromotive force developed at the thermocouple junction can be measured directly on a microammeter; it ranges from 0 to 14 mV for the range of 1 to 10^{-3} mm Hg. At high residual pressures the rate of cooling of the junction by conduction is large, and thus its temperature is low; at low residual pressures the junction temperature rises.

A sketch of a thermocouple gauge† is given in Fig. 4.19a, while Fig. 4.19b shows the simple circuit used in this laboratory with such a gauge.

The ionization gauge, shown in Fig. 4.20, is widely used in medium-to-high vacuum systems. Like the ion pump, the gauge can be of either the cold-cathode or hot-filament type. In the filament ion gauge (much as in the Frank-Hertz tube), the emitted electrons are accelerated by a positive grid ($E_g = +150$ V); the ions produced by the electron beam are collected at a plate maintained at a negative potential ($E_p = -25$ V). The electron beam must be standardized, and this is achieved by measuring and maintaining a fixed grid current, usually $I_g = 5$ mA.

For pressures lower than 10^{-3} mm Hg, and for *dry air,* the ion current is of the order of 100 μ amp/10^{-3} mm Hg; this calibration, however, is different for other gases. The ion current varies to a good approximation linearly with pressure, so that depending on the sensitivity of the d-c amplifier used measurements can be made down to 10^{-7} mm Hg (10^{-10} amp). In order to obtain accurate readings at low pressures, it is necessary to "degas" the grid by heating it electrically.

Unfortunately, if the gauge is operated at pressures above 10^{-3} mm at full heater current, the filament burns out (due to oxidation) in a few seconds.

† For example, the NRC 501, manufactured by the National Research Corporation, Cambridge, Mass.

4.4 Some Experimental Measurements of Pumping Speeds

To demonstrate some of the properties of vacuum systems discussed previously, we present data obtained by a student.† The vacuum system was made of glass with a two-stage oil diffusion pump and no cold traps.

Figure 4.21 gives the pressure-against-time curve, with $t = 0$ chosen not as soon as the diffusion pump heaters were turned on, but a short time thereafter; the heater voltage was set at 55 V. From the graph we infer an equilibrium pressure of 4×10^{-5} mm and a fairly constant pumping rate $S/C \simeq 0.12$ sec^{-1}. Due to the complicated tubing and other constrictions,

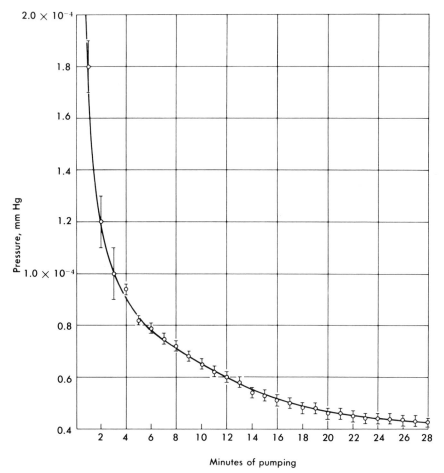

Minutes of pumping

FIG. 4.21   Experimental data on the pumping speed of a two-stage oil diffusion pump.

† D. Statt, class of 1963.

it is difficult to evaluate exactly the volume under evacuation; we assume that it is of the order of 10 liters (10^4 cm^3).

Even so, from the manufacturer's data on Fig. 4.16, the ultimate pressure should be of the order of 10^{-6} mm Hg, and the pumping speed an order of magnitude higher. Such differences are not uncommon; the low value of S is due partly to the reduced heater voltage, and partly due to degassing of the pump oil (heaters just turned on) and leakage from gases trapped in the vacuum grease of the stopcocks. Such gas release at low pressures and at the early stages of pumping can be approximated by a constant leak Q_L (micron-cm^3/sec); hence the differential equation given by Eq. 4.1 takes the form

$$\frac{dP}{dt} = -\frac{S}{C}(P - P_s) + \frac{Q_L}{C} \tag{4.1a}$$

and has the solution

$$\left[P - \left(P_s + \frac{Q_L}{S} \right) \right] = \left[P_0 - \left(P_s + \frac{Q_L}{S} \right) \right] \exp\left[-(s/c)t \right] \tag{4.2a}$$

where again P_0 is the initial pressure. Thus the ultimate pressure is determined by the largest of Q_L/S or P_s, or by both. If the leak Q_L cannot be considered constant, but is a function of pressure, then also the determination of S is affected.

5. Radiation Safety and Handling of Radioactive Materials

5.1 INTRODUCTION

In a series of experiments on quantum physics, the student necessarily comes in contact with radioactive sources, either while studying the properties of the nucleus itself or when using the sources to obtain energetic beams of alpha or beta particles or gamma radiation. As is well known, radiation can be harmful to humans and therefore precautions must be taken against undue exposure to it, and in the handling of radioactive materials.

In addition to the naturally occurring radioisotopes (which have long lifetimes), a great variety have been produced artificially and many of them can be purchased. Table 4.2 is a listing of the radioactive materials primarily suggested for use in this laboratory, some of which, like Co^{60}, Na^{22}, and Po^{210}, are quite standard for training, testing, and calibration purposes. The table gives the type and energy of the radiation, the half-life, the main decay scheme, and the maximum permissible burden in the body; in the last column a few remarks are given on the experiments for which the isotope is most useful.

TABLE 4.2

PROPERTIES OF SOME RADIOACTIVE MATERIALS

Element	Energy of radiation (MeV)		Half-life	Main decay scheme	Maximum permissible burden in total body (μCi)	Remarks
				β, γ Emitters		
Cesium 137 ($_{55}Cs^{137}$ & $_{56}Ba^{137}$)	β^-	0.514	27 yr	Cs^{137} (27 yr) Ba^{137} — decay scheme: $7/2+$, β^- 92% (2.6m), $11/2-$ 8%, 0, $3/2+$, γ 0.662, e_k/γ 10%, 0	30	Compton scattering β, γ spectra
	γ	0.662				
Cobalt 60 ($_{27}Co^{60}$ & $_{28}Ni^{60}$)	β^-	0.312	5.2 yr	Co^{60} (5.2 yr) Ni^{60} — decay scheme: $5+$, β^- Prompt, $4+$, γ 2.505, 8×10^{-13} sec., $2+$ 1.333, γ, 0	10	γ–γ correlation, γ spectra
	γ_1	1.172				
	γ_2	1.333				

TABLE 4.2 Continued

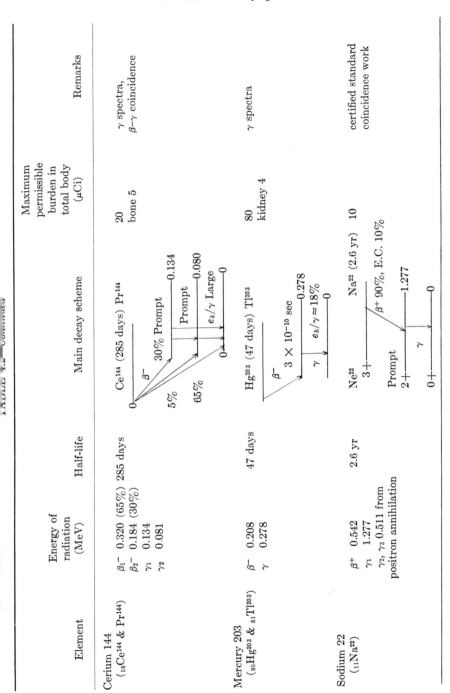

Element	Energy of radiation (MeV)	Half-life	Main decay scheme	Maximum permissible burden in total body (µCi)	Remarks
Cerium 144 ($_{58}$Ce^{144} & Pr^{144})	β_1^- 0.320 (65%), β_2^- 0.184 (30%), γ_1 0.134, γ_2 0.081	285 days	Ce^{144} (285 days) Pr^{144}	20, bone 5	γ spectra, β–γ coincidence
Mercury 203 ($_{80}$Hg^{203} & $_{81}$Tl^{203})	β^- 0.208, γ 0.278	47 days	Hg^{203} (47 days) Tl^{203}	80, kidney 4	γ spectra
Sodium 22 ($_{11}$Na^{22})	β^+ 0.542, γ_1 1.277, γ_2, γ_3 0.511 from positron annihilation	2.6 yr	Na^{22} (2.6 yr) β^+ 90%, E.C. 10%	10	certified standard coincidence work

TABLE 4.2—*Continued*

Element	Energy of radiation (MeV)	Half-life	Main decay scheme	Maximum permissible burden in total body (μCi)	Remarks
			Pure β Emitters		
Yttrium 91 ($_{39}$Y^{91})	β^- 1.537	58 days	Y^{91} (58 days) Zr^{91}; $1/2-$, β^- 98%, β^- 2%, $5/2+$, 1.19, 0	30 bone 5	β spectra
Thallium 204 ($_{81}$Tl^{204})	β^- 0.764	3.6 yr	Tl^{204} (3.6 yr); $2-$, E.C. 2%, β^- 98%, Hg^{204}, $0+$, Pb^{204}	80 kidney 10	β spectra
Phosporus 32 ($_{15}$P^{32})	β^- 1.707	14 days	P^{32} (14 days); $1+$, β^-, $0+$, S^{32}	bone 6	lifetime measurements

TABLE 4.2—*Continued*

Element	Energy of radiation (MeV)	Half-life	Main decay scheme	Maximum permissible burden in total body (µCi)	Remarks
			α Emitters		
Polonium 210 (Po^{210})	α 5.300	138 days	Po^{210} (138 days) 0+ —— α ——> Pb^{206} 0+	0.4 kidney 0.04 lethal 40µCi/kg	Rutherford scattering range in air

In the handling of radioactive materials the following regulations should always be observed:

1. Wear a film badge when using radioisotopes.
2. Refrain from eating and smoking while using radioisotopes.
3. Check hands for activity after completing work with radioisotopes (use the appropriate detector, that is, for alphas, betas, etc.).
4. Use gloves when danger of contamination exists.
5. Use tongs for handling strong samples (but only if you can do so safely).
6. In case of a spill, wash it off immediately.
7. *Report all accidents and mishaps connected with radioisotopes.*
8. Do not take radioactive sources out of the laboratory.

Radiation is harmful to living organisms because by ionization it destroys individual cells, and also because it may induce genetic changes. It seems established that low levels of radiation do not produce permanent injury, but the effect is assumed to be cumulative. A genetic change, on the other hand, can be produced by low-level radiation as well as by high-level radiation, but it should not be forgotten that human beings have always been exposed to cosmic rays and natural radioisotopes.

In all establishments where some potential radiation hazard might prevail there must exist an agency (the health physics group) which is responsible for personnel and area monitoring, and for source custody. The health physics group keeps a record of radioactive sources and other hazards, and of radiation accidents, and in general helps in the enforcement of safe procedures. It should be clear, however, that the sole responsibility for enforcement of proper practices rests with the individual who has been granted the privilege to work with a radioactive source. The aversion of many scientists to observe strict rules is a common phenomenon, but it must not be imitated by the student.

Two peculiar aspects of harm from radiation need special mention and warning: (a) radiation is neither visible nor painful, hence one may not be aware of having been exposed, unless proper detectors are used; (b) in general it is *too late* to do anything after one has been exposed.

Excluding nuclear reactors and particle accelerators, the most serious radiation hazards come from x-ray machines and from taking internally a small amount of radioactive material from a source used in a laboratory.

Following is a brief discussion of radiation units and conversion factors, of the maximum permissible radiation levels, and of shielding considerations.

5.2 Radiation Units and Conversion Factors

The broad designation *radiation* is understood to cover electromagnetic radiation from soft x-rays to gamma rays, all types of charged particle beams, such as alpha and beta particles, and protons and neutron beams as well. Thus a very wide spectrum of energies and types of interaction is included.

The units for the quantitative description of radiation are based on (a) the number of quanta of the particular radiation, or (b) the effects produced by that radiation, such as ionization or biological damage.

Group (a) includes:

1. The *Curie*, which is defined as 3.7×10^{10} disintegrations/sec (applicable in the case of a radioactive source); this is the activity of 1 gm of radium in equilibrium with its decay products.

2. *Flux*, which is defined as the number of quanta incident on $1 \text{ cm}^2/\text{sec}$; it is applicable mainly to (neutron) particle beams, but can also be used to specify the strength of a radioactive source.

Group (b) includes:

3. The *roentgen*, which is defined as the amount of gamma radiation that produces in 1 cm^3 of air ionization equal to 1 esu. Now 1 esu = 3.3×10^{-10} coulombs = 2×10^9 ion pairs/cm³ of air. Since the density of air is

$$1.293 \times 10^{-3} \text{ gm/cm}^3$$

1 roentgen produces 1.61×10^{12} ion-pairs/gm of air; the average ionization potential for air was taken to be 30 eV = 48×10^{-12} ergs. Hence,

1 Roentgen = the amount of gamma radiation that loses in air 78 ergs/gm

4. The *rad*, which is defined as that amount of radiation that deposits (loses) 100 ergs in each gram of tissue it traverses. Two different types of radiation may, however, produce different degrees of biological damage even though they were both rated as 1 rad.

Therefore, the unit most commonly used in health radiation is

5. The rem (*roentgen equivalent man*), defined as that amount of any radiation which when absorbed by man, will produce the same biological effects as the absorption of 1 roentgen of x-ray or gamma ray radiation. To define the *rem* quantitatively, a *relative biological effectiveness* (RBE) has been established and

(number of rem) = (number of rad) × RBE

Table 4.3 gives RBE for the usual types of radiation.

TABLE 4.3
RELATIVE BIOLOGICAL EFFECTIVENESS

Type of radiation	rad	× RBE	= rem
X-rays and gamma rays	1	1	1
Betas	1	1	1
Protons	0.1	10	1
Alphas	0.05	20	1
Fast neutrons	0.1	10	1
Slow neutrons	0.3	3	1

As the reader is probably aware, one rad or one rem represent fairly high levels of radiation; the corresponding flux of quanta is so high that it jams a counter or scaling device. Such levels of radiation are therefore measured by integrating devices such as ionization chambers or electroscopes. "Survey" instruments of this type read "mr/hr" and are always calibrated against standard sources of known strength.

Checks of calibration and linearity can be easily made by taking advantage of the inverse square law, which holds for any point source (source dimensions small as compared to the distance).

$$\frac{I(r_1)}{I(r_2)} = \frac{(r_2)^2}{(r_1)^2} \tag{5.1}$$

It follows that the number of quanta emitted into a solid angle $d\Omega$ by a source of strength C curies is

$$I = \frac{3.7 \times 10^{10}}{4\pi} \times C \times d\Omega \quad \text{(counts/sr)} \tag{5.2}$$

or the flux at a distance r(cm) from the source is

$$F \approx \frac{3 \times 10^9}{r^2} \times C \quad \text{(counts/cm}^2\text{)} \tag{5.3}$$

independent of the area of the detector.

To connect the above expression for the flux with the units of group (b), the ionization of the Co^{60} gamma rays has been taken into account, and the following conversion factor is obtained:

1 Curie of Co^{60} produces at a distance of 1 meter radiation of 1300 mr/hr

TABLE 4.4
MAXIMUM PERMISSIBLE EXPOSURES

Whole body	$(A - 18) \times 5$ rem but not over 3,000 mrem in 13 weeks nor over 12 rem in 12 months	approx 100 mrem/week
Hands and forearms	75 rem/year but not over 25 rem in 13 weeks	approx 1500 mrem/week
Skin of whole body	30 rem/year but not over 7 rem in 13 weeks	

5.3 RADIATION LEVELS AND PERMISSIBLE EXPOSURES

The direct effects of radiation are not well known, neither do extensive statistics exist on the genetic effects. From existing experience the United States government (AEC) has set maximum permissible exposures which are indicated in Table 4.4; these are occasionally revised.

It is believed that radiation effects are cumulative, and therefore the whole body maximum permissible radiation dosage over the whole life span of an individual is calculated through the formula $(A - 18) \times 5$ rem, where A is his age. Thus the need for continuous monitoring is evident (a badge should always be worn), and the keeping of accurate records is clearly indicated. Further, if an overexposure occurs at some time, the individual should not work with radiation until his over-all dosage is again below the permissible maximum.

Correspondingly, Table 4.5 gives the designation of areas as related to the dosage a human being would receive in that area.

TABLE 4.5
PERMISSIBLE DOSE FOR UNRESTRICTED AND WORK AREAS

Designation of area	Dose to any part of the body
Unrestricted	Less than 2 mrem in one hour or 100 mrem in seven consecutive days or 500 mrem in one year
Radiation area	More than 5 mrem in one hour or 100 mrem in five days
High radiation area	More than 100 mrem in one hour
Work benches and surface areas	100 counts/min-100 cm^2 or 1 mr/hr at 2 cm

Similar restrictions exist for the concentration of radioisotopes in air, waste water, etc. As an example, the maximum permissible concentration of Sr^{90} in the atmosphere is 3×10^{-10} $\mu Ci/ml$. This indicates that radioactive wastes and contaminated equipment or clothing must be treated carefully and disposed of in a systematic fashion.

The most serious danger, however, to a person working in a laboratory is that of taking internally even a very minute amount of a radioactive substance. An internal source will subject the human body to 4π radiation, and can reach sensitive organs which are otherwise protected by the skin; furthermore, there may be no way of removing the source even after its detection. Potentially the most dangerous sources in this category are long-lived isotopes which are chemically retained in the body and deposited in essential organs. The maximum permissible total body burden is indeed very small (consult Table 4.2) and of the order of a few microcuries; for dangerous isotopes such as Po^{210} it is only a few *hundredths* of a microcurie.

To avoid contamination, sources are hermetically sealed or embedded in plastic whenever possible. For certain experiments, however, involving alpha rays or low-energy beta rays, it is not possible to protect the source. This is the reason for enforcing the safety procedures given at the beginning of Section 5. Especially in preparing sources, all work should be done under a fume hood to avoid inhalation, and persons with cuts or breaks in the skin of their hands should avoid handling radioisotopes.

It should be kept in mind that the best-known industrial radiation accidents, like the death of the first workers who painted luminous radium dials on watches, and that of the uranium ore miners in Central Europe,

TABLE 4.6

RADIATION LEVELS DUE TO NORMAL EXPOSURE (WHOLE BODY)

1. Internal natural sources	K^{40}	25 mrem/year
	C^{14}	1 mrem/year
	Ra^{226}	5 to 350 mrem/year†
Man-made sources	Sr^{90}	1–5 mrem/year
2. External natural sources	Cosmic rays	35 mrem/year (at sea level)
		60 mrem/year (at 5000 ft)
	Radioactivity	35 to 70 mrem/year
Man-made sources	Luminous watch	40 mrem/year
	Medical x-rays	100 mrem/year
	Fall-out	1 to 6 mrem/year
	TV screen	less than 1 mrem/year

† The concentration of Ra^{226} is different in different areas of the world.

TABLE 4.7
EFFECTS OF WHOLE BODY EXPOSURE RECEIVED IN A FEW HOURS

1 rem	No detectable change
10 rem	Blood changes detectable
100 rem	Some injury; no disability
200 rem	Injury and some disability
300 rem	Injury and disability
400 rem	50% deaths in 30 days
600 rem	100% deaths in 30 days
10,000 rem	50% deaths in 4 days
100,000 rem	Quick deaths

were due to internal acceptance—of radium in the first instance, and of radon gas in the second. In the explosions of fission bombs during World War II, however, 15 to 20 percent of the injuries were due to gamma rays or neutrons.

Finally, for comparison reasons, we include Tables 4.6 and 4.7. Table 4.6 gives the radiation levels to which a human is normally exposed, both internally and externally. Table 4.7, on the other hand, gives the most probable effects of whole-body radiation if it is received in a short time span.

5.4 SHIELDING AGAINST RADIATION

The purpose of shielding is to attenuate the radiation beam. If the beam consists of charged particles, they do lose energy as they cross matter, and if the shield is sufficiently thick the beam will be completely *stopped*. Since the energy loss is proportional to the number of atomic electrons Z of the shielding material, low Z materials have a larger stopping power per (nucleon) *gram*. On the other hand, the higher the density, the higher the stopping power per unit *length* of shielding.

The attenuation of a gamma-ray beam, however, is different; no gradual energy loss occurs, but there exists a finite probability (cross section) for an interaction. Interactions (electromagnetic) of a gamma-ray beam with matter are either the photoelectric effect, Compton scattering, or pair production, depending on the energy of the beam. As explained in detail in the following Chapter 5, through a series of such processes a fraction of the beam becomes completely *absorbed* in the material used for shielding. Since the interaction probability is proportional to the amount of material present, we have

$$-\frac{dI}{dx} = I\mu$$

hence

$$I = I_0 e^{-\kappa x}$$

where x is the length of the shield, $\kappa = 1/L = \sigma_t \rho N_0$ is the absorption coefficient, and L the radiation length ($L = 0.51$ cm for lead).

If the beam consists of particles with strong interactions, such as neutrons or protons, the formalism is similar, but now $\kappa = 1/\lambda$ where λ is the mean free path, which can be roughly taken as 60 gm/cm^2.

In spite of these considerations, still the best shielding against a radioactive source is distance; since the inverse square law holds, keeping at

FIG. 4.22 Photograph of a facility adequate for handling and preparation of small samples of radioactive material for a teaching laboratory. Note the fume hood and storage bins.

a 10-m distance dilutes the flux over the value it had at contact with the source (assuming an extent of 5 cm) by a factor of 40,000; for gamma rays such attenuation is equivalent to shielding by 7 cm of lead.

In conclusion, Fig. 4.22 shows a typical installation for safe storage of radioactive materials and preparation of sources. The student should not necessarily be afraid of handling radioisotopes, but he should treat them with due respect.

REFERENCES

Much of the material presented here on radiation safety can be found in H. W. Patterson's article in UCRL Report 8816, *Lectures on Physics, Biophysics and Chemistry*, available from the Office of Technical Services, Washington 25, D. C.

Also the U. S. Government AEC Act, Title 10 may be consulted.

║5

║RADIATION AND PARTICLE DETECTORS

1. General Considerations

The terms *radiation* and *particle* used in the chapter title require clarification. The term *radiation* here designates electromagnetic energy propagating in space (crossing a given area in unit time), but specifically of a frequency higher than that of the visual spectrum; namely, x-rays and gamma rays. Visible, infrared, microwaves, and radiofrequency waves are not included. Because of the quantum-mechanical aspects of the electromagnetic field, such radiation can be described by a flux of (neutral) quanta, the photons, with an energy $E = h\nu$ and a momentum $p = h\nu/c$, where ν is the frequency of the radiation. These quanta interact with electric charges, and the probability for such interactions is of the same order as that for the interaction of two charges.

The term *particle* here encompasses all entities of matter (energy) to which can be assigned discrete classical and quantum-mechanical properties, such as rest mass, spin, charge, lifetime, and so on. The use of the term "particle" is not always clear: for example, we speak of a hydrogen molecule, whereas we refer to the nucleus of the hydrogen atom, the proton, as a particle. Similarly, the electron, the neutron, the massless neutrino, the π meson, etc., are referred to as particles; the same term is frequently used for a fission fragment, a helium nucleus, or a heavy ion. The visualization of a particle is that of a massive point describing a certain trajectory under the influence of external forces and initial conditions; this provides a useful model for many calculations.

Since particles have dimensions of the order of fermis (10^{-13} cm), they cannot be "seen" even by electron microscopes[†], but their impact on certain materials, or passage through them, can be noticed readily. Even more remarkably, in certain substances and under specific conditions the whole trajectory of a charged particle can become visible and be permanently recorded. Thus a particle detector, or radiation detector, is a device that produces a signal (intelligible to the experimenter) when a particle or photon arrives; if the device reveals to the experimenter the whole trajectory of the particle, it is called an image-forming detector.

All detectors known today are based on the electromagnetic interaction of the charge of the incoming particle with the atoms or molecules of the detector. The different types of interaction (ionization is the most common) and the different principles of amplification of this interaction distinguish the different types of detector. Neutrons, however, are detected through the interaction of the charged particles of the detector to which they transfer energy. This occurs either through elastic collisions of the neutrons with protons (hydrogenous materials), or through neutron capture in certain nuclei, or through the production of fission by the neutron: for example, $n + B^{10} \rightarrow Li^7 + \alpha$. Image-forming devices are not treated in this chapter, but we mention those that are currently used: (a) the nuclear emulsion, (b) the bubble chamber, and (c) the spark chamber.

In the following discussion we will be concerned with signal-producing devices, which we classify as follows:

(a) Gaseous ionization instruments, encompassing the ionization chamber, the proportional counter, and the Geiger counter
(b) Scintillation counters
(c) Solid-state detectors
(d) Cherenkov counters

All these detectors can be designed so as to respond to the passage or arrival of a *single* particle or quantum. They can also be used as integrating devices (as is frequently done with ionization chambers) giving a signal proportional to $N\bar{E}$, where N is the total number of particles crossing the instrument per unit time and $\bar{E}$ the average energy deposited by each particle.

---

† High-energy electron-scattering experiments (which serve as a sort of microscope) have, however, revealed much about the electromagnetic structure of the proton and neutron.

In evaluating a detector, the following properties are taken into consideration:

(a) Sensitivity, which defines the minimum energy that must be deposited in the detector so as to produce a signal; related to it is the signal-to-noise ratio at the system's output.

(b) Energy resolution, in certain detectors, which are large enough to stop the particle, the signal may be proportional to the initial energy of the particle. In other cases the velocity of the traversing particle can be measured, as in Cherenkov counters, or in dE/dx (ionization per unit length) detectors.

(c) Time resolution, which characterizes the time lag and time jitter from the arrival of the particle until the appearance of the signal, and the distribution in time (duration) of the output pulse; related to it is the dead time of the device, that is, the period during which no (correct) signal will be generated for the arrival of a second particle.

(d) Efficiency, which specifies the fraction of the flux incident on the counter that is detected. It usually is fairly high for charged particles, but can be as low as a few percent for neutral particles and for photons.

Particle detectors play a most important role in nuclear physics, and in many of the experiments described in this text some type of particle detector is used. Just as the spectrograph was the paramount instrument of atomic physics, so the Geiger counter, and later, the NaI scintillation counter have been the paramount instruments of nuclear physics.†

In Section 2 is presented a brief discussion of the interaction of charged particles and of photons with matter. In Section 3 gaseous ionization instruments are described with specific emphasis on the Geiger counter and on an experimental determination of the time distribution of nuclear radiation. This is followed in Section 4 by description of the scintillation counter and the measurement of nuclear gamma-ray spectra. Section 5 deals with solid-state detectors and the measurement of the specific ionization of polonium alpha rays in air. The chapter concludes with some remarks on Cherenkov counters and neutron detectors.

2. Interaction of Charged Particles and Photons with Matter

2.1 GENERAL REMARKS

As mentioned in the previous section the interaction of charged particles and photons with matter is electromagnetic and results either in a gradual

---

† It is interesting that the first particle detector ever to be used (by Lord Rutherford in his alpha-particle scattering experiments in 1910) was a scintillating screen, a technique which came again into prominence after 40 years.

reduction of energy of the incoming particle (with a change of its direction) or in the absorption of the photon. Particles coupled with the nuclear field, such as nuclei, protons, neutrons, and π-mesons, are subject to a nuclear interaction as well, which is, however, of much shorter range than the electromagnetic one. The nuclear interaction may become predominant only when the particles have enough energy to overcome Coulomb-barrier effects, and when the amount of matter traversed is large enough to be of the order of a nuclear mean free path—which is approximately 60 gm/cm². Heavy charged particles lose energy through collisions with the atomic electrons of the material, while electrons lose energy both through collisions with atomic electrons and through radiation when their trajectory is altered by the field of a nucleus (*bremsstrahlung*—see Sec. 2.5). Photons lose energy through collisions with the atomic electrons of the material, either through the photoelectric or Compton effect; at higher energies photons interact by creating electron-positron pairs in the field of a nucleus.

Since by necessity the ensuing discussion will be brief, the reader is referred to the excellent description of these phenomena in Fermi's *Nuclear Physics*,† Chapter II, or to one of the other references given at the end of the chapter.

A brief review of definitions will be helpful.

Cross Section. We define the cross section, σ, for scattering from a single particle

$$\sigma = \frac{\text{scattered flux}}{\text{incident flux per unit area}} \qquad (2.1a)$$

Thus σ has dimensions of area (usually cm²) and can be thought of as the area of the scattering center projected on the plane normal to the incoming beam. If the density of scatterers is n (particles/cm³), there will be $n\,dx$ scatterers per unit area in a thickness dx of material, and the probability $dP = I_s/I_0$ of an interaction in the thickness dx is

$$dP = \frac{\sigma I_0/S}{I_0} \,(Sn\,dx) = \sigma n\,dx \qquad (2.1b)$$

where S is the area covered by the scattering material and I_0 is the total flux incident on the target; thus I_0/S is the flux per unit area as shown‡

---

† Enrico Fermi, *Nuclear Physics*. Notes compiled by Orear, Rosenfeld, and Shluter. Univ. of Chicago Press, 1950.

‡ Occasionally confusion arises because the area of the incoming beam may be smaller than the area presented by the target as shown in Fig. 5.1b. Clearly the definition of Eq. 2.1a is valid in either case and always leads back to Eq. 2.1b.

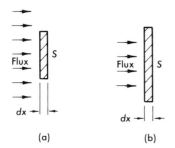

in Fig. 5.1a. The result of Eq. 2.1b is not surprising since dP *must* be proportional to n and dx;

$$dP \propto n\, dx$$

σ is then the factor that transforms this proportionality into an equality. Atomic and nuclear cross sections are of the order of 10^{-24} cm^2 (one barn), which is not surprising, given the geometrical size (cross section) of the nucleus

$$\sigma_{\text{geom}} = \pi R^2 = 3.14 \times 10^{-26}\, A^{2/3}\ \text{cm}^2$$

Differential Cross Section. For a single scatterer we define

$$\frac{d\sigma(\theta, \phi)}{d\Omega} = \frac{\text{flux scattered into element } d\Omega \text{ at angles } \theta, \phi}{\text{incident flux per unit area}}$$

It follows that

$$\int_0^{2\pi} d\phi \int_0^{\pi} \frac{d\sigma}{d\Omega}\, d(\cos\theta) = \sigma$$

where the integration is over all angles. If after the scattering process a variety of energies is possible,

$$\frac{d^2\sigma(\theta, \phi, E)}{d\Omega\, dE} = \frac{\text{flux with energy } E, \text{ within } dE, \text{ scattered into } d\Omega \text{ at } \theta, \phi}{\text{incident flux per unit area}}$$

It follows that

$$\int_0^{\infty} \frac{d^2\sigma(\theta, \phi, E)}{d\Omega\, dE}\, dE = \frac{d\sigma(\theta, \phi)}{d\Omega}$$

where the integration is over all possible energies of the scattered flux.

Absorption Coefficient. To obtain the probability for scattering in a length x of some material, we consider an incident flux per unit area I_0; $I(x)$ represents the flux at a distance x into the material. According to Eq. 2.1b

$$-dI(x) = dP\, I(x) = I(x)\sigma n\, dx \qquad (2.1c)$$

thus

$$\frac{dI}{I} = -\sigma n \, dx \qquad I(x) = I_0 \, e^{-\sigma n x}$$

If we designate $P(x)$ the probability for scattering in a length x, we have

$$P(x) = 1 - \text{(probability for survival in a length } x) = 1 - e^{-\sigma n x} = 1 - e^{-\kappa x}$$

where $\kappa = \sigma n$ is the absorption coefficient. Similarly $\lambda = 1/\sigma n$, which has dimensions of length, is called the absorption length, or, mean free path.

The density of scattering centers n is given by

$$\begin{aligned} n &= \rho N_0/A & &\text{if we consider scattering by nuclei} \\ n_e &= \rho N_0 Z/A & &\text{if we consider scattering by electrons} & (2.2) \\ n_n &= \rho N_0 & &\text{if we consider scattering by nucleons} \end{aligned}$$

where N_0 is Avogadro's number 6.023×10^{23} and ρ is the density of the material in gm/cm^3; Z and A are the atomic and mass number respectively.

Often we wish to express the absorption in terms of the equivalent matter traversed, namely, $\xi = gm/cm^2$. Then the thickness of the material can be expressed by $d\xi$, where

$$d\xi = \rho \, dx$$

The *mass* absorption coefficient is defined by

$$\mu = \frac{\kappa}{\rho} \qquad (2.3)$$

so that the fraction of a beam *not* absorbed is

$$\frac{I}{I_0} = e^{-\mu \xi} \qquad (2.4)$$

Similarly, if the region of interaction is very thin, the scattered flux is given directly by

$$I_s = I_0 \sigma n \, dx \qquad \text{for example, for nuclei,} \qquad I_s = I_0 \frac{N_0}{A} \sigma \, d\xi$$

2.2 Energy Loss of a Charged Particle

When a charged particle collides with atomic electrons, as we have already seen in the Frank-Hertz experiment, it can transfer energy to them only in discrete amounts. It can either excite an electron to a higher atomic quantum state or impart to the electron enough energy so that it

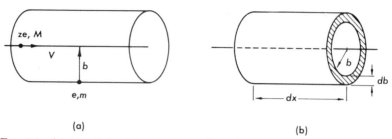

(a)   (b)

FIG. 5.2   (a) A particle of charge ze, mass M, and velocity v passes by an electron
with an impact parameter b. (b) The differential number of electrons with an
impact parameter b in the interval db is given by the volume of the cylindrical shell
$2\pi b\ db\ dx$.

will leave the atom; the latter process is the ionization of the atom. Since
in our present considerations the incoming particles have considerable
energy, the process of ionization is by far the prevailing one and we will
use this term in the discussion.

Let us consider then an atomic electron at a distance b from the path
of a charged heavy particle, of charge ze, mass M, and velocity v, as shown
in Fig. 5.2a. If we assume that the electron does not move appreciably
during the passage of the heavy particle, we can easily obtain the impulse
transferred to it due to the electric field, $\mathbf{E}$, of the passing heavy particle:

$$I_\perp = \int_{-\infty}^{+\infty} F_\perp(t)\ dt = e \int_{-\infty}^{+\infty} E_\perp(t)\ dt$$

$$= e \int_{-\infty}^{+\infty} E_\perp(x)\ \frac{dt}{dx}\ dx = \frac{e}{v} \int_{-\infty}^{+\infty} E_\perp(x)\ dx$$

We use only the component of the electric field normal to the particle's
trajectory since the longitudinal component averages to zero when summed
from $-\infty$ to $+\infty$. However, from Gauss's law, integrating over a cylinder
of radius b, coaxial with the trajectory (see Fig. 5.2a) we have

$$4\pi ze = \oint \mathbf{E}\cdot\mathbf{dS} = \int_{-\infty}^{+\infty} E_\perp 2\pi b\ dx \qquad \text{and} \qquad \int_{-\infty}^{+\infty} E_\perp\ dx = \frac{2ze}{b}$$

hence

$$I_\perp = \frac{2ze^2}{vb}$$

Since the electron was originally at rest, its momentum after the collision,

$\eta = I_\perp$, and the energy transferred is

$$E(b) = \frac{p^2}{2m} = \frac{2z^2e^4}{mv^2b^2} \tag{2.5}$$

Thus E is a function of the impact parameter b. To obtain the total energy lost by the heavy particle per unit path length, we must count how many electrons it encounters and average over the impact parameters.

From Fig. 5.2b we see that in a cylindrical ring of radius b and width db, and of unit height dx, there are contained† $n_e 2\pi b \, db \, dx$ electrons; hence

$$dE(b) = \frac{4\pi n_e \, dx z^2 e^4}{mv^2} \frac{db}{b}$$

and

$$-\frac{dE}{dx} = \frac{4\pi z^2 e^4}{mv^2} n_e \ln\left[\frac{b_{\max}}{b_{\min}}\right] \tag{2.6}$$

where because of the logarithm we had to use finite limits on b rather than 0 and ∞. The finite limits are imposed by physical considerations: for $b_{\max}$ we consider the distance where the time of passage of the heavy particle's field becomes of the same order as the period of rotation of the atomic electron in its orbit. Thus

$$\tau = \frac{b}{v} = \frac{1}{\nu} \quad \text{or} \quad b_{\max} = \frac{v}{\nu} \tag{2.7}$$

For the minimum value‡ we equate b to the DeBroglie wavelength of the electron

$$b_{\min} = \lambda = \frac{\hbar}{p} = \frac{\hbar}{mv} \tag{2.8}$$

We then obtain

$$-\frac{dE}{dx} = \frac{4\pi z^2 e^4}{mv^2} n_e \ln\left[\frac{mv^2}{\hbar\nu}\right] \tag{2.9}$$

The frequencies of the atomic electrons ν are, however, different for each orbit, so that a suitable average must be taken; we thus replace $\langle \hbar\nu \rangle$ with an average ionization potential $\bar{I}$. Finally, inclusion of relativistic effects

---

† n_e is the electron density as also given by Eq. 2.2.

‡ An alternate approach is to set $b_{\min}$ such that maximum energy is transferred to the electron. Because of momentum conservation we have $p_{\max} = 2mv$ and $E_{\max} = 2mv^2$, thus $b_{\min} = ze^2/mv^2$ using Eq. 2.5.

and a precise calculation give

$$-\frac{dE}{dx} = \frac{4\pi z^2 e^4}{mv^2}\, n_e \left[\ln \frac{2v^2 m}{\overline{I}(1-\beta^2)} - \beta^2 \right] \tag{2.10}$$

for the energy loss of heavy particles due to ionization.

In Eq. 2.10, $\beta = v/c$, and we see that the energy loss is only a function of the velocity, v, of the charge ze of the incoming particle, and of the electron density, n_e, of the scattering material. Note that in Eq. 2.10, m is the mass of the *electron* while the mass of the incoming particle does not appear at all.

Before further investigating Eq. 2.10, we should note the following effects:

(a) Equation 2.10 was derived on the assumption that the incoming particle is not deflected, and thus it is valid only for *heavy particles*; for electrons the term in the parentheses must be slightly modified.

(b) Electrons also lose energy through their interaction with the nucleus, and this is the prevailing mechanism at high energies. That is the electron's trajectory is bent by the field of the nucleus, which implies an acceleration (since the velocity vector changes), and from electrodynamics we know that accelerated charges radiate. This radiation, called "bremsstrahlung," is discussed in Section 2.5.

(c) For extremely relativistic particles, $v \approx c$, $\beta \approx 1$, Eq. 2.10 predicts a continuous rise in dE/dx proportional to $\ln \gamma^2$ where $\gamma^2 = (E/mc^2)^2 = 1/(1-\beta^2)$. Such a fast rise, however, is not observed experimentally. This is due to polarization of the medium: the electrons that are being set into motion by the field of the incoming particle move so as to reduce the effect of the external field. Consequently a much slower rise with energy results; the correct expression is†

$$-\frac{dE}{dx}\bigg]_{\text{ion}} = \frac{4\pi z^2 e^4}{mc^2}\, n_e \left[\ln \left(\frac{2\gamma c^2 m}{I'} \right) + \frac{1}{2} \right] \tag{2.11}$$

where

$$I' = \hbar\omega_p = \sqrt{\frac{4\pi n_e z e^2}{m}}$$

For silver bromide $I' \simeq 48$ eV

(d) For low-energy particles we obtain from Eq. 2.10

$$-\frac{dE}{dx} \propto \frac{z^2}{v^2} = \frac{z^2 M}{2E}$$

where M is the mass of the incoming particle and E its kinetic energy.

---

† See J. D. Jackson, *Classical Electrodynamics*, Wiley, 1962, Chap. 13.4.

The above expression (when applicable) is useful since a measurement of dE/dx and of E identifies the incoming particle

$$E\left(\frac{dE}{dx}\right) \propto z^2 M$$

(e) In image-forming devices and particularly in nuclear emulsions, the density of developed silver bromide grains can be used as a measure of the particle's velocity because of the dependence of Eq. 2.10 or Eq. 2.11 on β. However, the density of the track depends only on energy transfers ≤ 5 keV, since when an atomic electron acquires more energy, its own track becomes visible and separated from the primary particle's track: such electrons are called knock-ons or delta rays. The energy-loss expression for energy transfers ≤ 5 keV does not exhibit at all the relativistic rise of Eq. 2.11, but for high values of γ, stabilizes at a plateau 1.2 times the minimum value.

The energy loss of a heavy particle in a typical absorber, such as nuclear emulsion, as a function of the logarithm of its kinetic energy (in units of rest energy) is given in Fig. 5.3. Strictly speaking, this curve holds only for a given absorber and all singly charged particles, since we know from Eqs. 2.10 and 2.11 that dE/dx is a function only of the *velocity* of the incoming particle and its charge. (Note that $T/mc^2 = \gamma - 1$, which has a one-to-one correspondence to β). However, the general behavior of this curve holds for all absorbers.

We do recognize four regions of interest: (a) near the stopping point where a Bragg curve is applicable (see Fig. 5.42); (b) the low-energy region where the $1/v^2$ dependence of Eq. 2.10 dominates, and tends asymptotically toward the value $1/c^2$; (c) the relativistic region, where because

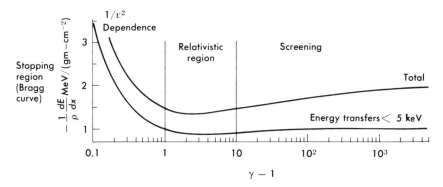

FIG. 5.3 The universal energy-loss curve for a singly charged particle plotted in Mev/(gm $-$ cm^{-2}) against $\gamma - 1$. Note the upper curve for the total energy loss and the lower curve for energy loss involving only energy transfers smaller than 5 keV.

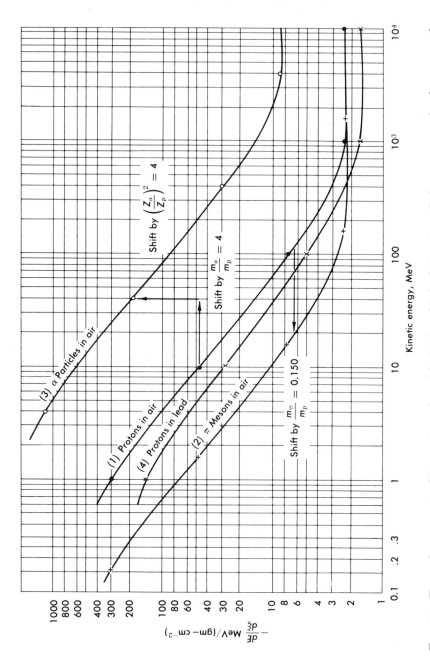

FIG. 5.4   Energy-loss curves for different charged particles in air and in lead. Note how all the curves are related to one another.

of the rise of the logarithmic term, a minimum appears approximately at $\gamma = 1$; and (d) the screened region in which Eq. 2.11 becomes applicable. Had polarization effects not been included, the rise of the dE/dx curve in this last region would be steeper than indicated in Fig. 5.3. The lower curve in Fig. 5.3 (energy transfers ≤ 5 keV) gives the grain density in nuclear emulsions.

If we choose to calibrate the abscissa of Fig. 5.3 in units of energy (MeV) of the particle rather than by $\gamma - 1$, we will not have a universal curve any more, but for each particle the energy loss curve will be shifted horizontally by m_1/m_2, in such fashion that $dE/dx]_{m_1} = dE/dx]_{m_2}$ when the corresponding energies T_1 and T_2 result in the same value of $\gamma - 1$:

$$\frac{T_1}{m_1} = \frac{T_2}{m_2} = c^2(\gamma - 1)$$

This is shown in Fig. 5.4 which gives the absolute value of energy loss $-dE/d\xi$ in MeV/(gm-cm^{-2}) in air for protons (curve 1) and π mesons ($m_\pi = 140$ MeV; curve 2), where the latter is shifted to the left by a factor $m_\pi/m_p = 0.150$.

Further, if we consider particles of different z, the energy loss will differ by the ratio $(z_1/z_2)^2$. In this fashion we obtain curve 3 in Fig. 5.4 for the energy loss of alpha particles in air, which is shifted (with respect to curve 1) to the right by a factor of $m_\alpha/m_p = 4$ and upwards by a factor of $(z_\alpha/z_p)^2 = 4$.

If we now turn our attention to the dependence of dE/dx on the absorber material, it is clear that it will vary rapidly, due to its dependence on n_e. If instead we use $-dE/d\xi$ (the energy loss per gm/cm^2 of material) the variation is much slower, since

$$n_e = \rho N_0 \frac{Z}{A}$$

and

$$d\xi = \rho \, dx$$

Thus

$$-\frac{dE}{d\xi} = N_0 \frac{Z}{A} z^2 f(\beta, \bar{I})$$

so that the energy loss per gm/cm^2 is larger for low Z materials, neglecting the small dependence on $\bar{I}$, the average ionization potential. Curve 4 of Fig. 5.4 gives $dE/d\xi$ for protons in lead, which is indeed lower than that in air, but not by a large amount.

An approximate universal figure for the energy loss of a relativistic singly charged particle in any material is 2 MeV/(gm-cm^{-2}).

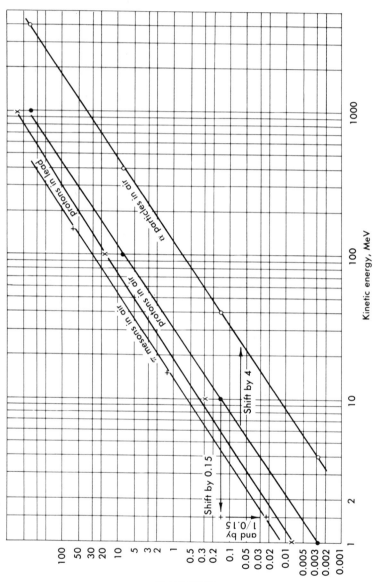

Fig. 5.5   Range curves for different particles in air and in lead. Note how the different curves are related to one another.

2.3 RANGE OF A CHARGED PARTICLE

Since the exact expression for the energy loss of a charged particle is known, it is possible by integration to find what total length of material an incoming particle of given energy will traverse before coming to rest; this is called its range R, and we can set

$$E_0 = \int_0^R \frac{dE}{dx}\, dx$$

or conversely, since $dE/dx = z^2 n_e f_1(\beta)$ and† $dE = M f_2(\beta)\, d\beta$ (M is the mass of the incoming particle),

$$R = \int_0^R dx = -\frac{1}{z^2 n_e}\int_{E_0}^0 \frac{dE}{f_1(\beta)} = \frac{M}{z^2 n_e}\int_0^{\beta_0} \frac{f_2(\beta)}{f_1(\beta)}\, d\beta = \frac{M}{z^2 n_e} F(\beta_0) \qquad (2.12)$$

That is, for the same velocity the range is proportional to the mass of the incoming particle, inversely proportional to the square of its charge, and inversely proportional to the electron density of the stopping material. Extensive tabulations of range curves for different particles, and different absorbers are available.‡ Also various empirical formulas have been devised; for example, for electrons, Feather's expression gives for the range of electrons in gm/cm² of aluminum

$$R = 0.543E - 0.160 \qquad E > 0.8 \text{ MeV} \qquad (2.13)$$

where E is the initial kinetic energy of the electron in MeV.

As suggested above, it is highly preferable to express the range in gm/cm², because then the dependence on the absorber material is slow (as $n_e/\rho = N_0 Z/A$), resulting in larger range (in gm/cm²) in heavy elements.

Figure 5.5 gives the range of protons, π mesons alpha particles as a function of their kinetic energy for air, in gm/cm². As explained for Fig. 5.4, the π-meson curve is obtained from the proton curve by shifting to the left by the factor $m_\pi/m_p = 0.15$ to reach the same β_0, but also multiplying the ordinate values by $m_\pi/m_p = 0.15$; for the alpha particles the curve is obtained by shifting to the right by the factor $m_\alpha/m_p = 4$ and multiplying the ordinate (first) by $m_\alpha/m_p = 4$ (due to the mass change) and then by $(z_p/z_\alpha)^2 = \frac{1}{4}$, hence leaving it unshifted.

Finally, the range of protons in lead is also given. The concept of range loses its meaning, however, when the amount of material that the particle must traverse before coming to rest is of the order of a nuclear mean free path as explained in the introduction to this section.§

---

† Nonrelativistically we obviously have the simple relation $dE = Mc^2\beta\, d\beta$.

‡ U.C.R.L. Report 2426 Rev, Vol II, *High Energy Particle Data*, available from O.T.S., U. S. Dept. of Commerce, Washington 25, D. C.

§ For heavy ions, energy loss due to collisions with the nuclei must also be considered.

2.4 MULTIPLE SCATTERING

In discussing the passage of a charged particle through matter, we have neglected up to now its interaction with the electric field of the nucleus, because indeed the energy transfer to the nucleus is minimal. However, when a particle of charge ze, mass m, and velocity v passes by the vicinity of a nucleus of charge Ze, it will be scattered (Fig. 5.6) with the Rutherford cross section (see Chapter 6).

$$\frac{d\sigma}{d\Omega} = \frac{1}{4}\left(\frac{e^2 Zz}{mv^2}\right)^2 \frac{1}{\sin^4 \theta/2} \tag{2.14}$$

showing that the probability for small-angle scattering is predominant.

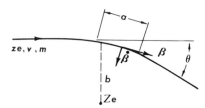

FIG. 5.6 Deflection of a charged particle when passing in the vicinity of a nucleus. Note the scattering angle θ.

For such small angles we approximate the angle of deflection by

$$\theta = \frac{\Delta p}{p} = \frac{2Zze^2}{pvb} \tag{2.15}$$

where p is the momentum of the particle and b is the impact parameter.

During its traversal of the material, the incoming particle suffers many small angle scatterings. It can be shown that the resultant scattering angle θ, after traversal of a finite thickness of material D, has a Gaussian† distribution about the mean $\theta = 0$; the probability for a scattering through an angle Θ within the interval $d\Theta$ is

$$P(\Theta)\, d\Theta = \frac{1}{\sigma\sqrt{2\pi}} \exp\left[-\frac{1}{2}\left(\frac{\Theta}{\sigma}\right)^2\right]$$

The standard deviation $\sigma = \sqrt{\overline{\theta^2}}$ (the root mean square scattering angle).

For the mean square scattering angle we have

$$\overline{\theta^2} = \frac{8\pi z^2 Z^2 e^4}{v^2 p^2}\, nD \ln\left(\frac{a_0 vp}{2Z^{4/3} ze^2}\right) \tag{2.16}$$

We further simplify Eq. 2.16 in order to exhibit the dependence of $\overline{\theta^2}$ on the incoming particle's charge z, velocity β, and momentum p; and on the

† See Chap. 10.

material's thickness D, the density of nuclei $n = \rho N_0/A$, and the atomic number Z: we obtain

$$\sqrt{\overline{\theta^2}} = \frac{z}{p\beta} \sqrt{DZ^2 n} F$$

where F is a slowly varying function of the parameters of the incoming particle and the scattering material (it contains the logarithmic term and constants). Furthermore $1/(Z^2 n)$ is proportional to the "radiation length" L_{rad} of the material (defined in Section 2.5 below) which is frequently tabulated, so that we finally write

$$\sqrt{\overline{\theta^2}} = |\theta|_{rms} = z \frac{21.2(\text{MeV}/c)}{p\beta} \sqrt{\frac{D}{L_{rad}}} (1 + \epsilon) \qquad (2.17)$$

where $|\theta|_{rms}$ is in radians, and p must be expressed in MeV/c; ϵ is a small correction† depending both on the scattering material and on β/z of the incoming particle. When we are interested in the rms *projected* angle, the numerical factor in Eq. 2.17 must be replaced by 15.

2.5 Passage of Electromagnetic Radiation (Photons) through Matter

As mentioned in the introduction to this section photons lose energy or are absorbed in matter by one of the following three mechanisms:

(a) Photoelectric effect, which predominates at low energies
(b) Compton effect, which predominates at medium energies (below a few MeV)
(c) Pair production of electrons and positrons, which predominates in the high-energy region

The relative importance of these processes and the energies at which they set in are best seen in Fig. 5.7, which gives the cross section for the interaction of a photon as a function of its energy (in units of the electron's rest mass). We will now briefly consider each process separately.

(a) *Photoelectric effect.* We speak of the photoelectric effect when the photon is completely absorbed and all its energy is transferred to an atomic electron. Consequently the photon must have enough energy to excite the bound electron from its quantum state to a higher state or into the continuum; the latter process (ionization of the atom) is much more probable.

---

† Calculated from Moliére theory; see U.C.R.L. Report 8030 by W. Barkas and A. H. Rosenfeld for tables of ϵ.

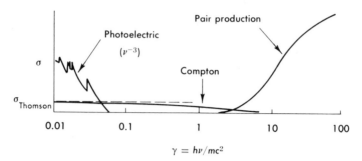

Fɪɢ. 5.7   The cross section for the interaction of photons with matter as a function of their energy (expressed in units of the electron's rest mass).

Since the binding energy of the inner electrons in atoms is of the order of keV, as the frequency of the photon is increased and it reaches the value of the binding energy of a particular shell,† a new "channel" opens, and we expect a sudden rise in the absorption cross section. Apart from the onset of new channels, the over-all variation of the photoeffect is a rapid decrease as the third power of the photon frequency (as $v^{-7/2}$) thus resulting in the curve shown on the left in Fig. 5.7. The cross sections for photoeffect are derived in Heitler,‡ from where we give the nonrelativistic value for the ejection of *one* electron from the K shell, when the photon energy is not too close to the absorption edge:

$$\sigma_P = \sigma_T \frac{Z^5}{(137)^4} 2\sqrt{2} \left[\frac{h\nu}{mc^2} \right]^{-7/2} \quad (\text{cm}^2) \qquad (2.18)$$

where we note the dependence on the Z of the nucleus, indicating that L-shell and higher-shell ejection is less probable because of the screening of the nuclear charge. Here σ_T is the classical Thomson cross section, which is derived on the simplified assumption of a plane polarized electromagnetic wave scattering from a free electron (it is assumed that the displacement of the electron is much smaller than the wavelength); we obtain

$$\sigma_T = \frac{8\pi}{3} \left[\frac{e^2}{mc^2} \right]^2 = \frac{8\pi}{3} r_0^2 \quad (\text{cm}^2) \qquad (2.19)$$

where $r_0 \equiv e^2/mc^2$ is the classical radius of the electron $= 2.8 \times 10^{-13}$ cm. Note that the Thomson cross section is independent of the frequency of the incoming photon. Near the absorption edge, Eq. 2.18 is to be modified

† Note that $n = 1$ electrons are said to be in the K shell, $n = 2$ in the L shell, $n = 3$ in the M shell, etc.

‡ W. Heitler, *The Quantum Theory of Radiation*, 3rd ed., Oxford Univ. Press, 1954, pp. 207 and 208.

as follows:

$$\sigma_P = \frac{128\pi}{3} \frac{e^2}{mc} \frac{\nu_K^3}{\nu^4} \frac{\exp\left(-4\epsilon \cot^{-1}\epsilon\right)}{1 - e^{-2\pi\epsilon}} \quad \text{(cm}^2\text{)} \qquad (2.20)$$

where ν_K is the frequency of the K-shell absorption edge and $\nu > \nu_K$ the frequency of the photon; $\epsilon = \sqrt{\nu_K/(\nu - \nu_K)}$.

(b) *Compton effect.* In the Compton effect, the photon scatters off an atomic electron and loses only part of its energy. This phenomenon, which is one of the most striking quantum effects, is described in detail in Chapter 6; the cross section for Compton scattering is given by the Klein–Nishina (K–N) formula, shown in an expanded scale in Fig. 5.8. The energy of the photon is given on the abscissa in units of the electron rest mass $\gamma = h\nu/mc^2$, and the ordinate gives the ratio of the Compton cross section σ_C to the classical Thomson cross section σ_T.

We give below the asymptotic approximations to the (K–N) Compton scattering cross section:

For low energies:

$$\sigma_C = \sigma_T(1 - 2\gamma + \tfrac{26}{5}\gamma^2 + \cdots) \quad \gamma \ll 1$$

For high energies: $\qquad\qquad\qquad\qquad\qquad\qquad\qquad\qquad (2.21)$

$$\sigma_C = \frac{3}{8}\sigma_T \frac{1}{\gamma}\left(\ln 2\gamma + \frac{1}{2}\right) \qquad \gamma \gg 1$$

(c) *Pair production.* In pair production a photon of sufficiently high energy is annihilated and an electron-positron pair is created. For a free

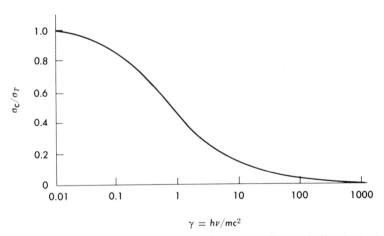

$\gamma = h\nu/mc^2$

Fig. 5.8 The ratio of the Compton scattering cross section, σ_C, to the constant Thomson cross section, σ_T, as a function of photon energy expressed in units of the electron's rest mass.

photon conservation of energy and momentum would not be possible in this process, so pair production must take place in the field of a nucleus (or of another electron) which will take up the balance of momentum. Clearly the threshold for this process is $2mc^2$ (where m is the mass of the electron), hence 1022 keV. The cross section for pair production rises rapidly beyond the threshold, and reaches a limiting value for $h\nu/mc^2 \approx 1000$ given by the following equation:†

$$\sigma_{\text{pair}} = \frac{Z^2}{137} r_0^2 \left[\frac{28}{9} \ln \frac{183}{Z^{1/3}} - \frac{2}{27} \right] \quad (\text{cm}^2) \qquad (2.22)$$

Since both the photoelectric and Compton effect cross sections decrease as the photon energy rises, pair production is the predominant interaction mechanism for very high-energy photons.

It is advantageous to introduce the mean free path (L_{pair}) for pair production; when a photon traverses a material with density of nuclei n,

$$L_{\text{pair}} = \frac{1}{n\sigma_{\text{pair}}} = \frac{1}{(28/9)(Z^2 n/137) r_0^2 \ln(183/Z^{1/3})} \quad (\text{cm}) \qquad (2.23)$$

where we have dropped the small term 2/27. Thus, the attenuation of a beam of I_0 photons will proceed as

$$I(x) = I_0 e^{-x/L_{\text{pair}}} \qquad (2.24)$$

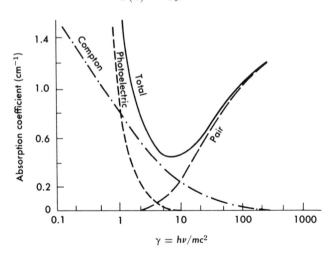

FIG. 5.9   The relative contribution of the three effects responsible for the interaction of photons with matter. The absorption coefficient in lead is plotted against the logarith of photon energy (in units of the electron's rest mass).

† See Heitler, *loc. cit.* p. 260.

In conclusion, Fig. 5.9 gives the total absorption coefficient for a photon traversing lead as a function of its energy (in units of the electron rest mass). Note that

$$\kappa_p = \sigma_p \, 2n \qquad \text{because there are 2 } K\text{-shell electrons per nucleus}$$
$$\kappa_c = \sigma_c \, n_e \qquad \text{electron density}$$
$$\kappa_{\text{pair}} = \sigma_{\text{pair}} \, n \qquad \text{density of nuclei}$$

The dotted curves in Fig. 5.9 indicate the relative contributions of each of the three interaction mechanisms.

2.6 Interaction of Electrons with Matter (Bremsstrahlung)

Since electrons carry charge, their interaction with matter must follow along the lines given in Section 5.2. Because of their small mass, however, their interaction with the nucleus results in significant energy loss by radiation; this process, called "bremsstrahlung," becomes the dominant mode of energy loss for high-energy electrons.

We can obtain an estimate of the cross section for "bremsstrahlung" from a classical nonrelativistic model. Consider an electron (charge e, mass m, and velocity v) passing by the vicinity of a nucleus of charge Ze, and let us assume that in the collision process the nucleus does not move (Fig. 5.6). The scattering angle of the electron is given by Eq. 2.15, and the change in the velocity vector of the electron is

$$\Delta v = \frac{2Ze^2}{mvb} \tag{2.25}$$

The radiation formula for an accelerated charge† is

$$P(t) = \frac{dE}{dt} = \frac{2}{3} \frac{e^2}{c} \left[(\dot{\beta})^2 - (\beta \times \dot{\beta})^2 \right] \tag{2.26}$$

So for our case, since $\dot{\beta}$ is normal to β,

$$dE(t) = \frac{2}{3} \frac{e^2}{c} \mid \dot{\beta}(t) \mid^2 dt \tag{2.27}$$

By a general theorem of Fourier analysis, if

$$E = \frac{2}{3} \frac{e^2}{c} \int_{-\infty}^{+\infty} \mid A(t) \mid^2 dt$$

---

† See Jackson, *loc. cit.*, p. 470. In Eq. 2.26 γ was set equal to 1; similarly Eq. 2.27 should include a term $(1 - \beta^2) = 1/\gamma^2$ which was also set equal to 1.

then also

$$E = \frac{2}{3}\frac{e^2}{c}\int_{-\infty}^{+\infty} | A(\omega) |^2 \, d\omega$$

where

$$A(\omega) = \frac{1}{\sqrt{2\pi}}\int_{-\infty}^{+\infty} A(t) \, e^{i\omega t} \, dt \qquad (2.28)$$

is the Fourier transform amplitude of $A(t)$.

Using then Eq. 2.28, we obtain in analogy with Eq. 2.27 the frequency spectrum of the radiation†

$$dE(\omega) = \frac{2}{3}\frac{e^2}{c} \left[| A(\omega) |^2 + | A(-\omega) |^2 \right] d\omega = \frac{4}{3}\frac{e^2}{c} | A(\omega) |^2 \, d\omega \qquad (2.29)\,†$$

To obtain $dE(\omega)$ we must perform the integral of Eq. 2.28 with $A(t) = | \dot{\beta}(t) |$. We assume that the acceleration $\Delta\beta$ occurs in a very brief interval of time, of the order of $\tau = a/v$, where a is the characteristic distance over which the force is appreciable‡; then

$$A(\omega) = \frac{1}{\sqrt{2\pi}}\int_{-\infty}^{+\infty} | \dot{\beta}(t) | \, e^{i\omega t} \, dt = \begin{cases} \dfrac{1}{\sqrt{2\pi}}\,\Delta\beta & \omega\tau < 1 \\[2ex] 0 & \omega\tau > 1 \end{cases} \qquad (2.30)$$

If $\omega\tau > 1$, there will be several oscillations of the exponential term over the region where $| \dot{\beta}(t) |$ is different from zero, and the integral will average to zero.

The integral results in a rectangular spectrum for the emitted radiation, as shown in Fig. 5.10, with

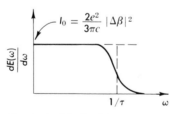

FIG. 5.10 Idealized bremsstrahlung spectrum resulting from the sudden acceleration $| \Delta\beta |$ of a charged particle.

$$\frac{dE(\omega)}{d\omega} = \begin{cases} \dfrac{2e^2}{3\pi c}\dfrac{4Z^2e^4}{c^2m^2v^2b^2} & \omega\tau < 1 \\[2ex] = 0 & \omega\tau > 1 \end{cases} \qquad (2.31)$$

---

† Because $A(t)$ is real, $A(\omega) = A^*(-\omega)$.

‡ See, for example, W. K. H. Panofsky and M. Phillips, *Classical Electricity and Magnetism*, Addison-Wesley, 1955, p. 304.

Next we integrate over all impact parameters b

$$\chi(\omega) = \int_{b_{\min}}^{b_{\max}} \frac{dE(\omega)}{d\omega} 2\pi b \, db$$

where we can set $b_{\max} = a = \tau v$ and in view of $\omega \tau \sim 1$ we also let $b_{\max} \sim v/w$; from classical considerations†

$$b_{\min} = \frac{Ze^2}{mv^2}$$

The cross section σ_{brems}, giving the probability of emission of a photon of energy $\hbar\omega$ in the interval $d(\hbar\omega)$, is related to $\chi(\omega)$ through

$$(\hbar\omega)\sigma_{\text{brems}}(\hbar\omega) \, d(\hbar\omega) = \chi(\omega) \, d\omega$$

Resulting in the classical nonrelativistic bremsstrahlung cross section

$$\sigma_{\text{brems}}(\hbar\omega) = \frac{16}{3} \frac{Z^2 e^2}{\hbar c} \left(\frac{e^2}{mc^2}\right)^2 \left(\frac{c}{v}\right)^2 \frac{1}{\hbar\omega} \ln \left(\frac{mv^3}{Ze^2\omega}\right) \qquad (2.32)$$

The average energy loss per path length, $-dE/dx$, is obtained by integrating over all photon energies (the square pulse) and multiplying by the density of nuclei:

$$-\frac{dE}{dx} = \int n(\hbar\omega)\sigma_{\text{brems}} \, d(\hbar\omega) = n(\hbar\omega)(\hbar\omega_{\max})\sigma_{\text{brems}}$$

Substituting $1/137 = e^2/\hbar c$, $r_0 = e^2/mc^2$ and $(\hbar\omega_{\max}) = E_0$, the energy of the electron, we obtain

$$-\left(\frac{dE}{dx}\right)_{\text{av}} = \frac{16}{3} \frac{Z^2 n}{137} E_0 r_0^2 \left(\frac{c}{v}\right)^2 \ln \left(\frac{mv^3}{Ze^2\langle\omega\rangle_{\text{av}}}\right) \qquad (2.33)$$

Equation 2.32 is a fair approximation; the correct quantum-mechanical result, including the screening of the nucleus by the atomic electrons, is given by the following:‡

$$-\left(\frac{dE}{dx}\right)_{\text{av}} = \frac{Z^2 n}{137} E_0 r_0^2 \left(4 \ln \frac{183}{Z^{1/3}} + \frac{2}{9}\right) \qquad (2.34)$$

The mean free path for bremsstrahlung by an *electron*, called the "radiation length," is defined through Eq. 2.35:

$$L_{\text{rad}} = \frac{1}{n\sigma_{\text{brems}}} = \frac{1}{4(Z^2 n/137)r_0^2 \ln (183/Z^{1/3})} \qquad (2.35)$$

---

† See footnote on page 157.

‡ See Heitler, *loc. cit.*, p. 253.

which is obtained from Eq. 2.34 by setting $L_{rad} = dx$, when $-dE/E_0 = 1$; the term $\frac{2}{9}$ (small as compared to the ln) was dropped.

To show at what electron energies bremsstrahlung becomes important, we note that

$$\frac{(dE/dx)_{rad}}{(dE/dx)_{ioniz}} = \frac{ZE(\mathrm{MeV})}{800}$$

This is shown in Table 5.1, where we give for some common absorbers, L_{rad}, as well as the electron energy at which bremsstrahlung loss becomes equal to ionization loss.

TABLE 5.1

RADIATION LENGTH OF ELECTRONS IN DIFFERENT MATERIALS

Material	L_{rad}	Electron energy for $(dE/dx)_{rad} = (dE/dx)_{ion}$
Air	330 m	120 MeV
Aluminum	9.7 cm	52 MeV
Lead	0.52 cm	7 MeV

Equation 2.35 is amazingly similar to Eq. 2.23, by which we defined L_{pair}. We have

$$L_{rad} = \tfrac{7}{9} L_{pair}$$

indicating that in matter the mean free path of a high-energy *electron* is of the same order as the mean free path of a high-energy gamma ray; this is the reason for the phenomenon of the electromagnetic cascade, first observed in cosmic rays.

If a very high-energy electron is incident in the atmosphere, it will soon (after approximately 330 m) emit one or more high-energy gamma rays. These gamma rays will soon again (after approximately 330 m) produce

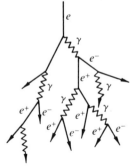

FIG. 5.11 Formation of an electromagnetic cascade. Note that high-energy electrons (positrons) radiate gamma rays and the gamma rays later convert into electron-positron pairs and so forth.

electron-positron pairs. Each of the secondary electrons and positrons will again radiate, and so on, until most of the energy of the primary electron (or gamma ray) has been transferred to many less energetic electrons (Fig. 5.11).

In another connection we have already used L_{rad} in Eq. 2.17 for multiple scattering; from Table 5.1 we see that in heavy materials scattering will be much more pronounced. Note that multiple scattering is the same for particles of the same momentum. However, for low energies a light particle will scatter much more than a heavier particle of the same kinetic energy ($p = \sqrt{2Tm}$). This is clearly seen when observing the tracks of low-energy protons and electrons in an image-forming device; the former ones are in general straight whereas the latter ones suffer multiple scattering through large angles.

2.7 CHERENKOV RADIATION

We know from classical electrodynamics that a uniformly moving charge does not radiate (you may consider the Lorentz transformation to the frame where the charge is at rest). If, however, the charge moves with a velocity $v = \beta c$ larger than the velocity of propagation of electromagnetic radiation (signals) in that medium, $c' = c/n$, detectable radiation covering the visible spectrum is emitted. This is a consequence of the retarded solutions of the wave equation,† and here we will only give a qualitative argument which leads to the correct conclusions.

Consider a charge e moving uniformly along the x axis, with velocity βc, in a medium of refractive index n. In Fig. 5.12a, let 1, 2, 3,··· denote the position of the charge at times t_1, t_2, t_3, ···; then the spheres of radii $r_1 = c'(t - t_1)$, $r_2 = c'(t - t_2)$, $r_3 = c'(t - t_3)$ denote the position in space, at a time t, where the potential due to this charge can manifest itself. We note then that the surface of the cone $A'OA$ is the locus of the extreme points that information about the particle can have reached at time t. We can think of this cone as representing the wave front of an electromagnetic wave propagating in a direction normal to the cone surface, in analogy (two-dimensional, however) to the wake generated by a ship.

From the geometry of Fig. 5.12a it is clear that the angle θ is given by

$$\cos \theta = \frac{(c/n)(t - t_1)}{\beta c(t - t_1)} = \frac{1}{n\beta} \qquad (2.36a)$$

The angle θ is real only if

$$n\beta > 1 \qquad (2.36b)$$

---

† See, for example, Jackson, *loc. cit.*, p. 494.

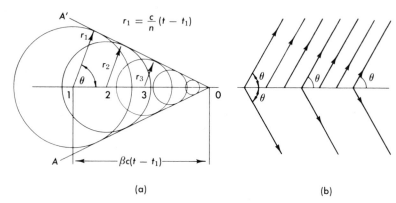

(a)   (b)

FIG. 5.12   Schematic representation of the Cherenkov effect. (a) The information
spheres propagating with velocity $c' = c/n$ are shown at a time t; the charged
particle propagates with velocity $v = \beta c > c'$. (b) Direction of emission of Cherenkov light as the charged particle propagates along the x axis.

Then radiation is emitted in a cone of angle θ about the instantaneous
direction of the velocity of the charged particle and with its apex at the
instantaneous position of the particle; Fig. 5.12b shows this in the form of
the emitted rays. This phenomenon is called the Cherenkov effect, after
the man who first observed it.

The energy radiated per unit frequency interval per unit path length is
given by

$$\frac{dI}{d\omega\,dx} = \frac{e^2\omega}{c^2}\sin^2\theta = \frac{e^2\omega}{c^2}\left[1 - \frac{1}{\beta^2 n^2(\omega)}\right] \tag{2.37}$$

indicating an increase in radiation with frequency. However, a cutoff sets
in because of the dispersive behavior of the refractive index $n(\omega)$ at high
frequencies. A numerical expression, giving the total number of photons
emitted in the *visible* spectrum, is given by

$$N = 400L\sin^2\theta \tag{2.38}$$

where L is the length of the radiator in centimeters.

The Cherenkov effect has been widely used for the detection of very
high-energy particles. Cherenkov detectors are of two main types: threshold
counters that give a signal for all particles with a β larger than some threshold value $\beta_0 = 1/n_0$, and differential counters that give a signal for particles of a given β within the interval $\Delta\beta$. It is possible to design counters
with a resolving power of $\Delta\beta/\beta = 0.0005$. The radiators may be either solid,
liquid, or gaseous, but must be transparent to the radiation; the advantage
of gases is that refractive indices very close to unity can be achieved and
readily controlled by adjusting the gas pressure.

3. Gaseous Ionization Instruments; the Geiger Counter

3.1 GENERAL

As mentioned earlier, most particle detectors are based in one form or another on the energy lost by ionization by the charged particle. In a large class of instruments the detecting material is a gas; the ionization potentials are of the order of 10 eV, but on the average, for example in air, the charged particle loses 30 to 35 eV for each electron-ion pair formed.† By collecting the free charges that were thus created, it is possible to obtain an electrical pulse signaling the passage of the charged particle.

The simplest type of gaseous detector consists of a cylindrical chamber with a wire stretched along its center, as shown in Fig. 5.13. The chamber walls act as the negative electrode, and positive voltage is applied to the central electrode. Under the influence of the electric field, the electrons are collected at the center while the positive ions move toward the walls. Clearly, it is desirable to collect the free charges before they recombine in the gas; this is mainly a function of the pressure of the gas and of the applied voltage‡

If, however, the voltage is sufficiently raised, the electrons gain enough energy to ionize through collision further atoms of the gas, so that there is a significant multiplication of the free charges originally created by the passage of the particles. In Fig. 5.14, Curve 1 gives the number of electron-ion pairs collected as a function of applied voltage when an electron (minimum ionizing) traverses the counter; Curve 2 gives the same data, but for a much more heavily ionizing particle. Thus the ordinate is proportional to the pulse height of the signal that will appear after the coupling condenser C (of Fig. 5.13).

Referring to Fig. 5.14, we see the following regions of operation of a gaseous counter: in region II the voltage is large enough to collect all the electron-ion pairs, yet not so large as to produce any multiplication. A detector operated in this region is called an *ionization chamber*. As the voltage is further raised, region III is reached, where multiplication of the

FIG. 5.13 Diagrammatic arrangement of a cylindrical Geiger counter; the central wire is charged to B^+ through R_C while the cylindrical envelope is held at ground. The output signal appears across R_L.

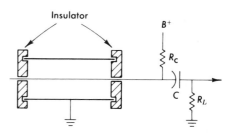

† This is due to other types of interaction as excitation and elastic scattering.
‡ It is also, of course, a function of the specific gas or mixture of gases used.

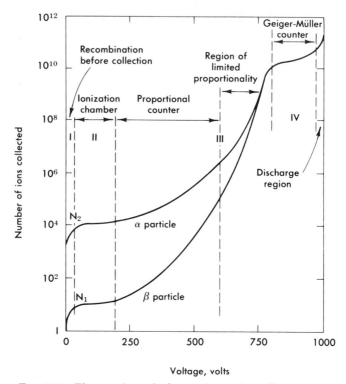

FIG. 5.14 The number of electron-ion pairs collected when a charged particle traverses a gaseous counter of average size plotted against the voltage applied between electrons. Curve 1 is for a minimum ionizing particle, whereas curve 2 refers to a heavily ionizing particle. Note the three possible regions of operation as an (a) ionization counter, (b) proportional counter, and (c) Geiger counter.

original free charges takes place through the interaction of the electrons as they move through the gas towards the collecting electrode. However, over a considerable range of voltage, the total number of collected electron-ion pairs is fairly proportional to the original ionization caused by the traversal of the charged particle.† A detector operated in this region is called a *proportional counter*; it has an advantage over the ionization counter in that the signals are much stronger, achievable gains being of the order of 10^2 to 10^4. Finally, further increase of the high voltage leads to region IV, where very large multiplications are observed, and where the number of collected electron-ion pairs is independent of the original ionization. This is the region of the *Geiger-Müller counter*, which has the great ad-

† The proportionality does not have to be a linear function of the applied voltage.

vantage of a very large output pulse, so that its operation is simple and reliable. Indeed, at such high voltages, once a few electron-ion pairs are formed the electrons produce more ionization at such a rapid rate that regenerative action sets in, the whole gas becomes ionized, and a discharge takes place. At that point, the resistance between the central electrode and the chamber walls becomes negligible, and the counter acts as a switch that has been closed between the high-voltage source and ground; this discharges condenser C through resistor R_L (Fig. 5.13). Since C was charged at B^+ (of the order of 1000 V), very large output signals may be obtained. For example, if the number of electron-ion pairs collected is 10^{10} (as given by Fig. 5.14) and $C = 0.001$ μF, we obtain

$$V = \frac{Q}{C} = \frac{1.6 \times 10^{-19} \times 10^{10}}{10^{-9}} = 1.6 \text{ V} \tag{3.1}$$

By scaling this figure, it is easy to appreciate the difficulties involved in the amplification of proportional-counter and ionization-counter signals.

The disadvantages of the Geiger counter are the loss of all information on the ionizing power of the charged particle that traversed the counter, and the long time necessary for restoring the gas to its neutral state after a discharge has taken place. However, the simplicity and good efficiency of the device for single-particle detection have made it a very common nuclear radiation detector. We therefore will elaborate on the Geiger counter and only briefly give results pertaining to ionization and proportional counters.

3.2 THE IONIZATION CHAMBER

The main difficulty of ionization counters is their very low signal output. If they are used, however, in a high flux of radiation as an integrating device, high signal levels can be reached; in that case the output signal corresponds to the total number of electron-ion pairs formed (per unit time) by the radiation. In this fashion ionization chambers are frequently used for monitoring x-ray radiation or high levels of radioactivity; in such applications they are far superior to Geiger counters, since the rates are so high that a Geiger would be completely jammed.

When an absolute measurement of the created free charges is made, as with an electrometer, ionization chambers may also serve as standards of ionizing radiation. Most commercial instruments, however, amplify the output pulse and are directly calibrated in roentgens (or fractions of roentgens) per hour. For use in the laboratory an ionization counter Model 2526 ("cutie-pie") manufactured by the Nuclear-Chicago Company is sug-

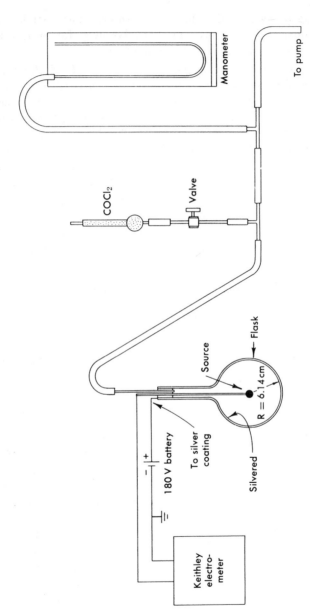

FIG. 5.15   A simple arrangement for the determination of the range of alpha particles in air by measuring the ionization current as a function of chamber pressure.

gested for radioactivity surveys and as an x-ray monitor in the range of 0–2500 mr/hr.

The following paragraphs describe a very rudimentary "student-type" ionization chamber that was used in this laboratory for measuring the range of alpha particles emitted by polonium 210.

Figure 5.15 is a sketch of the apparatus; it consists of a flask 5 in. outer diameter, its inside wall having been coated with a conducting material (such as aqua-dag or silver). A rubber stopper inserted at the mouth of the flask acts as a support, electrical insulator, and vacuum lock. Through the stopper is fastened a brass rod at the tip of which has been attached a 20-μCi Po^{210} source, which is thus located at the center of the flask. A 180-V battery is connected between the flask walls and the rod supporting the source, and the ionization current is measured with a Keithley electrometer.

The energy† of the Po^{210} alpha rays is 5.25 MeV, and their range in air at stp is 3.93 cm; thus the alphas stop before reaching the walls of the flask and deposit all their energy in the gas. By using the number of approximately 30 eV per electron-ion pair, mentioned on page 175, we would expect per alpha particle a total of

$$5.25 \times 10^6/30 = 170,000 \text{ electron-ion pairs}$$

(the true number in this case being closer to 110,000).

Since

$$20 \ \mu\text{Ci} = 20 \times 10^{-6} \times 3.7 \times 10^{10} = 7.4 \times 10^5 \text{ alpha particles/sec,}$$

if all the electrons were collected, the ionization current should be

$$I = 1.6 \times 10^{-19} \times 7.4 \times 10^5 \times 1.1 \times 10^5 = 1.3 \times 10^{-8} \text{ amp} \qquad (3.2)$$

which is readily measurable.

If now the flask is slowly evacuated, the alpha particles will traverse a longer path before stopping; however, as long as the alphas stop in the gas, the same number of electron-ion pairs is formed and the ionization current should remain flat and independent of pressure. When the density of the air in the flask becomes so low that the alphas reach the wall *before* losing *all* their energy in the gas, fewer electron-ion pairs are formed and the ionization current will drop monotonically with decreasing pressure.

Data obtained in this fashion by a student‡ are shown in Fig. 5.16. Indeed, the expected qualitative behavior of the ionization current is observed; from the breaking point we conclude that at a pressure of 51.5 ± 1

---

† See Table 4.2.

‡ R. Nebel, class of 1962.

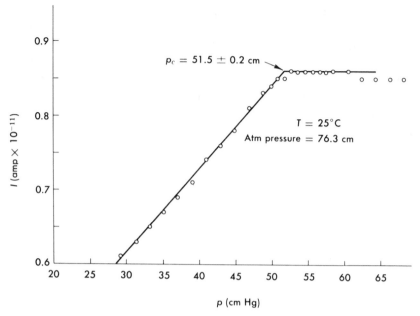

$$p_c = 51.5 \pm 0.2 \text{ cm}$$

$$T = 25°C$$
Atm pressure $= 76.3$ cm

I (amp $\times 10^{-11}$)

p (cm Hg)

FIG. 5.16 The results of the measurement referred to in Fig. 5.15. The ionization current is plotted against residual air pressure and a decrease in current begins at $P = 51.5$ cm Hg. This corresponds to a range of 4.02 cm in air at stp.

cm Hg, the range of Po²¹⁰ alpha rays in air is $R = 6.14$ cm. Hence at stp (760 mm Hg; 15° C)

$$R_{\text{stp}} = R \times \frac{P}{P_{\text{stp}}} \times \frac{T_{\text{stp}}}{T} = 6.14 \times \frac{51.5}{76.0} \times \frac{288}{298} = 4.02 \pm 0.1 \text{ cm}$$

in good agreement with the accepted value of $R_{\text{stp}} = 3.93$ cm.

From the ordinate of Fig. 5.16 we note, however, that the ionization current is three orders of magnitude lower than the estimate given by Eq. 3.2; this is due to the recombination of the electron-ion pairs, which proceeds at a fast rate because of the long path in the air, the high pressure, and the low value of the electric field accelerating the electrons.†

With such elementary apparatus it is difficult to perform a more detailed study of the specific ionization of the Po²¹⁰ alphas or of other charged particles; this will be done in Section 5 of the present chapter. We have, however, given here an example of a low-efficiency integrating ionization chamber.

---

† Some loss is also due to self-absorption in the source, and the geometrical solid angle is only 2π.

3.3 THE PROPORTIONAL COUNTER

Again we will not describe in detail the proportional counter† but only give the results obtained by a student‡ using such a detector in connection with the experiment on the Mössbauer effect (see Chapter 6). The advantage of proportional counters lies in the detection of very low-energy x-rays or gamma rays which can hardly penetrate a scintillation crystal, and when in addition good energy resolution is required. This is the case in the Mössbauer experiment from Fe^{57}, where it is necessary to identify a 14.4-keV gamma ray in a strong background of 123-keV gamma rays and 5-keV x-rays.

The proportional counter used was§ Amperex type 300-PC. It was filled with a xenon methane mixture at a pressure of 38 cm Hg. The equipment used for amplification and pulse-height measurement¶ is shown in Fig. 5.17, and the counter was operated at 2100 V. Fig. 5.18 gives the results

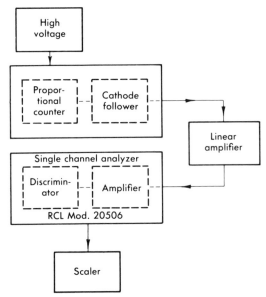

FIG. 5.17 Block diagram for pulse height measurements using a proportional counter.

† For an extensive discussion of proportional and ionization counters see the *Encyclopedia of Physics*, Vol. 45, *Nuclear Instrumentation II*. Berlin: Springer, 1958; articles by H. W. Fulbright, pp. 1–50, and by S. C. Curran, pp. 174–221.

‡ R. Harris, class of 1963.

§ Manufactured by the Amperex Corporation and obtainable from Scientific Sales, Inc., L. I., N. Y.

¶ For a more detailed discussion of pulse-height spectra see Section 4 of this chapter.

obtained, where the number of pulses is plotted against the discriminator channel. The large peak at Channel 12 is the 5-keV x-ray; the small peak at Channel 26 represents the sought-after 14.4-keV gamma ray.

As we know from Section 2.5 (Fig. 5.7) the predominant interaction of low-energy gamma rays in the gas is the photoelectric effect. The cross section for 5-keV quanta is of the order of 6×10^{-24} cm², so that if the counter represents approximately 50 milligrams/cm² of material, the ef-

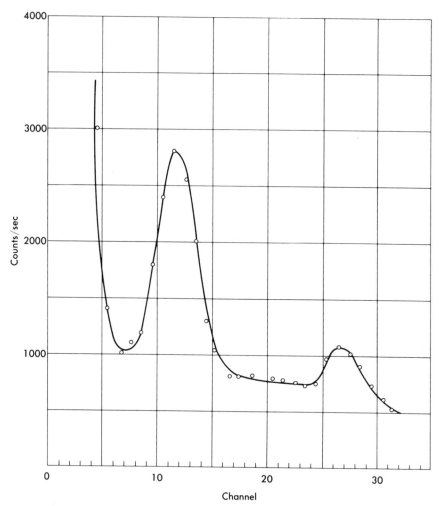

FIG. 5.18   Pulse height spectrum of the low-energy gamma radiation from Fe⁵⁷ as obtained with a commercial proportional counter. The pronounced peak at channel 12 is the 5 keV x-ray while the smaller peak is the 14.4 keV gamma-ray line used in the Mössbauer effect.

ficiency for gamma-ray detection might be as high as

$$6 \times 10^{23} \times 50 \times 10^{-3} \times \frac{Z}{A} \times 6 \times 10^{-24} \simeq 10 \text{ percent}$$

Using the data from the 5 keV peak, we obtain for the resolution of this proportional counter,

$$\Delta E/E = 1.7/12 = 14 \text{ percent}$$

where for ΔE we chose the half-width of the peak at half maximum (after background subtraction).

3.4 The Geiger Counter; Plateau and Dead Time

It has been pointed out in Section 3.1 that a gaseous counter operates in the Geiger region when the voltage between electrodes is sufficiently large; that is, the traversal of a charged particle initiates a discharge in the gas, and as a result a pulse appears at the output which is independent of the original ionization. If the voltage is further increased, spontaneous discharges occur, making the device useless as a particle detector.

Because the principle of operation is simple, Geiger counters are simply constructed, the geometry of Fig. 5.13 being typical. Fig. 5.19 is a photograph of some types of "end window" Geiger counters made by the Anton Laboratories of Skokie, Illinois. For certain applications, the thickness of the walls is an important consideration, and Geiger counters may be built with special thin windows (usually mica of a few mg/cm²). In the early days glass envelopes for Geiger counters were fairly common, and various pressures as well as mixtures of gases have been used.

Another important consideration for Geiger counters is the "quenching" of the discharge initiated by the traversal of a charged particle. Until the gas is returned to its neutral state, the passage of a charged particle will not produce an output pulse; this is the period of time during which the

Fig. 5.19 Photograph of three types of "end window" Geiger counters manufactured by Radiation Counter Laboratories of Skokie, Illinois.

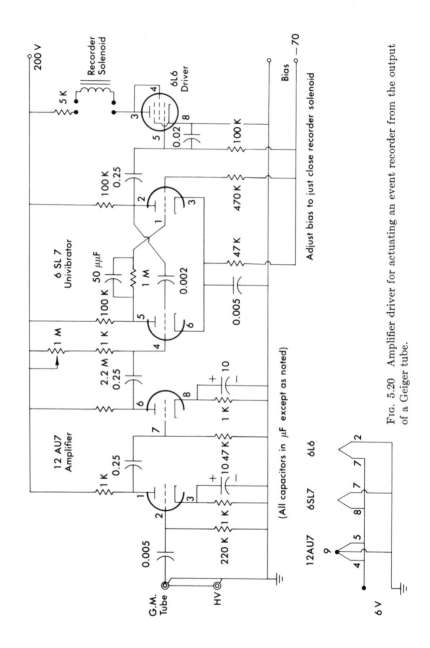

FIG. 5.20 Amplifier driver for actuating an event recorder from the output of a Geiger tube.

counter is "dead." The quenching of the discharge can be achieved through the external circuit (for example in Fig. 5.13 the charging resistor R_C will introduce such a voltage drop that the discharge will extinguish itself) ; or through the addition of special impurities (such as alcohol) to the gas of the counter, or by both methods used together.

The circuitry necessary for the operation of a Geiger counter is also extremely simple. A single stage of amplification and pulse shaping is usually sufficient to drive any scaler. As an example, Fig. 5.20 gives a circuit that has been used in the laboratory for driving the pen of an Esterline-Angus recorder, (or mechanical register), whenever a pulse appears at the counter output.

In order to operate a Geiger counter properly, the high-voltage source must be set in the "plateau" region (Fig. 5.14, region IV), where a similar output is consistently obtained for all charged particles traversing the counter. We may then define the *efficiency* of the detector as the ratio of the number of output pulses over the total flux traversing the counter; since the pulse heights are all equal in the plateau region, we do expect the efficiency to remain constant in that same region. Clearly any particle detector should be operated in a region where the efficiency is "flat" with respect to variation of operating parameters. The efficiency of Geiger counters is 90 percent or higher for charged particles, but for photons it is much lower, being only of the order of 1–2 percent.

It is difficult to make absolute efficiency measurements for Geiger counters. A "standard" calibrated source of radioactive material may be used, and its output count compared with the expected flux from a knowledge of the solid angle subtended by the Geiger counter. If the counter is placed at several distances from the source, the consistency of the measurements may also be checked through the $1/r^2$ dependence. However, a *relative* measurement of the efficiency as a function of the high voltage is easy to make; if it yields a *flat plateau*, this is an indication that the detector operates at high efficiency (close to 100 percent) for the particular type of radiation that is incident. Geiger-counter plateaus are usually a few hundred volts wide and have a small slope, of the order of 1–2 percent per 100 V.

To determine the plateau, either a radioactive source or the cosmic-ray flux may be used; since this flux is of the order of 10^{-2} particles/cm^2-sec, it takes several minutes to accumulate 1000 counts for a counter of average size. As explained in Chapter 10, the emission of radiation is a random process, so that the standard deviation† of any measurement is given by

---

† If this measurement is repeated many times, in 68 percent of the cases we will obtain $\bar{N} - \sigma < N < \bar{N} + \sigma$, where $\bar{N}$ is the average of all measurements. See Chapter 10 for the definition of σ.

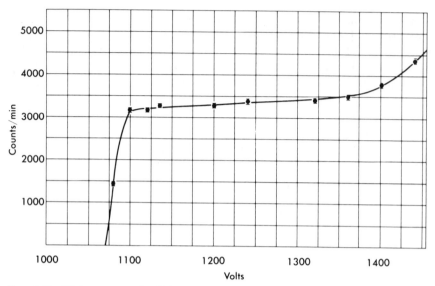

FIG. 5.21   Plateau curve of a Geiger counter. Note that the plateau region extends for 250 V and has a slope of the order of 5 percent per 100 V.

the square root of the number of counts, and thus the measurement should be interpreted as

$$1000 \pm 31 = 1000 \times (1 \pm .03) \text{ counts}$$

or in common parlance, 1000 counts give 3 percent statistics. The high voltage should be well stabilized, usually to a few parts in one thousand.

Figure 5.21 gives the plateau found by a student[†] for the RCL[‡] type 10104 Geiger counter. A 10-μCi Co[60] source was used for the measurements, and the standard deviation at each point is shown by the error flags. The plateau begins at 1100 V and is approximately 250 V wide; the discharge region begins at 1400 V.

The slope of the plateau, from Fig. 5.21, is

$$150/3200 = 5 \text{ percent per } 100 \text{ V}$$

Next we will turn our attention to the dead time of the Geiger counter already mentioned. Indeed, once a discharge has been initiated, the counter will not register another pulse unless the discharge has extinguished itself, and until, in addition, the counter has "recovered"—that is, returned to a neutral state. During the recovery period, the counter will generate an output pulse, but of a smaller-than-normal amplitude depending on the stage of recovery.

---

† D. Owen, class of 1963.
‡ Radiation Counter Laboratories, Inc., 5121 West Grove Street, Skokie, Ill.

This phenomenon of recovery can be clearly seen in Fig. 5.22, obtained by a student.† The Geiger counter was exposed to a high flux of radiation; the trace of a Tektronix 545 oscilloscope is triggered when the output pulse appears. The horizontal scale is 100 μsec/cm so that the shape of the output pulse and its exponentially decaying tail can be seen in detail. If now a second particle arrives within 1 msec of the previous one, it will appear on the same oscilloscope trace since the scope will not trigger again unless the spot has reached the end of the screen (which is 10 cm wide). The picture shown in Fig. 5.22 was obtained by making a multiple exposure of such traces. The correlation of pulse height against delay in arrival time and the exponential dependence of the recovery are noticeable. If we consider that the counter is inoperative until the output is restored to 63 percent of its original value $(1 - 1/e)$, the data of Fig. 5.22 give a value for the dead time τ of the order of

$$\tau = 400 \ \mu\text{sec} \tag{3.3}$$

Pulses, however, seem to appear after an interval

$$\tau \simeq 300 \ \mu\text{sec} \tag{3.4}$$

The dead time of a counter may also be obtained by an "operational" technique, such as by measuring the counting loss when the detector is

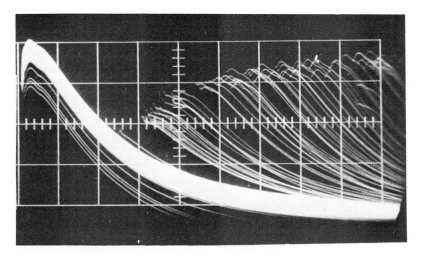

Horizontal scale	100 μsec/cm
Vertical scale	5 V/cm

Fig. 5.22　Multiple exposure photograph of oscilloscope traces obtained from a Geiger counter exposed to a high flux of radiation. Note the effect of the "dead-time" of the counter and the gradual build up (recovery) of the output pulses.

† R. Harris, class of 1963.

subjected to high flux. If the dead time is τ (sec), and the counting rate R (counts/sec), the detector is inoperative for a fraction $R\tau$ of a second; the true counting efficiency is then $1 - R\tau$.

Consider two sources S_1 and S_2 which when placed at distances from the counter D_1 and D_2 give a true rate (counts/unit time) R_1, R_2. The counter, however, registers rates $R_1' < R_1$, $R_2' < R_2$ due to dead-time losses, and when both sources are simultaneously present, it registers $R_{12}' < R_1' + R_2'$ due to the additional loss accompanying the higher flux. Now,

$$R_1' = R_1(1 - R_1'\tau)$$
$$R_2' = R_2(1 - R_2'\tau)$$
$$R_{12}' = (R_1 + R_2)(1 - R_{12}'\tau)$$

We solve by writing

$$\frac{R_{12}'}{1 - R_{12}'\tau} = \frac{R_1'}{1 - R_1'\tau} + \frac{R_2'}{1 - R_2'\tau}$$

which reduces to a quadratic equation in τ with the solution

$$\tau = \frac{1 \pm \sqrt{1 - R_{12}'(R_1' + R_2' - R_{12}')/R_1'R_2'}}{R_{12}'}$$

This can be expanded in the small quantity $(R_1' + R_2' - R_{12}')$ to give the approximate expression

$$\tau \simeq \frac{(R_1' + R_2' - R_{12}')}{2R_1'R_2'} \tag{3.5}$$

We now apply Eq. 3.5 to data obtained by students† with the same counter used for Fig. 5.22. In practice, source S_1 is first brought to the vicinity of the counter and R_1' is obtained; next S_2 is also brought in the area and R_{12}' is obtained and finally S_1 is removed and R_2' is measured: thus no uncertainties due to source position can arise. They obtain

$$R_1' = 395 \pm 3 \text{ counts/sec}$$
$$R_{12}' = 655 \pm 3 \text{ counts/sec}$$
$$R_2' = 334 \pm 3 \text{ counts/sec}$$

yielding $\tau = 282 \pm 20$ μsec, in better agreement with Eq. 3.4 than with Eq. 3.3.

The rather long dead time of the Geiger counter is a serious limitation restricting its use when high counting rates are involved; the ionization counter and proportional counter, however, have dead times several orders of magnitude shorter.

---

† D. Owen and D. Sawyer, class of 1963.

3.5 Time Distribution of Nuclear Radiation†

It has already been mentioned that nuclear radiation and cosmic rays are random processes and therefore follow a Poisson distribution. One consequence of this distribution is that when a large number of counts, N, is considered, the distribution becomes Gaussian, with a standard deviation $\sigma = \sqrt{N}$. To test the Gaussian nature of the frequency distribution of counts, we may take several measurements of the same source and plot the results. Such data are shown in Fig. 5.23a, where the Gaussian curve, normalized to the same area as the histrogram and with the appropriate σ, has also been plotted; the fit is quite satisfactory (that is, it has a low χ^2 probability as explained in Chapter 10).

The data for this plot were taken from 100 repeated measurements of approximately 100 counts each; the parameters of the sample are

$$\text{mean} \qquad \bar{N} = 85.34 \text{ counts/min}$$

$$\text{standard deviation} \qquad \sigma = \sqrt{N} = 9.23 \text{ counts/min}$$

(3.6)

Table 5.2 gives the frequency distribution of these 100 measurements in small "bins" of a width of 2 counts/min. Note that for the histrogram of Fig. 5.23a, the bin size was taken twice as large.

The procedure for testing if the distribution of a sample, as in Table 5.2 above, is indeed Gaussian may be facilitated by use of a special graphing technique shown in Fig. 5.23b. Here probability paper‡ is used, where on the abscissa is given the integrated probability of obtaining a count smaller than the value shown on the ordinate.

The advantage of this plot is that a Gaussian distribution appears as a straight line with a slope determined by the standard deviation. Indeed this is the case in Fig. 5.23b, where the same data of Table 5.2 have been plotted; the line is the prediction for a Gaussian distribution with a mean and standard deviation as obtained from the sample (Eq. 3.6).

We know, however, that the frequency distribution of nuclear radiation is of the Poisson type,§ which in the limit of many counts tends towards the Gaussian. To test the applicability of the Poisson distribution, we have to examine parameters which are related to *few counts*, for example, the distribution of the time intervals between any two successive counts (namely, the time elapsed after the arrival of one count until the next one). Similarly we may test the distribution of the time intervals from any count to the second next, and so on; however, as the number of counts contained

---

† To follow the content of this section the student must understand the material of par. 3, Chapter 10, and especially Section 5.

‡ O. Riedel, "Statistical Purity in Nuclear Counting," *Nucleonics*, **12**, 64 (1954).

§ Chapter 10, Section 5.

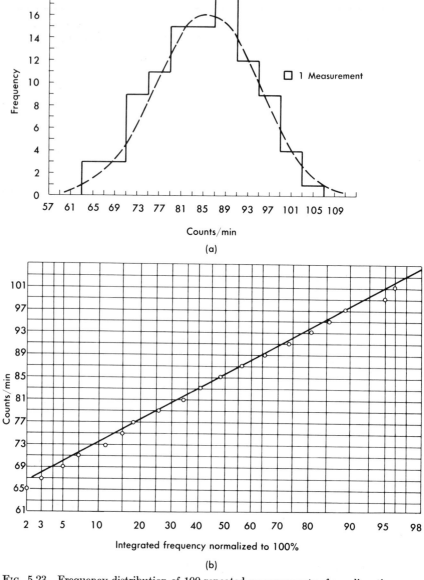

FIG. 5.23 Frequency distribution of 100 repeated measurements of a radioactive sample. (a) Histogram of the measurements; the dotted curve is the best fit Gaussian. (b) The integrated frequency distribution plotted on a special "probability scale"; in such a representation a Gaussian curve appears as a straight line.

TABLE 5.2

Frequency Distribution of 100 Repeated Measurements of a
Radioactive Sample

Interval (counts/min)	Number of measurements (yielding a result in this interval)	Cumulative % probability (of obtaining this result)
64–65	2	2
66–67	1	3
68–69	2	5
70–71	1	6
72–73	6	12
74–75	3	15
76–77	3	18
78–79	8	26
80–81	8	34
82–83	7	41
84–85	8	49
86–87	7	56
88–89	11	67
90–91	7	74
92–93	8	82
94–95	4	86
96–97	3	89
98–99	6	95
100–101	1	96
102–103	3	99
104–105	1	100

in the time intervals is increased, the distribution tends more and more to a Gaussian. These frequency distributions have been calculated in Section 5 of Chapter 10; the result is

$$q_m(t) \;=\; r\,\frac{(rt)^{m-1}e^{-rt}}{(m-1)!} \tag{3.7}$$

where $q_m(t)$ gives the probability of finding a time interval between t and $t + dt$; r is the true mean rate of counts (that is, counts per unit time, where the same units are to be used for measuring t), and m is the number of counts between which the time intervals are measured. If we measure the intervals between every count, $m = 1$; if between every second count, $m = 2$, etc.

We note that the distribution for $m = 1$ is simply

$$q_1(t) \;=\; re^{-rt} \tag{3.8}$$

it has no maximum and is peaked at $t = 0$. This is true for any random

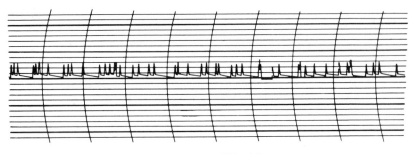

Increasing *t* ➤ 12 in./min

FIG. 5.24   Reproduction of the tape of the event recorder indicating the arrival of single cosmic-ray counts; the time scale is 5 sec/in.

process, and seems to justify an old German proverb that one calamity is always followed by a second one.† For $m \neq 1$, however, Eq. 3.7 has a maximum at

$$t = \frac{m - 1}{r}$$

which for large m becomes more pronounced and also approaches the mean time interval between m counts, m/r.

To obtain experimental data on such distributions of nuclear radiation, the output of the Geiger-counter amplifier is connected to a pen recorder (Esterline Angus-Model AW). It is convenient to use the cosmic rays as a source of radiation, in which case a recorder paper speed of 12 in./min provides adequate spacing between counts; Fig. 5.24 is a reproduction of such a record.

All required information is then contained on the recording, and we may measure off (as with a ruler) the time intervals between each count. The same recording may be used for measuring off time intervals between any second or tenth count, etc., but this becomes tedious. If such distributions are desired, it is preferable to connect the recorder to the second, or tenth stage, etc., of a scaler which is driven by the Geiger counter.

In Fig. 5.25 are presented data obtained in this fashion by students.‡ Figure 5.25a gives the histogram of the frequency distribution of the intervals between every count; the solid curve is Eq. 3.8 properly normalized to equal areas and indicates a very satisfactory fit. The histogram in Fig. 5.25b is the frequency distribution of the time intervals between every second count for the same sample; again the solid curve is obtained from Eq. 3.7 with $m = 2$, and also shows a satisfactory fit.

---

† W. Bothe, *Physikalishe Zeitschrift*, **37**, 520 (1936).

‡ D. Owen and D. Sawyer, class of 1963.

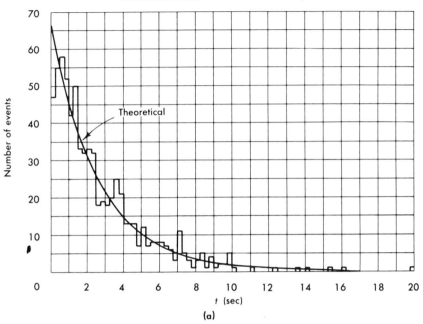

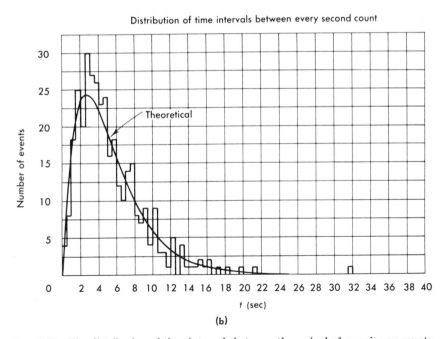

Fig. 5.25   The distribution of time intervals between the arrival of cosmic-ray counts as obtained from the record of Fig. 5.24. The solid curves are the predictions of Eq. 3.7. (a) Time intervals between every count ($m = 1$). (b) Time intervals between every second count ($m = 2$).

4. The Scintillation Counter

4.1 GENERAL

As we saw in gaseous-ionization instruments, the electron-ion pairs were directly collected; in the scintillation counter the ionization produced by the passage of a charged particle is detected by the emission of weak scintillations as the excited molecules of the detector return to the ground state. The fact that certain materials emit scintillations when traversed or struck by charged particles has been known for a long time, Lord Rutherford being the first to use a ZnS screen in his alpha-scattering experiments.

The scintillation counters used presently were developed in the nineteen-fifties and consist of an organic or inorganic crystal coupled to a sensitive photomultiplier which responds to the light pulses. Anthracene or stilbene crystals make excellent scintillators, but organic compounds embedded in transparent plastic, such as polystyrene, are now widely used because of ease in handling and machining and availability in large sizes. Such materials are commercially available† under the general designation of "plastic scintillators." The active materials are compounds, such as "PPO" 2–5–diphenyloxazole or diphenylstilbene, or others, and are also available in liquid form.

Organic scintillators have an extremely fast response, of the order of 10^{-9} sec, which can be matched by good photomultipliers. On the other hand, because of the low density and low Z, their efficiency for gamma-ray conversion is not high. To detect gamma rays, inorganic crystals are used instead, such as NaI or CsI, activated with some impurity such as Tl (1 part in 10^3). Inorganic crystals have an excellent efficiency for gamma-ray conversion, due to their high Z; from Eq. 2.18 we recall that the photo-electric effect is proportional to Z^5 and from Eq. 2.22 pair production is proportional to Z^2. However, the light output from inorganic crystals is spread over a much longer time interval, of the order of 10^{-6} sec. Such inorganic crystals are also available commercially,‡ appropriately encased since they are damaged by humidity; they come in sizes up to several cubic inches.

The light output of scintillators is proportional (as a matter of fact, linear) to the energy lost by the particle that traverses the detector; thus, by pulse-height analyzing the electrical output of the photomultiplier, the scintillation counter may be used as a spectrometer. This procedure is discussed in detail in the following section, where it is seen that energy resolution of the order of 10 percent is achievable.

The mechanism of emission of the photons in the scintillator material is rather involved. Table 5.3 gives a chart of the processes involved in the

---

† For example, from Pilot Chemicals Inc., 36 Pleasant St., Watertown 72, Mass.

‡ For example, from Harshaw Chemical Corp., Cleveland, Ohio.

TABLE 5.3†

THE SERIES OF PROCESSES LEADING TO THE EMISSION OF LIGHT WHEN A CHARGED PARTICLE
TRAVERSES A SCINTILLATOR MATERIAL

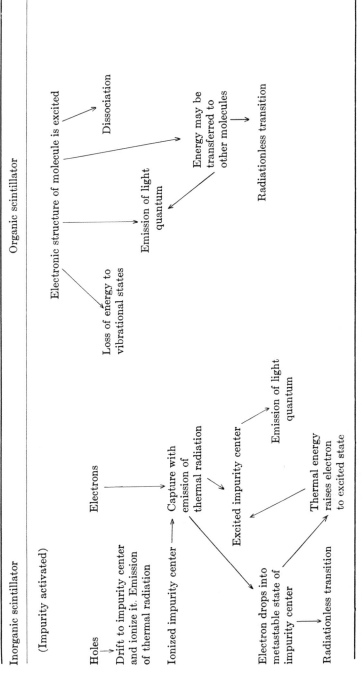

† After J. Sharpe, Nuclear Radiation Detectors, Methuen Co. (Courtesy of the Publishers)

emission of light in organic and inorganic crystals. In inorganic materials it is the migration of the electrons through the lattice (until they excite an impurity center) that is responsible for the long duration of the light pulse.

Even though the efficiency for transferring the energy lost by ionization to the photons in the visible region is on the average low, $\epsilon \approx 1.5$ percent, still a scintillator provides ample light output. Consider the case of a plastic scintillator 1 cm thick, traversed by a minimum-ionizing particle: $dE/dx = 2 \times 10^6$ eV per gm/cm^2; if we take the average photon energy as 3 eV, we obtain 10^4 photons. The efficiency of a photomultiplier cathode for converting photons into electrons is of the order of $\frac{1}{10}$, and the geometric efficiency for collecting the photons onto the photocathode is usually high, so that of the order of 1000 electrons are released. With modern techniques, however, it is possible to detect the release of a few photoelectrons, or even of a single one.

Clearly the scintillator material must be transparent to the visible radiation and optical coupling to the photomultiplier must be provided. This is achieved either directly or through a "light pipe"; light pipes are appropriately shaped pieces of lucite or other medium of high refractive index which traps and guides the light due to total internal reflection at its surfaces. At the surfaces where the light pipe is joined to the scintillator or to the photomultiplier, optical contact is achieved by the use of either viscous fluids or special glues.† Obviously the whole assembly must be light tight; this is frequently achieved by wrapping black electrical tape around the scintillator, light pipe, and phototube.

Because of its great stability and ease of operation, as well as because of its time and energy resolution, the scintillation counter has become the most frequently used detector in nuclear physics, especially for high-energy particles.

4.2 EXPERIMENT ON THE DETERMINATION OF THE ENERGY OF GAMMA
     RAYS WITH A SCINTILLATION COUNTER

. From the previous section we know that the appropriate detector for measurements of the energy of gamma rays is an inorganic crystal. When a gamma ray of energy <1 MeV enters the detector, it will interact either by the photoelectric effect or the Compton effect. In the former case it is fair to assume that the ejected photoelectron will deposit all its energy in the scintillator; in the Compton effect, however, the scattered photon may or may not convert in the scintillator (depending on the size and geometry of the detector).

---

† In the first category: Corning 200,000 centipoise fluid or clear vacuum grease; in the latter, R 363, PS 28 acrylic glue, etc.

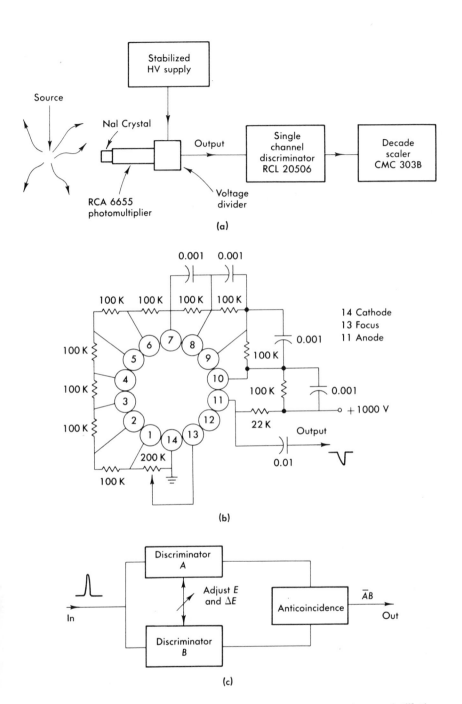

FIG. 5.26   Measurement of a gamma-ray pulse-height spectrum using a scintillation counter. (a) Block diagram of the apparatus. (b) Wiring for the base of an RCA 6655 photomultiplier in order to provide the appropriate HV to each stage and coupling of the output signal. (c) Diagrammatic arrangement for a single-channel pulse-height analyzer.

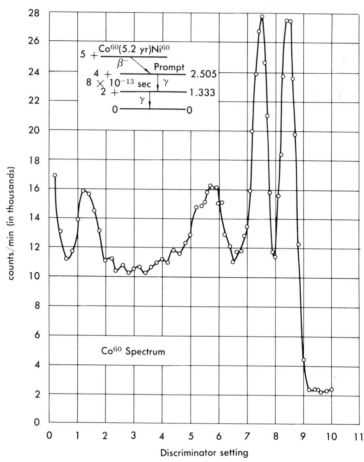

FIG. 5.27 Pulse-height spectrum of Co^{60} gamma rays obtained with a NaI crystal.

The pulse-height spectrum for gamma rays of a given energy will consist of a peak at an energy corresponding to that of the gamma ray, and a continuum below the peak, corresponding to Compton-scattered gamma rays which escaped from the crystal before totally converting. This can be seen in Fig. 5.27 and the following ones. Clearly the larger the size of the crystal, the larger the percentage of the output counts that will lie in the photopeak; thus, the gamma-ray line will become more pronounced.

The data reported here were obtained with a NaI-Tl activated crystal,† 2 in. in diameter and 1 in. wide, coupled directly to an RCA 6655 photo-

† Hawshaw Chemical Corp., Cleveland, Ohio.

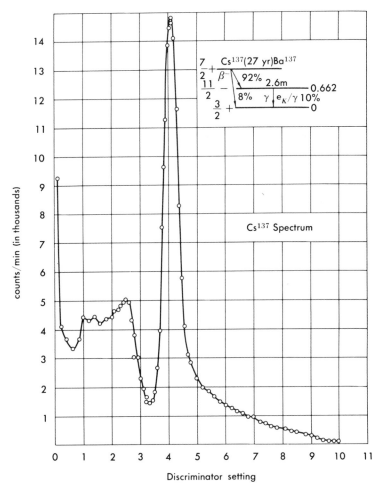

Fig. 5.28   Pulse-height spectrum of Cs^{137} gamma rays obtained with a NaI crystal.

multiplier tube.† Figure 5.26a shows the arrangement and block diagram of the electronics. A stabilized high-voltage supply feeds the divider chain to provide the appropriate voltage for each stage of the photomultiplier (Fig. 5.26b). The output pulse is fed to a preamplifier and single-channel discriminator (Radiation Counter Laboratories Model 20506), and its output is fed to a decade scaler (Chapter 4, Section 3.1). The single-channel analyzer consists essentially of two Schmidt triggers, as shown in

† For the characteristics and ratings of this tube see the specifications sheet of the Radio Corporation of America.

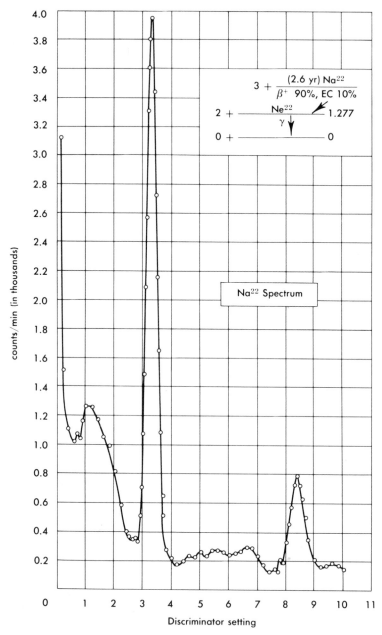

FIG. 5.29   Pulse-height spectrum of Na²² gamma rays obtained with a NaI crystal. Note that the 511-KeV line is due to positron annihilation.

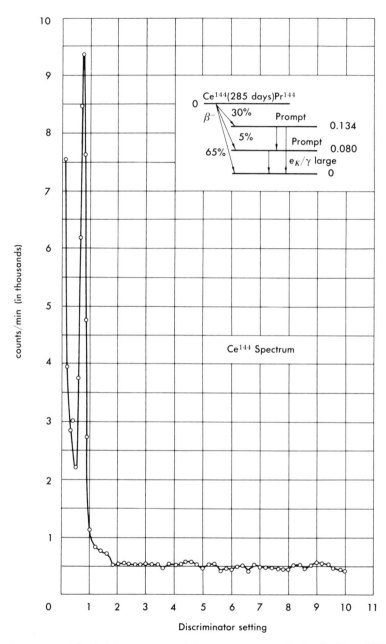

FIG. 5.30   Pulse-height spectrum of Ce¹⁴⁴ gamma rays obtained with a NaI crystal.

Fig. 5.26c, where the threshold of trigger A is always set higher than the threshold of B by an amount ΔE, the window width. If then an input signal passes through trigger B, which is set at threshold E, but *not* through A, set at $E + \Delta E$, an output is obtained. Both E and ΔE can be manually adjusted, and by taking a count for various settings of E it is possible to scan the whole spectrum.

The following figures give the results obtained by a student[†]. Figure 5.27 gives the spectrum of Co^{60} and shows two distinct peaks, which we attribute to gamma rays emitted in the de-excitation of Ni^{60} from its 2.505 MeV level to the 1.333 level, and from that level to the ground state according to the decay scheme also shown in the figure. As a measure of the energy resolution, we may consider the full width of the peak at half maximum, which is of the order of 0.8 V; hence a resolution of $0.8/8 \simeq 10$ percent. We also notice a significant background for pulse heights lower than that of the peaks, which is due to Compton scattered gamma rays which subsequently escaped from the crystal.

Figure 5.28 gives similar data for a sample of Cs^{137}; here the 0.662 MeV gamma ray represents the de-excitation of Ba^{137}. Again we notice some Compton background and an energy resolution of the order of 10 percent. Figures 5.29 and 5.30 give the pulse-height spectra from Na^{22} and Ce^{144}, respectively. For Na^{22}, the peak at 1.277 MeV arises from the de-excitation of Ne^{22}; the larger peak at 0.511 MeV arises from annihilation radiation. Indeed, from the level diagram of Na^{22} decay, we notice that positrons are emitted; the positrons are usually stopped in the walls of the source container, or in the crystal face, and as they come close enough to an electron they annihilate into two gamma rays, each gamma ray sharing the energy of the electron-positron pair[‡]. It is *one*[§] of these gamma rays that is then converted in the crystal and gives rise to the 0.511 peak.

Finally, in Fig. 5.31 is given a plot of all the observed peaks against channel number, showing the linearity of pulse height against energy; some of these data were referred to already in Chapter 2 when the line spectra of quantum-mechanical systems were discussed. In addition to the gamma rays, the nuclei that were investigated do also emit beta rays, and one would expect to see the corresponding peaks in the pulse-height spectrum. This, however, is not true because the beta spectrum is continuous instead of being a sharp line as is the case with gamma-ray spectra; in addition, electrons may lose variable amounts of energy before reaching the scintillation crystal, so that unless special precautions are taken, the energy resolution is usually poor.

† R. Stevens, class of 1963.
‡ See also the detailed discussion in Chap. 9.
§ Note that they are emitted with a relative angle of 180 degrees.

In interpreting gamma-ray spectra some care must be taken since spurious peaks due to instrumental effects or physical effects do appear. First, there can be peaks arising from the emission of x-rays, following photoejection of K-shell electrons either in the source or in the shielding. Also, a peak may appear due to photons that backscatter (by 180°) in the photomultiplier window or elsewhere; then the Compton-scattered electron escapes, but the scattered photon becomes converted in the crystal. For Cs¹³⁷ with its 0.662-MeV gamma ray, the backscattering peak appears at 0.185 MeV and can be identified in a carefully measured spectrum.

Another spurious effect occurs when an incoming photon of energy E ejects a K-shell electron from the iodine of the crystal, but the x-ray that is

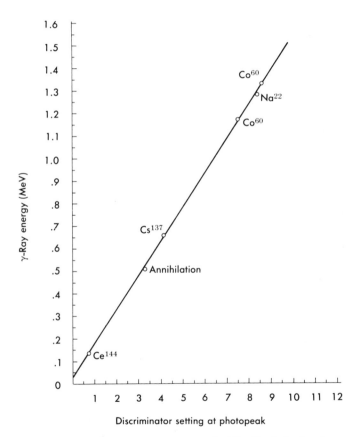

FIG. 5.31 Plot of gamma-ray energy against the discriminator channel of the photopeaks appearing in the spectra of Figs. 5.27 through 5.30 above. Note the linearity of the plot as well as its intersection with the origin.

emitted escapes without converting in the detector. The ejected photoelectron has an energy

$$E - E_K$$

where E_K is the energy of the K shell of iodine, namely, 29 keV, and will give rise to a peak not coinciding with the true photopeak. This so-called "escape-peak" can be identified because it is located 29 keV below the photopeak; it is most pronounced in the pulse-height spectra of low-energy gamma rays.

The relative ratio of counts in the photopeak as compared to the counts in the Compton background depends on the crystal and source geometry and on the gamma-ray energy. Usually the relative counts in the photopeak give sufficient information, but when the absolute number of gamma rays is required, we can calculate the efficiency of the crystal for the particular geometry and gamma-ray energy. Extensive tables of efficiency for most combinations of the relevant parameters have been calculated.†

4.3 RANGE AND ABSORPTION MEASUREMENTS; APPLICATION TO THE ABSORPTION OF GAMMA RAYS IN Pb AND Al

We saw in Section 2.3 of this chapter that the range of a charged particle in an absorber provides a measure of its energy. Therefore by using a detector (which in itself is not sensitive to energy) we may obtain a range curve by inserting between source and detector absorbers of varying thickness.

It is by this technique, and using a Geiger counter, that beta-ray energy spectra were first measured. The absorbers are thin aluminum foils (a few mg/cm² thick) and the corresponding energy is obtained from calibration against particles of known energy or from empirical formulae such as Eq. 2.13. Since the beta-ray (electron) spectra from radioactive nuclei are continuous, the range curve does not have a simple shape but shows a monotonic decrease in counting rate with increasing absorber thickness. Further difficulties arise from the scattering of the electrons either out or *into* the detector‡; scattering from the air, table tops, counter shielding, the absorber itself,§ and the foil supporting the source (back-scattering) does take place. Therefore good collimation and geometry must be carefully selected, and corrections for counter-window thickness and absorp-

---

† See the *Encyclopedia of Physics*, Vol. 45, *Nuclear Instrumentation II*, p. 110.

‡ We note from Eq. 2.17 that multiple scattering is much more severe for electrons than for heavy charged particles of the same (nonrelativistic) energy, since p is much smaller.

§ In poor geometry cases it is often possible that the counting rate might increase (!) when the first absorber is introduced.

tion in the air and the source must be applied. If gamma radiation is also present, it will give a continuous background beyond the maximum range of the electrons.

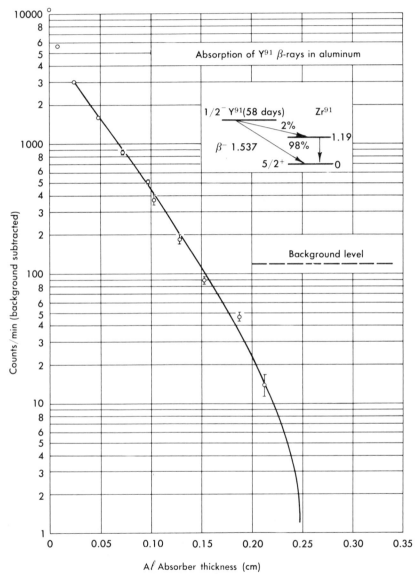

Fig. 5.32   Data on the absorption of Y^{91} beta rays in aluminum. Note that to observe the range of the betas, we need consider also data below background level. The range corresponding to the betas of maximum energy is 0.25 cm of aluminum.

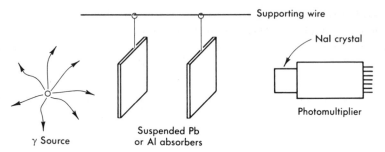

FₐG. 5.33   A suggested arrangement for the insertion of lead absorbers when measuring the attenuation of gamma rays. This special mounting is used in order to minimize scattering.

Since by now more precise techniques have been developed for measuring charged-particle energy spectra (electromagnetic spectrometers), and the scintillation counter for determining gamma-ray spectra, range and absorption measurements are not commonly used. In Fig. 5.32 is given such a range curve obtained by a student[†] using Y^{91} beta rays which have a maximum energy of 1.54 MeV.

From Fig. 5.32 we obtain

$$R_{\max} \simeq 0.25 \text{ cm Al} = 0.680 \text{ gm/cm}^2 \text{ Al}$$

and using Feather's relation, Eq. 2.13,

$$E = 1.55 \text{ MeV}$$

which is the correct order of magnitude, and coincidentally, also in exact agreement with the accepted value.

In interpreting the data we should keep in mind the limitations previously mentioned, and this can be seen from the data of Fig. 5.32. Neither end of the curve (low absorber or high absorber side) can be easily fitted; to find the maximum range one always has to take into account the background which might be an order of magnitude higher than the true counting rate. For these reasons a comparison technique, developed by Feather, is more frequently used.[‡]

If we consider absorption measurements with gamma rays, we know from Section 2 that no distinct range exists but that instead an exponential attentuation of the beam will take place. The absorption coefficient is slightly dependent on energy (see Eqs. 2.18 and 2.21), so that we can in principle obtain the gamma-ray energy from the slope of the counting rate against

---

[†] H. Nebel, class of 1962.

[‡] See, for example, G. Friedlander and J. W. Kennedy, *Nuclear and Radiochemistry*, John Wiley, 1957.

absorber-thickness curve; it must be a straight line when presented on a semilog plot (Fig. 5.34). This method, however, is far less sensitive than a direct measurement of the gamma-ray energy in a scintillation crystal as discussed in the previous section. Even though absorption measurements

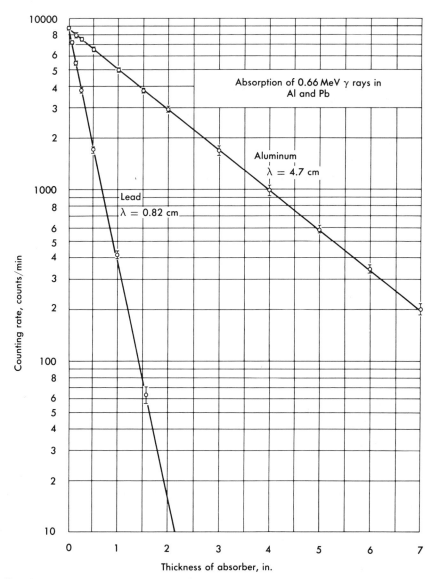

FIG. 5.34 Data on the absorption of Cs[137] gamma rays in lead and aluminum. Note the exponential nature of the absorption; also the difference in the absorption lengths of the two materials.

are not very useful for the determination of the energy of gamma rays, the exponential absorption of gamma rays by matter, and the dependence of the absorption coefficient on the material of the absorber can be demonstrated. Further, it is easier to collimate a gamma-ray beam, and scattering effects are less pronounced than for electrons, so that it is possible to obtain "clean" data with only a reasonable amount of care.

The detector for measuring gamma-ray absorption could be a Geiger counter, but a scintillator is preferable because of its higher efficiency. The absorbers have now a thickness of the order of grams per square centimeter and are usually lead or aluminum; they should be placed so as to minimize scattering, a possible arrangement being shown in Fig. 5.33.

The data on Fig. 5.34 were obtained by a student[†] using a Cs^{137} source. The resulting values for the absorption coefficients are

for Pb $\kappa = 1/0.82$ cm^{-1} and $\mu = 0.107$ cm^2/gm

for Al $\kappa = 1/4.7$ cm^{-1} $\mu = 0.079$ cm^2/gm

in good agreement with the accepted values for gamma rays of 0.662 MeV energy[‡].

Pb $\mu = 0.11$ cm^2/gm

Al $\mu = 0.075$ cm^2/gm

The exponential nature of the absorption process is clearly demonstrated.

5. Solid-State Particle Detectors

5.1 GENERAL

We have seen how the gaseous ionization counters and the scintillation counters are widely used for the detection of radiation and charged particles. Recently it has been possible to use semiconductor materials for the detection of charged particles, especially those of low energy; such detectors are called "solid-state counters" after the name of the field of physics responsible for the development and understanding of semiconductors.[§]

In a general sense, we can think of this type of detector as a solid-state ionization chamber, having two basic advantages over a gas-filled ionization chamber:

(a) The energy required for the creation of an electron-ion pair is 3 eV (as compared to approximately 30 eV in a gas) so that stronger signals and better statistics can be achieved.

[†] K. Douglass, class of 1964.

[‡] See, for example, *Nucleonics*, **19:** 6, p. 62 (1961).

[§] The scintillation counter is also a detector in the solid state!

(b) The stopping power is approximately 10^3 times that of a gas-filled device (since the detector material is so much denser), and thus it becomes possible to stop, in the *detector*, particles with energies typical of nuclear interactions. Consequently a very large number of electron-ion pairs is formed, leading to very good energy resolution. A 1-MeV proton stopping in a solid-state detector will create 300,000 electron-ion pairs, while the same proton traversing a proportional counter of 2-cm thickness would only release approximately 30 pairs.

In practice, however, it must be possible to collect the free charges (those created by the passage of the charged particle) before they recombine; this might be done, for example, by the application of an electric field in the detector material. This requirement is very difficult to meet with any of the ordinary crystals. Clearly, the material must have a high resistivity, since otherwise current will flow under the influence of the field, masking the effect of the pulse produced by the passage of the particle; on the other hand, in high-resistivity materials, the mobility of the free carriers is very low and the recombination probability high.

Even though some results have been obtained by using diamond as a detector, semiconductor materials come much closer to fulfilling the requirements mentioned above. Very pure material (an intrinsic semiconductor) is used to achieve the necessary high resistivity of the order of 10^7 ohm-cm, and the detector is operated at low temperatures. Such devices are called "bulk semiconductor detectors" and have yielded only moderate results.

A great improvement occurs when a semiconductor junction† is used as the detector volume; a device of this kind is called a barrier-layer detector. The junction is made by either of the following methods:

(a) Diffusing a high concentration of *donor* impurities on a p-type material, usually silicon, thus creating an n–p junction

(b) Utilizing a thin p-type surface that is formed by oxidizing when n-type silicon or germanium are exposed to the air. This surface is so thin that it is usually coated with gold to provide a good electrical contact; thus we have a p-n junction.

In either case the operation is similar, but the junction is always reverse biased.

Below we will briefly discuss the diffused junction (n-p) type of detector; Fig. 5.35a is a reproduction of Fig. 3.29, and gives the configuration of the energy bands at an n-p junction, electrons being the majority carriers in the left, or n, region, and holes the majority carriers in the right, or p,

---

† Semiconductor junctions were discussed in Chapter 3, Section 4, and the reader may find it useful to review that material, as well as Chapter 3, Section 3.

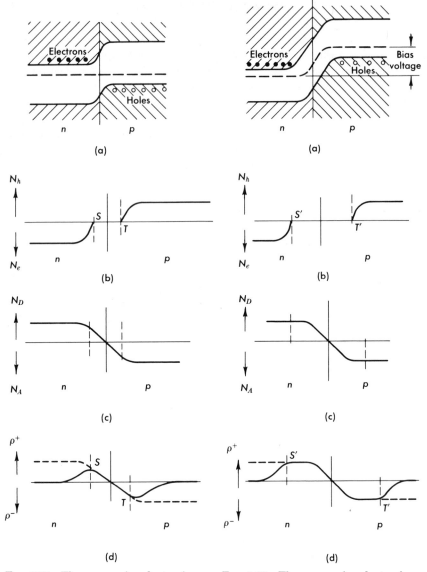

Fig. 5.35   The n-p semiconductor junction. (a) Position of conduction and valence bands and of the Fermi level across the junction; note the majority carriers for each region. (b) Density distribution of majority carriers on the two sides of the junction. (c) Density distribution of impurity centers on the two sides of the junction. (d) Distribution of space charge on the two sides of the junction.

Fig. 5.36   The n-p semiconductor junction under reverse bias. The plots in (a), (b), (c), and (d) pertain to the same distributions as in Fig. 5.35 but under reverse bias. Note the increase of the "depletion zone," $S'T'$.

region. Clearly, electrons may not move to the right, since the conduction band is at a higher (negative) potential, and holes may not move to the left, since the valence band is now at a higher (positive) potential; as a consequence there is some *repulsion* of *majority* carriers from the junction; Fig. 5.35b shows their density distribution. We note a "depletion zone" in the region marked $S–T$.

Next, Fig. 5.35c shows the density of impurity centers on the two sides of the junction; that is, these centers which may be *expected* to be ionized by the passage of a charged particle. To the left the donors have given electrons to the conduction band and are left positive; to the right the acceptors have acquired electrons from the valence band and are left negative. However, these impurity centers are neutralized by the majority carriers, so that the free (space) charge distribution is the sum of Figs. 5.35b and 5.35c, as shown in Fig. 5.35d.

Thus we see that space charge exists in the region of the junction, and as a consequence an electric field (the so-called barrier) exists as well, and extends over the depletion zone. Clearly if an electron-ion pair is created in the *depletion zone*, the electric field is such as to accelerate the

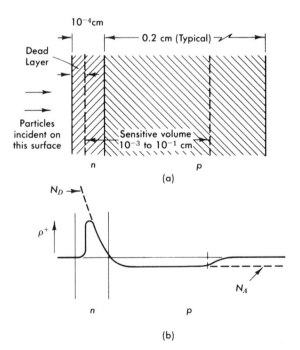

FIG. 5.37 Arrangement of an n-p semiconductor junction for use in a solid state detector. (a) Actual dimensions. (b) Distribution of space charge.

negative charge towards the n region, where it will have high mobility, (being a majority carrier); similarly, the hole will be accelerated towards the p region. Thus good collection efficiency is achieved.

Under reverse bias, Fig. 5.36 shows the same junction, 5.36a being the same as Fig. 3.30. Figure 5.36b gives, as before, the density distribution of majority carriers, which are now further removed from the junction, and Fig. 5.36c is exactly the same as 5.35c, giving the density of impurity centers. Figure 5.36d, however, which gives the space-charge distribution, shows that the *ionized* impurity centers have reached saturation and extend beyond the junction. Thus, most of the applied bias voltage appears across the depletion zone, which now is much more extended; the limit to this increase in *sensitive detector depth* is set by the breakdown voltage of the semiconductor material itself.

In a diffused junction, as used for a detector, the concentration of donors in the n-type material is much larger (about 10^3) than the concentration of acceptors in the p-type material. Since the total free charge must be the same on both sides of the junction, the space-charge distribution is asymmetric, as shown in Fig. 5.37b; Fig. 5.37a gives some of the physical dimensions in a realistic diffused junction; we note that most of the "sensitive volume" is in the p-type material.

5.2 PRACTICAL CONSIDERATIONS IN SOLID-STATE DETECTORS

From the previous discussion we have seen how a semiconductor junction may provide the appropriate electric field within a solid so as to collect electron-hole pairs produced by the passage of a charged particle. Multiplication such as occurs in the proportional or Geiger counter never takes place in a solid. The only way to achieve good resolution in a solid state detector is to *always* collect *all* the electron-hole pairs produced. Clearly, then, the sensitive volume of the detector must be longer than the range of the particle that is detected; it is also desirable that the dead layer at the entrance side be as thin as possible.

Since in recent detectors sensitive volumes† typical of a length of 3 mm have been achieved, the use of solid state detectors has been extended to particles of energies as high as 30 MeV. The resolution in energy is usually extremely good—that is, of the order of 0.25 percent for alpha particles (see also Fig. 5.41). The over-all size of the detector is at present restricted to a few cubic centimeters, due to the available semiconductor crystals; on the other hand, small size and the absence of need for a photomultiplier tube can be quite advantageous.

---

† The sensitive volume or barrier depth can be obtained from a nomograph, as given by J. L. Blankenship, "Proceedings of the Seventh Scintillation Counter Symposium, Institute of Radio Engineers, N.Y.," *Nuclear Science* 7: 2, 3 (Sept. 1960), p. 190.

It is also possible to use solid-state detectors, not as total absorption counters, but as dE/dx devices, in which case the p region is also made thin and no electrodes are placed in the path of the particle. Such detectors have been made to respond to high-energy (minimum-ionizing) particles as well. Semiconductor devices are also very useful for the detection of gamma rays. In general due to their small size, the ratio of counts in the photopeak as compared to background counts is smaller than for a scintillation crystal; however, the resolution is excellent, reaching one part in thousand.†

In practice, the construction of a solid-state detector is an art, and the attachment of electrodes to insure good ohmic contacts may be quite difficult. When germanium is used, cooling to liquid nitrogen temperatures may be required, while silicon gives good resolution at ambient temperature. The output signals are small, the voltage being determined by the capacities of the junction and of the amplifier input; the former depends on the length of the depletion zone and the area of the detector. If we assume a typical capacity of 200 $\mu\mu$F, then for 1 MeV energy loss, the signal voltage is

$$V = \frac{Q}{C} = \frac{1.6 \times 10^{-19} \times (10^6/3)}{200 \times 10^{-12}} \approx 2.5 \times 10^{-4} \text{ V} \qquad (5.1)$$

It is necessary to use a charge-sensitive preamplifier because the capacity C depends on the applied bias; thus if voltage is directly measured, severe variations in gain occur when the bias is changed. Leakage current in the crystal, and amplifier noise, set the limits of the smallest detectable signals.

Most of the hardware for solid-state detectors as well as the detectors themselves are now commercially available‡; Fig. 5.38 shows a typical setup with a feedback preamplifier. A surface-barrier silicon detector is used and operated at room temperature. Figure 5.41 gives the response ob-

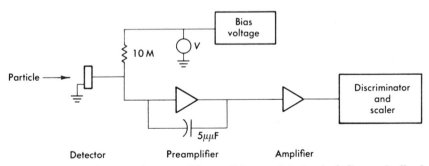

FIG. 5.38 Typical set-up for use with a solid-state detector including a feedback preamplifier.

---

† G. T. Ewan and A. J. Tavendale, Can. J. Phys., **42**, 2286, (1964).

‡ For example, from Oak Ridge Technical Enterprises, Oak Ridge, Tennessee.

tained from polonium alpha particles of different energies (after attenuation in air).

Another recent type of solid-state detector, called p-i-n (positive-intrinsic-negative material), consists of a layer of intrinsic crystal placed between p-type and n-type material. Having the advantage of a much longer sensitive volume, it holds better promise for high-energy particle detection.†

5.3 RANGE AND ENERGY LOSS OF Po²¹⁰ ALPHA PARTICLES IN AIR

In Section 3 a description has been given of the method of obtaining an estimate of the range (and hence energy) of Po²¹⁰ alpha particles in air, by means of a crude ionization chamber. With solid-state detectors, it is possible to improve on these measurements, as well as to study the rate of energy loss of the alpha particles as a function of their energy.

A collimated Po²¹⁰ source and the detector are both placed in an evacuated vessel at a fixed distance of 15 cm, as shown in Fig. 5.39. Then air is allowed into the vessel, and as a function of the pressure we measure

(a) The number of particles counted in the detector
(b) The pulse height distribution of the output signals, namely, the energy of the alpha particles when they reach the detector

In measurements of type (a), the *same* number of alpha particles should be reaching the detector until the pressure is raised to the point where the amount of material (gm/cm² of air) between source and detector is equal to the range of the alpha particles; beyond that pressure the counting rate should abruptly fall to zero. Note that since the relative position of source and detector is not altered, the solid angle $\Delta\Omega$ does not change, and the

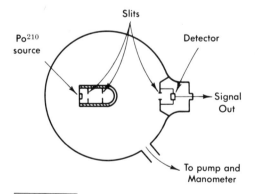

FIG. 5.39 Arrangement for the measurement of the range in air of Po²¹⁰ alpha particles. Note mounting of the solid–state detector and source inside an evacuated chamber (see also Fig. 6.4a).

---

† For more details see J. M. Taylor, *Semiconductor Particle Detectors*. London and Washington, D. C.: Butterworth, 1963; also consult the current literature.

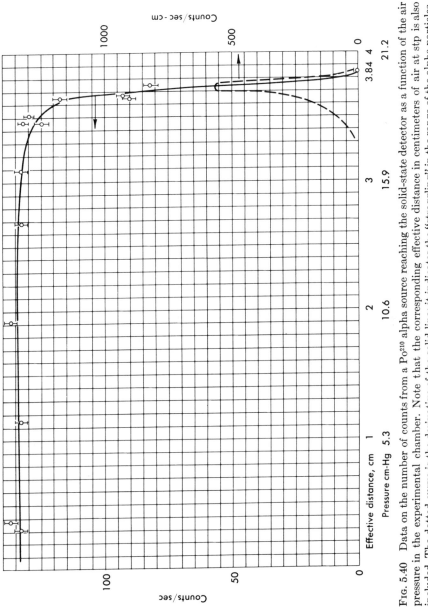

Fig. 5.40 Data on the number of counts from a Po²¹⁰ alpha source reaching the solid-state detector as a function of the air pressure in the experimental chamber. Note that the corresponding effective distance in centimeters of air at stp is also included. The dotted curve is the derivative of the solid line; it indicates the "straggling" in the range of the alpha particles.

215

only variation arises from the increase in multiple scattering; this, in turn, may result in some loss of particles from the beam.

These considerations are indeed borne out by the results obtained by a student† and shown in Fig. 5.40. Here the ordinate to the left gives the counts per second while the abscissa gives the pressure of air in centimeters of mercury, or, equivalently, the effective distance of air at stp. The dotted curve to the right is the derivative with respect to distance of the counting curve and gives the range (and so-called range straggling) of Po^{210} alpha particles. We obtain a mean range of

$$R = 3.72 \pm .06 \text{ cm}$$

and an extrapolated range

$$R = 3.82 \pm .06 \text{ cm}$$

which might indicate some systematic discrepancy from the accepted value for the extrapolated range of 3.93 cm.

Turning now to the measurements of type (b), Fig. 5.41 shows the distribution of the detector pulse heights as obtained with the single channel discriminator (described in connection with the scintillation counter). Each peak corresponds to a different pressure, and we thus note that the alpha particles reach the detector with progressively less energy when they have traversed more gm/cm^2 of air. We set the pulse height obtained in vacuum equal to the full energy of the Po^{210} alpha particle, namely, 5.25 MeV, and use the linear characteristic of the solid-state detector to obtain

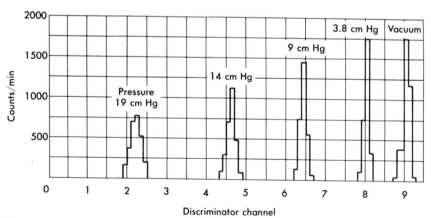

FIG. 5.41 Distribution of output pulse height of the solid-state detector for five different pressures. Note the gradual decrease of the energy of the alpha particle.

† K. Douglass, class of 1964.

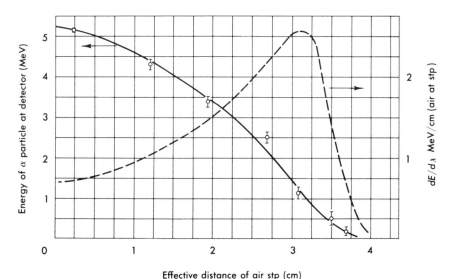

Fig. 5.42 Plot of the residual energy of a polonium alpha particle when it reaches the detector as a function of air pressure (plotted, however, in terms of the equivalent amount of air (stp) traversed). These data are obtained from distributions such as shown in Fig. 5.41. The dotted curve represents the derivative of the solid (energy) curve; thus it gives the energy loss per unit length. It is called the "Bragg curve."

the energy of the alphas as a function of material traversed. The results, obtained by a student† are given in Fig. 5.42 (solid curve).

If the derivative of the energy curve is taken with respect to distance, we obtain the energy-loss curve, dE/dx, as a function of distance, as shown by the dotted curve in Fig. 5.42. Such a curve is called a Bragg curve, and shows a $1/E$ dependence‡ as predicted by Eq. 2.10; for these very slow particles $E = \frac{1}{2}Mv^2$ and the influence of the logarithmic term of Eq. 2.10 is minimal. As the particle reaches the end of its range the energy loss dE/dx drops rapidly to zero.

From the energy curve of Fig. 5.42, we note that in air at stp the polonium alpha particle produces at the end of its range approximately 67,000 electron-ion pairs per centimeter, whereas at its full energy it produces only 20,000 pairs per centimeter; these numbers were obtained by using an average loss of 36 eV for the production of one electron-ion pair in air. More accurate results, especially close to the stopping point, can be best obtained with special ionization chambers.

---

† K. Douglass, class of 1964.

‡ We might plot the dE/dx curve against energy by making use of the data of the energy curve to express the distance from the stopping point in energy units.

6. Some Other Detectors

In this concluding section the basic features used for the detection of neutrons will be discussed briefly. Also some data obtained with Cherenkov counters are presented. We wish to point out, however, that in order to obtain the latter data particle beams produced by high-energy accelerators are required; such beams are usually not available in a teaching laboratory.

6.1 NEUTRON DETECTORS

It has been noted how charged particles may be detected through the ionization they produce in matter; similarly, efficient techniques for detecting gamma rays have been discussed. Short-lived neutral particles such as the π^0 meson, K^0 meson, and the Λ^0 hyperon can usually be detected and identified through their decay into electromagnetic quanta or charged particles. Thus special techniques are required only for the detection of neutrons (lifetime 17 sec) and the stable neutrinos.

The detection of neutrinos is quite difficult and was not achieved until 1956, but neutrons can be readily detected by observation of the secondary particles that emerge from the interaction of neutrons with nuclei.

Several such interactions are used for neutron detection, the efficiency of each depending on the neutron energy. They are

(a) Nuclear reactions following neutron capture, such as

$$n + X_Z{}^A \rightarrow X_{Z-2}{}^{A-3} + \alpha \qquad (n, \alpha) \text{ reaction}$$

$$n + X_Z{}^A \rightarrow X_{Z-1}{}^A + p \qquad (n, p) \text{ reaction}$$

$$n + X_Z{}^A \rightarrow X + X' \qquad (n, \text{ fission})$$

where the alpha particle, proton, or fission fragment may be detected. This type of process is of great use with slow neutrons, since for such neutrons the capture cross section is very large.

(b) Elastic scattering of neutrons from a proton, at such momentum transfer that enough energy is imparted to the recoil proton so that it can be detected. This technique is applicable for higher-energy neutrons.

(c) Nuclear transmutations, in which the neutrons produce a long-lived isotope or nuclear excited state in the detector; the decay of this radioactive product with the emission of gamma rays or beta rays is then a measure of the total flux of neutrons that was incident on the detector.

Turning our attention to the nuclear reactions, the process

$$n_0^1 + B_5^{10} \rightarrow Li_3^7 + \alpha \tag{6.1}$$

is very frequently used. The reaction is exothermic with a Q value of 2.79 MeV; so that the alpha particle is released with an energy of 1.77 MeV (following capture at rest). The cross section for the reaction of Eq. 6.1 is 4×10^{-20} cm^2 for neutrons of 10^{-3} eV. Any gaseous detector may be used for observing the alpha particles; because of their limited range, however, the alphas must be produced inside the sensitive volume of the detector. This is achieved by using boron trifluoride as the filling gas and operating the counter in the Geiger region; when the neutron flux is high, it is possible to operate in the ionization chamber region as well. Another technique consists in lining the inside wall of the detector with boron so as to use a filling gas more appropriate than BF$_3$.

If it is desired to detect with a B^{10} counter, neutrons of a few electron volts energy, the neutrons must be slowed down (thermalized) before entering the detector; this is necessary because the cross section for the reaction of Eq. 6.1 drops by a factor of 500 when the neutron energy reaches 100 eV. Neutrons are slowed down by passing through a moderator, which is usually a material rich in hydrogen, such as water or paraffin.

In the category of fission detectors for thermal neutrons, typically reactions such as the following are used:

$$n_0^1 + U_{92}^{235} \rightarrow \text{fission fragments, for example, } \rightarrow Sr_{38}^{90} + Xe_{54}^{143} + 3n$$

For higher-energy neutrons, many more materials, such as U^{238} and Th^{232}, become fissionable at various thresholds.

The elastic scattering of neutrons from protons, as mentioned before, is used for the detection of neutrons with energies of at least a few hundred keV. Since organic scintillating materials contain a substantial amount of hydrogen, it is possible to detect neutrons through the proton recoil and obtain pulse-height information as well.

A general consideration with neutron detectors is that they should be insensitive to gamma radiation; this is especially important since emission of neutrons is usually accompanied by gamma-ray emission. Such discrimination may be achieved by pulse-height selection, in gaseous neutron detectors, but becomes much more difficult when scintillation counters are used for the detection of neutrons.

Finally, we mention the "foil activation techniques" in which a thin indium foil is exposed to the neutron flux, and after a certain irradiation period the foil is removed from the beam and its activity is measured with a Geiger counter. The reaction that takes place is

$$n_0^1 + In_{49}^{115} \rightarrow In_{49}^{116} \tag{6.2}$$

the In^{116} subsequently decaying with a 54.1-min lifetime by the emission of an electron. The reaction of Eq. 6.2 has a very high cross section for thermal neutrons, and the 54.1-min activity may be conveniently measured after other prompt radiations have decayed.

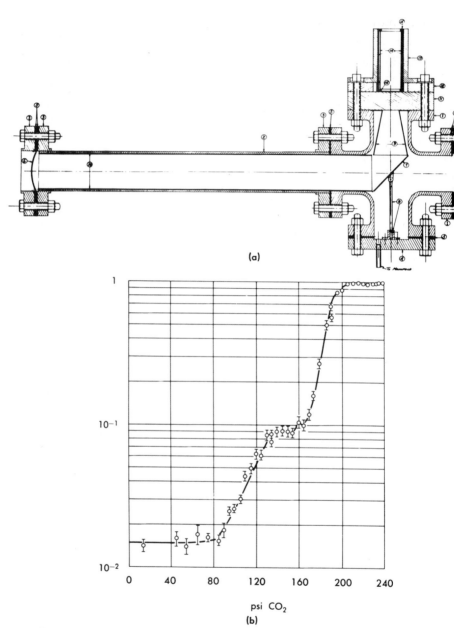

FIG. 5.43   A threshold Cherenkov counter for detection of high-energy particles; the radiating material is CO_2 or SF_6 gas. (a) Mechanical arrangement of a 4-ft-long counter; the particles travel along its axis from left to right; (7) plane mirror, (11) Lucite window, (14) photomultiplier, (9) and (10) reflecting liners, (6) thin stainless steel entrance and exit windows. (b) Results obtained showing number of counts against pressure. Note three distinct plateaus indicating electrons, μ mesons, and π mesons.

6.2 CHERENKOV COUNTERS

In Section 2.6 the basic facts of Cherenkov radiation were given; their application in a practical detector will now be described.

First we consider a "threshold" counter, as shown in Fig. 5.43a; it consists of a cylinder in which the radiating material (in this case SF_6 up to pressures of 400 psi) is contained; a 45° mirror reflects the Cherenkov light into a photomultiplier tube through an appropriate pressure window. This counter was placed in a beam of negative particles emerging at 32° from the Cosmotron.† By means of a bending magnet and a collimator, the momentum of the beam was selected to be 920 MeV/c; the beam contained electrons, μ mesons, and π mesons. Table 5.4 gives the mass of these particles, their velocity (for the specific momentum of 920 MeV/c), and the (threshold) value of the refractive index a radiator must have before any Cherenkov light is emitted; in the fifth column is tabulated the pressure at which SF_6 reaches this index.

TABLE 5.4

MASS ANALYSIS OF 920-MeV/c, 32°-NEGATIVE BEAM FROM THE COSMOTRON

Particle	Mass (MeV/c^2)	β (velocity/c)	n (threshold)	psi of SF_6	Percentage in beam
e	0.511	1.0000†	—	—	1.5
μ^-	105.7	0.9935	1.0065	103	7.5
π^-	139.6	0.9885	1.0115	176	91.0

† To within 10^{-6}.

The data obtained are shown in Fig. 5.43b, where the ordinate gives the fraction of the total flux (traversing the detector) that was registered by the Cherenkov counter, against gas pressure. We note that even at a pressure of a few psi a 1.5-percent count is obtained, and we attribute this to the electrons in the beam. At 80 psi the count begins to rise and reaches a plateau of 9 percent at 140 psi, which we attribute to μ mesons. Finally at 170 psi we notice a new rise in the count, which levels off at 200 psi with an efficiency close to 100 percent; it is due to the π mesons. From the level of the three plateaus we obtain the relative abundance of each type of particle in the beam; these are given in the last column of Table 5.4.

Next we consider a "differential" Cherenkov counter, in which use is made of the specific angle at which the Cherenkov light is emitted; a schematic diagram is shown in Fig. 5.44a. The spherical mirror focuses the Cherenkov light emitted at an angle θ into a ring of radius $R = f \tan \theta$,

† The 3-BeV proton synchrotron at Brookhaven National Laboratory.

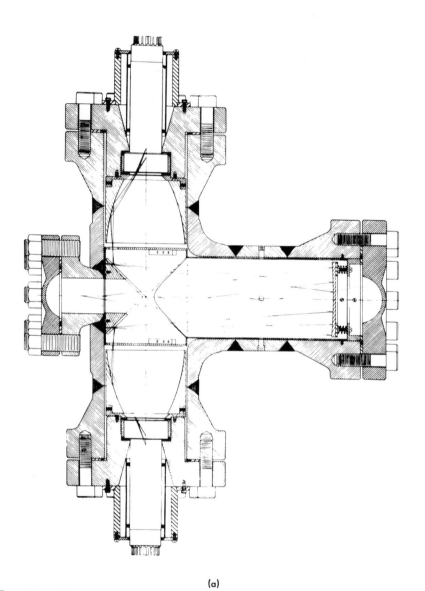

(a)

FIG. 5.44   A differential Cherenkov counter for high-energy particles; the radiating material is CO_2 gas. (a) (*Above*) Mechanical arrangement of the 18 in. long, 5 in.diameter, high-pressure counter. The particles enter from the left; the path of the optical rays is indicated. (b) (*Opposite page*). Results obtained showing number of counts against pressure. Note the two distinct peaks due to π mesons and K mesons.

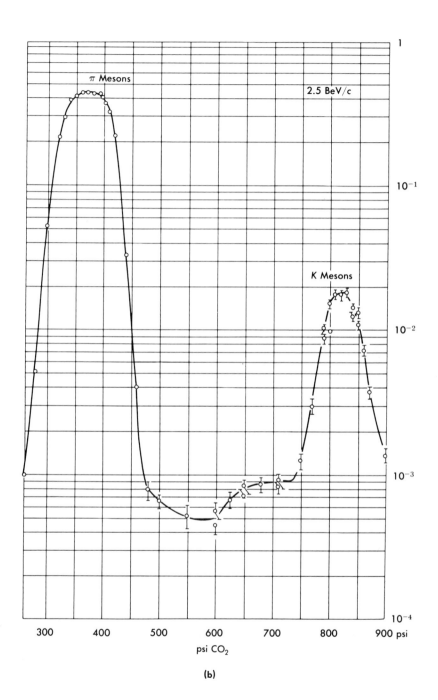

(b)

where f is the focal length; if then a circular slit of mean radius R is placed in the focal plane of the mirror, only Cherenkov light emitted at the angle θ (within some $\Delta\theta$) may reach the photomultiplier. Thus for a given value of the refractive index of the radiator, and since θ is fixed, only particles with β satisfying

$$\cos\theta = \frac{1}{n\beta}$$

give a count in the detector. By varying the refractive index (that is, the pressure in the case of a gas radiator) one may cover the range of β close to unity.

Results so obtained are shown in Fig. 5.44b, where the counter was exposed to a 2.5-BeV/c beam of positive particles from the A.G.S.†, at 17° from the target. The beam consists of some μ mesons, π mesons, K mesons, and protons. The slit position and focal length were chosen so as to select $\theta = 9°$ for the angle of acceptable Cherenkov light; the radiator material was CO_2 gas heated above the critical temperature. We note a peak due to the π mesons, extending (with a flat top) from 350 to 400 psi, and the K-meson peak from 810 to 830 psi. The proton peak is not shown in Fig. 5.44b but it corresponds to a pressure of 1230 psi; the μ-meson contamination is too small and too close to the π-meson peak to be resolved. These results are summarized in Table 5.5.

TABLE 5.5
2.5 BeV/c POSITIVE BEAM FROM AGS

Particle	Mass (MeV/c^2)	β (velocity/c)	n (for $\theta = 9°$)	psi of CO_2 at 40° C	Percentage in beam
π^+	139.6	0.9984	1.0141	400	43.5
K^+	494.0	0.9803	1.032	860	1.7
p	938.2	0.9307	1.082	approx 1200	approx 55.0

The observed resolution is in good agreement with the calculated one, and the background level of the detector appears to be of the order of 5×10^{-4}.

REFERENCES

By necessity the discussion presented in this chapter is not complete. Below is a selective list of references (including those already mentioned

† The 33-BeV alternating gradient synchrotron at Brookhaven National Laboratory.

in the footnotes to the chapter) which the reader may consult for additional information.

On interaction of radiation and particles with matter:

1. E. Fermi, *Nuclear Physics*. Chicago: University of Chicago Press, 1950.
2. J. D. Jackson, *Classical Electrodynamics*. New York: Wiley, 1962.
3. W. Heitler, *The Quantum Theory of Radiation*. London and New York: Oxford Univ. Press, 1954.

On gaseous and scintillation detectors; neutron detectors:

4. W. J. Price, *Nuclear Radiation Detectors*. New York: McGraw-Hill, 1958.
5. B. Rossi and H. Staub, *Ionization Chambers and Counters*. New York: McGraw-Hill, 1949.
6. J. Sharpe, *Nuclear Radiation Detectors*. Methuen and Company, 1955.
7. *Encyclopedia of Physics*, Vol. 45, *Nuclear Instrumentation II*. Berlin: Springer, 1958.
8. J. B. Birks, *The Theory and Practice of Scintillation Counting*. New York: Pergamon, 1964.

On solid state detectors:

9. J. M. Taylor, *Semiconductor Particle Detectors*. London and Washington, D. C.: Butterworth, 1963.

On Cherenkov detectors:

10. J. V. Jelley, *Cherenkov Radiation*. New York: Pergamon, 1958.

6

SCATTERING EXPERIMENTS

1. Introduction

Ever since Rutherford performed his original experiments on the scattering of energetic alpha particles from atomic nuclei, scattering has become increasingly more powerful as a tool for investigating the forces between elementary particles. By now it is familiar to the reader that an electron, under the influence of the attractive electromagnetic force of the nucleus, may be found in a bound state. The classical analogue of this situation is the motion of the planets around the sun under the influence of the gravitational force; they describe elliptical orbits.

According to Newton's laws, in the absence of a force, electrons (atoms, planets) will travel in a straight line,† since $(d\mathbf{p}/dt) = 0$. If, however, the trajectory of the electron passes by a nucleus, under the influence of the electromagnetic force, two things may happen: either the electron will fall into a bound state, or, if its total energy is larger than the ionization potential, its trajectory will be altered but it will not become bound to the

† A remarkable experimental confirmation of this statement is obtained in atomic-beam experiments.

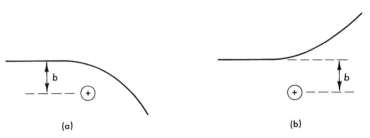

FIG. 6.1 Scattering of a particle due to the presence of a force center. (a) Attractive force. (b) Repulsive force.

nucleus. The latter is shown in Fig. 6.1 and we speak of "scattering" of the electron by the nucleus. The classical analogue of this situation is the motion of "comets" under the influence of the gravitational force; they describe hyperbolic orbits.

Scattering is electromagnetic in the electron-nucleus system, and gravitational in the comet-sun system. In general by investigating scattering processes we may obtain information on the nature and strength of any force that may exist between projectile and target. In particular, we already know the form of both the gravitational and electromagnetic force

$$ F_e = k\,\frac{e_1 e_2}{r^2} \qquad F_g = G\,\frac{m_1 m_2}{r^2} \tag{1.1} $$

Here k, and G are constants, $e_1 e_2$ and $m_1 m_2$ are the electric charges or gravitational masses, and r the distance between projectile and target. From Eq. 1.1, we see that the shorter the distance, the stronger the force. Strong forces lead to *large-angle scattering*, which therefore must be a consequence of a close approach of the projectile to the target. The distance of closest approach is related to the *impact parameter b*, which is the distance from the target normal to the extension of the undeflected trajectory of the projectile as shown in Fig. 6.1.

Equation 1.1 is correct under the condition that the charges (masses) are concentrated at one point, or at least that the smallest distance r is always much larger than the extent over which the charge is distributed; if this is not true, the force will be given by an appropriate extension of Eq. 1.1

$$ F = k e_1 \int \frac{\rho(r')}{|\,\mathbf{r} - \mathbf{r}'\,|^2}\, d\mathbf{r}' \tag{1.2} $$

where $\rho(r')$ is now the charge density of the target and we have assumed that the projectile is a point charge. Thus, if the force law is known, scatterings with small impact parameters may provide information on the structure of the target. As a matter of fact, the process of human vision consists of the detection by the eye of quanta of electromagnetic radiation, scattered by the objects that are seen. In this process, however, the distance of closest approach is very much smaller than the dimensions of the object and, consequently, the resolution is good. Radar detection is similar, the only difference being that the wavelength of the radiation is longer and consequently the resolution is poorer.

Since the scattered particle moves in the field of force of the target, work is done, with consequent changes in potential and kinetic energy. However, when the projectile and target are considered together as a system (isolated), there can be no change in its total energy: what takes place is a transfer of energy and momentum from the projectile to target; if the projectile is scattered by an angle θ, the momentum transfer is

$$q = |\mathbf{p}_i - \mathbf{p}_f| = \sqrt{p_i^2 + p_f^2 - 2p_i p_f \cos \theta} \qquad (1.3)$$

and if the energy of the projectile is not much altered in the scattering process, $|p_i| = |p_f|$

$$q = 2p \sin \frac{\theta}{2} \qquad (1.4)$$

The momentum transfer is the basic physical quantity that characterizes all scattering experiments; it depends on the momentum (rigidity) of the projectile and the angle of scattering. The probability for scattering is, to first order, a function only of the momentum transfer (and not of the angle), large momentum transfer scatters being, in general, less probable.

In the process of scattering, if the target is very massive, as compared to the momentum transferred, it is clear that the energy transferred to it, ΔE, will be small, since

$$\Delta E = T = \sqrt{m^2 + q^2} - m \simeq \frac{1}{2} \frac{q^2}{m} \qquad (1.5)$$

In Eq. 1.5 we assumed that all the transferred energy appeared as kinetic energy of the target; in this case we speak of *elastic* scattering. However, it is possible that in the scattering process the target may be excited to a state of different energy (we may write this as $m_f \neq m_i$), or that part of the transferred energy is converted to mass in the form of new particles, or that the target dissociates itself with the absorption or liberation of energy; in these cases we speak of *inelastic* scattering.

Besides gravitational and electromagnetic forces, today we know of two other types of force fields: the *nuclear force* (or *strong interaction force*) and the *weak interaction force*. The former is responsible for holding nuclei

TABLE 6.1

THE FOUR FORCE FIELDS AND THEIR RELATIVE STRENGTH

Force field	Strength	Main manifestation
Nuclear force	1.0	Binding of nuclei
Electromagnetic forces	1/137	Binding of atoms, molecules; also macroscopic effects
Weak interaction	10^{-14}	Decay of elementary particles
Gravitational force	10^{-42}	Everyday experience

together, and the latter is responsible for the radioactive decay of nuclei with the emission of an electron (or positron). In Table 6.1 are given the relative strengths of the four force fields presently known to man.

In scattering between elementary particles, the probability for scattering (cross section) is related to the strength of the interaction as given in Table 6.1. Therefore, strong-interaction scattering (as of a neutron from a proton) is easier to observe than electromagnetic scattering (as of an electron from a proton) at the same momentum transfer. Weak interaction scattering has a much smaller cross section and was only recently (1962) observed in the inelastic scattering of neutrinos, $\nu_\mu + n \rightarrow \mu^- + p$. Gravitational scattering, while absurdly small at the elementary particle level, is, however, observed macroscopically.

The three scattering experiments described in this chapter are all due to the electromagnetic interaction. In the first experiment, alpha particles (nuclei of helium) of sufficient energy (5.2 MeV) are scattered from the nucleus of gold by virtue of the interaction of the alpha-particle charge, $Z = +2$, with the charge of the gold nucleus $Z = 79$. It is true that a nuclear-force interaction also exists between these two particles. Even though the nuclear force is stronger, its range is short, so that unless the alpha particles have sufficient energy to overcome the repulsive Coulomb potential, they cannot approach close enough to the nucleus to be affected by the nuclear force. If we take for the range of the nuclear force 10^{-13} cm, the Coulomb potential will be

$$\Phi \simeq \frac{e^2}{4\pi\epsilon_0} \frac{Z(\text{He})Z(\text{Au})}{10^{-15}} \text{ (MKS units)} = 230 \text{ MeV}\dagger$$

which a 5.2-MeV alpha particle can clearly not overcome (see Fig. 6.2).

$\dagger$ $230 \text{ MeV} = \dfrac{(1.6 \times 10^{-19})^2}{1.6 \times 10^{-19}} \times 9 \times 10^9 \times \dfrac{2 \times 79}{10^{-15}} \text{ eV}$

Such alpha-particle scattering experiments were first performed by E. Rutherford in 1910 and are named after him.

In the second experiment, photons of an energy of 662 keV are scattered from atomic electrons. At these energies, the momentum transfer is large enough so that the electrons may be considered free and, consequently, the energy transfer also becomes considerable; therefore, the energy of the scattered photon is decreased as a function of the angle of scattering. As a result, a continuous shift is observed in the frequency of the scattered radiation, a result inexplainable on the basis of a classical theory of radiation. The first experiments of this type were performed with x-rays by A. H. Compton in 1923 and are now named after him. The simple assumption that the scattering of the flux of electromagnetic radiation can be described by the scattering of individual quanta carrying energy and momentum leads to the correct explanation of the effect. It is the electromagnetic interaction, but now taking place between the photons and charged particles, that is responsible for the scattering.

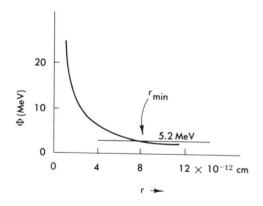

FIG. 6.2 The coulomb (electrostatic) potential of a gold nucleus as a function of the distance from it; r_{min} is the closest distance that a 5.2 MeV alpha particle can reach.

The third experiment described concerns again the electromagnetic interaction of photons with an electric charge, now the charge of the nucleus. However, the scattering is completely inelastic, resulting in the absorption of the photon by the nucleus; to satisfy energy and momentum conservation, and in order for this absorption process to take place, the photons must have exactly the correct energy: we speak of resonant absorption of gamma rays. This effect was first observed by R. Mössbauer (1960) and is now named after him; since the very narrow band of gamma-ray energies over which the absorption takes place is of the order of 1 part in 10^{13}, the Mössbauer effect is the most precise measurement of frequency (or for that matter, the most precise measurement of any kind) that man has ever performed.

2. Rutherford Scattering

2.1 DERIVATION OF THE SCATTERING CROSS SECTION

We shall now derive the *differential cross section* $d\sigma/d\Omega(\theta, \varphi)$, for an alpha particle scattered by the Coulomb field of the nucleus into the element of solid angle $d\Omega$ centered about the angles θ and φ. To do this we will first find an expression for the angle of the scattered particle as a function of the impact parameter b.

Consider therefore Fig. 6.3a, where the initial position and velocity of the alpha particle is $\mathbf{v}_i$, and the impact parameter b. Since the force is always directed along the line joining the projectile and target, the motion will be in the plane defined by $\mathbf{v}_i$ and the position of the scattering center. We will first obtain the equation for the orbit†. The equations of motion can be had either through Lagrangian formalism, or from Newton's equations (CGS units are used)

$$m\ddot{r} - mr\dot{\theta}^2 = \frac{ee'}{r^2} \tag{2.1}$$

$$\frac{d}{dt}(mr^2\dot{\theta}) = 0 \tag{2.2}$$

Here the notation of Fig. 6.3a has been used and e and e' are the charges of target and projectile, respectively.

We can immediately obtain a "first integral" of Eq. 2.2, namely,

$$mr^2\dot{\theta} = \text{constant} = l \tag{2.3}$$

where l, the angular momentum about the scattering center, is conserved for all central force fields. Using Eq. 2.3 in Eq. 2.1 we obtain

$$m\ddot{r} - \frac{l^2}{mr^3} - \frac{ee'}{r^2} = 0$$

or

$$m\ddot{r} = -\frac{\partial}{\partial r}\left[\frac{l^2}{2mr^2} + \frac{ee'}{r}\right] \tag{2.4}$$

and multiplying by $\dot{r}$ both sides of Eq. 2.4

$$m\ddot{r}\dot{r} = \frac{m}{2}\frac{d}{dt}(\dot{r}^2) = -\frac{dr}{dt}\frac{dg}{dr} = -\frac{dg}{dt} \tag{2.5}$$

---

† For a detailed treatment of the two-body central force problem, see H. Goldstein, *Classical Mechanics*, Addison-Wesley, Chapter III.

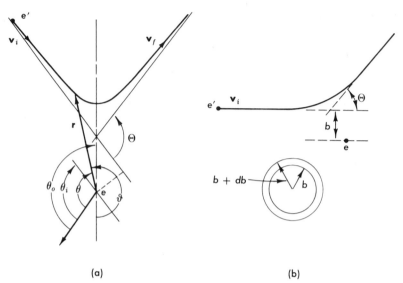

(a)   (b)

FIG. 6.3   The scattering of a particle of charge e' from a particle of charge e; v_i and v_f are the initial and final velocities. The scattering angle is Θ. (a) Definition of the angles appearing in the derivation of the cross section. (b) The range of impact parameters b in db that contribute to scattering at Θ into $d\Theta$.

with

$$g = \frac{l^2}{2mr^2} + \frac{ee'}{r}$$

Integration of Eq. 2.5 yields

$$\frac{1}{2} m\dot{r}^2 + \frac{1}{2} \frac{l^2}{mr^2} + \frac{ee'}{r} = \text{constant} = E \tag{2.6}$$

The second "first integral" is the total energy of the alpha particle, which is conserved since the scattering is assumed to be elastic (we also assumed that the scattering center at the origin does not move). Note that Eq. 2.6 reduces correctly to $E = \frac{1}{2}mv^2$ for $r \to \infty$. Solving Eq. 2.3 and Eq. 2.6, we have

$$\frac{d\theta}{dt} = \frac{l}{mr^2} \tag{2.6a}$$

and

$$\frac{dr}{dt} = \sqrt{\frac{2}{m}\left[E - \frac{l^2}{2mr^2} - \frac{ee'}{r}\right]} \tag{2.6b}$$

Equations 2.6a and 2.6b may be combined to give the differential equation of the orbit

$$\frac{d\theta}{dr} = \frac{l}{m} \frac{1/r^2}{\sqrt{(2/m)[E - (l^2/2mr^2) - (ee'/r)]}} \tag{2.7}$$

In order to integrate Eq. 2.7, we make the change of variable

$$u = \frac{1}{r} \qquad du = \frac{-dr}{r^2}$$

so that

$$d\theta = \frac{-du}{\sqrt{-u^2 - (2ee'm/l^2)u + (2Em/l^2)}} \tag{2.8}$$

hence

$$\theta = \text{arc cos}\left[\frac{(l^2/ee'm)u + 1}{\sqrt{(2El^2/(ee')^2m) + 1}}\right] \tag{2.9}$$

and finally the equation of the orbit is

$$u = \frac{1}{r} = \frac{ee'm}{l^2}[\epsilon \cos(\theta - \theta_0) - 1] \tag{2.10}$$

where

$$\epsilon = \sqrt{\frac{2El^2}{(ee')^2m} + 1} > 1, \qquad \text{since } E > 0 \tag{2.11}$$

θ_0 is a constant of integration, such that at θ_i, $\cos(\theta - \theta_0) = 1/\epsilon$, as shown in Fig. 6.3a. If instead we choose to express the angles in terms of ϑ, we have

$$\cos(\theta - \theta_0) = \cos(\pi - \vartheta) = -\cos\vartheta$$

and

$$\frac{1}{r} = -\frac{ee'm}{l^2}(1 + \epsilon \cos\vartheta) \tag{2.12}$$

which is the equation of a hyperbola, with the focus at the origin, eccentricity ϵ, and center at $\vartheta_0 = \pi$, $r_0 = l^2/[ee'm(\epsilon - 1)]$. The limits on the angle ϑ are from

$$\cos\vartheta_1 = -\frac{1}{\epsilon} \qquad \frac{3\pi}{2} < \vartheta < \pi$$

$$\tag{2.13}$$

to

$$\cos\vartheta_2 = -\frac{1}{\epsilon} \qquad \pi < \vartheta < \frac{\pi}{2}$$

The angle Θ through which the alpha particle was scattered is given by

$$\Theta = \pi - (\vartheta_1 - \vartheta_2) = 2\left(\frac{\pi}{2} - \cos^{-1}\frac{1}{\epsilon}\right) = 2\sin^{-1}\frac{1}{\epsilon}$$

$$\sin\frac{\Theta}{2} = \frac{ee'}{\sqrt{(2E/m)l^2 + (ee')^2}} \tag{2.14}$$

It is convenient to use in Eq. 2.14 the impact parameter b; we have

$$l^2 = (mv_i b)^2$$

and since

$$E = \tfrac{1}{2}mv_i^2$$

$$\sin\frac{\Theta}{2} = \frac{1}{\sqrt{(2Eb/ee')^2 + 1}} \tag{2.15}$$

$$\cot\frac{\Theta}{2} = b\,\frac{2E}{ee'}$$

Next we wish to obtain the probability for scattering through an angle Θ, at the azimuth φ; from Eq. 2.15 we see that there is no dependence on φ and the angle of scattering Θ is uniquely determined by the impact parameter b. To obtain the differential cross section for scattering into an angle (as defined in Chapter 5, Section 2) we note that the probability of the incident particle having impact parameter b in db is proportional to the element of area that can contribute to b. From Fig. 6.3b this is evidently

$$P(b)\,db = 2\pi b\,db$$

And the probability that a particle with b in db is scattered by the angle Θ in $d\Theta$ is

$$P(\Theta)\,d\Theta = P(b)\left(\frac{db}{d\Theta}\right)d\Theta = (2\pi b)\left(\frac{db}{d\Theta}\right)d\Theta$$

Substituting expressions for b and $db/d\Theta$ obtained from Eq. 2.15,

$$db = -\frac{ee'}{4E}\frac{1}{\sin^2\Theta/2}\,d\Theta$$

we have

$$dP(\Theta) = \frac{1}{8}\left(\frac{ee'}{E}\right)^2\frac{2\pi\cos\Theta/2}{\sin^3\Theta/2}\,d\Theta$$

$$dP(\Theta) = \left(\frac{ee'}{4E}\right)^2\frac{1}{\sin^4\Theta/2}\,2\pi\sin\Theta\,d\Theta \tag{2.16}$$

Since the scattering into φ is isotropic

$$\int_0^{2\pi} d\varphi = 2\pi$$

and we may write

$$d\sigma(\Theta, \varphi) = \frac{P(\Theta)\, d\Theta\, d\phi}{2\pi} = \left(\frac{ee'}{4E}\right)^2 \frac{1}{\sin^4 \Theta/2} \sin \Theta\, d\Theta\, d\varphi$$

And since the element of solid angle is $d\Omega = \sin \Theta\, d\Theta\, d\varphi$,

$$\frac{d\sigma}{d\Omega} = \left(\frac{ee'}{2}\right)^2 \frac{1}{4E^2} \frac{1}{\sin^4 \Theta/2} \tag{2.17}$$

which is the famous Rutherford scattering cross section.

Since our derivation was nonrelativistic, we may write $E = p^2/2m$, in which case

$$\frac{d\sigma}{d\Omega} = (2mZZ'e^2)^2 \frac{1}{q^4} \tag{2.18}$$

where we used Ze and $Z'e$ for the charge of scatterer and projectile and $q = 2p \sin \Theta/2$ for the momentum transfer as defined by Eq. 1.4. Equation 2.18 shows clearly that the scattering cross section is a function of q *only* and in this case a very simple one.

A typical feature of Eq. 2.17 is that it increases very rapidly for small angles and even becomes infinite at $\Theta \to 0$; the relevant point is that even the total cross section

$$2\pi\left(\frac{ZZ'e^2}{4E}\right)^2 \int_{\Theta_0}^{\pi} \frac{\sin \Theta}{\sin^4 \Theta/2}\, d\Theta = (ZZ'e^2)^2 \frac{\pi}{2}\left[\frac{1}{\sin^2 \Theta_0/2} - 1\right]_{\text{as }\Theta_0 \to 0} \to \infty \tag{2.19}$$

goes to infinity if scattering through small angles is included. This fact is a reflection of the long-range nature of the electromagnetic force†; small-angle scattering corresponds to very large impact parameters. It is this property of electromagnetic scattering that gave rise to "multiple scattering" as discussed in the Chapter 5. It is easy, however, to see that a cutoff in the smallest possible scattering angle Θ_0 (Eq. 2.19) must be imposed from physical considerations. That is, the largest permissible impact parameter b_0 is determined by the transverse size of the beam, or of the scatterer (whichever one is the largest). But even long before that limit is reached, the impact parameter cannot be larger than the size of the atom, since outside it the charge of the nucleus is "screened" by the atomic electrons.

---

† As compared to the nuclear force.

It is interesting that even though Eq. 2.17 was derived on classical arguments, it is also correct quantum mechanically. The same result is obtained by an exact solution of the Schrödinger equation for a Coulomb potential[†], as well as by a first-order perturbation theory (Born approximation) treatment. We remark also that the same scattering cross section is obtained when the potential is attractive, the two cases being experimentally indistinguishable (as shown in Fig. 6.1). We know, however, that for the electromagnetic force the potential is "attractive" for particles of opposite charge and "repulsive" for particles of the same charge.

2.2 SCATTERING OF ALPHA PARTICLES BY THE NUCLEUS OF GOLD

We will now describe in detail a measurement of the scattering of polonium 210 alpha particles from a very thin foil of gold. The apparatus used is, in essence, similar to that of Rutherford's except for the detection technique. As in any scattering experiment we need:

(1) *The beam of particles to be scattered.* The alpha particles (He^4 nuclei) from Po^{210} decay are ideally suited for this purpose, since (a) they have sufficient energy (5.2 MeV) to traverse a thin target; (b) the beam is monoenergetic and does not contain electrons; (c) the high intensity of a radioactive source permits adequate collimation to yield a narrow beam; (d) they are readily available.[‡]

(2) *The target, or scattering material.* This needs to be sufficiently thick to produce enough scattering events for the available incident-beam intensity; but it should not be so thick as to change appreciably the energy of the primary beam or to affect the scattered alpha particles.[§] The target thickness used is of the order of a few mg/cm^2.

(3) *The detector.* In this apparatus a thin piece of scintillating material (organic) mounted onto a photomultiplier was used.[¶] Since the range of 5.2-MeV alphas in air is only approximately 4 cm, the beam, scatterer, and detector must all be enclosed in a vacuum. The detector can be moved to different angles with respect to the beam line, so that the angular distribution of the scattered alpha particles may be obtained.

The apparatus is shown in Fig. 6.4a; it consists of the cylindrical vessel A, containing the beam source, target holder, and detector; vessel A can

---

† See, for example, M. Born, *Atomic Physics*, Hafner Publishing Co., 1957, Appendix XX, p. 360.

‡ To accelerate an alpha particle to this energy one would need a potential difference of 2.6×10^6 eV (Van de Graaf generator) or would have to use a cyclotron.

§ It should not be thicker than a fraction of a mean free path.

¶ A solid-state counter (as described in Chapter 5, Section 5.2) may also be used to detect the alpha particles. It has the advantage of simplicity and better discrimination against background.

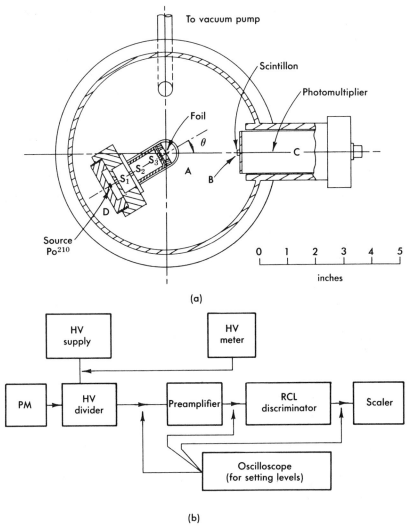

FIG. 6.4 Apparatus used for Rutherford scattering experiment. (a) The scattering chamber, source, target and detector. (b) Block diagram of the electronics.

be readily evacuated. The thin slab of scintillon† B, of width 0.565 cm, height 1.55 cm, and thickness 0.15 cm, is the detecting element, and is glued onto a 6192 DuMont photomultiplier tube C; the photomultiplier is permanently set into the scattering chamber with the vacuum seal made at its plastic base (one should ascertain that the tube does not get drawn

† The material used was Pilot B (Pilot Chemicals, Inc., 36 Pleasant St., Watertown, Mass.).

into the chamber when it is first pumped down). For the beam source, 0.5 millicuries of Po^{210} were plated on the inside of the round cap D, which can be unscrewed easily for removal. Three slits, S_1, S_2, and S_3, 1 mm by 1 cm, are used to collimate the flux of alpha particles emitted by the source. Beyond the third slit, the target holder can be positioned; it consists of a small 2 cm by 2 cm frame on which the scattering foil (gold, aluminum, etc.) is mounted.

Usually the detector is swung around the incident-beam direction. In the present setup, however, it is more convenient to have the detector fixed and to rotate the beam and scattering foil. This latter assembly is mounted as a whole on a shaft coaxial with the scattering center, and can be moved without breaking the vacuum; a pointer outside the chamber indicates the angle of rotation. The associated circuitry (Fig. 6.4b) is the standard one consisting of the high-voltage supply and divider for the photomultiplier, a preamplifier, and an RCL 20506 single-channel discriminator driving a scaler, as described in Chapter 5, Section 4.2.

The counts that register on the scaler do not all come from scattered alpha particles but contain "background" of two types:

(a) A counting rate R'', which is present even when the source is removed from the chamber. This is due mainly to contamination of the chamber with Po^{210}, and to noise in the detector or electronics. To measure R'' the source is removed and a count is taken at different angles; it is usually independent of angle.

(b) A counting rate R' due to the source, but not produced by scattering in the target material itself. The rate R' is mainly due to poor beam collimation, slit scattering, scattering off residual air molecules, and so on. To measure R', the source is placed in position, but the scattering foil is removed and again a count is taken at different angles. This time an angular dependent background may be expected.

Since R' contains R'',† the true rate is given by

$$R_{\text{true}}(\theta) = R(\theta) - R'(\theta) \qquad (2.20)$$

where R is the counting rate with both source and target in place. It is necessary to know R' and R'' separately in order to understand the causes of the background and thus reduce it as much as possible.

Let us next make some quantitative estimates on the expected counting rates. The defining beam slit is 1 mm by 1 cm at a distance of 5 cm, and hence subtends a solid angle

$$\Delta\Omega(\text{beam}) = 4 \times 10^{-3} \text{ sr}$$

---

† Unless the scattering foil is contaminated, which can be readily ascertained.

We thus obtain for the beam intensity

$$0.5 \times 10^{-3} \times 3.7 \times 10^{10} \times \frac{.004}{4\pi} = 6000 \text{ counts/sec}$$

The observed beam intensity, however, is 110,000 counts/min, the difference from our simple estimate being due in part to the extent of the source but mainly to self-absorption in the source.

Next we consider the detector solid angle. The size of the scintillon is 0.873 cm² at a distance of 6.66 cm, hence

$$\Delta\Omega \sim 0.02 \text{ sr}$$

If we use a gold foil of thickness 0.0001 in., and

$$Z' = 2; \qquad Z = 79; \qquad \text{and} \qquad E = 5.2 \text{ MeV}$$

we obtain †

$$\frac{d\sigma}{d\Omega} = \left[\frac{2 \times 79}{4\pi\epsilon_0} \frac{(1.6 \times 10^{-19})^2}{5.2 \times 10^6 \times 1.6 \times 10^{-19}} \frac{1}{4} \right]^2 \frac{1}{\sin^4 \theta/2} = \frac{1.20 \times 10^{-20}}{\sin^4 \theta/2} \text{ m}^2$$

Thus for scattering through 15°

$$\frac{d\sigma}{d\Omega} (\theta = 15°) = 4.17 \times 10^{-21} \text{ cm}^2$$

The number of alphas scattered into the detector is given by

$$I_s = I_0 N \frac{d\sigma}{d\Omega} d\Omega$$

where

$I_0 = 1.1 \times 10^5$ counts/min in the incident beam

$d\Omega = 2 \times 10^{-2}$ sr

$N = t \times \rho \times (N_0/A)$, the area density of scatterers, where

  $N_0 = 6 \times 10^{23}$, Avogadro's number

$\rho = 19.3$ gr/cm³, the density of the scatterer (gold)

$t = 0.00025$ cm, the thickness of the foil

$A = 197$, the atomic weight of gold

---

† Note that we calculate in the MKS system and that dimensionally

$$[(Z^2 e^2)/(4\pi\epsilon_0)]^2 = [F]^2 [L]^4 \qquad \text{while} \qquad E^2 = [F]^2 [L]^2$$

This yields

$$N = 1.48 \times 10^{19} \text{ gold nuclei/cm}^2$$

and

$$I_s(\theta = 15°) = 132 \text{ counts/min} \dagger$$

This seems to be a sizable rate; however, the pertinent question is how this rate compares to the background rate R'; that is, what is the signal-to-noise ratio (S/N). In the present experiment, the background (mainly due to the contamination of the vessel) was high, and of the order of 130 counts/min; thus already at 15°, $S/N = 1$.

To improve the S/N ratio, we could increase I_s by increasing the solid angle (which is impractical), or by increasing the beam intensity (which might raise the noise level as well) or most simply, by increasing the scattering-foil thickness. If we increase the foil thickness, however, we are limited by the energy loss of the beam particles in the target. If we wish, for example, to determine the cross section to 25 percent, then since

$$\frac{d\sigma}{d\Omega} \propto E^{-2}$$

and

$$\frac{\Delta(d\sigma/d\Omega)}{d\sigma/d\Omega} = -2 \frac{\Delta E}{E}$$

the energy loss must not exceed 12 percent. By referring to the Bragg curve (Fig. 5.42) we note that a 5.2-MeV alpha particle will lose 1.5 MeV of its energy after traversing the equivalent of 1.2 cm of air at stp, namely, approximately 2 mg/cm², which corresponds to a gold foil of thickness $t = 0.00012$ cm.‡

Multiple scattering in the foil is not significant for the alpha particles. We use Eq. 2.17, Chapter 5, and $L_{\text{rad}} \approx 6$ gr/cm² for gold; then with the above value $t = 0.00025$ cm, we obtain

$$\theta_{\text{rms}} = \frac{21.2}{\sqrt{2mE}} Z^2 \sqrt{\frac{10^{-3}}{6}} \text{ rad} \simeq 0.25°$$

---

† $I_s(\theta = 15°) = 1.1 \times 10^5 \times 1.5 \times 10^{19} \times 2 \times 10^{-2} \times 4 \times 10^{-21}$.

‡ From Chapter 5 we see that energy loss/(gm/cm²), $dE/d\xi = N_0(Z/A)z^2 f(I, \beta)$, yielding the equivalent thickness of gold $t = 0.00012$ cm. However, at these low velocities a more detailed treatment of the energy loss is required, and as also observed experimentally the alpha particle loses 1.5 MeV after traversing a gold foil of thickness $t = 0.00025$ cm (see also Fig. 5.4).

2.3 RESULTS AND DISCUSSION

We now will give results for Rutherford scattering obtained by students†
with the apparatus described in Section 2.2.

It is important to be extremely careful when handling the radioactive
source for this experiment; polonium, while very convenient for Rutherford
scattering, is a "nasty" isotope. As noted in Chapter 4, it can be lethal when
taken internally, and due to the recoil following alpha emission, small
parts of the source break off and contaminate the vessel in which it is
enclosed. Further, alpha-particle contamination cannot be detected with
a Geiger counter, but only with special alpha detectors, such as a gas-flow
counter‡. *Gloves must always be worn* when handling the source, and the
cap must be replaced whenever the source is removed from the apparatus.

First the chamber is evacuated and the detection system is adjusted
with the source in place, but without the scattering foil. The detector is
placed at 0° and the photomultiplier output is observed on an oscilloscope;
the high voltage is then raised until clean pulses of a few volts amplitude
are obtained. Next the discriminator is adjusted by taking a plateau curve
in the integral mode; it is also possible to operate the discriminator in the
differential mode, but in either case attention must be paid to the energy
loss of the alphas when the foil is inserted.

We are now in a position to measure the beam profile when the scatterer
is *not* in place; the results of counting rate against angle are shown on a
linear scale in Fig. 6.5a and on a logarithmic scale in Fig. 6.5b. This measurement serves three purposes:

(a) It determines the background rate R' and gives the extent of the
beam, namely, the detector angles beyond which the counts will be due to
scattered alphas. From Fig. 6.5b we see that for $\theta \gtrsim 6°$ there are no beam
counts; also the value of the background is 130 counts/min. (As noted
earlier, this rate was due almost entirely to contamination of the chamber,
as evidenced by a separate measurement with the source removed).

(b) It provides the information on the incoming beam intensity, and
for this purpose the linear plot of Fig. 6.5a is more useful. If the over-all
beam dimensions are smaller than the dimensions of the detector, then
the peak count simply gives the beam intensity and the profile of Fig. 6.5a
should have a flat top.

This is not always true, however. Let us first consider the distribution of
the beam in the θ direction (horizontal); this may be uniform, or Gaussian,
or of another type. Let the interval $\Delta\theta = x$ contain 90 percent of the beam

† R. Dockerty and S. McColl, class of 1962.
‡ PAC 3G Gas Proportional Counter; may be purchased from Eberlein Instruments
Corp., Sante Fé, N. M.

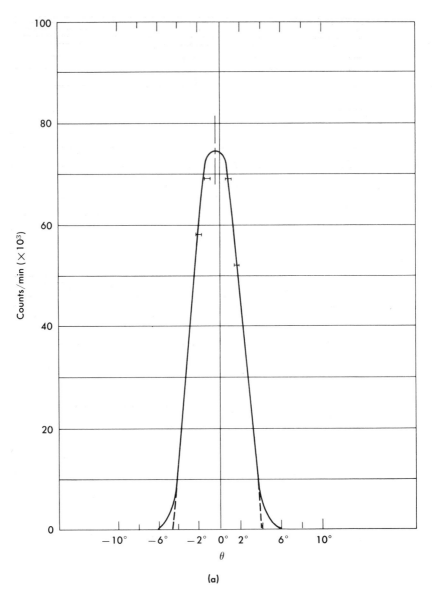

(a)

FIG. 6.5   The profile of the alpha-particle beam as measured in the scattering chamber with the scattering foil removed. (a) (*Above*) Linear plot that is used for obtaining
the total flux, and the beam center. (b) (*Opposite page*) Semilogarithmic plot giving the
background level outside the beam.

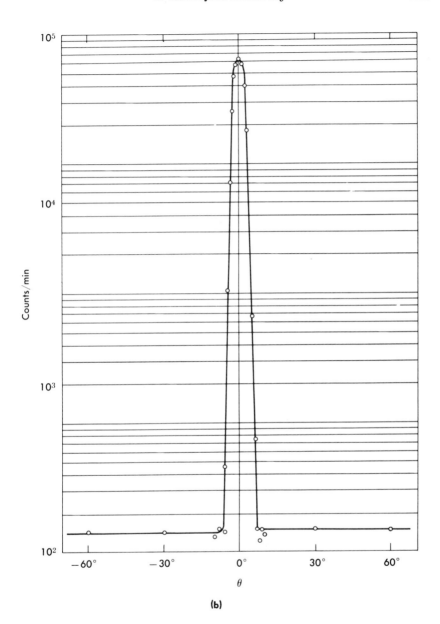

(b)

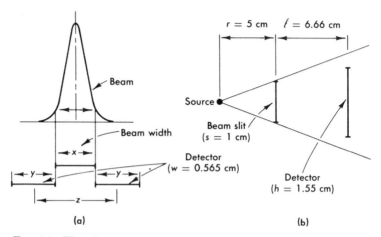

FIG. 6.6   The effects of the finite dimensions of the beam and of the detector. (a) In the horizontal plane. (b) In the vertical plane.

(see Fig. 6.6a). The angular width of the detector is

$$\Delta\theta = y = \frac{0.565}{6.66} \times \frac{180}{\pi} = 4°9$$

Further, from Fig. 6.5a we observe that the beam counts drop to 10 percent of their peak value in $\Delta\theta = z = 8°5$ (by extrapolating to zero beam count, we obtain $\Delta\theta = z' = 9°$). As seen in Fig. 6.6a, $z = x + y/2 + y/2$, so that we find $x \simeq 3°6$, which is smaller than the detector width; consequently we must expect in the profile of Fig. 6.5a a flat top of width $\Delta\theta = y - x \approx 1°3$.

Unfortunately in the vertical direction, the beam size is larger than the dimension of the detector, as is seen in Fig. 6.6b, where the dimensions of the beam-defining slit and the detector are shown. We note that only a fraction F of the incident beam reaches the detector, where

$$F = \left(\frac{h}{s}\right)\left(\frac{r}{r + l}\right) = 0.665$$

Thus the total incident beam is given by

$$I_0 = \frac{I_\theta}{F} = 110,000 \text{ counts/min}$$

where I_θ is the peak counting rate obtained from the beam profile, which we took as $I_\theta = 74,000$ counts/min. The above value of I_0 is subject to at least a ± 20 percent error in view of the approximations used and the nonuniformities in beam density and direction.

(c) Finally, the beam profile gives information on the true position of the beam axis. From Fig. 6.5a we find that the axis is located at $\theta_0 = -0°25$, and all the scattering angles must be corrected accordingly.

Now data may be taken. The chamber is opened, the scattering foil inserted, and the chamber again evacuated. The counting rate is measured as a function of angle first to the one side and then to the other side of the beam. These raw data are given in column 2 of Table 6.2. Column 3 gives the counts after background subtraction and column 4 the probable error; in column 5 are shown the corrected angles. At each angle enough data are accumulated so that the statistical accuracy is of the order of 3 percent.

In evaluating columns 3 and 4, the background rate R' was taken as 130 ± 10 counts/min (see Fig. 6.5b). The large error on the background is not due to a statistical uncertainty (which could be reduced) but to fluctuations in R' over the period that the experiment was in progress. As R becomes comparable to R', the error in the true rate $R_t = R - R'$ increases, reaching $\Delta R_t/R_t \approx 0.5$, which sets a limit to the largest useful scattering angle.

From the observed yields of scattered particles, we can obtain the differential cross section, from the expression

$$\frac{d\sigma}{d\Omega} = \frac{I_s}{(\Delta\Omega)I_0N}$$

where the symbols are defined as on page 239 and have the same values

$I_0 = 110{,}000$ counts/min

$N = 1.48 \times 10^{19}$ gold nuclei/cm^2 (for the given thickness of the foil $t = 0.00025$ cm)

$\Delta\Omega = 0.02$ sr (but see next paragraph)

$I_s = $ value given by column 3 of Table 6.2

The differential cross section so obtained is shown in column 6 of Table 6.2 and is also plotted against the scattering angle in Fig. 6.7a.

The process of dividing the yield by $\Delta\Omega$ to obtain the cross section needs some further discussion. Two points are of special importance:

(a) In evaluating $\Delta\Omega$ we use the approximation

$$\Delta\Omega = \frac{hw}{l^2} = \frac{1.555 \times 0.565}{(6.66)^2} = 0.0197 \text{ sr} \tag{2.21}$$

where w and h are the width and height of the rectangular detector and l the distance from the target (see also Fig. 6.6). The approximation is valid because the detector area is always normal to the scattered beam, and

TABLE 6.2
RUTHERFORD SCATTERING DATA

θ_{dial} (1)	R (counts/min) (2)	$R - R'$ (3)	$\Delta(R - R')$ (4)	$\theta_{correct}$ (5)	$d\sigma/d\Omega$ (observed) (6) $\times10^{-21}$ cm^2/sr	$(\sin^4 \theta/2)^{-1}$ (7) $\times10^3$	k (8) $\times10^{-24}$ cm^2/sr
−70°	144.3						
−60°	147.2						
−50°	139.8						
−40°	148.6						
−35°	145.2						
−30°	144.2						
−25°	166.6	37	±15	24°45'	1.13	0.460	2.46
−20°	228.6	99	±15	19°45'	3.04	1.16	2.62
−15°	443	313	±17	14°45'	9.60	3.72	2.59
−10°	1868	1738	±40	9°45'	53.4	19.2	2.78
−8°	5336	5206	±72	7°45'	160	47.9	3.34
0°	41662						
+6°	8576	8146	±90	6°15'	250	115	2.17
+8°	3663	3533	±60	8°15'	107	37.5	2.85
+10°	1681	1551	±40	10°15'	47.6	15.7	3.04
+15°	380	250	±17	15°15'	7.68	3.24	2.37
+20°	215.7	86	±15	20°15'	2.54	1.05	2.42
+25°	182.2	52	±15	25°15'	1.60	0.437	3.66
+30°	161.6	32	±15	30°15'	0.98	0.190	
+35°	163.2						
+40°	132.7						
+50°	142.1						

it becomes better as l increases, and the beam spot on the target decreases. More accurately, we must integrate the element of solid angle

$$d\Omega = \sin\theta\, d\theta\, d\varphi$$

over the area of the detector. Clearly, if we approximate and assume θ to be constant, $d\theta = \Delta\theta = w/l$, and $d\varphi = \Delta\varphi = h/(l\sin\theta)$, we obtain Eq. 2.21, which is independent of θ.

(b) In dividing the yield by $\Delta\Omega$ to obtain the differential cross section we must assume that $d\sigma/d\Omega$ does not change appreciably over the angular range subtended by the detector. This assumption is not very good, especially at the smaller scattering angles. Correctly, we should integrate $d\sigma/d\Omega$ over $d\Omega$ to obtain the yield

$$I = \int_{\varphi_1}^{\varphi_2} d\phi \int_{\theta_1}^{\theta_2} \frac{1}{\sin^4\theta/2} \sin\theta\, d\theta = \frac{4}{2}(\varphi_1 - \varphi_2)\left. \frac{1}{\sin^2\theta/2}\right]_{\theta_1}^{\theta_2}$$

but we may set $\varphi_1 - \varphi_2 \simeq h/(l\sin\bar\theta)$ as before, and

$$I = \frac{h}{l\sin\bar\theta}\left[\frac{\sin^2(\theta_1/2) - \sin^2(\theta_2/2)}{\sin^2(\theta_1/2)\sin^2(\theta_2/2)}\right]$$

which approximates but is not equal to the result obtained by using Eq. 2.21

$$I = \frac{hw}{l^2}\frac{1}{\sin^4\bar\theta/2}$$

In order to compare the results with the theoretical prediction of the Rutherford cross section (Eq. 2.17) we give in column 7 of Table 6.2 the factor $(1/\sin^4\theta/2)$ evaluated at the appropriate angle. The observed cross section should be proportional to this factor, and column 8 gives the ratio $k = $ (column 6)/(column 7), which should be equal to

$$k = \left(\frac{zZe^2}{4\pi\epsilon_0}\frac{1}{4E}\right)^2 \tag{2.22}$$

In Fig. 6.7b is a log-log plot[†] of yield against $1/(\sin^4\theta/2)$; horizontal error bars correspond to $\pm 0°.25$ uncertainty in the scattering angle,[‡] while the vertical ones correspond to the errors given in column 4 of Table 6.2. Angles to the right of the beam axis are indicated by a cross, to the left of the axis by a circle.

[†] On a linear scale, the plot should yield a straight line of slope k. However, on the log-log plot we cover a much larger range of values; the slope of the line must be 1 and the intercept gives k.

[‡] Remember, however, that the detector angular width is $\pm 2°.5$.

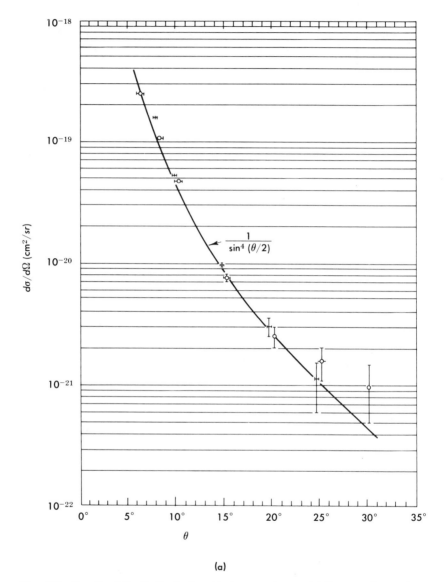

(a)

FIG. 6.7   Results of the Rutherford scattering experiment. (a) Cross section against angle; note that the measurement is extended over three decades. (b) Cross section against $1/(\sin^4 \theta/2)$; note the straight line fit and the unit slope of the line.

The straight line on Fig. 6.7b is the theoretical prediction that has slope $= 1.00$. We note that it provides a very good fit to the experimental points over more than two decades, and therefore these data confirm the

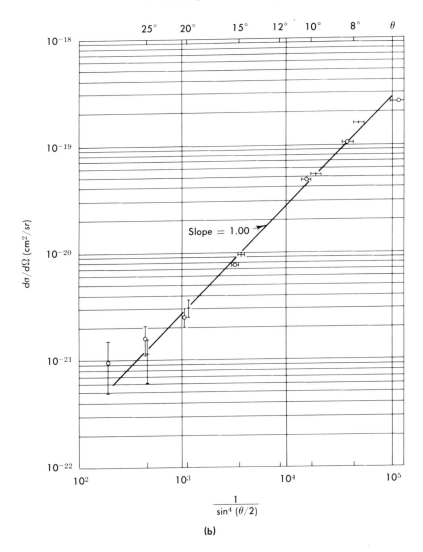

(b)

hypothesis of the nuclear atom and the angular dependence of the Rutherford scattering cross section as obtained in Eq. 2.17. The small deviations
from the fit are due to experimental difficulties which will be discussed
below.

While the straight line in Fig. 6.7b was constrained to have slope = 1.00,
the intercept (that is, the normalization) was obtained by a least-squares
fit to the data points. It yields a value

$$k = 2.70 \times 10^{-24} \text{ cm}^2$$

Evaluation of Eq. 2.22 with $E = 5.2$ MeV yields $k = 1.20 \times 10^{-24}$ cm^2. However, the alpha particle loses a considerable amount of energy when traversing the target; it is therefore more appropriate to average $1/E^2$ over this energy range, that is,

$$\left\langle \frac{1}{E^2} \right\rangle = \frac{\int_{x_1}^{x_2} E^{-2}\, dx}{\int_{x_1}^{x_2} dx}$$

We assume for simplicity that the energy loss is proportional to the thickness

$$E = E_0 - \left(\frac{dE}{dx}\right) x$$

hence

$$dE = -\left(\frac{dE}{dx}\right) dx$$

and we can write

$$\left\langle \frac{1}{E^2} \right\rangle = \frac{\int_{E_1}^{E_2} E^{-2}\, dE}{\int_{E_1}^{E_2} dE} = \frac{1/E_1 - 1/E_2}{E_2 - E_1} = \frac{1}{E_1 E_2}$$

We now use $E_1 = 5.2$ MeV, $E_2 = 3.7$ MeV, and obtain from Eq. 2.22

$$k = 1.67 \times 10^{-24} \text{ cm}^2 \text{ (theory)}$$

whereas

$$k = 2.70 \times 10^{-24} \text{ cm}^2 \text{ (experiment)}$$

The difference between the observed and theoretical constants, while at first sight large, can be traced to the limited sensitivity of the apparatus and mainly to

(a) Uncertainty in incoming flux
(b) Uncertainty in foil thickness

and to a lesser extent to

(c) Extended size of the beam and lack of parallelism
(d) Extended angular size of the detector
(e) Plural scattering in the foil (for the data at small angles)
(f) Background (for the data at large angles)

The reader should keep in mind that the main purpose of the experiment was to prove the $1/(\sin^4 \theta/2)$ dependence. Further, the observed value of k is of the correct order of magnitude and if we used it to find the charge of the gold nucleus, we would obtain

$$Z' = 99 \qquad \text{instead of} \qquad Z = 79$$

It is interesting to note that when the foil was inserted, the counting rate at $0°$ dropped from $I_0 = 74{,}000$ to $I_0' = 41{,}660$ counts/min (see Table 6.2). This reduction is a measure of the total cross section, or more precisely of

$$I_0 - I_0' = \sigma_t N I_0 = N I_0 \int_{\theta_0}^{\pi} \frac{d\sigma}{d\Omega}(\theta)\, d\Omega$$

where for θ_0 we use the angular limits of the detector.† Then we obtain for the probability of interaction (see Eq. 2.19)

$$\frac{I_0 - I_0'}{I_0} = Nk \frac{2\pi}{\sin^2 \theta_0/2}$$

With $\theta_0 = 2°.5$, $N = 1.48 \times 10^{19}$, and the observed value of

$$\frac{I_0 - I_0'}{I_0} = 0.44$$

we obtain

$$k = 2.26 \times 10^{-24}\,\text{cm}^2$$

which is of the correct order of magnitude. However, in view of the crude approximations made in evaluating the total cross section, the agreement with the previously discussed values of k is fortuitous.

The large value for $(I_0 - I_0')/I_0$ indicates that the probability for scatters $\geq 2°.5$ is considerable, and that therefore it is probable that an alpha particle may suffer in traversing the foil more than one (small angle) scattering from a nucleus.

We conclude this section with two further remarks:

(a) If we changed the scattering material, the cross section would also change, as $(Z/Z')^2$, while maintaining the same angular dependence. We can thus obtain information on the charge of the nucleus and confirm that it is equal to the atomic number Z of the material. Convenient target materials are silver, $Z = 47$, aluminum, $Z = 13$, and others.

---

† This discussion is really applicable to a beam of circular cross section and to a circular detector.

(b) For angles $\theta > \pi/2$ the cross section does not vary as rapidly, and at the limit $\theta \approx \pi$ it has the value $d\sigma/d\Omega = k = 1.2 \times 10^{-24}$ cm². Precise measurements at these large angles reveal deviations from the Rutherford equation and are due to the strong interaction (nuclear interaction, see Table 6.1) between the alpha particle and the gold nucleus.

As explained in detail in the introduction to this chapter, the nuclear interaction will manifest itself only at short distances—that is, at high-momentum transfers q (Eq. 1.4), where $q = 2p \sin \theta/2$. Clearly the maximum momentum transfer in this experiment is

$$q = 2p = 2\sqrt{2Em} = 200 \text{ MeV}/c$$

The recent experiments on Coulomb scattering of electrons from protons have been extended to $q \simeq 2.2$ BeV/c.

3. Compton Scattering

3.1 FREQUENCY SHIFT AND CROSS SECTION

This section deals with the scattering of electromagnetic radiation by free electrons. As mentioned in the introduction to this chapter, it is the scattering of electromagnetic radiation from various objects that makes it possible for us to "see" them. However, as the frequency of the radiation is increased beyond the visible region, the light quanta have energies comparable to, or larger than the binding energy of the electrons in atoms, and the electrons can therefore be considered as free.

In 1920 A. H. Compton investigated the scattering of monochromatic x-rays from various materials. He observed that after the scattering, the energy (frequency) of the x-rays had changed, and had always decreased. From the point of view of classical electromagnetic theory, this frequency shift cannot be explained,† since the frequency is a property of the incoming electromagnetic wave (field) and cannot be altered by the change of direction implied by the scattering. If, on the other hand, we think of the incoming radiation as being represented by a beam of photons, we need only consider the scattering of a quantum of energy $E = h\nu$ from a free electron; then, because of energy-momentum conservation, the scattered quantum has energy $E' = h\nu' < E$, in complete agreement with the experiments of Compton.

The frequency shift will depend on the angle of scattering and can be easily calculated from the kinematics. Consider an incoming photon of energy $E = h\nu$ and momentum $p = h\nu/c$ (Fig. 6.8) scattering from an electron of mass m; $\mathbf{p}$ is the momentum of the electron after scattering and

---

† See, for example, J. D. Jackson, *Classical Electrodynamics*, John Wiley, p. 488.

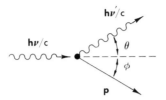

FIG. 6.8 Compton scattering of
a photon from a free electron.

$h\nu'$, $\mathbf{h\nu'}/c$ the energy and momentum of the photon after the scattering. The three vectors $\mathbf{h\nu}/c$, $\mathbf{h\nu'}/c$, and $\mathbf{p}$ must lie on the same plane, and energy conservation yields

$$h\nu + mc^2 = h\nu' + \sqrt{p^2c^2 + m^2c^4} \qquad (3.1)$$

From momentum conservation we obtain

$$h\nu = h\nu' \cos\theta + cp \cos\phi \qquad (3.2)$$

$$0 = h\nu' \sin\theta + cp \sin\phi \qquad (3.3)$$

Here θ is the photon scattering angle, and ϕ the electron recoil angle. To solve the above equations we transpose appropriately, square, and add Eq. 3.2 and Eq. 3.3 to obtain

$$h^2\nu^2 - 2h^2\nu\nu' \cos\theta + h^2\nu'^2 = c^2p^2$$

while by squaring Eq. 3.1,

$$h^2\nu^2 + h^2\nu'^2 - 2h^2\nu\nu' + 2hmc^2(\nu - \nu') = c^2p^2$$

which by subtraction yields

$$\frac{\nu - \nu'}{\nu\nu'} = \frac{h}{mc^2}(1 - \cos\theta) \qquad (3.4)$$

We can recast Eq. 3.4 into two more familiar forms: (a) to give the shift in wavelength of the scattered x-ray beam:

$$\Delta\lambda = \lambda' - \lambda = \frac{h}{mc}(1 - \cos\theta) \qquad (3.5)$$

or (b) to give the energy of the scattered photon:

$$E' = \frac{E}{1 + (E/mc^2)(1 - \cos\theta)} \qquad (3.6)$$

From Eq. 3.5 we see that the shift in wavelength, except for the angular dependence, is a constant, the Compton wavelength[†]

$$h/mc = 2.42 \times 10^{-10} \text{ cm} = 0.0242 \text{ Å}$$

For low-energy photons, with $\lambda \gg 0.02$ Å, the Compton shift is very small, whereas for high-energy photons with $\lambda \ll 0.02$ Å, the wavelength of the *scattered* radiation is always of the order of 0.02 Å, the Compton wavelength. These conclusions can equally well be obtained from Eq. 3.6, where the energy shift increases when E/mc^2 becomes large. For $E/mc^2 \gg 1$, E' is independent of E and of the order $E' \approx mc^2$. [Hence $\lambda' = c/\nu' = c/(E'/h) \simeq c/(mc^2/h) = h/mc$ as stated before.]

As an example, in this laboratory gamma rays from Cs^{137} are scattered from an aluminum target; since $E = 0.662$ MeV, we have $E/mc^2 = 1.29$, so that back-scattered gamma rays ($\theta = 180°$) will have $E' = E/3.6$, which is less than 30 percent of their original energy. It thus becomes quite easy to observe the Compton energy shift as compared to x-ray scattering, where, if we assume $\lambda = 2$ Å, $\Delta\lambda/\lambda = \Delta E/E = 0.01$.

In the original experiments Compton and his collaborators observed (especially for high Z materials) in addition to the frequency shifted x-rays, scattered radiation *not shifted* in frequency. The unshifted x-rays are due to scattering from electrons that remained bound in the atom[‡]: in this process the recoiling system is the entire atom, and we replace in Eq. 3.5 m by m_A (where $m_A \simeq 2{,}000 \times A \times m_e$) resulting in an undetectable wavelength shift, $\Delta\lambda' \approx 10^{-7}$ Å.

Next we are interested in the differential cross section for the scattering of the radiation from the electrons. Classically this is given by the Thomson cross section,[§] which can be easily derived: consider a plane wave propagating in the z direction with the E vector linearly polarized along the x direction. This is incident on an electron of mass m, as shown in Fig. 6.9 The electron will experience a force $F = eE = eE_0 \cos \omega t$, and its acceleration will be

FIG. 6.9 Classical picture of the scattering of electromagnetic radiation by an electron; this leads to the Thomson cross section.

$$\dot{v} = \frac{eE_0}{m} \cos \omega t$$

[†] The mass of the electron m_e was used in evaluating h/mc; by using the mass of the pion, or other particle, we obtain the pion Compton wavelength, and so forth.

[‡] A similar situation is discussed in the following section on the Mössbauer effect, where the nucleus remains bound in the lattice and the recoiling system is the entire crystal.

[§] Already discussed in Chapter 5, Section 2.5.

According to Eq. 2.26, Chapter 5, the power radiated by this accelerated electron will be (nonrelativistically, in MKS units)

$$\frac{dP}{d\Omega} = \frac{e^2}{(4\pi\epsilon_0)^2\mu_0 c^5}\, \dot{v}^2 \sin^2 \Theta \tag{3.7}$$

where Θ is the angle between the direction of observation and the E vector of the incoming wave. Using the expression for $\dot{v}$, we can write for Eq. 3.7 averaged over one cycle

$$\left\langle \frac{dP}{d\Omega} \right\rangle = \frac{1}{2}\left(\frac{e^2}{4\pi\epsilon_0 mc^2}\right)^2 \epsilon_0 E_0^2 c \sin^2 \Theta$$

Finally, from the definition for a cross section given in Chapter 5, Section 2.1, we have

$$\frac{d\sigma}{d\Omega} = \frac{\text{energy radiated/(unit time-unit solid angle)}}{\text{incident energy/(unit area-unit time)}}$$

Here the denominator is clearly given by the Poynting vector

$$\langle P \rangle = \frac{1}{2}\sqrt{\frac{\epsilon_0}{\mu_0}}\, E_0^2 = \tfrac{1}{2}\epsilon_0 c E_0^2$$

Thus we obtain

$$\frac{d\sigma}{d\Omega} = \left(\frac{e^2}{4\pi\epsilon_0 mc^2}\right)^2 \sin^2 \Theta \tag{3.8}$$

Where

$$\frac{e^2}{4\pi\epsilon_0 mc^2} = r_0$$

has dimensions of length, the so-called "classical electron radius"

$$r_0 = 2.82 \times 10^{-13}\,\text{cm}$$

Finally, we average over all possible directions of polarization of the incoming wave and use the angle θ measured from the direction of propagation of the wave to obtain

$$\frac{d\sigma}{d\Omega} = r_0^2 \left(\frac{1 + \cos^2\theta}{2}\right) \text{cm}^2 \tag{3.9}$$

When integrated over all angles, Eq. 3.9 yields the Thomson cross section

$$\sigma_T = \frac{8\pi}{3} r_0^2 \tag{3.10}$$

(also given in Chapter 5, Eq. 2.19).

Several objections can be raised to the simple cross section given by Eq. 3.9 or Eq. 3.10: (a) it does not depend on frequency, a fact not supported by experiment; (b) the electron, even though free, is assumed not to recoil; (c) the treatment is nonrelativistic; and (d) quantum effects are not taken into account. Indeed, the correct quantum-mechanical calculation for Compton scattering yields the so called Klein-Nishina formula[†]

$$\frac{d\sigma}{d\Omega} = r_0{}^2 \frac{1 + \cos^2 \theta}{2} \frac{1}{[1 + \gamma(1 - \cos \theta)]^2}$$

$$\times \left[1 + \frac{\gamma^2(1 - \cos \theta)^2}{(1 + \cos^2 \theta)[1 + \gamma(1 - \cos \theta)]} \right] \quad (3.11)$$

where r_0 and θ were defined previously, and $\gamma = h\nu/mc^2$. The cross section has been averaged over all incoming polarizations. By integrating Eq. 3.11, the total cross section can be had. We will not give the complete result here, but the assymptotic expressions have already been presented in Chapter 5, Eq. 2.21.

A comparison of the Thomson (Eq. 3.9) and Klein-Nishina cross sections, including the results obtained in this laboratory for $\gamma = 1.29$, is shown in Fig. 6.14. We remark that although the Thomson cross section is symmetric about $90°$, the Klein–Nishina cross section is peaked forward strongly as γ increases. This is due to a great extent to kinematical factors associated with the Lorentz transformation from the center of mass to the laboratory; note that the center of mass velocity of the (quantum + free electron) system is

$$v = \beta c = c\gamma/(1 + \gamma)$$

where as before $\gamma = h\nu/mc^2$.

The experimental data are in perfect agreement with the results of Eqs. 3.6 and 3.11, which are among the most impressive and convincing successes of quantum theory. In the following two sections we will describe the experimental verification of these predictions.

3.2 THE COMPTON SCATTERING EXPERIMENT

As with any scattering experiment, the apparatus will consist of:

(a) The beam of incident particles, in this case photons.
(b) The target (containing the electrons from which the photons scatter).
(c) The detector of the scattered photons.

[†] See W. Heitler, *The Quantum Theory of Radiation*, 3rd ed., Oxford University Press, p. 219.

The beam of photons is obtained by collimating the gamma radiation from a Cs^{137} source. As we know (Table 4.2) Cs^{137} (Ba^{137}) emits a gamma ray of energy 0.662 MeV, and the detection techniques have been discussed in Chapter 5. In Fig. 5.28 is shown the pulse-height spectrum of the gamma radiation from Cs^{137}, as obtained with standard equipment; the same detection equipment is used in this experiment with the only difference that the scintillation crystal is placed in a heavy shield.

A schematic of the apparatus is shown in Fig. 6.10. The lead pig A is fixed and holds the source, which can be introduced through the vertical hole (v). Another lead shield B contains the detector and can be rotated about the center, where the target is located. The lead assemblies are rather heavy (approximately 200 lb) and some provisions must be taken for adequate mounting.

For the source, a 35-mCi Cs^{137} sample was used, which was properly encapsulated before being shipped to the laboratory. It should always be transported in a lead container, and when transferred into the lead pig A, it must be handled only by the attached string. The source holder (A) has a collimator (h) drilled horizontally, subtending a solid angle of the order of 0.03 sr. Of interest to us will be the density of the photon beam at the target, and the *expected* value is

$$\frac{3.7 \times 10^{10} \times 0.035}{4\pi} \frac{1}{r^2} = 8.8 \times 10^4 \text{ photons/cm}^2\text{-sec}$$

Where we use $r = (13.5 \times 2.54) = 34.3$ cm, as read off Fig. 6.10; indeed the observed density of 41×10^3 photon/cm²-sec is of the predicted order of magnitude.

In contrast to the situation in Rutherford scattering, there is no need to enclose the beam and detector in vacuum or to use a very thin target. We know that gamma rays do not gradually lose energy when traversing matter as a charged particle does, but their interaction can be characterized by a mean free path. For the Cs^{137} gamma ray we find from Fig. 5.34

$$\lambda = 4.7 \text{ cm in Al}; \quad \lambda = 0.92 \text{ in Pb}$$

this corresponds to 10^4 cm of air, so that the interaction of the photon beam in the air of the apparatus (approximately 100 cm) is indeed negligible. Also, the target thickness can safely be a fraction of a mean free path before the probability for multiple interactions becomes considerable. Aluminum targets $\frac{1}{2}$-in. thick are quite adequate for this experiment.

Some special mention must be made of the geometrical shape of the target. We may use a flat target (such as an aluminum plate), in which event the cross section is obtained by considering the interaction of the total beam with the number of electrons per square centimeter of the

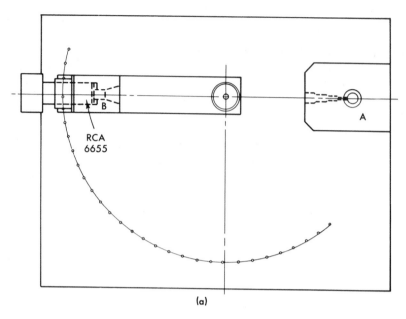

(a)

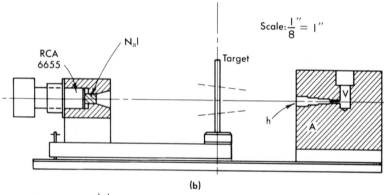

Scale: $\frac{1}{8}'' = 1''$

(b)

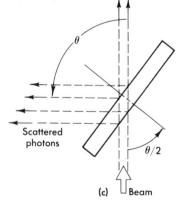

(c)

FIG. 6.10 Apparatus used for measuring the Compton scattering of Cs^{137} gamma rays from different targets; note that the detector is movable. (a) Top view. (b) Elevation. (c) Use of a flat target when measuring Compton scattering at large angles. By such placement the scattered photons do not have to traverse very large amounts of the target material.

target†; alternatively, we may use a target of circular cross section (such as a rod), in which event the cross section is obtained by considering the interaction of the beam density (photons per square centimeter) with the total number of electrons in the target.† When using a plate, it is advisable to rotate it so that it always bisects the angle between beam and detector, since otherwise the scattered photons may have to traverse a very large amount of material before leaving the target (see Fig. 6.10c). In that case, however, the amount of scattering material in the beam path varies as $1/\cos{(\theta/2)}$, and this correction must be applied to the yield of scattered particles. These effects are obviously eliminated when a target of circular cross section is used. In addition, the scattering point is better defined even if the beam is only poorly collimated. On the other hand, accurate evaluation of the flux density at the target is difficult. The results presented here were obtained by using a $\frac{1}{2}$-in. aluminum rod as the target.

An interesting refinement of the technique is made by observing the recoil electrons in time coincidence with the scattered photon. However, the kinetic energy of the recoil electron is

$$T_e = E - E' = E \frac{\gamma(1 - \cos\theta')}{1 + \gamma(1 - \cos\theta)}$$

which at its maximum value ($\theta = 180°$) is

$$T \text{ (electron)} = 0.662 \times (2.58/3.58) = 475 \text{ keV}$$

The range of such an electron in aluminum is only 150 mg/cm² (see, for example, Feather's rule, Chapter 5, Eq. 2.13), which corresponds to approximately 0.06 cm. Thus, the recoil electrons will, in almost all cases, stop in the target. On the other hand, if a plastic scintillator is used as the target, and is viewed with a photomultiplier, the recoil electrons do produce a signal that can be easily detected.

As mentioned before, the detection system consisted of a NaI crystal mounted on a 6655 RCA photomultiplier. The dimensions of the crystal were 1-in. diameter and 1-in. thick. In Fig. 6.11 are shown two typical pulse-height spectra: one at 0° which obviously is dominated by the primary beam and does not change whether the target is in place or removed; and one at 120°, where the photopeak is again clearly observable but appears at a much lower discriminator setting.

By measuring the pulse-height distribution at various angles, we obtain the energy of the scattered photons as it is given by the position of the photopeak, and under the assumption that the discriminator is linear (see Fig. 5.31). To obtain the yield of scattered photons, we may either (a) accept all counts above noise level and apply a correction for the efficiency

---

† See Fig. 5.1.

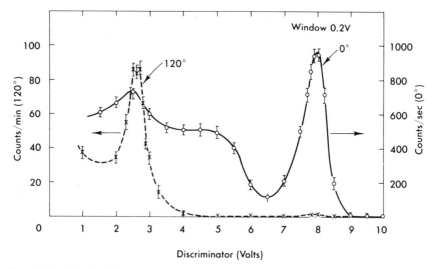

FIG. 6.11   Pulse-height spectrum of scattered gamma rays, at 0° (solid curve) and at 120° dotted curve; note the shift of the photopeak indicating a shift in the energy of the incident gammas.

of the crystal, or (b) integrate the counts in the photopeak only and apply a correction for the "photofraction" as well as for crystal efficiency. These corrections depend on the crystal size and on the photon energy (which varies with angle); Fig. 6.12 gives the efficiency against energy for a 1-in. diameter, 1-in. thick NaI crystal.† In the present experiment we have chosen the former method, that is, of accepting all counts above noise level.

In this experiment, unlike the Rutherford scattering experiment the background is quite low. When the detector is moved outside the primary beam, which has an angular width of $\theta_0 \approx 8°5$, the background counts are less than 0.01 of the scattered counts. Further, the cross section for Compton scattering falls off much more slowly than $1/(\sin^4 \theta/2)$, and therefore a higher background can be tolerated; thus useful data can be obtained at large angles. We also note that Cs^{137} emits mainly the 0.662-MeV gamma ray and 0.514-MeV electrons; these electrons, however, can hardly penetrate the target, much less reach the detector.

3.3 RESULTS AND DISCUSSION

The results presented below were obtained by students‡ using the apparatus described in the previous section.

† From *Scintillation Phosphors*, Harshaw Chemical Company, Cleveland, Ohio.
‡ D. Kohler and A. Rosen, class of 1962.

At first the beam is surveyed by measuring the profile (see Section 2.3 on Rutherford scattering) and it is ascertained that the pulse-height spectrum has the correct shape, as shown in Fig. 6.11. The beam is found to have an angular spread† of $\theta_0 = \pm 8°.5$, consistent with the dimensions of the source collimator; from Fig. 6.10 we find $\tan \theta_0 = 1.5/13.5 = 0.110$, thus $\theta_0 \approx 7°$. The peak beam-rate, which is fairly flat near the center is 25.4×10^3 counts/sec, which corresponds to a density at the detector of

$$\text{flux} = 5 \times 10^3 \text{ counts/cm}^2\text{-sec}$$

We have to correct this rate for the efficiency of the crystal, which at this energy (Fig. 6.12) is $\epsilon = 0.47$, and to extrapolate from the density at the detector to the flux density at the target. Assuming a $1/r^2$ dependence and since r (source detector) $= 2r'$ (source-target) as shown in Fig. 6.10, we obtain

$$\text{flux (at target)} = \frac{5 \times 10^3}{\epsilon} \times \left(\frac{r}{r'}\right)^2 = 42.6 \times 10^3 \text{ photons/cm}^2\text{-sec}$$

Next, a pulse-height spectrum is taken at different angles and the position of the photopeak carefully measured. Two such spectra have already been shown in Fig. 6.11; these were taken with a discriminator window 0.2 V wide (the abscissa is calibrated in volts). To avoid drifts of the detection system, after each pulse-height spectrum is taken, the detector as-

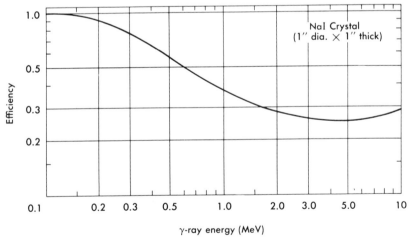

FIG. 6.12   Efficiency of a NaI crystal for the detection of gamma rays, as a function of their energy.

---

† Full width at half maximum, 17°.

TABLE 6.3

PHOTOPEAK POSITION AS A FUNCTION OF ANGLE

Angle (degrees)	Discriminator channel	Energy	$(1 - \cos\theta)$
0	7.95	0.662	0
20	7.45	0.620	0.06
50	5.45	0.455	0.36
60	4.80	0.400	0.50
70	4.30	0.358	0.66
90	3.45	0.287	1.00
120	2.60	0.218	1.50

sembly is returned to 0° and the center of the photopeak is scanned. The photopeak values are summarized in the following Table 6.3 and are also shown in Fig. 6.13. We plot the inverse of the photon energy, $1/E'$, against

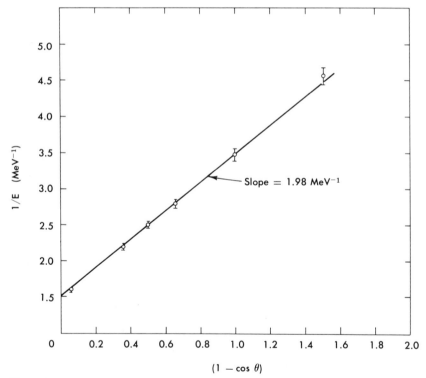

FIG. 6.13   The results obtained for the energy (frequency shift) of the Compton scattered gamma rays. Note that $1/E$ is plotted against $(1 - \cos\theta)$ leading to a linear dependence; the slope of the line gives the mass of the electron.

TABLE 6.4

DIFFERENTIAL CROSS SECTION FOR COMPTON SCATTERING

Angle (degrees) (1)	E (MeV) (2)	Yield (counts/sec) (3)	Efficiency ϵ (4)	Corrected yield (5)	$d\sigma/d\Omega$ (experimental) (6)	Klein-Nishina (7)
					$\times 10^{-26}$ cm^2	$\times 10^{-26}$ cm^2
0	0.662					6.65
20	0.614	52.2	0.47	111.0	6.06	5.39
30	0.564	35.1	0.49	71.5	3.90	4.29
40	0.503	27.5	0.52	51.0	2.78	3.19
50	0.452	20.7	0.55	37.7	2.06	2.42
60	0.402	17.7	0.60	29.5	1.61	1.85
70	0.353	15.3	0.65	23.5	1.28	1.50
80	0.320	13.7	0.69	20.0	1.09	1.23
90	0.289	13.6	0.72	19.0	1.04	1.09
100	0.263	12.5	0.75	16.6	0.91	1.02
110	0.242	13.4	0.78	17.2	0.94	1.01
120	0.225	14.8	0.81	18.3	1.00	0.97
130	0.211	14.8	0.84	17.6	0.96	0.97
140	0.201	16.6	0.86	19.3	1.05	0.99
150	0.193	17.1	0.88	19.4	1.06	0.98

$(1 - \cos \theta)$; according to Eq. 3.6, a straight line should be obtained, since

$$\frac{1}{E'} - \frac{1}{E} = \frac{1}{mc^2} (1 - \cos \theta)$$

This is indeed the result, and the slope of the line gives $1/mc^2$; from a least-squares fit we obtain

$$mc^2 = 505 \pm 12 \text{ keV}$$

in very good agreement with the known value of the electron mass.

We thus conclude that Eq. 3.6 is very well verified and that our explanation of the Compton frequency shift is firmly supported by these data.

We next turn to the evaluation of the differential cross section. As explained before, all counts above the 1 V level were accepted, and data were taken at several angles. The results are summarized in Table 6.4: here column 1 gives the angle and column 2 the corresponding photon energy; column 3 gives the raw yield in counts per second at that particular angle; column 4 gives the efficiency of the detector for the photon energy (of column 2) as obtained from Fig. 6.12, and the corrected counting rate appears in column 5.

To obtain the cross section we note that

$$\frac{d\sigma}{d\Omega} = \frac{\text{yield}}{(d\Omega)\, N I_0}$$

From Fig. 6.10 we see that the detector solid angle is given by

$$d\Omega = \frac{\text{crystal area}}{r^2} = \frac{5.07 \text{ cm}^2}{(34.3)^2 \text{ cm}^2} = 4.3 \times 10^{-3} \text{ sr}$$

For the total number of electrons in the target, we have

$$N = \pi \left(\frac{d}{2}\right)^2 h\rho \frac{N_0}{A} Z$$

Where

d = diameter of target = 1.27 cm
h = height of target† = 10 ± 2 cm
ρ = density of aluminum = 2.7 gm/cm³
N_0 = Avogadro's number = 6 × 10²³
A = atomic weight of aluminum = 27
Z = atomic number of aluminum = 13

thus

$$N = 10^{25} \text{ electrons}$$

For I_0, the flux density at the target, we use the previously obtained value

$$I_0 = 42.6 \times 10^3 \text{ photons/cm}^2\text{-sec}$$

so that finally

$$\frac{d\sigma}{d\Omega} = \frac{\text{corrected yield}}{4.3 \times 10^{-3} \times 10^{25} \times 42.6 \times 10^3} = \frac{\text{corrected yield}}{1.83 \times 10^{27}}$$

The values of the differential cross section obtained in this fashion are given in column 6 of Table 6.4, and are also plotted in Fig. 6.14. The solid line in Fig. 6.14 gives the theoretical values for $d\sigma/d\Omega$ derived from the Klein–Nishina formula (Eq. 3.11) for $\gamma = 1.29$, while the dashed curve represents the Thomson cross section.

The agreement of the *angular dependence* of the experimental points with the theoretical curve is indeed quite good and clearly indicates the inadequacy of the Thomson cross section for the description of the scatter-

---

† This is obtained by estimating the length of target intercepted by the beam. We assume that the angular spread in the vertical direction is the same as in the horizontal, $\theta_0 = \pm 8°5$ and use $r = 34.3$ cm; see also Fig. 6.10.

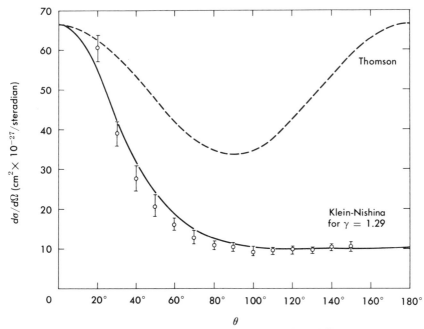

FIG. 6.14   The results obtained for the scattering cross section of Cs^{137} gamma rays as a function of angle. The solid line is the prediction of the Klein-Nishina formula for that particular energy; the dotted line is the Thomson cross section.

ing of high-energy photons, while confirming the Klein–Nishina formula. On the other hand the *absolute value* of the experimental cross section is subject to considerable error due to the way in which the flux density I_0 and total number of electrons N were estimated. Therefore, in spite of the close agreement of the absolute values of the experimental data with the theoretical curve, we may only conclude that the observed cross section is of the correct order of magnitude.

4. Mössbauer Effect

4.1 GENERAL CONSIDERATIONS

In the two experiments previously described, we could visualize the scattering process as if it were a collision of two billiard balls in which the incoming alpha particle or photon maintained its identity but suffered a change in momentum and energy. The phenomenon of scattering can, however, also be visualized as the absorption by the target of quanta of the incoming beam, with the subsequent re-emission of these quanta; this was the model we used in the derivation of the Thomson scattering cross section in the previous chapter.

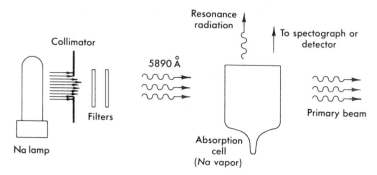

FIG. 6.15   The arrangement of an optical (atomic) resonance radiation experiment. Here the sodium D lines are incident on a cell containing sodium vapor; it is then possible to observe, at right angles to the incident beam, the appearance of the D lines.

Since we know that emission of quanta of energy $h(\nu_\beta - \nu_\alpha)$ in the visible spectrum is due to transitions of atoms from a state of $\beta \rightarrow \alpha$ we must also expect that when quanta of this energy $h(\nu_\beta - \nu_\alpha)$ are incident on an atomic system in the state α, they may be strongly absorbed, with the consequent raising of the atom from the state α, to the state β. Evidence for such strong absorption is obtained by detecting radiation of frequency $(\nu_\beta - \nu_\alpha)$ emitted from the absorber in all directions; it is due to the atoms that, having absorbed a quantum from the beam, were raised to the state β and then underwent a spontaneous transition back to the state α, emitting the quantum $h(\nu_\beta - \nu_\alpha)$, but with equal probability into all directions. Such radiation is called "resonance radiation" and was first observed by R. W. Wood in sodium vapor in 1904. A schematic of the apparatus is shown in Fig. 6.15. An absorption cell was illuminated by sodium light, and at right angles to the incident beam the sodium D lines were observed.

Let us note two facts: (1) Since the atom must be in the state α when the radiation is incident, α is usually the ground state of the atom† (2) The incident radiation must be exactly of the correct energy $h(\nu_\beta - \nu_\alpha)$ corresponding to the separation of the levels α and β.

If we now try to observe in a similar manner resonance radiation, using a nuclear gamma ray (instead of sodium D lines), we will obtain a negative result. This is a simple consequence of energy and momentum conservation which produces a negligible effect in the case of an atomic line. To understand this, consider a system R originally at rest; R undergoes a tran-

---

† The available intensities of visible radiation, the absorption cross section, and the density of the absorbers are all such that the entire sample must participate to yield observable results. In very special cases, a metastable state, to which a large fraction of the atoms can be transferred (by some other means), can serve as the state α.

sition from $\beta \rightarrow \alpha$, where the energy difference between states α and β is

$$E_\beta - E_\alpha = h\nu \qquad (4.1)$$

As a result of the transition, a quantum is emitted, which will carry away energy $h\nu_e$ and momentum $h\nu_e/c$; ν_e is to be determined. From Fig. 6.16a we see that to conserve momentum, the emitting system R must recoil with momentum $h\nu_e/c$; therefore it will have energy (nonrelativistically)

$$E_R = \frac{(h\nu_e)^2}{2mc^2} \qquad (4.2)$$

To balance energy, we must have

$$E_R + h\nu_e = h\nu$$

leading to

$$h\nu_e = h\nu(1 - x + 2x^2 + \cdots) \qquad (4.3a)$$

where

$$x = \frac{h\nu}{2mc^2}$$

Similarly a system R' originally at rest in order to be raised from the level $\alpha \rightarrow \beta$, where $E_\beta - E_\alpha = h\nu$, it must absorb a quantum of energy

$$h\nu_a = h\nu(1 + x - 2x^2 + \cdots) \qquad (4.3b)$$

If the emitted quanta were strictly monochromatic, then it is clearly not possible for a free system R to absorb a quantum $h\nu_e$ emitted by a similar free system R', since $h\nu_a \neq h\nu_e$ (Fig. 6.17a).

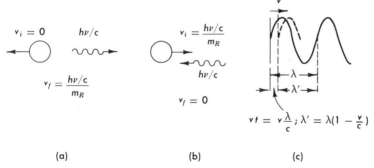

(a)   (b)   (c)

Fig. 6.16 The effect of momentum conservation (recoil effects) in the emission and absorption of nuclear gamma rays. (a) A system R originally at rest emits a gamma ray $h\nu$; it must recoil with a velocity $v_f = (h\nu/c)/m_R$. (b) A system R moving originally with a velocity $v_i = (h\nu/c)/m_R$ absorbs a gamma ray $h\nu$; after the absorption the system will be at rest. (c) Derivation of the first-order Doppler shift for an observer moving with velocity v.

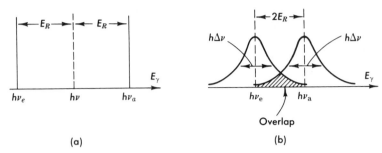

(a)   (b)

FIG. 6.17   Indication of the energy shift of an emitted or absorbed gamma
ray due to the recoil of the nucleus. (a) The situation when the line width is
very narrow in comparison to the recoil energy; no resonant absorption can then
take place under normal conditions. (b) The situation when the line width is of
the same order as the recoil energy; note that resonance absorption can now
take place and it will be proportional to the amount of overlap of the two lines.

We know, however, that spectral lines have a certain width† $\Delta\nu$; in
Fig. 6.17b the emission and absorption lines are shown appropriately centered about $h\nu_e$ and $h\nu_a$, but with a width $\Delta\nu$. If then the two line-shapes
overlap, it is possible to have resonant absorption.

This is true for atomic systems: here $h\nu \approx 2$ eV, and for hydrogen
$mc^2 \approx 10^9$ eV; thus $x \approx 10^{-9}$. The width of atomic spectral lines, however,†
is of the order of $\Delta\nu/\nu \approx 10^{-6}$. Thus

$$\left(\frac{\Delta\nu}{\nu} \approx 10^{-6}\right) \gg \left(\frac{h\nu}{2mc^2} \approx 10^{-9}\right)$$

For nuclear gamma rays, $h\nu \approx 10^4 - 10^6$ eV; also, in general, nuclear
lifetimes are longer than for atomic systems, so that

$$\frac{\Delta\nu}{\nu} \approx 10^{-10} - 10^{-15}$$

Thus we see, contrary to the situation for atomic systems,‡

$$\left(\frac{\Delta\nu}{\nu} \approx 10^{-10}\right) \ll \left(\frac{h\nu}{2mc^2} \approx 10^{-7}\right)$$

making resonance radiation impossible.

---

† The minimum or "natural width" of a line is determined by the lifetime τ for the
transition $\beta \to \alpha$; from the uncertainty principle $\Delta E \Delta t \approx \hbar$, and thus $\Delta\nu \approx 1/\tau$. Other
contributions are the "Doppler broadening" due to the thermal motion of the atom or
nucleus, collisions, and external perturbations or imperfections in a crystal lattice, etc.

‡ For example if $\tau \approx 10^{-9}$ sec, then $\Delta E \approx 6 \times 10^{-7}$ eV. Further, nuclear gamma rays
are subject to broadening influences much less than atomic lines.

In the preceding discussion we assumed that the emitting and absorbing nuclei were at rest. We could, on the other hand, think of imparting to the absorbing nucleus (by some means) enough velocity in a direction opposite to that of the quantum (Fig. 6.16b) so as to satisfy Eqs. 4.3. For example, if $h\nu \approx 10^4$ eV, and the nucleus has $A \approx 100$

$$\frac{h\nu}{c} = mv \qquad h\nu c = (mc^2)v \qquad (4.4)$$

and

$$v = \frac{3 \times 10^{10} \times 10^4}{100 \times 10^9} = 3 \times 10^3 \text{ cm/sec}$$

Such velocities can be obtained in the laboratory by placing the samples on the rim of a centrifuge and orienting the incoming beam towards one of the tangents. It then becomes possible to observe nuclear resonant absorption.

Nuclear resonant absorption would also occur if both the emitter and absorber were so massive that momentum could be balanced with negligible energy being given to the recoiling system; that is, if the denominator m in Eq. 4.2 became infinite. Indeed, R. Mössbauer showed in 1958 that for atoms bound in a crystal lattice, a nucleus does not recoil individually† but the momentum of the nuclear gamma ray is shared by the whole crystal. This can be seen, if we consider that the binding energies of the atoms in a lattice site are of the order of 10 eV, whereas the recoil energies, given by Eq. 4.2, are always less than 1 eV.

Since, however, the nucleus is now part of a larger quantum-mechanical system, there exists the possibility that the energy available from the deexcitation of the nucleus $\beta \to \alpha$ might not all be given to the gamma ray, but might be shared between the gamma ray and the lattice, in the form of vibrational energy. Lattice vibrations—the so-called emission of *phonons*—are a quantized process, and the lowest energy phonon that a single nucleus can emit has

$$E = kT$$

where $T = \Theta_D$ is a characteristic temperature for the crystal, the *Debye temperature*. Thus, if the recoil energy of the free nucleus, as given by Eq. 4.2, is $E_R < k\Theta_D$, it is not possible for the lattice to become excited into a vibrational mode, and the total energy of the transition is taken by the gamma ray. The probability of recoilless emission of the gamma ray is

---

† It is customary to say that "the nucleus does not *always* recoil individually," in order to account for the instances where the nucleus *transfers energy* to the lattice as explained in the following paragraph.

then given by

$$f = \exp\left(-\frac{3}{2}\frac{E_R}{k\Theta_D}\right) \qquad (4.5)$$

Eq. 4.5 holds at absolute zero, and for finite temperatures we may use

$$f = \exp\left(-\frac{\langle x^2\rangle}{\lambdabar^2}\right) \qquad (4.6)$$

Here $1/\lambdabar^2 = (2\pi\nu/c)^2$ is the wave number of the emitted gamma ray and $\langle x^2\rangle$ is the mean square deviation of the atoms from their equilibrium position and is proportional to T. As an example, for the 14.4 keV line of Fe^{57},

$$E_R = 0.002 \text{ eV} \quad \text{and} \quad \Theta_D = 490° \text{ K}$$

hence

$$f = e^{-0.08} = 92 \text{ percent}$$

We therefore see that in certain materials (Fe^{57} being the most suitable) the Mössbauer conditions are met; recoilless emission and absorption can take place, and consequently nuclear *resonance radiation* can be observed.

It has been explained earlier (Eq. 4.4) that we could compensate for the recoil of the nucleus by moving the absorber in a direction opposite to the incoming gamma ray (so as to make the total momentum of the nucleus-plus-gamma-ray system zero). It follows then that if the absorption is recoilless, such motion of the absorber would *destroy* the resonance condition. In recoilless emission (absorption) the gamma ray has energy $E_\gamma = h\nu_0$ in the system, which is at rest with respect to the nucleus; if the nucleus is moving in the laboratory with a velocity v in the direction of the gamma ray, the laboratory energy of the gamma ray E_γ' is given through a Lorentz transformation

$$E_\gamma' = \frac{1}{\sqrt{1-\beta^2}}(E_\gamma - vp_\gamma) = E_\gamma\frac{1-\beta}{\sqrt{1-\beta^2}}$$

where $\beta = v/c$. For $\beta \ll 1$ we obtain to first order

$$\Delta E = E_\gamma' - E_\gamma = \beta E_\gamma \quad \text{or} \quad \frac{\Delta E}{E} = \beta = \frac{v}{c}$$

which, written as $\Delta\nu/\nu = v/c$, is the first order Doppler shift of a wave emitted (absorbed) by a moving observer (Fig. 6.16c). To obtain a quantitative estimate we consider again the 14.4-keV line of Fe^{57}, which has a lifetime $\tau \simeq 10^{-7}$ sec and hence $\Delta\nu/\nu = 4.5 \times 10^{-13}$. Thus, velocities of the order of $v = c(\Delta\nu/\nu) \simeq 1.5 \times 10^{-2}$ cm/sec will be sufficient to destroy

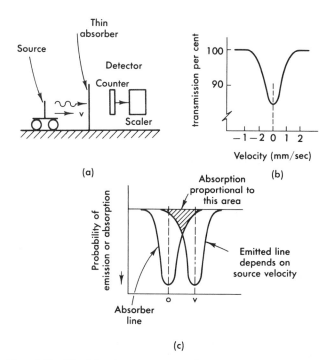

FIG. 6.18   The Mössbauer resonant absorption experiment. (a)
Diagrammatic view of the equipment. (b) The probability for
transmission of a gamma ray as a function of the source (or absorber) velocity when no hyperfine structure is present. (c) The
width of the transmission curve is a combination of the shape of
both the source and absorber lines.

the resonant absorption. Such velocities are easy to achieve and control
in the laboratory, and we measure the transmission of the 14.4-keV gamma
ray through an Fe^{57} absorber as a function of its velocity. Alternatively
we can leave the absorber stationary and move the source.

A possible experimental arrangement, indicated in Fig. 6.18a, consists
of an Fe^{57} source, an Fe^{57} absorber which can be moved at a constant ve-
locity†, and a detector for the 14.4-keV gamma rays; we measure the rate
of transmitted gamma rays. At zero velocity the transmission is low because of resonant absorption; as the velocity of the absorber is increased,
however, the resonance is destroyed and the transmission increases, leading
to a typical curve as shown in Fig. 6.18b. We may think of the incoming
gamma ray as scanning over the absorption line as a function of the velocity, and therefore the observed absorption is a measure of the overlap

† Which velocity, however, is varied in the course of the experiment.

of the two lines as shown in Fig. 6.18c. In this way we "trace out" the natural line width for this nuclear gamma ray, and measure energy deviations of 1 part in 10^{13} ($v \approx 0.06$ mm/sec). This represents the most precise measurement performed by man and is one of the reasons the Mössbauer effect is such an important tool of physics.

4.2 THE APPARATUS AND SOME EXPERIMENTAL CONSIDERATIONS

In this laboratory the Mössbauer effect was observed using the 14.4-keV gamma ray of Fe^{57} which follows the decay, by electron capture, of Co^{57} (see Fig. 6.19). Basically the apparatus required for the experiment consists of (Fig. 6.18) (1) The source (with or without appropriate collimation), (2) the absorber and a mechanism for moving the absorber or the source at constant speed, (3) the detector for the 14.4-keV gamma ray. From Fig. 6.19 we note that the 14.4-keV line of interest will be accompanied by a 122-keV gamma ray as well as by a weaker 136-keV line. There is also a strong background present from the 6.5-keV x-ray of Co^{57}, which follows the electron capture from the K shell. The source used was 1 mCi of Co^{57} plated and annealed onto an ordinary iron backing†.

The detector is chosen so as to provide good efficiency and discrimination for the 14.4-keV gamma ray. A xenon-methane proportional counter, followed by a single-channel discriminator as described in Chapter 5, Section 3.3, was used. In Fig. 6.20, curve (a) gives the pulse-height spectrum of the gamma rays emitted by the source, while curve (b) gives the same spectrum after the gamma rays have traversed a 0.001-in. absorber. The shaded area represents the "window" selected on the discriminator, so that only gamma rays within these energy limits were recorded by the scaler.

The absorber in this case is usually a thin steel foil, but it should not exceed 0.001 in., since nonresonant scattering increases so much as to

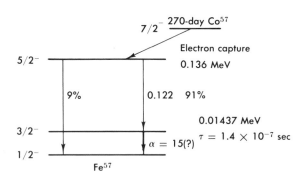

FIG. 6.19   The energy-level diagram of the Fe^{57} nucleus.

† Purchased from Nuclear Science and Engineering Co., P.O. Box 1091, Pittsburgh, Pa.

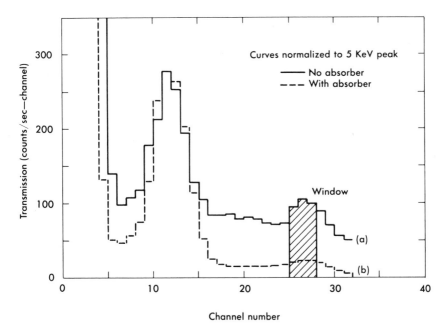

Fɪɢ. 6.20 Pulse-height spectrum of the low-energy gamma rays of Fe^{57} as obtained with a proportional counter. The solid curve has been taken without the absorber in place whereas the dotted one with the absorber in place. The shaded region indicates the discriminator window used for observing the Mössbauer effect.

smear out the 14.4-keV line. Further, natural iron contains only 2.17 percent of Fe^{57}, so that poor signal-to-noise ratios result. It is possible, however, to obtain absorber samples enriched in Fe^{57}, and in the present experiment, such a foil (of 1 cm² area) was used; the Fe^{57} concentration was 91.2 percent and the thickness 1.9 mg/cm² (approximately 0.0001 in.).

The motion of the absorber can be achieved either by purely mechanical arrangements, or by a transducer of some type. Examples in the former category are a plunger driven by an appropriately shaped cam (logarithmic spiral $r = k\theta$) or the rim of a wheel rotating about an axis that is not normal to the surface of the wheel. In all cases of mechanical motion, special attention must be paid to decoupling the vibrations of the driving motor from the absorber.

For the present experiment, a device of the latter category was chosen, namely, a loudspeaker† driven by a sawtooth current. The source was mounted on the core of the speaker and the absorber was kept stationary. The driving waveform was obtained from the horizontal sweep of a

---

† University C-8HC, available from any electronics distributor.

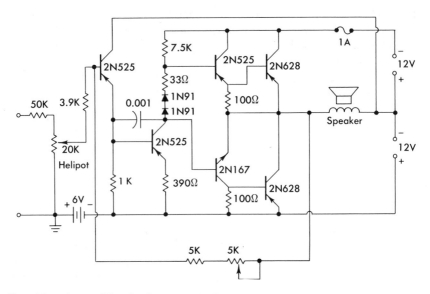

FIG. 6.21   An amplifier circuit capable for driving a speaker coil for use in the Mössbauer experiment.

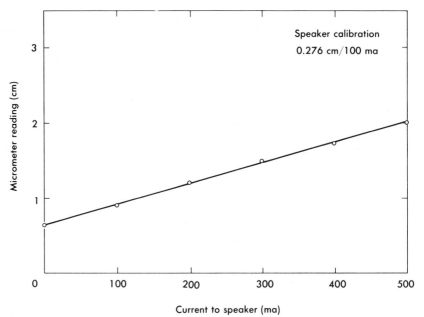

Current to speaker (ma)

FIG. 6.22   Velocity calibration of the speaker used to provide the motion of the source in the Mössbauer experiment.

Tektronix 545A oscilloscope after amplification (Fig. 6.21)†; by varying both amplitude and frequency the desired velocities were obtained.

To calibrate the speaker, a micrometer screw was mounted in a special manner above the speaker. By listening, the experimenter could discern when the screw touched the speaker, giving results to within ±0.003 cm out of a maximum travel of 0.2 cm. Assuming that the speaker is linear with current, the calibration shown in Fig. 6.22 was obtained. The small variation in solid angle with the change of source-detector distance does not affect the results obtained. It is also advisable to gate the scalers so as to count only during the linear part of the motion (and in the desired direction).

4.3 RESULTS AND DISCUSSION

In Fig. 6.23 the results obtained by a student‡ are given; the abscissa gives the velocity of the source in millimeters per second, and the ordinate the counting rate at the detector. It is clear that maximum absorption occurs at zero velocity, in accordance with the hypothesis of recoilless

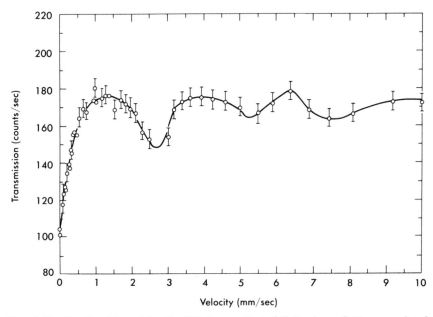

FIG. 6.23   Results obtained for the Mössbauer effect of Fe^{57} using a Co^{57} source plated on ordinary iron backing, and an enriched Fe^{57} absorber.

---

† Argonne National Laboratory internal report by J. Baumgardner.
‡ J. Harris, class of 1963.

emission (and absorption) of the gamma ray, and the conclusions reached in the previous sections.

The full width at half maximum for the zero-velocity peak as obtained from Fig. 6.23 is $\Gamma_{app} = 0.70$ mm/sec. If the two curves shown in Fig. 6.18c are assumed to have a Lorentzian shape, then the apparent width Γ_{app} can be related to the true line width Γ through

$$\Gamma_{app}/\Gamma = 2.00 + \text{term correcting for absorber thickness}$$

Thus we find

$$\Gamma(14.4 \text{ keV Fe}^{57}) \approx 0.30 \text{ mm/sec}$$

and

$$\frac{\Delta\nu}{\nu} = \frac{\Gamma}{c} \approx 10^{-12}$$

which is in fair agreement† with the accepted value of $\Delta\nu/\nu = 3 \times 10^{-13}$.

It is clear that in Fig. 6.23, apart from the zero velocity peak, there also appear subsidary peaks at $v = 2.5$ mm/sec, 5.5 mm/sec, and possibly also at 7.5 mm/sec. What is the origin of these peaks, so reminiscent of the hyperfine structure of atomic spectral lines?

Indeed this structure of the Mössbauer line is greatly dependent on the type of host material in which the absorber (or source) nuclei are embedded. In natural iron, there exist strong magnetic fields at the site of the nuclei; as a result, the nuclear energy levels are split, giving rise to a "Zeeman effect" for the nucleus.‡ Figure 6.24b shows the splitting of the 3/2 excited state and the 1/2 ground state of Fe^{57}, and consequently the 14.4-keV line has six hyperfine structure components. Figure 6.24a shows the same levels for stainless steel, where no splitting occurs.

If both the source and absorber are not split, then clearly only a single peak will be observed, as in Fig. 6.18b. If the source is not split, but the absorber is, then as a function of velocity we will "scan" with the single

---

† Most of the discrepancy can be traced to the considerable thickness of the absorber. The probability for interaction is given by

$$P = \sigma_0 f a (N_0/A) t$$

where

t = absorber thickness in gr/cm² $\approx 2 \times 10^{-3}$
$N_0/A = 6 \times 10^{23}/57 \approx 10^{22}$
σ_0 = is the Mössbauer absorption cross section in cm² 1.5×10^{-18} cm²
f = probability for recoilless absorption, approximately 1
a = concentration of the resonantly absorbing nuclei in the sample, approximately 1

Hence, for the present case, $P \approx 30!$

‡ See Chapter 7 for a detailed discussion of the Zeeman effect.

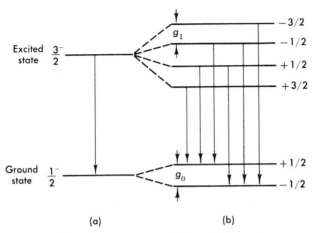

Fig. 6.24 Hyperfine structure splitting of the nuclear energy
levels of Fe^{57}. (a) When stainless steel is used, the levels are not
split. (b) In ordinary iron, however, both levels are split, giving
rise to a hyperfine structure with six components.

line over the hyperfine structure pattern of the absorber. In this case there
is no absorption at zero velocity (see Fig. 6.25a). Finally, if both the source
and absorber are split, a complicated pattern emerges, depending on the

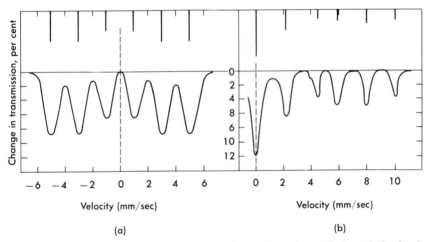

Fig. 6.25 The expected pattern of the Mössbauer line when splitting of the levels
takes place. (a) Only one of the source or absorber are split; note that the Mössbauer
line also is split into six components and no absorption takes place for zero velocity.
(b) When both source and absorber are split a complicated pattern results with maximum
absorption at zero velocity.

TABLE 6.5

POSITION AND AMPLITUDE OF MÖSSBAUER PEAKS IN Fe⁵⁷, INCLUDING THE
EXPERIMENTAL RESULTS

Peak	Amplitude	Position (mm/sec)	Observed position (mm/sec)
0	7	0	0
1	4	2.2	2.75
2–3	1.5	4.3	5.5
4	2.5	6	7.6 (?)
5	3	8	—
6	2	10	—

degree of overlap of the individual components as the two hyperfine structure patterns are shifted one over the other; however, maximum absorption occurs at zero velocity (see Fig. 6.25b).

In the experiment that yielded the data† of Fig. 6.23, both the source and the absorber were split, so that a pattern of the type shown in Fig. 6.25b was obtained. Table 6.5 gives the relative intensities and known positions of the peaks as well as the positions obtainable from the results of Fig. 6.23.

The apparent discrepancies in the known and observed positions are due in part to a small velocity calibration error. Materials like stainless steel, or potassium ferrocyanide, or sources made by diffusing Co⁵⁷ into chronium metal do not exhibit structure in the 14.4-keV line and give simple patterns. In Table 6.6 we summarize some of the numerical values pertinent to the Mössbauer effect in Fe⁵⁷.

TABLE 6.6

SOME NUMERICAL VALUES PERTINENT TO THE Fe⁵⁷ MÖSSBAUER LINE

Transition energy	$E_\gamma = 14.4 \times 10^3$ eV
Internal conversion coefficient	$\alpha = e/\gamma = 15$
Lifetime	$t = 1.4 \times 10^{-7}$ sec
Relative width	$\Delta\nu/\nu = 3 \times 10^{-13}$
Recoil energy of free nucleus	$E_R = 0.19 \times 10^{-2}$ eV
Debye Temperature (Mössbauer)	$\Theta_D = 490°$ K
Probability for recoilless transition at room temperature	$f = 0.80$
Cross section for resonant absorption	$\sigma_0 = 15 \times 10^{-19}$ cm²
Abundance of Fe⁵⁷	2.17 percent

† J. Harris, class of 1963.

A very complete description of the Mössbauer effect, including reprints of the most important papers, will be found in H. Franenfelder's *The Mössbauer Effect* (W. A. Benjamin, Inc., New York, 1962); this reference should be fully adequate until the student finds it necessary to consult the current literature.

7

HIGH-RESOLUTION SPECTROSCOPY

1. Introduction

In 1896, P. Zeeman observed that when a sodium source was placed in a strong magnetic field, the yellow D lines were split into several components. Faraday had performed the same experiment some thirty years earlier but had failed to observe an effect because of the low resolution of his spectrograph. We also know from Chapter 2 (4.3) that even in the absence of a magnetic field the atomic spectral lines have a fine structure which was easily observed with the small grating spectrometer; with a high-resolution instrument, however, it becomes possible to observe that each of these fine structure lines may again be resolved into closely spaced components, which form the so-called "hyperfine structure" (hfs) of atomic lines.†

The splitting of a spectral line obviously is a consequence of a splitting of the initial state, or of the final state, or of both states between which

† To set the reader at ease, no further splitting beyond the hyperfine structure has been observed, nor can it be expected for free atoms; in the hyperfine structure we include both the splitting due to *nuclear spin* and that due to *isotope shift*.

the transition takes place. The energy-level splittings produced by the application of an external magnetic field **H** (Zeeman effect) are of the order of

$$\Delta E = \mathbf{\mu} \cdot \mathbf{H} = \frac{e}{2m_e c} \mathbf{L} \cdot \mathbf{H} \sim \frac{e\hbar}{2m_e c} H \qquad (1.1)$$

where $\mathbf{\mu}$ is the magnetic moment of the state (see Section 2 of this chapter). The constant $\mu_0 = e\hbar/2mc = 0.579 \times 10^{-14}$ MeV/gauss is called the Bohr magneton, so that in units of wave numbers the displacement for *one* Bohr magneton is

$$\Delta \bar{\nu} = \frac{\Delta \nu}{c} = \frac{\Delta E}{hc} = \frac{e}{4\pi m_e c^2} H = 4.669 \times 10^{-5} \times H \text{ cm}^{-1} \qquad (1.2)$$

with H in gauss.

The hyperfine structure splitting is due to the interaction of the magnetic-dipole, electric-quadrupole, etc., moment of the nucleus, with the electromagnetic field produced by the electrons at the nucleus. The interaction energy for the magnetic dipole term is of the order of

$$\Delta E = \mathbf{\mu}_N \cdot \langle \mathbf{H}_J(0) \rangle \sim \mu_N \mu_0 \left\langle \frac{1}{r^3} \right\rangle = \frac{\mu_0^2}{1837} \left\langle \frac{1}{r^3} \right\rangle \qquad (1.3)$$

where μ_N is the nuclear magneton

$$\mu_N = \frac{e\hbar}{2mc} = \frac{\mu_0}{1837}$$

and $\langle \mathbf{H}_J(0) \rangle$ is the expectation value for the magnetic field of the electrons at the origin; it is equal to $\mu_0 \langle 1/r^3 \rangle$ (except for configurations with $l = 0$). Instead of evaluating $\langle 1/r^3 \rangle$, we recall that the fine structure splitting is due to an $\mathbf{L} \cdot \mathbf{S}$ coupling of the electrons, and therefore is of the order of $\mu_0^2 \langle 1/r^3 \rangle$, so that we expect

$$\Delta E(hfs) \sim \frac{\Delta E(\text{fs})}{1837} \qquad (1.4)$$

Let us substitute reasonable numbers in Eqs. 1.2 and 1.4; for example, $H \approx 10$ kilogauss yields

$$\Delta \bar{\nu}(\text{Zeeman}) \sim 0.460 \text{ cm}^{-1}$$

and† $\Delta \bar{\nu}$ (fine structure) ~ 100 cm^{-1} yields

$$\Delta \bar{\nu}(\text{hfs}) = 0.050 \text{ cm}^{-1}$$

---

† See for example Chapter 2, p. 47.

Thus the splitting of the lines is very small and can be observed only with a high-resolution instrument. Assuming $\lambda \approx 5000$ Å and $\Delta\bar{\nu} \sim 0.050$ cm^{-1}, we find that the required resolving power is

$$\frac{\lambda}{\Delta\lambda} = \frac{\bar{\nu}}{\Delta\bar{\nu}} = 4 \times 10^5$$

Such resolution may be achieved in two ways:

(a) With a large grating used in a high order; the resolving power of a grating is given by

$$\frac{\lambda}{\Delta\lambda} = Nn$$

where n, the diffraction order, can be as large as 20, and for a 10-in. grating with 7000 rulings to the inch, the number of rulings is $N = 7 \times 10^4$, so that

$$\frac{\lambda}{\Delta\lambda} \sim 10^6$$

Such gratings, are, however, very difficult to construct, but can now be obtained commercially.

(b) With a "multiple beam" interferometer; the most common one today and easiest to use being the Fabry-Perot. It consists of two reflecting glass plates of high quality which are kept parallel and at a distance t. The interference pattern appears as a series of concentric rings, and the resolution, the so-called "instrument width" is

$$\delta\bar{\nu} = \frac{1}{2t}\frac{1-R}{\pi\sqrt{R}} \tag{1.5a}$$

where R is the reflectivity of the plates. It is more helpful to use the expression

$$\Delta\bar{\nu} = \Delta n \frac{1}{2t} \tag{1.5b}$$

where Δn is the fraction of an order by which the ring has been shifted. It is easy to distinguish $\Delta n \sim 1/10$, and t can be as large† as 2 cm, so that

$$\Delta\bar{\nu} = 25 \times 10^{-3} \text{ cm}^{-1}$$

which for $\lambda = 4000$ Å corresponds to $\lambda/\Delta\lambda \approx 10^6$. An additional require-

---

† In special cases connected with the observation of lasers, spacings as large as 1m have been used.

ment for the observation of such small structure of spectral lines is that the lines themselves be narrower than the spacing between the components of the structure. For that purpose special light sources have been constructed which emit lines that are as narrow as possible.

In this chapter we first present an elementary discussion of the theory of the Zeeman effect and of the theory of hyperfine structure. A brief section is devoted to light sources, and next the Fabry-Perot interferometer is described. In Section 6, experimental data on the Zeeman effect of the 5461-Å green line of Hg^{198}, obtained with a Fabry-Perot, are treated in detail. Next, the medium-resolution grating spectrograph used in this laboratory is described and data on the Zeeman effect obtained with it, are presented; finally data on the hyperfine structure of mercury obtained with a grating of high quality are also presented.

REFERENCES

By necessity the discussion of certain topics in this chapter is very brief. However the bibliography on atomic spectroscopy is excellent and extensive. The following texts and monographs should be very useful to the student.

E. U. Condon and G. H. Shortley, *The Theory of Atomic Spectra*, Oxford Univ. Press. This is one of the most complete theoretical treatments on atomic spectroscopy, but at an advanced level.
H. E. White, *Introduction to Atomic Spectra*, McGraw-Hill. This book contains extensive data on atomic spectra and the treatment of the theory is based on the semi-classical approach of the vector model.
H. Kuhn, *Atomic Spectra*, Longman's, London, 1962. A recent good book on a slightly more advanced level than White's book, referred to above.
S. Tolansky, *High Resolution Spectroscopy*, Methuen, London. A very comprehensive and clear treatise on the instruments and techniques of high-resolution spectroscopy.
H. Kopferrnan, *Nuclear Moments*, Academic Press. This book contains a very complete discussion of atomic hyperfine structure, of analysis methods, and of the conclusions that are obtained from it.

2. The Zeeman Effect

2.1 THE NORMAL ZEEMAN EFFECT

As already discussed in Chapter 2, Section 4, the solution of the Schrödinger equation† yields "stationary states" labeled by three integer indices, n, l, and m, where $l < n$ and $m = -l, -l + 1, \cdots, l - 1, l$. For the screened Coulomb potential, the energy of these states depends on n and l but not on m; we therefore say that the $(2l + 1)$ states with the same

† See, for example, E. Fermi, *Notes on Quantum Mechanics*, Univ. of Chicago Press, or any other text on quantum mechanics.

n and l index are "degenerate" in the m quantum number. Classically we can attribute this degeneracy to the fact that the plane of the "orbit" of the electron may be oriented in any direction without affecting the energy of the state, since the potential is spherically symmetric.

If now a magnetic field H is switched on in the region of the atom, we should expect that the electrons (and the nucleus†) will interact with it. We need only consider the electrons outside closed shells, and let us assume there is one such electron; indeed the interaction of the magnetic field with this electron yields for each state an additional energy ΔE, given by

$$\Delta E = m\mu_0 H \tag{2.1}$$

Thus, the total energy of a state depends now on n, l, and m, and the degeneracy has been removed.

FIG. 7.1   Magnetic moment due to a current circulating in a closed loop.

To see how this additional energy arises we consider again the classical analogy. Then the orbiting electron is equivalent to a current density‡

$$\mathbf{J}(\mathbf{x}) = -e\mathbf{v}\,\delta(\mathbf{x} - \mathbf{r})$$

where $\mathbf{r}$ is the equation of the orbit and $\mathbf{x}$ gives the position of the electron; the negative sign obviously arises from the negative charge of the electron. Such a current density gives rise to a magnetic dipole moment

$$\mathbf{\mu} = \frac{1}{2c}\int \mathbf{x} \times \mathbf{J}(\mathbf{x})\,d^3x = -\frac{1}{2c}\,e(\mathbf{r} \times \mathbf{v})$$

---

† For our present discussion this interaction of the nucleus with the external field is so small that we will neglect it.

‡ For a circular orbit, the electron is equivalent to a current $I = \Delta Q/\Delta T = e/T = e\omega/2\pi$ where ω is the angular frequency $\omega = v/a$; a is the radius of the orbit. But a plane closed loop of current gives rise to a magnetic moment $\mu = (I/c)A$ where A is the area enclosed by the loop; in our case $A = \pi a^2$, hence

$$\mu = \frac{ev}{2\pi ac}\,\pi a^2 = \frac{eva}{2c}$$

The angular momentum for the circular orbit is $L = m_e va$, hence

$$\mu = \frac{e}{2m_e c}\,L$$

as in Eq. 2.2.

But the angular momentum of the orbit is given by

$$\mathbf{L} = \mathbf{r} \times \mathbf{p} = m_e(\mathbf{r} \times \mathbf{v})$$

so that

$$\mathbf{\mu} = -\frac{e}{2m_ec}\,\mathbf{L} = -\frac{e\hbar}{2m_ec}\,l\mathbf{u}_L \qquad (2.2)$$

where we expressed the angular momentum of the electron in terms of its quantized value (according to Bohr) $\mathbf{L} = lh/2\pi$ and $\mathbf{u}_L$ is a unit vector along the direction of $\mathbf{L}$. Now, the energy of a magnetic dipole in a homogeneous field is

$$E = -\mathbf{\mu}\cdot\mathbf{H} = \frac{e}{2m_ec}\,\mathbf{L}\cdot\mathbf{H} \qquad (2.3)$$

but the angle between $\mathbf{L}$ and the external field $\mathbf{H}$ cannot take all possible values.† We know that it is quantized, so that the projection of $\mathbf{L}$ on the z axis (which we can take to coincide with the direction of $\mathbf{H}$ since no other preferred direction exists) can only take the values $m = -l, -l + 1, \cdots, l - 1, l$. Thus, the energy of a particular state n, l, m in the presence of a magnetic field will be given by‡

$$E_{n,l,m} = -E_{n,l} + mH\mu_0 \qquad (2.4)$$

where§

$$\mu_0 = \frac{e\hbar}{2m_ec}$$

In Fig. 7.2 is shown the energy-level diagram for the five states with given n and $l = 2$, before and after the application of a magnetic field H. We

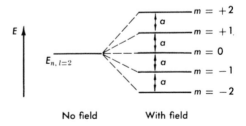

FIG. 7.2 Splitting of an energy level under the influence of an external magnetic field. The level is assumed to have $l = 2$ and therefore is split into five equidistant sublevels.

No field   With field

---

† This was first clearly shown in the Stern-Gerlach experiment.

‡ The energy in the field is $+(e\hbar/2m_ec)m$ because the electron charge is taken as negative.

§ In the MKS system $\mu_0 = (e\hbar/2m_e)$; m_e in this expression is obviously the mass of the electron, not to be confused with the magnetic quantum number m.

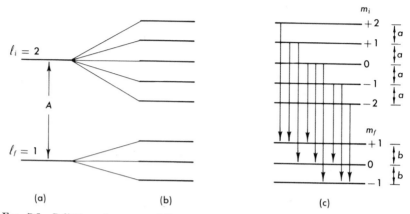

FIG. 7.3 Splitting of a spectral line under the influence of an external magnetic field. (a) The initial level ($l = 2$) and the final level ($l = 1$) with no magnetic field are shown. A transition between these levels gives rise to the spectra line. (b) The two levels after the magnetic field has been applied. (c) The nine allowed transitions between the eight sublevels of the initial and final state.

note that all the levels are equidistantly spaced, the energy difference between them being

$$\Delta E = \mu_0 H$$

Let us next consider the transition between a state with n_i, l_i, m_i and one with n_f, l_f, m_f. As an example we choose $l_i = 2$ and $l_f = 1$, and the energy-level diagram is shown in Fig. 7.3; without a magnetic field in Fig. 7.3a, and when the magnetic field is present in Fig. 7.3b.

However, we know that for an electric dipole transition to take place between two levels, certain selection rules must be fulfilled: in particular,

$$\Delta l = \pm 1 \tag{2.5}$$

Thus, when the field is turned on, we cannot expect transitions between the m sublevels with the same l, since they do not satisfy Eq. 2.5. Further, also the transitions between the sublevels with $l_i = 2$ to the sublevels with $l_f = 1$ which do satisfy Eq. 2.5 are now governed by the *additional* selection rule†

$$\Delta m = 0, \qquad \pm 1 \tag{2.6}$$

and thus only the transitions shown in Fig. 7.3c are allowed.

Let the energy splitting in the initial level be a, and in the final level be b, and let A be the energy difference between the two levels when no

---

† The selection rules of atomic spectroscopy are a consequence of the addition of angular momenta. In this specific case the selection rules indicate that we consider only *electric dipole* radiation.

TABLE 7.1

ALLOWED TRANSITIONS FROM $l_i = 2$ TO $l_f = 1$ AND THE CORRESPONDING ENERGIES

m of final state	m of initial state				
	$+2$	$+1$	0	-1	-2
$+1$	$A + 2a - b$	$A + a - b$	$A - b$	$\times$	$\times$
0	$\times$	$A + a$	A	$A - a$	$\times$
-1	$\times$	$\times$	$A + b$	$A - a + b$	$A - 2a + b$

Note: An X indicates that this particular transition may not take place.

magnetic field is applied. Then the energy released in a transition $i \rightarrow f$ is given by

$$E_i - E_f = A_{if} + m_i a - m_f b \qquad (2.7)$$

These energy differences for the nine possible transitions shown in Fig. 7.3c, are given in matrix form in Table 7.1; an $\times$ indicates that the transition is forbidden and will not take place.

At this point the reader must be concerned about the use of a and b; according to our previous argument (Eq. 2.4), as long as all levels are subject to the same magnetic field H, their splitting must also be the same, and

$$a = b = \mu_0 H$$

Thus, we see from Eq. 2.7 (or Table 7.1) that only *three* energy differences are possible

$$E_i - E_f = A + a(m_f - m_i) = A + a\Delta m$$

where Δm is limited by the selection rule, Eq. 2.6, to the *three* values $+1$, 0, -1. Consequently, in the presence of a magnetic field H, the single spectral line of frequency $\nu = A/h$ is split into three components with frequencies

$$\nu_- = (A - \mu_0 H)/h, \qquad \nu_0 = A/h, \qquad \text{and} \qquad \nu_+ = (A + \mu_0 H)/h$$

irrespective of the values of l_i and l_f. Furthermore, these spectral lines are polarized, as shown in Fig. 7.4. When the Zeeman effect is viewed in a direction *normal* to the axis of the magnetic field, the central component is polarized parallel to the axis whereas the two outer ones normal to the axis of the field. When the Zeeman effect is observed *along* the axis of the the field (by making a hole in the pole face, or using a mirror), only the two outer components appear, circularly polarized. The lines from

$\Delta m = +1$ transitions appear with right-hand circular polarization; from $\Delta m = -1$ transitions, with left-hand circular polarization. The central line does not appear, since the electromagnetic field must always have the field vectors ($\mathbf{E}$ and $\mathbf{B}$) normal to the direction of propagation.

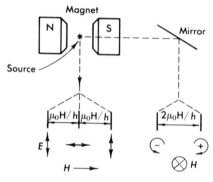

FIG. 7.4 The polarization and separation of the components of a normal Zeeman multiplet when viewed in a direction normal to, and in a direction parallel to, the magnetic field.

The splitting of a spectral line into a triplet under the influence of a magnetic field is called the *"normal" Zeeman effect*, and is occasionally observed experimentally, as, for example, in the 5790 Å line of mercury arising in a transition† from $^1D_2$ to $^1P_1$. But in most cases the lines are split into more components, and even where a triplet appears it does not always show the spacing predicted by Eq. 2.4. This is due to the intrinsic magnetic moment of the electron (associated with its spin) and will be discussed in the following sections.

2.2 THE INFLUENCE OF THE MAGNETIC MOMENT OF THE ELECTRON

In Chapter 2 it was discussed how the intrinsic angular momentum (spin) of the electron $\mathbf{S}$ couples with the orbital angular momentum of the electron $\mathbf{L}$ to give a resultant $\mathbf{J}$; this coupling gave rise to the "fine structure" of the spectra‡. The projections of $\mathbf{J}$ on the z axis are given by m_J and we could expect (on the basis of our previous discussion) that the *total* magnetic moment of the electron will be given by

$$\mathbf{\mu} = \frac{\mu_0}{\hbar} \mathbf{J} \tag{2.8}$$

---

† Note that both initial and final state have $S = 0$.

‡ We will use the following notation: $\mathbf{L}$, $\mathbf{S}$, $\mathbf{J}$ represent the angular momentum vectors which have magnitude $\hbar\sqrt{l(l+1)}$, $\hbar\sqrt{s(s+1)}$, $\hbar\sqrt{j(j+1)}$.

The symbols l, j, etc. (s is always $s = \frac{1}{2}$) are the *quantum numbers* which label a one-electron state and appear in the above square root expressions.

The symbols L, S, J, etc., are the *quantum numbers* which label a state with more than one electron and are then used instead of l, s, j.

Consequently, the energy-level splitting in a magnetic field H would be in analogy to Eq. 2.4,

$$\Delta E = -m_J \mu_0 H \qquad (2.9)$$

These conclusions, however, are not correct because the *intrinsic* magnetic moment of the electron is related to the *intrinsic* angular momentum of the electron (the spin) through

$$\mathbf{\mu}_e = 2\frac{e}{2m_e c}\mathbf{S} = 2\frac{e\hbar}{2m_e c}s\mathbf{u}_s \qquad (2.10)$$

and *not* according to Eq. 2.2.† Consequently, the *total* magnetic moment of the electron is given by the operator

$$\mathbf{\mu} = (\mu_0/\hbar)[\mathbf{L} + 2\mathbf{S}] \qquad (2.11)$$

We can think of it as a vector oriented along $\mathbf{J}$ but of magnitude

$$\mu = \mu_0 g J \qquad (2.12)$$

The numerical factor g is called the Landé g factor and a correct quantum-mechanical calculation gives‡

$$g = 1 + \frac{j(j+1) + s(s+1) - l(l+1)}{2j(j+1)} \qquad (2.13)$$

---

† The result of Eq. 2.10 is obtained in a natural way from the solution of the Dirac equation; also from the classical relativistic calculation of the "Thomas precession."

‡ This result can also be obtained from the vector model for the atomic electron. In Fig. 6.5 the three vectors $\mathbf{J}$, $\mathbf{L}$, and $\mathbf{S}$ are shown, and $\mathbf{L}$ and $\mathbf{S}$ couple into the resultant $\mathbf{J}$, so that

$$\mathbf{J} = \mathbf{L} + \mathbf{S}$$

By taking the squares of the vectors, we obtain the following values for the cosines

$$\cos(\mathbf{L}, \mathbf{J}) = \frac{j^2 + l^2 - s^2}{2lj} \qquad \cos(\mathbf{S}, \mathbf{J}) = \frac{j^2 + s^2 - l^2}{2sj}$$

From Eq. 2.11 we see that

$$\mu/\mu_0 = l\cos(\mathbf{L}, \mathbf{J}) + 2s\cos(\mathbf{S}, \mathbf{J})$$

Thus

$$g = \frac{\mu}{\mu_0 j} = \frac{j^2 + l^2 - s^2}{2j^2} + \frac{2j^2 + 2s^2 - 2l^2}{2j^2} = 1 + \frac{j^2 + s^2 - l^2}{2j^2}$$

Fig. 7.5 Addition of the orbital angular momentum $\mathbf{L}$, and of the spin angular momentum $\mathbf{S}$ into the total angular momentum $\mathbf{J}$, according to the "vector model."

Finally we must replace j^2, s^2, and l^2 by their quantum mechanical expectation values $j(j+1)$, etc., and we obtain Eq. 2.13.

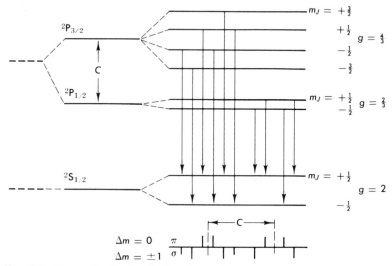

FIG. 7.6   Energy levels of a single valence electron atom showing a P state and an S state. Due to the fine structure, the P state is split into a doublet with $j = \frac{3}{2}$ and $j = \frac{1}{2}$. Further, under the influence of an external magnetic field each of the three levels is split into sublevels as shown in the figure where account has been taken of the magnetic moment of the electron. The magnetic quantum number m_j for each sublevel is also shown as is the g factor for each level. The arrows indicate the allowed transitions between the initial and final states, and the structure of the line is shown in the lower part of the figure.

The interesting consequence of Eqs. 2.12 and 2.13 is that now the splitting of a level due to an external field H is

$$\Delta E = (g\mu_0 H) m_j \qquad (2.14)$$

and in contrast to Eq. 2.4 is *not* the same for all levels; it depends on the j and l of the level ($s = \frac{1}{2}$ always when one electron is considered). The sublevels are still equidistantly spaced but by an amount

$$\Delta E = g\mu_0 H$$

Let us then consider again the transitions between sublevels belonging to two states with different l (in order to satisfy Eq. 2.5). Since, however, we are taking into account the electron spin, l is not a good quantum number, and instead the j values of the initial and final level must be specified. If we choose for this example $l_i = 1$ and $l_f = 0$, we have the choice of $j_i = \frac{3}{2}$ or $j_i = \frac{1}{2}$, whereas $j_f = \frac{1}{2}$. Transitions may occur only if they satisfy, in addition to Eq. 2.5, *also* the selection rules for j

$$\Delta j = 0, \pm 1 \qquad \text{not} \qquad j = 0 \to j = 0 \qquad (2.5a)$$

Furthermore the selection rules for m_j must also be satisfied; they are the same as given by Eq. 2.6

$$\Delta m_j = 0, \pm 1 \tag{2.6a}$$

In Fig. 7.6 the energy-level diagram is given without and with a magnetic field for the doublet initial state with $l = 1$, and the singlet final state, $l = 0$. Six possible transitions between the initial states with $j = \frac{3}{2}$ to the final state with $j = \frac{1}{2}$ are shown (as well as the four possible transitions from $j = \frac{1}{2}$ to $j = \frac{1}{2}$). By using Eq. 2.13 we obtain the following g factors

$$l = 1 \qquad j = \tfrac{3}{2} \qquad s = \tfrac{1}{2} \qquad g = \tfrac{4}{3}$$

$$l = 1 \qquad j = \tfrac{1}{2} \qquad s = \tfrac{1}{2} \qquad g = \tfrac{2}{3}$$

$$l = 0 \qquad j = \tfrac{1}{2} \qquad s = \tfrac{1}{2} \qquad g = 2$$

The sublevels in Fig. 7.6 have been spaced accordingly.

TABLE 7.2

ALLOWED TRANSITIONS FROM $j_i = \frac{3}{2}$ TO $j_f = \frac{1}{2}$ AND THE CORRESPONDING ENERGIES

m_j of final state	m_j of initial state			
	$+\frac{3}{2}$	$+\frac{1}{2}$	$-\frac{1}{2}$	$-\frac{3}{2}$
$+\frac{1}{2}$	$A + \dfrac{3a}{2} - \dfrac{b}{2}$	$A + \dfrac{a}{2} - \dfrac{b}{2}$	$A - \dfrac{a}{2} - \dfrac{b}{2}$	$\times$
$-\frac{1}{2}$	$\times$	$A + \dfrac{a}{2} + \dfrac{b}{2}$	$A - \dfrac{a}{2} + \dfrac{b}{2}$	$A - \dfrac{3a}{2} + \dfrac{b}{2}$

In Table 7.2 are now listed the six transitions from $j = \frac{3}{2}$ to $j = \frac{1}{2}$ in analogy with Table 7-1. However, *since now a $\neq$ b*, the spectral line is split into a six-component (symmetric) pattern. This structure of the spectral line is indicated in the lower part of Fig. 7.6; following adopted convention, the components with polarization parallel to the field are indicated above the base line, and with normal polarization below.† As before the parallel components have $\Delta m = 0$, the normal ones $\Delta m \pm 1$.

The horizontal spacing between the components is proportional to the differences in the energy of the transition, and the vertical height is pro-

---

† It is also conventional to label the parallel components with π, and the normal ones by σ (from the German "Senkrecht").

portional to the intensity of the components; the relative intensity can be predicted exactly since it involves only the comparison of matrix elements between the angular parts of the wave function.

As the magnetic field is raised, the separation of the components continues to increase linearly with the field until the separation between Zeeman components becomes of the order of the fine-structure separation (spacing C in Fig. 7.6). At this point the Zeeman components from the $j = \frac{3}{2} \rightarrow \frac{1}{2}$ and $j = \frac{1}{2} \rightarrow \frac{1}{2}$ transition begin to overlap; clearly the perturbation caused by the external magnetic field is of the order of the $\mathbf{L \cdot S}$ energy and affects the coupling of $\mathbf{L}$ and $\mathbf{S}$ into $\mathbf{J}$; $\mathbf{J}$ ceases to be a "good quantum number."

For very strong fields, $\mathbf{L}$ and $\mathbf{S}$ become completely uncoupled, so that the orbital and intrinsic magnetic moments of the electron interact with the field independently, giving rise to an energy shift

$$\Delta E = -\frac{\mu_0}{\hbar} \mathbf{L \cdot H} - 2\frac{\mu_0}{\hbar} \mathbf{S \cdot H} - a\mathbf{L \cdot S}$$

$$= -\mu_0 H (m_l + 2m_s) - am_l m_s \qquad (2.15)$$

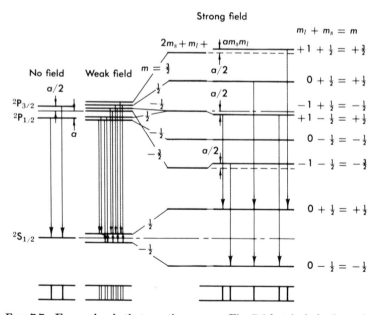

FIG. 7.7 Energy levels that are the same as Fig. 7.6 but include the region of very strong magnetic field. Note how the sublevels belonging to the $j = \frac{3}{2}$ and $j = \frac{1}{2}$ levels cannot any more be distinguished but have coalesced into five components typical of Fig. 7.2.

Here, m_l and m_s are the projections of **L** and **S** on the z axis, and in this case the selection rules for electric dipole transitions become

$$\Delta m_l = 0, \pm 1 \quad \text{and} \quad \Delta m_s = 0 \qquad (2.16)$$

In Fig. 7.7 the splitting of the six Zeeman sublevels for the $l = 1$, $s = \frac{1}{2}$ fine-structure doublet is shown for a weak and for a strong external magnetic field. For weak fields we have the situation already shown in Fig. 7.6 (Eq. 2.14) while in the limit of very strong fields the six sublevels coalesce into five components almost equidistantly split as predicted by Eq. 2.15. However, an additional term must be added to take account of the **L·S** interaction, which even though weak is still present. As a result, the structure of the spectral line becomes almost a normal Zeeman triplet, as can also be seen in Fig. 7.7. This phenomenon is called the Paschen-Back effect. It necessitates, however, magnetic fields that are beyond the reach of iron core magnets, and thus are difficult to achieve.

2.3 ATOMS WITH MORE THAN ONE VALENCE ELECTRON

In the previous section we discussed the Zeeman effect, including the electron's spin, but only for the case of one (valence) electron outside a closed shell. We have, however, considered in Chapter 2 (Section 4.4) the coupling of the orbital angular momenta l_1, l_2, $\cdots$, of the valence electrons into a resultant **L**, and of their spins s_1, s_2, $\cdots$, into a resultant **S**, which then couple to give **J** = **L** + **S**. This is the so-called Russell-Saunders coupling and is applicable to mercury; the energy-level diagram and a discussion of the mercury spectrum were given in Chapter 2 (Fig. 2.13).

Let us consider the familiar 5461 Å green line which arises from a transition between the $^3S_1(6s7s)$ state to the $^3P_2(6s6p)$ state. Using Eq. 2.13 we can obtain the g factors for the initial and final state:[†]

$$^3S_1(\; J = 1, \; L = 0, \; S = 1) \qquad g_i = 2$$
$$\qquad (2.16)$$
$$^3P_2(\; J = 2, \; L = 1, \; S = 1) \qquad g_f = \tfrac{3}{2}$$

In Fig. 7.8 is given the energy-level diagram for these two states without and with a magnetic field. The structure of the line is indicated in the lower part of the figure and exhibits a symmetric nine-component pattern.

Indeed, the example we are considering involves the same angular momenta $l_i = 1$, $l_f = 2$ as the example discussed in Section 2.1, where we concluded that only three components would appear because of the equality of the splitting a of the initial level, and of the splitting b, of the final level

---

[†] Note that we use in Eq. 2.13 L, S, and J, the quantum numbers for the coupled angular momenta of the two valence electrons.

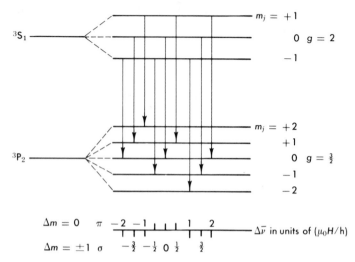

FIG. 7.8 Structure of the Zeeman multiplet arising in a transition from a $^3S_1$ to a $^3P_2$ line; the mercury green line at 5461 Å is an example of such a transition. Note that the situation here is similar to that indicated in Fig. 7.3 if the initial and final state are interchanged.

(Table 7.1). However, in the present instance, the different g factors (Eq. 2.16) make $a \neq b$ and give rise to *all* nine components; the experimental data obtained on the Zeeman effect of this spectral line and discussed in Section 6 below do confirm the structure predicted by Fig. 7.8.

3. Hyperfine Structure

3.1 MAGNETIC DIPOLE INTERACTION

As stated in the introduction, spectral lines, when examined under high resolution, do show a small structure even when no magnetic field is applied. It was also mentioned that such hyperfine structure is due to the interaction of the higher moments of the nucleus with the electromagnetic field of the valence electron or electrons. We will defer the more general treatment of the multipole moments of the nucleus to Section 3.4 and will first discuss only the interaction of the nuclear dipole magnetic moment.

If we consider a nucleus with an intrinsic angular momentum (spin) $\mathbf{I}$ different from zero ($I \geq \frac{1}{2}$), we can expect that the revolving charge of the nucleus will give rise to a magnetic moment (see Eq. 2.2) oriented along the spin axis.

$$\mathbf{\mu} = -\frac{e}{2Mc}\mathbf{I}$$

where M is the mass of the nucleus. In addition, nuclei exhibit an intrinsic magnetization†, so that in general we have

$$\mathbf{\mu} = -g_I \frac{e}{2m_p c} \mathbf{I} = g_I \mu_N I \mathbf{u}_I$$

where $\mathbf{u}_I$ is a unit vector along the spin direction, and

$$\mu_N = \frac{e\hbar}{2m_p c}$$

is the nuclear magneton; m_p is the proton mass. The numerical factor g_I includes all the effects of intrinsic and orbital magnetization of the nucleus and can be obtained only from a theory of nuclear structure.

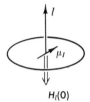

FIG. 7.9 Interaction of the nuclear magnetic moment with the magnetic field produced by the electrons at the nucleus.

Clearly, such a magnetic moment of the nucleus, $\mathbf{\mu}$, will interact with the magnetic field $\mathbf{H}_e(0)$ produced by the atomic electrons (at the nucleus; Fig. 7.9). This interaction then results in a shift of the energy levels of the atom (electrons + nucleus, system) by the amount

$$\Delta E = -\mathbf{\mu} \cdot \mathbf{H}_e(0) \tag{3.3}$$

The direction of $\mathbf{H}_e(0)$ is that given by the total angular momentum of the atomic electrons, namely, $\mathbf{J}$,‡ so that

$$\Delta E = +\left(\frac{\mu}{|I|}\right)\left(\frac{H_e(0)}{|J|}\right)\mathbf{I} \cdot \mathbf{J} \tag{3.4}$$

Thus, we expect the splitting of a level of given J according to the possible values of $(\mathbf{I} \cdot \mathbf{J})$ which as we know are quantized. As a matter of fact, the situation is analogous to that of the fine structure, where the interaction was proportional to the $(\mathbf{L} \cdot \mathbf{S})$ term. In that instance the two angular momenta coupled into a resultant $\mathbf{J} = (\mathbf{L} + \mathbf{S})$ according to the quantum-mechanical laws of addition of angular momentum. In the present situa-

---

† This gives rise to the so-called "anomalous" magnetic moment of the nucleon; for example, the neutron (an uncharged particle) has a magnetic moment of -1.91 μ_N.

‡ The direction of $\mathbf{H}_e(0)$ is really opposite to $\mathbf{J}$ because the electron has negative charge.

tion, $\mathbf{J}$ and $\mathbf{I}$ couple into a total angular momentum of the atom designated by $\mathbf{F}$:

$$\mathbf{F} = (\mathbf{I} + \mathbf{J}) \tag{3.5}$$

An energy level of given $\mathbf{J}$ is then split into sublevels having all possible values of F, namely, the integers (or half integers)

$$|J - I| \le F \le |J + I|$$

Thus if $I = \frac{1}{2}$, the level is split into two components, with $F_1 = J + \frac{1}{2}$ and $F_2 = J - \frac{1}{2}$ (provided $J \ge \frac{1}{2}$); if $I = 1$, the level is split into three components with $F_1 = J - 1$, $F_2 = J$, and $F_3 = (J + 1)$ (provided $J \ge 1$); etc. This situation is shown in Fig. 7.10 and we see that if J is known, the number of hyperfine structure components of a spectral line provides direct information on the spin of the nucleus.

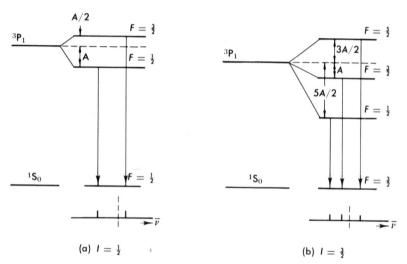

(a) $I = \frac{1}{2}$ (b) $I = \frac{3}{2}$

Fig. 7.10 Hyperfine structure splitting of a $^3P_1$ atomic energy level, and the allowed transitions between the hyperfine structure components of this level and a $^1S_0$ final state. (a) When the spin of the nucleus is $I = \frac{1}{2}$. (b) When the spin of the nucleus is $I = \frac{3}{2}$.

It should be clear from Eq. 3.5 that if either $I = 0$ or $J = 0$, no splitting of the energy levels can occur (the interaction energy specified by Eq. 3.4 vanishes). This is so because if $I = 0$, the nucleus cannot have a dipole moment,† and if $J = 0$, then by symmetry, the magnetic field at the origin $H_e(0) = 0$.

---

† Or any moment; see Section 3.4.

Using Eq. 3.5, we can now obtain the expectation value of the operator $(\mathbf{I} \cdot \mathbf{J})$ that appears in Eq. 3.4; referring again to the vector model (see p. 289) we write "classically"

$$\cos (\mathbf{I}, \mathbf{J}) = \frac{F^2 - I^2 - J^2}{2IJ}$$

and replacing F^2, etc., by the quantum-mechanical expectation values $F(F + 1)$ we obtain

$$\Delta E = \frac{A}{2} [F(F + 1) - I(I + 1) - J(J + 1)] \qquad (3.6)$$

where the constant A is given by

$$A = \frac{\mu}{|I|} \frac{\langle H_e(0) \rangle}{|J|} \qquad (3.7)$$

Note that the energy splitting between sublevels, as given by Eq. 3.6 (and shown in Fig. 7.10), is not symmetric. Further, if we succeed in extracting from the experimental data the constant A, we can obtain the nuclear magnetic moment if $\langle H_e(0) \rangle$ is known.

The calculation of the average value of the magnetic field of the electrons at the nucleus $\langle H_e(0) \rangle$, however, is not easy to perform, and depends on the orbital angular momentum of the valence electron or electrons. In the case of one electron in an s state $(l = 0)$

$$A = \frac{4}{3} \left(\frac{e^2}{4\pi\epsilon_0} \right) \left(\frac{h}{mc} \right)^2 \frac{m_e}{m_p} | \psi_s(0) |^2 g_I \qquad \text{(MKS)} \qquad (3.8)$$

where g_I was defined in Eq. 3.1, and $| \psi_s(0) |^2$ is the absolute square of the wave function at the origin. If a hydrogenlike wave function is used, $| \psi_s(0) |^2 = Z^3/(\pi a_0^3 n^3)$, we obtain

$$A = \frac{8}{3} \frac{hcR_\infty \alpha^3 Z^3}{n^3} \frac{m_e}{m_p} g_I \qquad (3.9)$$

where R_∞ is the Rydberg, α is the fine structure constant; Z is the charge of the nucleus and n the radial quantum number of the orbit. To obtain A in cm^{-1}, we divide Eqs. 3.9 and 3.10 by hc and express R_∞ in wave numbers, $R_\infty = 109737.31 \text{ cm}^{-1}$.

In the case of one electron with $l \neq 0$,

$$A = \frac{hcR_\infty \alpha^3 Z^3}{n^3(l + \frac{1}{2})j(j + 1)} \frac{m_e}{m_p} g_I \qquad (3.10)$$

Equations 3.9 and 3.10 are modified for the case of two valence electrons, and, in general, are subject to various corrections, especially in the heavier elements.†

If we now consider a spectral line arising from the transition between the initial state a and final state b, it will exhibit a hyperfine structure which will be a combination of the hyperfine structure of the two levels. The selection rules for a transition are $\Delta L = \pm 1$; $\Delta J = 0, \pm 1$ (not $0 \to 0$) and $\Delta F = 0, \pm 1$ (not $0 \to 0$).

As an example we examine the structure of the ultraviolet resonance line of Hg ($\lambda = 2537$ Å); as we have seen from Fig. 2.13, it arises from a transition between‡ the $^3P_1$ and $^1S_0$ states. If we assume that the spin of the nucleus of mercury is $\frac{1}{2}$, the line will be split into a doublet as in Fig. 7.10a; if we assume that the spin of the mercury nucleus is $\frac{3}{2}$, the line will be split into a triplet as in Fig. 7.10b. (Note that even if we assume $I > \frac{3}{2}$, the line can never be split into more than 3 components, since $J_i = 1$ and $J_f = 0$.)

However, when the hyperfine structure of the 2537 Å line of *natural* mercury is examined experimentally, it exhibits more than three components! This is because any sample of natural mercury contains several *isotopes*. Six such isotopes of mercury are found in nature in reasonable concentrations; these are listed in Table 7.3 with some of the properties of their nucleus.

From Table 7.3 we see that the two isotopes 199 and 201 will produce a splitting of the 2537 Å line into a doublet and a triplet, respectively. The exact position of these components is given in Fig. 7.11b, which shows the

TABLE 7.3
PROPERTIES OF THE ISOTOPES OF NATURAL Hg ($Z = 80$)

Isotope	Abundance (percent)	N (Neutrons)	I (Nuclear spin)	μ (units of μ_N)	Q (cm² $\times$ 10⁻²⁴)
198	10.1	118	0	0	
199	17.0	119	$\frac{1}{2}$	0.5	
200	23.3	120	0	0	
201	13.2	121	$\frac{3}{2}$	-0.6	$+0.45$
202	29.6	122	0	0	
204	6.7	124	0	0	

† For a detailed discussion of these topics, see the excellent book by H. Kopfeman, *Nuclear Moments*, Academic Press, 1958, Chapter I, Section III.

‡ Note that this is a so-called "intercombination" line (from a triplet state to a singlet; $\Delta S = 1$).

$$\lambda = 2537 \ \overset{\circ}{A} \quad \longrightarrow \ \bar{\nu}$$

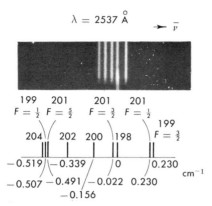

FIG. 7.11   High-resolution spectrogram of the 2537 Å line of natural mercury. In the lower part of the figure the various components are identified and their separation from the position of the Hg198 component is also indicated. (Note that the Hg198 component appears in the spectrogram as the longer line.)

structure of the line†; Fig. 7.11a is a spectrogram obtained with a high-resolution grating in perfect agreement with this structure.

Some important features of Fig. 7.11 are to be noted:

(a) The structure of the Hg201 line is inverted from that shown in Fig. 7.10b; this indicates that the magnetic moment of the Hg201 nucleus is negative, as shown in Table 7.3.

(b) The centers of gravity of the Hg199 and Hg201 levels do not coincide.

(c) There are additional lines in the structure which do not correspond to either Hg199 or Hg201 components.

Indeed, from (b) and (c) we conclude that the same spectral line originating in different isotopes is slightly displaced. This phenomenon is called *isotope shift*, and enables us to identify (correctly) all the components in the 2537 Å line; it is discussed briefly in Section 3.3.

3.2 ELECTRIC QUADRUPOLE INTERACTION

When the three components of the 2537 Å line that are due to Hg201 are isolated and carefully measured, it is found that the *relative* spacings do not agree with the intervals predicted by Eq. 3.6 (Note that this is independent of the value of A.) This deviation is due to the interaction of the atomic electrons with the next higher moment of the nucleus, the "electric quadrupole" moment. The quadrupole interaction is smaller

† To completely identify the structure of a line, information from several transitions is combined, along with Zeeman-effect measurements and spectrograms obtained from sources containing separated isotopes.

than the magnetic dipole interaction and depends on the angle $(\mathbf{I}, \mathbf{J})$ as a Legendre polynomial of $l = 2$, that is, as $\frac{1}{2}(3 \cos^2\theta - 1)$; the quantum-mechanical calculation yields for the energy shift of an F sublevel

$$\Delta E_Q = B \frac{\frac{3}{4}C(C + 1) - I(I + 1)J(J + 1)}{2I(2I - 1)J(2J - 1)} \tag{3.11}$$

where Eq. 3.11 must be added to the energy shift given by Eq. 3.6. Here C is the magnetic dipole spacing given by Eq. 3.6,

$$C = [F(F + 1) - I(I + 1) - J(J + 1)]$$

and B is the quadrupole interaction constant given by

$$B = eQ \left\langle \frac{\partial^2\phi}{\partial z^2}(0) \right\rangle \tag{3.12}$$

Here Q is the quadrupole moment of the nucleus (with respect to the nuclear spin axis, $\mathbf{I}$) and $\langle (\partial^2\phi/\partial z^2)(0) \rangle$ is the average value of the gradient of the electric field of the electrons at the nucleus; z is along the $\mathbf{J}$ direction about which the potential and field are symmetric.

It is shown in Section 3.4 that a nucleus may not have a quadrupole moment unless $I \geq 1$; we also note from Eq. 3.11 that if $J = 0$ or $\frac{1}{2}$, B must be zero, since otherwise $\Delta E_Q \to \infty$ (this is to be expected, because unless an energy level splits up into at least three components, clearly the quadrupole interaction cannot manifest itself.)

We will not discuss the quadrupole interaction any further because its effect can be revealed in the hyperfine structure only with very precise data. Such precision is difficult to achieve by the techniques of optical spectroscopy. Other techniques, however, which measure directly the spacing between hyperfine structure components† are capable of much higher precision; it then becomes possible to measure accurately the effects of the quadrupole and even of the *magnetic octupole* $(l = 3)$ moment.

When a magnetic field is applied, each hyperfine component (of a given F number) exhibits a Zeeman effect, splitting into $2F + 1$ components with quantum numbers m_F. The additional energy is given as before by

$$\Delta E = m_F H \mu_0 g_F \tag{3.13}$$

As the field H is increased, the energy shift given by Eq. 3.13 becomes very quickly of the order of the hyperfine splitting and F ceases to be a good quantum number. Then a Paschen-Back effect of the hyperfine structure takes place analogous to Eq. 2.15 and the energy shift is given by

$$\Delta E = m_J H \mu_0 g_J - m_I H \mu_N g_I + A m_I m_J \tag{3.14}$$

_____

† Resonance methods as described in Chapter 8.

3.3 ISOTOPE SHIFT

It has been noted in Section 3.1 that the complete structure of the 2537 Å line can be explained when we assume that the spectral lines of different istopes are shifted with respect to one another. This in turn means that the atomic energy levels are displaced slightly from the position calculated for a point nucleus; further, there must be a *different* displacement in the initial and in the final level if a shift is to be observed in the emitted spectral line (see Fig. 7.12a).

Clearly, it is not possible to measure the absolute value of the isotope shift of an energy level; we may only detect *differences* between the isotope shift of a level from one isotope to another.

The isotope shift arises from two conditions: (a) The finite mass of the nucleus (the nucleus is much heavier than the electron, but we can consider its mass as infinite only to a first approximation). (b) The finite size of the nucleus (the nuclear radius is much smaller than the orbit of the electron, but we can consider the nucleus as a point only to a first-order approximation). For light elements the isotope shift is mainly due to the effect of the finite mass, whereas for the heavy elements it is mainly due to the finite size effect.

We know that when solving the Schrödinger equation we cannot assume that the nucleus is stationary but instead must consider both the electron and nucleus as revolving about the center of mass of the electron-nucleus

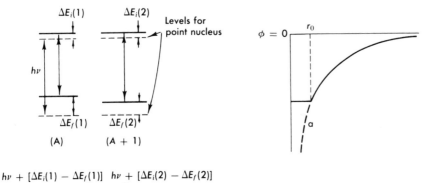

FIG. 7.12   The isotope shift of atomic spectral lines. (a) The energy levels of the initial and final state of two different isotopes with mass numbers A and $A + 1$ are shown. The dotted lines show the position the levels would have if the nucleus was an infinitely heavy point; the solid lines show the actual position of the levels which are shifted by a different amount for each isotope, and for each level. (b) Modification of the Coulomb potential of the nucleus due to its finite size.

system. This assumption leads back to the Schrödinger equation for a stationary attractive center (nucleus) if the mass of the electron is replaced by its *reduced mass*

$$m' = m \frac{M}{M + m} \tag{3.15}$$

Then the energy of a hydrogenlike level is given by

$$E_n = \frac{hcR_\infty Z^2}{n^2} \left(1 - \frac{m}{M} \right) \tag{3.16}$$

which depends on the mass of the nucleus. By differentiating Eq. 3.16 we obtain†

$$\Delta E_n = \frac{hcR_\infty Z^2}{n^2} \frac{m}{m_p} \frac{\Delta A}{A^2} \tag{3.17}$$

where A is the atomic number.

For the heavier elements, the finite mass effect given by Eq. 3.17 decreases as $1/A^2$, and the finite size of the nucleus becomes the dominant reason for a shift of the energy level. Consider Fig. 7.12b where curve (a) represents the Coulomb potential of a point charge. If it is assumed that the electric charge of the nucleus is distributed on a spherical surface of radius r_0, then the potential will not diverge at $r = 0$, but will be constant for all $r \leq r_0$. Thus the potential seen by an electron will be of the form shown by the solid curve of Fig. 7.12b.

It can be shown, then, that the energy of a level will be shifted due to changes Δr_0 in the nuclear radius r_0 by an amount

$$\Delta E_n \simeq \frac{hcR_\infty Z^2}{n^3} F(Z, r_0) \frac{\Delta r_0}{r_0} \tag{3.18}$$

The nuclear radius can be reasonably well expressed by

$$r_0 = A^{1/3} \times 1.2 \times 10^{-13} \text{ cm} \tag{3.19}$$

so that $\Delta r_0/r_0 = \Delta A/3A$, which for $\Delta A = 2$ and $A \sim 200$ yields a $1/300$ change in isotope shift.

Returning to the 2537 Å ultraviolet line of mercury in Fig. 7.11, we see that indeed the lines of the heavier elements (larger radius) are shifted toward lower energies, the spacings between 198–200, 200–202, and 202–204 being almost equal as expected. A similar displacement appears between the *centers of gravity* of the hyperfine structure multiplets of Hg^{199} and Hg^{201} (dotted lines in Fig. 7.10).

---

† Different expressions hold for atoms with more than one valence electron.

3.4 NUCLEAR MOMENTS AND THEIR INTERACTION WITH THE ATOMIC
ELECTRONS

This section contains a more general, but still not rigorous, discussion on nuclear moments. At a first reading it can be bypassed without affecting the continuity of the material presented in this chapter.

The term *nuclear moment* refers to the electromagnetic moments of the charge and current distribution of a nucleus. The moments arise from the deviation from spherical symmetry of these distributions, and are a property of the nucleus.

It is known that the potential due to a localized charge distribution can be expressed outside the charge by an expansion in spherical harmonics

$$\phi(\mathbf{r}) = \sum_{l=0}^{\infty} \sum_{m=-l}^{+l} \frac{q_{l,m}}{r^{l+1}} Y_{l,m}(\theta, \phi) \tag{3.20}$$

where the numerical coefficients $q_{l,m}$ are given by

$$q_{l,m} = \iiint \rho(r', \theta', \phi') r'^{l} Y_{l,m}^{*}(\theta', \phi') r'^{2} \sin \theta' \, d\theta' \, d\phi' \, dr' \tag{3.21}$$

and $Y_{l,m}$ are the spherical harmonics; ρ is the charge distribution of the nucleus and r', θ', ϕ' are the coordinates of the nucleus. The spherical harmonics in Eqs. 3.20 and 3.21 must be referred to some axis; this is chosen as the axis of the nuclear spin $\mathbf{I}$.

As usual, for the nuclear charge density $\rho(r', \theta', \phi')$ the absolute square of the nuclear wave function is substituted; hence,

$$q_{l,m} = e \int r'^{l+2} \, | \, \psi(r') \, |^{2} \, dr' \iint \psi(\theta', \phi') Y_{l,m}^{*} \psi^{*}(\theta', \phi') \, d\Omega' \tag{3.22}$$

where we have separated the radial and angular part of the wave function; $\psi(\theta', \phi') \propto Y_{l=I}^{m}(\theta', \phi')$. Thus we see that (1) $q_{l,m} = 0$ unless $l \leq 2I$, where I is the nuclear spin; and (2) assuming that $\psi(\theta', \phi')$ has a definite parity, then $| \, \psi(\theta', \phi') \, |^{2}$ has always even parity, and only Y_{lm}^{*} of even parity will give nonzero contributions to the angular integral. Thus l must be even: $l = 0, 2, 4, \cdots$.

Similarly, the nuclear current and magnetization distribution† give rise to the vector potential $\mathbf{A}$ which can also be expanded in powers of $(1/r)$; the expansion is more complicated than Eq. 3.20 because the vector spherical harmonics must be used.

---

† See, for example, J. D. Jackson, *Classical Electrodynamics*, John Wiley, Chapter 16.

The magnetic moments are defined in analogy with Eq. 3.21 as

$$p_{l,m} = -\frac{1}{l+1} \iiint r'^l \nabla \cdot \left(\frac{\mathbf{r'} \times \mathbf{J}(\mathbf{r'})}{c} \right) Y_{l,m}^* \, d^3r' \qquad (3.23)$$

However, the parity requirement on the angular integrals makes all even terms in l vanish (because ∇, $\mathbf{r}$ and $\mathbf{J}(\mathbf{r'})$ all have odd parity, so that $Y_{l,m}^*$ must also be odd).† Thus, for the magnetic moments, l must be odd: $l = 1, 3, 5, \cdots$.

For $l = 1$ there are three multipole moments‡ that can be thought of as the components of a vector; however, the $l = 1$, $m = 0$ term gives the projection onto the z axis (**I** axis) of the familiar magnetic moment vector,

$$p_{1,0} = \frac{1}{2c} \int [\mathbf{r'} \times \mathbf{J}(\mathbf{r'})] \cos \theta \, d^3r' \qquad (3.24)$$

In the present instance this is the only number of importance, since the p_x and p_y components will average out for a nucleus spinning about the z axis.

Similarly for $l = 2$ there are five multipole moments contributing to this quadrupole term; they can be thought of as the components of a 3×3 symmetric tensor. However, only the $l = 2$, $m = 0$ term is required in the calculation of the energy, and it is called the electric quadrupole moment Q, where

$$eQ = \frac{1}{2} \iiint (3 \cos^2 \theta - 1) r'^2 \rho(r', \theta', \phi') \, d^3r' \qquad (3.25)$$

As an example of an electric quadrupole moment, consider Fig. 7.13 where charge is uniformly distributed inside the equal ellipsoids (a) and (b);

Q > 0

(a)

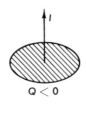

Q < 0

(b)

FIG. 7.13 The quadrupole moment of a nucleus with a charge distribution that is not spherically symmetric. Note that the value of Q depends on the distribution of the charge with respect to the nuclear spin axis.

---

† More precisely, $\mathbf{r}$ and $\mathbf{J}$ are both vectors, so that $\mathbf{r'} \times \mathbf{J}$ is a pseudovector. Similarly ∇ has transformation properties of a vector, so that $\nabla \cdot (\mathbf{r'} \times \mathbf{J})$ is a pseudoscalar.

‡ In general, $q_{l,m} = (-1)^m q_{l,-m}^*$.

the nuclear spin, however, is oriented in a different direction in the two cases. The quadrupole moment has the absolute value

$$Q \simeq \tfrac{5}{2} r_0{}^2 \epsilon$$

where r_0 and ϵ are the radius and eccentricity of the generating ellipse, but in case (a), Q is positive while in case (b), it is negative.

Having defined the moments of the nucleus, we wish to find the interaction energy with the atomic electrons. For a charge distribution $\rho(\mathbf{x})$ in a potential $\phi(\mathbf{x})$, the energy is

$$E = \iiint \rho(\mathbf{x})\phi(\mathbf{x}) \, d^3x \tag{3.26}$$

If we assume that the charge is localized at the vicinity of the origin and the potential $\phi(\mathbf{x})$ is slowly varying in that region, we can expand it in a Taylor series about the origin, obtaining for Eq. 3.26

$$E = e\phi(0) - \mathbf{p} \cdot \nabla\phi]_0 - \frac{e}{6} \sum_i \sum_j Q_{ij} \frac{\partial^2 \phi}{\partial x_i \partial x_j}\bigg]_0 + \cdots \tag{3.27}$$

However, since the potential of the atomic electrons is assumed to have cylindrical symmetry about $\mathbf{J}$,

$$\nabla\phi]_0 = \frac{\partial\phi}{\partial z}\bigg]_0 \mathbf{u}_J$$

and

$$\frac{\partial^2 \phi}{\partial x_i \partial x_j} \neq 0 \qquad \text{only for } i = j$$

further†

$$E = e\phi(0) - p_{1,0} \frac{\partial\phi}{\partial z}\bigg]_0 \cos(\mathbf{I}, \mathbf{J}) - \frac{q_{2,0}}{4} \frac{\partial^2 \phi}{\partial z^2}\bigg]_0 (\tfrac{3}{2}\cos^2(\mathbf{I}, \mathbf{J}) - \tfrac{1}{2}) \tag{3.28}$$

From Eq. 3.28 we can clearly see that the dipole moments interact with the field and show an angular dependence, typical of the $l = 1$ Legendre polynomial, whereas the quadrupole moments interact with the gradient of the field showing an angular dependence, typical of the $l = 2$ Legendre polynomial, and so on.

Referring now to Table 7.3, we can understand why the mercury nuclei that have $I = 0$ do not possess a magnetic moment. Similarly Hg[199] with

† Use is made of the fact $\nabla^2\phi]_0 = 0$ and of the circular symmetry, so that

$$\frac{\partial^2 \phi}{\partial z^2}\bigg]_0 = -2 \frac{\partial^2 \phi}{\partial x^2}\bigg]_0 = -2 \frac{\partial^2 \phi}{\partial y^2}\bigg]_0$$

$I = \frac{1}{2}$ cannot have a quadrupole moment, while Hg^{201} with $I = \frac{3}{2}$ may have one.

It is striking that all the *even* isotopes of mercury should have $I = 0$; we should also note that since for mercury $Z = 80$, these isotopes all have an even number of protons *and* an even number of neutrons. It is observed for the *ground* state of all nuclei that the even (neutrons)-even (protons) nuclei have $I = 0$, while the odd-even or even-odd ones have half-integral spin. This is then a strong indication that the spin of the nucleus is compounded from the spin and orbital angular momentum of its neutrons and protons (the orbital angular momentum being, in units of $\hbar$, of course, always an integer).

4. Line Width and Light Sources for High-Resolution Spectroscopy

It is obvious that if very small differences between components of a spectral line are to be resolved, the width of the component must be less than the separation. This is not easy to achieve without special precautions, but it constitutes the first step towards obtaining a high-resolution spectrogram.

Spectral lines, as also briefly discussed in Chapter 6, Section 4.1, have a natural width given by

$$\Delta\nu = \frac{\Delta E}{h} \simeq \frac{1}{2\pi\Delta\tau} \qquad (4.1)$$

where $\Delta\tau$ is the lifetime of the state; this is negligible, since atomic lifetimes are of the order of $\tau \gtrsim 10^{-8}$ sec; thus

$$\Delta\bar{\nu} = \frac{\Delta\nu}{c} = \frac{1}{2\pi \times 3 \times 10^{10} \times 10^{-8}} < 10^{-3} \text{ cm}^{-1}$$

However, external influences may and do broaden spectral lines considerably; the main causes are as follows:

(a) *Doppler broadening.* Due to their thermal energy, the atoms in the source move in random directions with a velocity given by the Maxwell distribution. Consequently, the wavelength emitted in a transition of the atom is Doppler-shifted†; this results in a broadening of the line, which can be shown to have a half-width

$$\Delta\bar{\nu} = 10^{-6} \sqrt{\frac{T}{A}} \, \bar{\nu} \text{ cm}^{-1} \qquad (4.2)$$

---

† See, for example, the discussion on the Mössbauer effect in Chapter 6, Section 4.

where $\bar{\nu}$ is the wave number of the line, T is the absolute temperature in degrees K, and A is the atomic number of the element.

Thus, Doppler broadening is most serious for the light elements and in sources that operate at high temperatures. For example, in an arc discharge operating at $T = 3600°$ K, a hydrogen line of $\lambda = 5000$ Å will have a Doppler width of 1.2 cm^{-1}, which will mask any hyperfine structure. For heavy elements, as in Hg ($A \sim 200$), $\Delta\bar{\nu} = 0.100$ cm^{-1}, which is still quite broad. However, by using a glow discharge that is cooled (sometimes even by liquid air) reasonable results can be obtained.

(b) *Pressure (or collision) broadening.* When the pressure in the source vapor is too high, the atoms are subject to frequent collisions, which in a way can be thought of as reducing the time interval $\Delta\tau$ entering into Eq. 4.1. To avoid this effect the pressure in the discharge tube is kept below 1 mm of mercury. Usually to facilitate the operation of the tube it is filled with some noble gas (such as argon or neon) while the concentration of atoms of the element under examination is kept much lower than 1 mm.

(c) *External fields.* Magnetic or electric fields produce Zeeman or Stark splitting of the components, resulting in effective broadening of the line. Electric fields of 1000 V/cm can cause a broadening of a few cm^{-1}.

(d) *Self-absorption and reversal.* This phenomenon is most pronounced with resonance lines.† As the radiation emitted from the atoms in the middle of the source travels through the vapor, it has a probability of being absorbed that is proportional to the path length it traverses and to the ab-

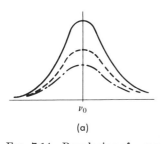

ν_0

(a)

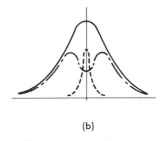

(b)

Fig. 7.14   Broadening of a spectral line due to self-absorption in the source. The solid curve is the emitted line, the dashed curve represents the part of the radiation that is absorbed and the dash-dot curve shows the transmitted line which is the difference of the two former curves. (a) Normal absorption, (b) strong absorption especially in the central region leading to self-reversal.

---

† The reader may again consult the discussion on resonant absorption of radiation presented in Chapter 6 in connection with the Mössbauer effect.

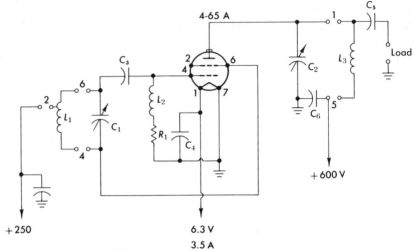

FIG. 7.15 A radio frequency oscillator suitable for exciting an electrodeless discharge tube. Such a source emits spectral lines of narrow width.

L_1 = 20 turns No. 16 wire $1\frac{1}{2}$ dia. $1\frac{5}{8}$L tapped $6\frac{1}{2}$ C_3 = 3800$\mu\mu$F
L_2 = 7 turns No. 16 wire $\frac{3}{8}$ dia. $1\frac{5}{8}$L C_4 = 0.02μF 600 V
L_3 = 7MC BUD-OE L-40 C_5 = 0.01μF 5000V
C_1 = 100$\mu\mu$F C_6 = 0.1μF 3000V
C_2 = 200$\mu\mu$F R_1 = 5K 20W

sorption cross section; this will be strongest in the center of the line and weaker in the wings. The result shown in Fig. 7.14a is that the line becomes "squashed" in the center; that is, it is broadened.

If now the outer layers of the source are much cooler than the middle ones, the width of the particular energy level (due to the Doppler effect) is smaller in the outer layers and absorption takes place only at the central frequency with almost none in the wings. The result is a "self-reversed" line as shown in Fig. 7.14b. This effect is very pronounced in the sodium D lines, and when it is viewed with a high-resolution instrument, the line exhibits a doublet structure which is frequently mistaken for hyperfine structure.

In this laboratory when a commercial high-pressure sodium light is used for the alignment of the Fabry-Perot interferometer, each D line gives rise to a double ring pattern. The separation between centers is of the order of approximately 0.100 cm^{-1} and is a function of the operating voltage, indicating that it is due to self reversal.†

In order to avoid these causes of spectral line broadening, special sources have been developed. They must radiate lines whose width is as close as

---

† The hyperfine structure pattern of the sodium D lines is 0.065 cm^{-1} wide.

possible to the natural one, but they must also be quite intense, since high-resolution instruments by necessity have small luminosity. The sources most commonly used are the hollow cathode and the electrodeless discharge. In this laboratory the electrodeless discharge is used. It consists of a narrow tube (pyrex or quartz) containing a separated mercury isotope (198) and filled with 10 mm of neon. A radiofrequency or microwave source is coupled to it by means of clip-leads or a wave guide and provides the power for sustaining the discharge; the source usually has to be started with a spark (Tesla) coil. The source may be cooled with an air blast, or by other means, to dissipate the power absorbed by the glass envelope; Fig. 7.15 shows the 7 Mc 50-watt radiofrequency oscillator used in this laboratory to excite the source; the tuning condenser is adjusted for maximum brilliance or as desired. The Hg^{198} source operated in this fashion yielded lines of half width at half maximum of the order of $\Delta \bar{\nu} = 0.030$ cm^{-1}.

If a quartz (or vycor) envelope is used for the source (as it must be when ultraviolet lines are investigated), the experimenter must always wear glasses. As is well known, even *short exposures* of the eye to ultraviolet light *cause damage*, sometimes permanent; strong ultraviolet mercury lines even cause sunburn of the skin.

5. The Fabry-Perot Interferometer

5.1 GENERAL DISCUSSION

As mentioned in the introduction, the Fabry-Perot is the most commonly used of the "multiple-beam" interferometers; other instruments in this category are the Lummer-Gehrcke plate and the Michelson échelon. The Fabry-Perot has found many uses, such as the measurement of refractive indices of gases, measurement of lengths, and so on, but here we will be mainly concerned with its application to the measurement of very small differences in wavelengths such as appear in the Zeeman effect and hyperfine structure.

The Fabry-Perot consists of two parallel flat glass plates, coated on the inner surface with a partially transmitting metallic layer. Consider

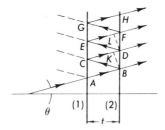

FIG. 7.16 Reflected and transmitted rays at the two parallel surfaces (1) and (2) of a Fabry-Perot étalon. The étalon spacing is t.

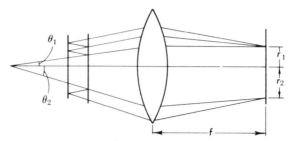

FIG. 7.17   Focusing of the light emerging from a Fabry-Perot étalon. Light entering the étalon at an angle θ is focused onto a ring of radius $r = r\theta$ where f is the focal length of the lens.

then, the two partially transmitting surfaces (1) and (2) shown in Fig. 7.16 and separated by a distance t. An incoming ray making an angle θ with the normal to the plates will be split into the rays AB, CD, EF, etc. The path difference between the wave fronts of two adjacent rays—for example, AB and CD—is

$$\delta = BC + CK$$

where obviously BK is normal to CD. Then†

$$\delta = 2t \cos \theta$$

and for constructive interference to occur

$$n\lambda = 2t \cos \theta \tag{5.1}$$

where n is an integer; Eq. 5.1 is the basic interferometer equation. If the refractive index of the medium between the plates is $\mu \neq 1$, we must modify Eq. 5.1 to

$$n\lambda = 2\mu t \cos \theta \tag{5.1a}$$

Let the parallel rays B, D, F, etc., be brought to a focus by the use of a lens of focal length f as shown in Fig. 7.17. Then when θ fulfills Eq. 5.1a, bright rings will appear in the focal plane, their radius being given by

$$r = f \tan \theta \simeq f\theta \tag{5.2}$$

Next we wish to find the intensity of the rings (fringes) and the contrast between bright fringes and background. Let the transmission coefficient be T, and the reflection coefficient be R (for no absorption at the surface $R = 1 - T$); and let the intensity of the incident radiation have an am-

† $CK = BC \cos 2\theta$,   and $BC \cos \theta = t$

hence

$$\delta = BCK = BC(1 + \cos 2\theta) = 2BC \cos^2 \theta = 2t \cos \theta$$

plitude A and therefore intensity I_0 where $I_0 = A^2$. The intensities of the rays that are transmitted through the second surface, B, D, etc., are obtained by squaring their amplitudes. Thus

$$I_B = (AT)^2 = I_0 T^2$$

$$I_D = (ARRT)^2 = I_0 R^4 T^2 \text{ etc.}$$

The amplitudes between adjacent rays decrease as R^2, and, therefore, unless R is close to 1, the interference maxima are not sharp. As R is increased, however, T decreases,† and one might suppose that the intensity of the rings would be greatly diminished; this is not true, since now the amplitudes of many rays can be added. The summation over all amplitudes, taking into account the change in phase, is given by Airy's formula:

$$I_T = \left[\sum_{N=1}^{\infty} A_N \right]^2 = I_0 \frac{T^2}{(1 - R)^2} \frac{1}{1 + [4R/(1 - R)^2] \sin^2 \delta/2} \quad (5.3)$$

with

$$\delta = 2\pi \frac{2t}{\lambda} \cos \theta$$

We note that at the maxima ($\delta = 0, 2\pi, \cdots$)

$$I_T = \frac{I_0 T^2}{(1 - R)^2} \quad (5.4)$$

which for no absorption yields $I_T = I_0$; at the minima, ($\delta = \pi, 3\pi, \cdots$)

$$I_T = I_0 \frac{T^2}{(1 + R)^2} \quad (5.5)$$

which for no absorption yields $I_T = I_0(1 - R)^2/(1 + R)^2$ showing that for R close to 1, a very good contrast can be achieved. The intensity distribution of the fringes for different values of R is shown in Fig. 7.18, and clearly high values of R, >0.90 are desirable.

It is important to note the following points:

(a) Since the interfering rays emerging from the interferometer are parallel, they must be focused by a lens of good quality (see Fig. 7.17).

(b) The order of interference n is, in general, very large $[n_0 = (2t)/\lambda]$ but the rings that are observed are only from the few orders‡

$$n = (n_0 - \epsilon) - (p - 1) \quad (5.6)$$

where $p = 1, 2, \cdots$ may be as high as 10.

---

† Typical values for a good interferometer are $R = 0.96$; $T = 0.04$.

‡ Equation 5.6 is discussed in Section 5.3.

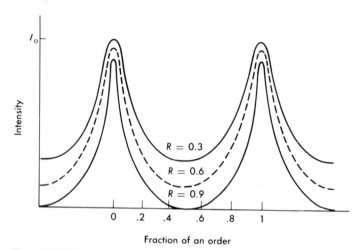

Fraction of an order

FIG. 7.18   The intensity of light from a Fabry-Perot étalon (viewed in the focal plane of the lens) as a function of radial displacement in units of separation between successive orders. R is the reflection coefficient of the interferometer plates; the contrast improves rapidly for increasing R.

(c) Since interference occurs for rays making an angle θ with the normal to the plates, a *perfectly* parallel beam may not produce fringes. Since we are interested only in small angles θ, an *almost* parallel beam is allowed to be incident on the interferometer; this is achieved by using an extended source at the focus of the first lens (see Fig. 7.21).

5.2 THE FABRY-PEROT ÉTALON

The multiple-beam interferometer discussed above is easily realized in practice. Two *optically flat* glass or quartz plates with one surface coated with an appropriate reflecting film are used. The plates are assembled in a holder (see Fig. 7.19) and held apart by three very accurately machined spacers. Three spring-mounted screws are used to apply pressure, and by

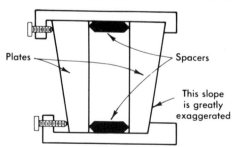

Plates

Spacers

This slope is greatly exaggerated

FIG. 7.19   Mounting of the interferometer plates into a Fabry-Perot étalon. Note that the slight slope of the two sides of a plate is usually of the order of 1/10 degree.

careful adjustment, the plates are made parallel. Such an assembly is frequently called a Fabry-Perot étalon.

The plates can be ground flat to $\lambda/20$ (approximately 250 Å), and
usually the outer surface is slightly inclined (about $0.1°$) with respect to
the inner one in order to avoid multiple reflections, which give rise to
"ghost" fringes. The reflecting coating is deposited by evaporation in
vacuum and is either silver or aluminum depending on the wavelength
that will be investigated. Also multilayers of dielectrics are now commonly
used. Figure 7.20 gives the reflectivity of these metals as a function of
wavelength, but for thick coatings. For thin layers, the reflectivity increases with the thickness of the metal film (but so does the absorption, A);
with a film 500 Å thick we may obtain $R = 0.94$, $T = 0.02$, and $A = 0.04$.
Note that the performance of the interferometer depends primarily on
the quality of these coatings.

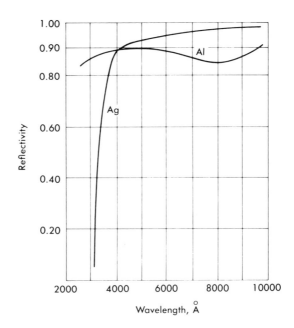

FIG. 7.20 The reflectivity of
aluminum and silver coatings
as a function of wavelength.

It is also important that the spacers be equal and accurately machined.
They can be made of quartz and should also be ground to a $\lambda/20$ precision.
It is possible to use several sets of spacers of different length with the same
pair of plates; this modifies the spectral range of the étalon, which is given
by $\Delta\bar{\nu} = 1/2t$ cm^{-1}.

When long exposures are taken, attention must be given to the stability
of the system. The air between the plates must be kept at constant tem-

perature in order that the refractive index† entering Eq. 5.1a does not change; this can be achieved by enclosing the étalon in an insulated box.

It is clear that when a source containing several wavelengths, for example, an atomic spectrum, is viewed with a Fabry–Perot, the ring patterns from all wavelengths are superimposed. To separate the patterns and view only the fringes of the line of interest, a low-dispersion spectrometer is used in conjunction with the Fabry–Perot. The ring pattern is focused onto the slit of the spectrometer, in the focal plane of which now appear vertical sections of each ring pattern, but these are dispersed according to the wavelength. We speak of "crossing" the Fabry–Perot with the spectrometer. Good grating spectrographs are occasionally used for that purpose, but a small prism spectrograph is usually adequate. There are basically two methods for introducing the Fabry–Perot étalon in the optical system: (a) a parallel beam arrangement as shown in Fig. 7.21a, or (b) a converging beam as shown in Fig. 7.21b.

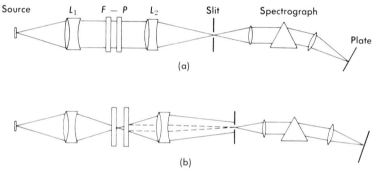

FIG. 7.21   Two possible arrangements for crossing a Fabry-Perot étalon with a low-resolution spectrograph; L_1 and L_2 are compensated lenses and F–P indicates the position of the étalon. (a) The Fabry-Perot is placed in a quasi-parallel beam originating from an extended source. (b) The image of the source is focused onto the Fabry-Perot.

In method (a), lens L_2 focuses the ring pattern onto the spectrometer slit; care must be taken that it be properly adjusted. Maximum light intensity is used with such an arrangement. In method (b), the light source is focused by means of lens L_1 halfway between the Fabry–Perot plates; and the ring pattern is focused by means of L_2 onto the slit. This method has the advantage of easier adjustment of the plates for parallelism and of producing sharper fringes. In both methods the diameter of the rings is

† The reader may verify that small changes in temperature may shift the ring pattern by a whole fringe.

determined by the focal length of L_2 and the magnification of the spectrograph, which is frequently 1:1 (the ratio of collimator and camera lens focal lengths).

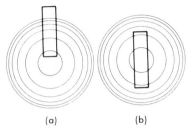

FIG. 7.22 Two possible arrangements of the spectrograph slit with respect to the Fabry-Perot ring pattern. (a) Off-center method, (b) the center of the pattern as well as two segments of each ring appear on the plate.

(a)   (b)

We may project onto the spectrometer slit the central portion of the ring pattern (Fig. 7.22b) or only part of it (Fig. 7.22a). The latter effect can be achieved by tilting the Fabry–Perot étalon (after adjusting the plates) with respect to the optical axis, and leaving the lenses L_1 and L_2 in their original positions. This "off-center" method is used when only a short slit is available, or when the source is so small (not extended) that only few rings can be formed with enough intensity.

5.3 REDUCTION OF THE DATA OBTAINED FROM A FABRY-PEROT

We have seen (Eq. 5.2) that the interference rings formed in the focal plane have radii

$$r_n = f\theta_n \tag{5.2}$$

where the angle θ_n is given by

$$n = \frac{2\mu t}{\lambda} \cos \theta_n = n_0 \cos \theta_n = n_0 \left(1 - 2 \sin^2 \frac{\theta_n}{2} \right)$$

and since θ is always small, we obtain

$$n = n_0 \left(1 - \frac{\theta_n^2}{2} \right) \quad \text{or} \quad \theta_n = \sqrt{\frac{2(n_0 - n)}{n_0}} \tag{5.7}$$

Now if θ_n is to correspond to a bright fringe, n must be an integer; however, n_0, which gives the interference at the center ($\cos \theta = 1$ or $\theta = 0$ in Eq. 5.1), is in general *not* an integer:

$$n_0 = \frac{2\mu t}{\lambda} \tag{5.8}$$

(There is no bright spot in the center of the pattern, in general.) If n_1 is the interference order of the first ring, clearly $n_1 < n_0$ since $n_1 = n_0 \cos \theta_1$. We then let $n_1 = n_0 - \epsilon$, with $0 < \epsilon < 1$ where n_1 is the closest integer to

n_0 (smaller than n_0). Thus, we have in general for the pth ring of the pattern, as measured from the center out,

$$n_p = (n_0 - \epsilon) - (p - 1) \tag{5.6}$$

Combining Eq. 5.6 with Eqs. 5.7 and 5.2, we obtain for the radii of the rings

$$r_p = \sqrt{\frac{2f^2}{n_0}} \sqrt{(p - 1) + \epsilon} \tag{5.9}$$

We note (a) that the difference between the squares of the radii of adjacent rings is a constant

$$r_{p+1}^2 - r_p^2 = \frac{2f^2}{n_0} \tag{5.10}$$

and (b) that the fraction of an order ϵ can be found by extrapolating to $r_p{}^2 = 0$ (according to the slope $2f^2/n_0$) †.

Now,‡ if there are two components of a spectral line with wavelengths λ_1 and λ_2, very close to one another, they will have fractional orders at the center ϵ_1 and ϵ_2:

$$\epsilon_1 = \frac{2t}{\lambda_1} - n_1(1) = 2t\bar{\nu}_1 - n_1(1)$$

$$\epsilon_2 = \frac{2t}{\lambda_2} - n_1(2) = 2t\bar{\nu}_2 - n_1(2)$$

where $n_1(1)$, $n_1(2)$ is the order of the first ring. Hence, if the rings do not overlap by a whole order ($n_1(1) = n_1(2)$), the difference in wave numbers between the two components is simply

$$\bar{\nu}_1 - \bar{\nu}_2 = \frac{\epsilon_1 - \epsilon_2}{2t} \tag{5.11}$$

If the orders are overlapped x times,

$$\bar{\nu}_1 - \bar{\nu}_2 = \frac{x + \epsilon_1 - \epsilon_2}{2t} \tag{5.11b}$$

From Eq. 5.11 we see that we do not need to know t much more accurately than $\epsilon_1 - \epsilon_2$. The fractional order $\epsilon_1, \epsilon_2, \cdots$, can hardly be measured to $1/1000$; therefore knowledge of t to this accuracy of $1/1000$ is amply adequate; this can be easily achieved with a micrometer or a microscope.

---

† See also Fig. 7.28.
‡ From here on we set $\mu = 1$.

The resolution† of the Fabry–Perot can be obtained from Eq. 5.1:

$$\bar{\nu} = \frac{1}{\lambda} = \frac{n}{2t \cos \theta}$$

and by differentiation

$$\Delta \bar{\nu} = \frac{\Delta n}{2t} \left[\frac{1}{\cos \theta} - \frac{n \sin \theta}{\cos^2 \theta} \right] \tag{5.12}$$

and since θ is always small

$$\Delta \bar{\nu} \simeq \frac{1}{2t} \Delta n \tag{5.13}$$

where Δn is the *fraction of order* by which one ring pattern is shifted with respect to an other. (Note that when this fraction of order is measured at the center, Eq. 5.13 becomes exact, and since $\Delta n (\theta = 0) \equiv \epsilon_1 - \epsilon_2$ we get back Eq. 5.11.)

The fraction of order Δn that can be measured experimentally depends on the quality of the plates (as shown by Eq. 5.5, contrast of fringes), on the proper alignment and *focusing* of the optical system, and on the width and relative intensity of the components that are being measured. Values of $\Delta n \approx 1/100$ are common, and with some care this value can be exceeded‡.

Using $\Delta n = 1/100$, we then find for the resolving power of the Fabry–Perot at a wavelength of $\lambda = 5000$ Å and a spacing $t = 0.5$ cm

$$\frac{\Delta \bar{\nu}}{\bar{\nu}} = \Delta n \left(\frac{\lambda}{2t} \right) = 10^{-2} \times 5 \times 10^{-5} = 5 \times 10^{-7} \tag{5.14}$$

which is quite satisfactory. In general, we see that the wave number interval between adjacent rings (called the free spectral range) is

$$\Delta \bar{\nu} \text{ (for } \Delta n = 1) \simeq \frac{1}{2t}$$

which can be used as a "scale factor" for the frequency difference of any components appearing in the spectral line.

In reducing the data our aim is to obtain the orders of fractional interference $\epsilon_1, \epsilon_2, \cdots$, for all the components of the line, and also to know if any of the components overlap in order, and in that case by how many orders.

---

† See also Eq. 1.5a.

‡ For example, for the data discussed in Section 6.2, and presented in Fig. 7.27, Δn is of the order of approximately 1/20.

The possibility of components overlapping in order increases when a large spacing is used ($\Delta\bar{\nu}$ between adjacent fringes is small). To clear up such ambiguity, exposures of the same line are obtained with two or three different sets of spacers, providing also consistency checks of the assignments that were made.

As stated before, we may obtain a Fabry-Perot pattern containing the center of the ring system as shown in Fig. 7.22b. In that case we work with rings close to the center so as to extract the fractional order ϵ for each component. Alternatively, we may obtain a pattern not containing the center as shown in Fig. 7.22a, in which case we use only the pth to the $(p + q)$th rings; this is known as the off-center method.

We will be concerned with the first method, and let R_p be the radius of the pth ring as measured on the photographic plate. Note that it is possible to measure R_p only if the center of the pattern is included on the plate. The fractional order ϵ is then given by Eq. 5.15 below, which follows from Eq. 5.9

$$\frac{R_{p+1}^2}{R_{p+1}^2 - R_p^2} - p = \epsilon \qquad (5.15)$$

We note that the denominator in Eq. 5.15 is a constant, and any adjacent pair of rings can yield a value for ϵ. However, since the squares of the radii of successive rings are linearly related (they form an arithmetic progression) in order to utilize all available information a least squares fit to Eq. 5.15 must be made.

A somewhat less tedious reduction technique, in which the squares of the radii and their differences are tabulated in a square array, is given by Tolansky.† Consider a line which has three components a, b, c, and let the respective radii be R_{1a}, R_{2a}, R_{3a}, $\cdots$, for component a; R_{1b}, R_{2b}, R_{3b}, $\cdots$, for component b, and similarly for c. From Eq. 5.10 it is clear that the difference between the squares of the radii of component a,

$$\Delta_a = R_{(p+1),a}^2 - R_{p,a}^2 = \frac{2f^2}{n_{0,a}}$$

is equal (to within a very small part) to the same difference for component b,

$$\Delta_b = R_{(p+1),b}^2 - R_{p,b}^2 = \frac{2f^2}{n_{0,b}}$$

or any other component of the same line; let these differences be designated

---

† S. Tolansky, *High Resolution Spectroscopy*, Methuen, London, 1947, p. 130.

by Δ. Now from Eq. 5.15,

$$\epsilon_a = \frac{R^2_{(p+1),a}}{\Delta} - p$$

$$\epsilon_b = \frac{R^2_{(p+1),b}}{\Delta} - p \text{ etc.}$$

and the required separation (in wave numbers) between the two components, a and b, is

$$\Delta \bar{\nu} = \frac{\epsilon_a - \epsilon_b}{2t} = \frac{R^2_{p,a} - R^2_{p,b}}{\Delta} \times \frac{1}{2t} \tag{5.16}$$

If we designate by $\delta^p_{a,b}$ the difference between the square of the radii of the pth order rings of components a and b, and by $\Delta^a_{p,p+1}$ the difference between the square of the radii of the pth and $(p+1)$th ring of component a, we can form the square array shown in Table 7.4. We note that all Δ's

TABLE 7.4

SQUARE ARRAY FOR REDUCTION OF FABRY–PEROT DATA

Component				Ring number					
	1		2		3		4		5
a	R^2_{1a}	Δ^a_{12}	R^2_{2a}	Δ^a_{23}	R^2_{3a}	Δ^a_{34}	R^2_{4a}	Δ^a_{45}	R^2_{5a} etc ...
	δ^1_{ab}		δ^2_{ab}		δ^3_{ab}		δ^4_{ab}		δ^5_{ab}
b	R^2_{1b}	Δ^b_{12}	R^2_{2b}	Δ^b_{23}	R^2_{3b}	Δ^b_{34}	R^2_{4b}	Δ^b_{45}	R^2_{5b} etc ...
	δ^1_{bc}		δ^2_{bc}		δ^3_{bc}		δ^4_{bc}		δ^5_{bc}
c	R^2_{1c}	Δ^c_{12}	R^2_{2c}	Δ^c_{23}	R^2_{3c}	Δ^c_{34}	R^2_{4c}	Δ^c_{45}	R^2_{5c} etc ...

must have the same value in any one row, and *also* in any column. Thus when an error in measurement has been made, it becomes immediately apparent, and that radius can be rejected. Similarly all δ's in one row must be equal and their average can be taken. To obtain $\Delta \bar{\nu}$, we first obtain the average value of Δ; however, we should not use every available Δ, but only alternate† ones to obtain the average. Then the average value of the

---

† Note that if we take

$$\langle \Delta \rangle = (1/k)(\Delta_{12} + \Delta_{23} + \cdots + \Delta_{k,k+1})$$

this is equivalent to

$$(1/k)[(R_2{}^2 - R_1{}^2) + (R_3{}^2 - R_2{}^2) + \cdots + (R^2_{k+1} - R_k{}^2)] = (1/k)(R^2_{k+1} - R_1{}^2)$$

so that only the information from the first and last ring is used.

δ's is obtained and

$$\Delta \bar{\nu}_{ab} = \frac{\langle \delta_{ab} \rangle}{2l \langle \Delta \rangle} \tag{5.16a}$$

This method of reduction of the data is quite satisfactory, and use of it will be made in the next section.

For an "off-center" pattern this technique is not applicable, since the radii are not known. The data reduction is based on the fact that now the difference in radii (not in the square of radii) of adjacent rings is almost a constant.†

6. The Zeeman Effect of the λ = 5461 Å Line of Hg^{198}

6.1 Equipment and Alignment

We now consider the observation in this laboratory of the Zeeman effect on the λ = 5461 Å line of Hg^{198}. The choice of the green line is due to its predominance in the mercury spectrum, and the ease with which it can be observed. In an external magnetic field, it is split into nine components, as discussed in detail in Section 2.3. In the present observations, a polarizer parallel to the magnetic field was used, so that only three of the nine components (the π light) appeared. Furthermore, natural mercury exhibits in the green line a large number of hyperfine structure components, and each of them forms a Zeeman pattern. To avoid a multiplicity of components in one spectral line, a separated isotope of mercury was used as the source. Hg^{198} is well suited for our purpose since $I = 0$, and therefore

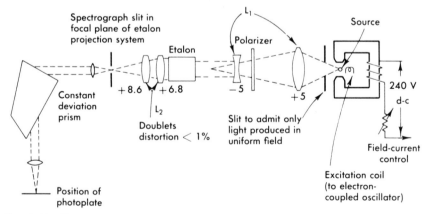

Fig. 7.23 Experimental arrangement used for observing the Zeeman effect with a Fabry-Perot étalon, crossed by a constant deviation prism spectrograph.

† See also Tolansky, loc. cit.

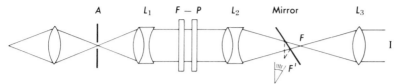

A L_1 F – P L_2 Mirror L_3

Fɪɢ. 7.24 Optical arrangement for aligning a Fabry–Perot étalon. Rough adjustment is made by viewing the image formed by L_3. Final adjustment is made by viewing the plates from the point F (or F'). Patterns as shown in Fig. 7.25 will be observed.

it exhibits no hyperfine structure; Hg^{198} is usually obtained by neutron capture in gold

$$_{79}Au^{197} + n \rightarrow {}_{80}Hg^{198} + e^-$$

and subsequent separation by chemical methods of the mercury from the gold.†

The optical system used for this investigation is shown in Fig. 7.23. The Fabry–Perot was crossed in the parallel-beam method with the small constant-deviation spectrograph described in Chapter 2. The étalon and lenses are all mounted on an optical bench to which the spectrograph is rigidly attached. The pair of lenses L_1 forms the light from the source into a parallel beam, while the pair L_2 focuses the Fabry–Perot ring pattern onto the spectrograph slit; the effective focal length of L_2 is 8 cm, and a further magnification of 2 takes place in the spectrograph.

The discharge tube is mounted vertically, as is the spectrograph slit; the slit width was 1 mm. It is clear that in this arrangement not only the ring pattern is focused onto the spectrometer slit but also the image of the source. A sheet of polaroid that could be rotated at will was used as a polarizer.

The spacing of the Fabry–Perot étalon is

$$t = 0.5002 \text{ cm}$$

and it is imperative to adjust the plates carefully for parallelism. This can be done either by viewing through the spectrograph with a frosted glass in the focal plane, and adjusting for the best quality of the pattern, or by a much more sensitive arrangement as shown in Fig. 7.24. A very small aperture (less than 1 mm in diameter) is placed at the position of the source and illuminated with an intense sodium lamp. The Fabry–Perot plates are adjusted to be normal to the optical axis by bringing the image of A reflected by the étalon back onto A. Next, L_3 is adjusted until a series of multiple images of A appears when the observer is located at I; the plates of the étalon can then be roughly adjusted for parallelism by bringing all

---

† Ready-made Hg^{198} electrodeless discharge tubes can be obtained from the Ryan, Velluto, and Anderson Glass Works, Inc., in Cambridge, Mass., or from other suppliers.

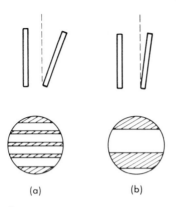

FIG. 7.25 Patterns seen on the axis in the focal plane of a Fabry-Perot when illuminated with a point source. (a) Poor plate parallelism, (b) improved plate parallelism. The tilt of the plates shown is greatly exaggerated.

(a)   (b)

the images into coincidence. The final adjustment is made by removing L_3 so that the observer locates his eye at F (or a mirror can be used); then, fringes of equal width do appear parallel to the base of the wedge that is formed by the two plates (Fig. 7.25). As the plates are moved into parallelism, the fringes become broader and finally the whole image of the aperture A seems to have a uniform illumination (bright or dark depending on the exact value of $n_0 = 2t/\lambda$). It is equally important that the ring pattern be in sharp focus at the plane of the photographic plate. For this experiment Kodak Royal-Pan film was used.

The electrodeless discharge was excited as described in Section 4, and to observe the Zeeman effect the discharge tube was placed in a magnetic field. A small iron core electromagnet powered by a 220-V d-c line was used to produce the field; it was controlled by a bank of resistors and a rheostat.

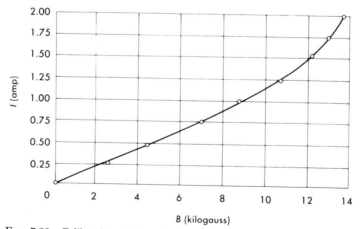

FIG. 7.26 Calibration of the electromagnet used in the Zeeman effect experiment. The magnetic field is plotted against current; note the saturation at high fields.

The diameter of the pole faces was only $1\frac{1}{2}$ in., and a small gap ($\frac{1}{2}$ in.) was used. By tapering the pole faces, higher magnetic fields can be achieved but this reduces the effective area of the field as well as the homogeneity. The magnetic field was measured with a "flip coil" and the calibration of field against current is given in Fig. 7.26. It is seen that field strengths of 12 kilogauss could be reached.

6.2 Data on the Zeeman Effect

The data presented below were obtained by students†. Figure 7.27 shows the Hg 5461 Å line photographed at various magnet settings. As explained earlier, the source contains a single isotope, and the polarizer allows only the observation of π light. We note that the fringes are rather broad, but it can clearly be seen that when the field is applied the single-line pattern (Fig. 7.27a) breaks up into a triplet, the separation between the components of the triplet becoming larger with increasing field.

Fig. 7.27   Fabry-Perot patterns showing the Zeeman effect of the green line of mercury. (See the text for additional details.) (a) No magnetic field applied. (b)—(e) A magnetic field of progressively greater strength is applied. Note the splitting of the original line into a triplet of increasing separation.

† E. Glover and T. Wagner, class of 1961.

We will first analyze the exposure with zero field in order to verify that the squares of the radii do indeed follow Eqs. 5.9 and 5.10, and we will find the fractional order at the origin by making a least-squares fit to the squares of the radii. Next we will analyze the 1.00-amp and 1.50-amp exposures by the square-array method.

The initial step in the reduction of the data is the measurement of the diameters (or radii) of the rings. To this effect a traveling microscope was used, and readings were taken directly off the plate; care must be taken to insure that the travel of the microscope is indeed along the diameter of the rings and that the crosshairs are properly oriented. When the fringes in the pattern are as broad as in Fig. 7.27, it is much more accurate to measure the two edges and take the average rather than try to set the crosshairs in the center of the fringe. A more advanced technique for obtaining the ring diameters from the plates is to use a microphotometer.

In Table 7.5 are tabulated the radii of the rings, their squares, and the differences of the squares for the exposure with no field. The same data are also plotted in Fig. 7.28, and it is seen that the straight-line fit is quite satisfactory. The fractional order at the center is found to be $\epsilon_0 = 0.595 \pm 0.024$. For the exposures at 1.00 amp (8.7 kilogauss) and 1.50 amp (12.1 kilogauss), the square array, as described in the Section 5.2, is presented in Tables 7.6 and 7.7; it yields the separation in wave numbers of component a from the central component b, and of c from b. Furthermore, if the fractional order of the central component b is calculated as before, it is found to be $\epsilon_1 = 0.649 \pm 0.024$ and $\epsilon_2 = 0.667 \pm 0.014$, respectively, in

TABLE 7.5

RADII OF FABRY–PEROT PATTERN FOR SINGLE LINE

Ring p	Radius R_p (cm)	R_p^2 (mm²)	$(R_{p+1}^2 - R_p^2)$ (mm²)	Ring p	Radius R_p (cm)	R_p^2 (mm²)	$(R_{p+1}^2 - R_p^2)$ (mm²)
							3.095
1	0.103	1.071		7	0.444	19.758	
			3.140				2.995
2	0.205	4.211		8	0.477	22.753	
			3.144				3.104
3	0.271	7.355		9	0.508	25.857	
			3.091				3.087
4	0.323	10.446		10	0.538	28.944	
			3.281				2.922
5	0.370	13.727		11	0.564	31.866	
			2.936				2.826
6	0.408	16.663		12	0.589	34.692	

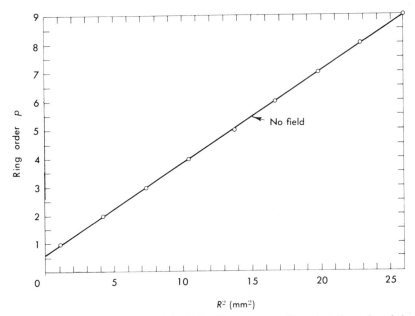

FIG. 7.28 Plot of the square of the Fabry-Perot ring radii against the order of the ring (as measured from the center out). This plot pertains to the data of Fig. 7.27(a) tabulated in Table 7.5 where no magnetic field is applied. The straight line is the least squares fit to the experimental points.

TABLE 7.6
FABRY–PEROT DATA SQUARE ARRAY (FIELD OF 8.7 KILOGAUSS)

Component	Ring order p								
	1	Δ	2	Δ	3	Δ	4	Δ	5
a	0.436	3.136	3.572	3.188	6.760	3.226	9.986	3.118	13.104
δ_{ab}	0.625		0.796		0.692		0.838		0.660
b	1.061	3.307	4.368	3.084	7.452	3.372	10.824	2.940	13.764
δ_{bc}	0.762		0.650		0.784		0.668		0.524
c	1.823	3.195	5.018	3.218	8.236	3.256	11.492	2.796	14.288

Note: All numbers are in mm²; to evaluate $\langle \Delta \rangle$ we used the six numbers between rings 1–2 and 3–4; to evaluate $\langle \delta \rangle$ we used the information from rings 1–4 only. We obtain

$$\langle \Delta \rangle = 3.249, \qquad \langle \delta_{ab} \rangle = 0.737, \qquad \langle \delta_{bc} \rangle = 0.716$$

and using $t = 0.5002$

$$\Delta \bar{\nu}(a - b) = 0.227 \text{ cm}^{-1}; \qquad \Delta \bar{\nu}(b - c) = 0.220 \text{ cm}^{-1}.$$

TABLE 7.7

FABRY–PEROT DATA SQUARE ARRAY (FIELD OF 12.1 KILOGAUSS)

Component	Ring order p								
	1	Δ	2	Δ	3	Δ	4	Δ	5
a	0.207	3.142	3.349	3.411	6.670	3.289	10.049	2.983	13.032
δ_{ab}	0.917		0.895		0.858		0.841		1.031
b	1.124	3.120	4.244	3.374	7.618	3.272	10.890	3.173	14.063
δ_{bc}	0.921		0.864		0.908		0.806		0.837
c	2.045	3.063	5.108	3.418	8.526	3.170	11.696	3.203	14.900

Note: All numbers are in mm²; to evaluate $\langle \Delta \rangle$ we used the six numbers between rings 1–2 and 3–4; to evaluate $\langle \delta \rangle$ we used all values except the $\delta_{ab} = 1.031$. We obtain

$$\langle \Delta \rangle = 3.176, \qquad \langle \delta_{ab} \rangle = 0.878, \qquad \langle \delta_{bc} \rangle = 0.867$$

and using $t = 0.5002$

$$\Delta \bar{\nu}(a - b) = 0.276 \text{ cm}^{-1}, \qquad \Delta \bar{\nu}(b - c) = 0.273 \text{ cm}^{-1}$$

agreement with the no-field value ϵ_0. Thus we conclude that the central component is not shifted by the application of the magnetic field.

The final data are summarized in Fig. 7.29 where the spacing of the components against magnetic field is shown. We see that the spacing varies linearly with the field, and the proportionality coefficient has the value

$$\frac{\Delta \bar{\nu}}{H} = 0.024 \text{ cm}^{-1}/\text{kilogauss} \tag{6.1}$$

From the discussion of Section 2.3, we know that the Hg 5461 Å line connects the $^3S_1$ and $^3P_2$ states; the structure of the line is indicated in Fig. 7.8. Since the polarizer was set to select only components arising in transitions with $\Delta m = 0$, we expect to observe only the three central components, which will be separated by (Eq. 2.16):

$$\Delta \bar{\nu} = \frac{\mu_0}{hc} (g_i - g_f) H = \frac{1}{2} \frac{\mu_0}{hc} H \tag{6.2}$$

By comparing Eq. 6.2 with the experimental result of Eq. 6.1, we obtain

$$\frac{\mu_0}{hc} = (4.80 \pm 0.5) \times 10^{-5} \quad \text{cm}^{-1}/\text{gauss}$$

in good agreement with the accepted value

$$\frac{\mu_0}{hc} = 4.669 \times 10^{-5} \quad \text{cm}^{-1}/\text{gauss}$$

From these data we conclude that indeed spectral lines are split into

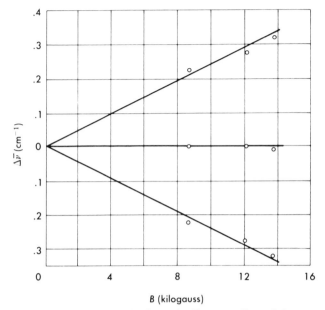

Fig. 7.29   Results obtained on the Zeeman effect of the green line of mercury (see text). The observed displacement of the three components from the zero field value (of the single line) is plotted against magnetic field.

components when the source is placed in a magnetic field. Further, the splitting observed was in excellent agreement with the theory of the anomalous Zeeman effect; the normal Zeeman effect can be excluded, since the energy difference between the components of the line was not $\mu_0 H$ (but instead $\frac{1}{2}\mu_0 H$).

Finally to illustrate the use of the Fabry–Perot in the off-center method, Fig. 7.30 has been included. This exposure was obtained with a high-quality interferometer by Dr. L. C. Bradley III, on the 4047 Å line of Hg^{197}; the orders shown are from $p = 5$ to $p = 8$. No magnetic field is applied but the hyperfine structure is clearly resolved.

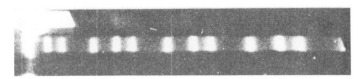

Fig. 7.30   Fabry-Perot pattern of good quality, obtained in the off-center method on the 4047 Å line of Hg^{197} (radioactive). The three distinct components observed are due to the hyperfine structure of this line.

7. The High-Resolution Grating; Observation of Hyperfine Structure and of the Zeeman Effect

7.1 THE GRATING SPECTROGRAPH

As discussed in the introduction, a large grating of good quality is a powerful-high resolution instrument. We will first discuss the medium resolution instrument installed and operated in this teaching laboratory; this provided enough resolution in order to observe the Zeeman effect of the yellow lines of the mercury spectrum. In the next section, data on the hyperfine structure of other lines of mercury will be presented; these data, however, were obtained with a research instrument.

The basic equation for the reflection grating (Chapter 2, Eq. 2.1) is

$$n\lambda = d(\sin \theta_r - \sin \theta_i) \qquad (7.1)$$

from which it follows that†

$$\frac{d\lambda}{d\theta} = \frac{d}{n} \cos \theta_r = \frac{A \cos \theta_r}{Nn} \qquad (7.2)$$

Here, n is the order of interference, $d = A/N$ is the spacing between rulings, A being the total width of the grating, and N the total number of rulings; θ_i and θ_r are the angles of incidence and refraction, respectively (see Fig. 7.31).

Two lines differing in wavelength by $\Delta\lambda$ will appear in the focal plane of the instrument (that is, on the plate) separated by a distance $\Delta x = \Delta\theta R$ where R is the focal length of the element that focuses the refracted ray onto the detector (plate). It is then clear from Eq. 7.2 that the "plate factor" is given by

$$\frac{d\lambda}{dx} = \frac{1}{n}\frac{d}{R} \cos \theta_r \qquad (7.3)$$

Clearly, to separate two lines which appear displaced by Δx, the image of the slit on the plate must be smaller than Δx. In practice‡ it is usually difficult to make $\Delta x < 10^{-2}$ cm. Even if the plate factor is extremely good (small), however, the smallest separation in wavelength $\Delta\lambda$ that can be observed is determined by the width of the diffracted line; namely, the resolution of the grating that is given by

$$\frac{d\lambda}{\lambda} = \frac{1}{Nn} \qquad (7.4)$$

Eqs. 7.3 and 7.4 are the basic equations for a grating spectrograph.

† Note that the angle of incidence is fixed for all wavelengths, so that $d\theta_i = 0$.

‡ This is due to very slight errors in the exact focal setting of the mirrors, imperfections of the slit, and other image-broadening effects.

A layout of the spectrograph is shown in Fig. 7.32. It consists of:

(1) A Bausch and Lomb 52 × 52-mm plane grating ruled to 30,000 lines/in.

(2) Two front-surface silvered 4-in. diameter spherical mirrors, with a 10-ft focal length.

(3) The entrance slit and the plate holder.

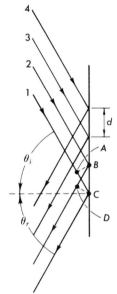

FIG. 7.31 Principle of operation of the reflection diffraction grating. θ_i is the incidence angle and θ_r the reflection angle; d is the spacing between rulings.

FIG. 7.32 A large focal-length spectrograph. The source and film holder are located in a different room from the mirrors and the grating. The separation between the two 4-in. diameter spherical mirrors can be adjusted up to 3 ft.

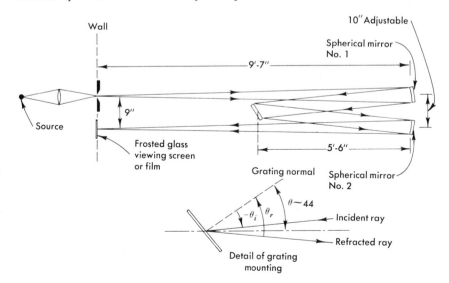

When such large focal lengths are used, it is advantageous to use a small room as the spectrometer camera; the room is made light tight and the grating and the mirrors are placed in it. The slit is mounted through the wall leading to a small "source room"; similarly, one or more plate holders are located at a convenient position through one wall or in the camera room. The grating and mirrors are mounted on rigid bases and have enough freedom of motion for adjustment, as well as for the selection of the appropriate order of the line that is to be examined.

For the particular grating under discussion we note that the 5790 Å yellow mercury line, for example, is refracted

through first order $(n = 1)$ if $\sin \theta_i - \sin \theta_r = 0.6948$ (7.5a)

through second order $(n = 2)$ if $\sin \theta_i - \sin \theta_r = 1.3896$ (7.5b)

and cannot be observed in the third or higher order. Thus if we wish to observe this particular line in second order, $\sin \theta_i$ and $\sin \theta_r$ must have opposite signs; that is, both the incident and reflected rays must be on the *same* side of the normal.

If we now look more carefully at Fig. 7.32, we note that from the relative positioning of the grating and of the spherical mirrors the angles θ_i and θ_r are constrained as follows.

$\theta_r + \theta_i \sim 8°6$ if r and i on opposite sides of the normal

$\theta_r + \theta_i = \theta_r - |\theta_i| \sim 8°6$ if r and i on same side of the normal

Using Eq. 7.5b and $\theta_i - \theta_r = 8°6$, we find that the grating must be inclined at $\theta = 44°$ (see again Fig. 7.32). Thus

$$|\theta_i| = 44° - 4°3 = 39°7$$

$$\theta_r = 44 + 4°3 = 48°3$$

which satisfy Eq. 7.5b.

Having obtained θ_r we can apply Eq. 7.3; using $R = 115'' = 292$ cm (see Fig. 7.32), we obtain a plate factor

$$\frac{d\lambda}{dx} = \frac{0.665}{2} \frac{(1/12000)}{292} = 9.5 \times 10^{-8} = 9.5 \text{ Å/cm} \qquad (7.6)$$

If we assume that the closest separation that can be resolved on the plate is 10^{-2} cm, we obtain for this instrument a resolving limit of $\Delta\lambda \sim 10^{-1}$ Å at $\lambda = 5790$ Å. We may compare this result with the theoretical resolving power of the grating as given by Eq. 7.4; in second order

$$\frac{\lambda}{d\lambda} = Nn = 2 \times 2 \times 30,000 = 1.2 \times 10^{-6}$$

Namely, at $\lambda = 5790$ Å,

$$\Delta\lambda \text{ (theoretical)} \sim 5 \times 10^{-2} \text{ Å}$$

Therefore we see that the grating as mounted in this arrangement can reach half its theoretical resolving power† at $\lambda = 5790$ Å. The above conclusions are indeed borne out by the experimental data as discussed below.

In order to decrease the plate factor, and therefore increase the effective resolution of the instrument, we may work in a higher order of diffraction. There is here, however, the difficulty that the intensity decreases with increasing order, and furthermore, every grating is constructed so as to give maximum intensity at a specific angle of refraction. That is, the grooves are cut in the grating in such a fashion that their normal bisects the angle of refraction, as shown in Fig. 7.33; this angle is called the *blaze* and for the particular grating used here, $\theta_b = 15°$, restricting its use to the first few orders.

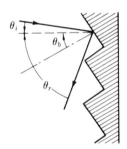

FIG. 7.33 Highly magnified view of the reflection grating shown in Fig. 7.31. Note how the grating rulings are grooves cut into a very flat surface. θ_b is the "blaze" angle, and for maximum intensity $\theta_i + \theta_b = \theta_r - \theta_b$.

One minor difficulty with grating spectrographs is the overlap of different orders, which becomes more confusing for large n. By visual observation, however, the color of the different lines is very helpful in separating the orders; also filters can be used to illuminate or isolate specific lines. Another difficulty is the appearance of "ghost" lines due to periodic imperfections of the grating.

In Fig. 7.34a is shown a portion of the mercury spectrum as obtained by students‡ with the medium resolution grating spectrograph described above. The two yellow lines at 5769.6 Å and 5790.7 Å are shown and are separated (on the plate) by $\Delta x = 2.350$ cm. Thus the plate factor is

$$\frac{\Delta\lambda}{\Delta x} = \frac{21.1}{2.35} = 9.0 \text{ Å/cm} \tag{7.7}$$

---

† Violet lines can be observed in third order with a consequent improvement in the plate factor, however, for the shorter wavelengths the theoretical value of $\Delta\lambda$ decreases as well.
‡ W. Lama and E. Holroyd Class of 1966.

in close agreement with the value calculated by Eq. 7.6. To appreciate the improvement in resolution the reader should compare Fig. 7.34a with the spacing between these same lines in Fig. 2.14. By referring to Fig. 2.13 we note that another yellow line appears at 5789.7 Å arising from the transition $6s6p\ ^3D_1 \rightarrow 6s6p\ ^1P_1$; however, this line is too faint and does not appear in this exposure. The irregular appearance of the image is due to an imperfection in the slit.

The resolution of this spectrograph is sufficient to observe the (gross structure) Zeeman effect in the yellow lines of mercury. The experimental

(a) No magnetic field

(b) B = 15.7 kilogauss

2.348 cm

Fig. 7.34   Spectrogram of the two yellow lines of natural mercury obtained in second order with the medium resolution spectragraph shown in Fig. 7.32. (a) No magnetic field. (b) A magnetic field of 15.7 kilogauss is applied with its axis normal to the direction of observation. No polarizer is used and the splitting of each line into a triplet can be clearly seen. The plate factor was 9 Å/cm yielding a separation of 2.35 cm (on the plate) between the two lines. The faint line to the left of λ = 5790.7 Å is probably the λ = 5789.7 Å line, known to be weak. The distorted shape of the lines is due to imperfections of the slit.

data obtained by the students† are shown in Fig. 7.34b, and in Fig. 7.35a, b, and c. Fig. 7.36 shows the initial and final states of the transition as they

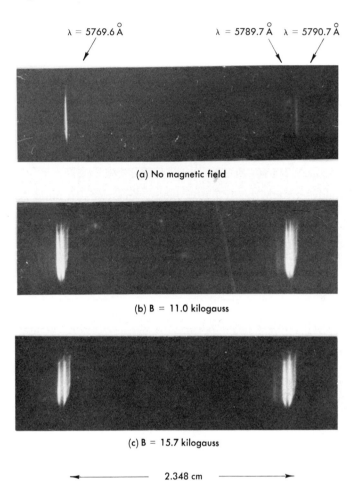

$\lambda = 5769.6$ Å $\lambda = 5789.7$ Å $\lambda = 5790.7$ Å

(a) No magnetic field

(b) B = 11.0 kilogauss

(c) B = 15.7 kilogauss

◄───────── 2.348 cm ─────────►

FIG. 7.35   Spectrogram of the two yellow lines of natural mercury obtained as in Fig. 7.34 above but with the insertion of a polarizer; the components polarized parallel to the magnetic field axis ($\Delta m = 0$) are eliminated. (a) No magnetic field. (b) Magnetic field $B = 11.0$ kilogauss. (c) Magnetic field $B = 15.7$ kilogauss. Note the increasing separation of the components and the appearance of ghosts. For a discussion of the structure of these lines see the text and Fig. 7.36.

† See footnote on p. 331.

are split in a magnetic field; the structure of the lines is shown in the lower part of the figure according to the adopted convention.

Let us first calculate according to Eq. 2.13 the g factors of the states involved in these transitions

$$g = 1 + \frac{j(j+1) + s(s+1) - \ell(\ell+1)}{2j(j+1)} \tag{2.13}$$

Since the lower state is a singlet ($^1P_1$) its g factor must be one. Similarly for the $^1D_2$. Thus the 5790.7 Å line will exhibit a normal Zeeman triplet structure. For the $^3D_2$ state we obtain

$$g(^3D_2) = \tfrac{7}{6} \tag{7.8}$$

Interestingly enough this g factor is also very close to unity so that the 5769.6 Å line will also appear almost as a normal Zeeman triplet; the split-

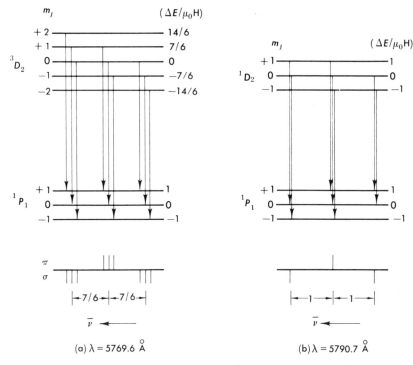

(a) $\lambda = 5769.6$ Å   (b) $\lambda = 5790.7$ Å

FIG. 7.36   The structure of the (a) $\lambda = 5769.6$ Å and (b) $\lambda = 5790.7$ Å yellow lines of mercury when placed in a magnetic field. Note that line (b) yields a normal Zeeman triplet pattern whereas line (a) when viewed under medium resolution will also appear as an almost normal Zeeman triplet.

ting between the centers of gravity of the three components, however, will be $\frac{7}{6} \mu_0 H$. These conclusions can be verified by the diagram of Fig. 7.36. In Fig. 7.34b is shown the mercury yellow doublet in a magnetic field of 15.7 kilogauss.† No polarizer was used and the axis of the field was normal to the direction of observation. The source was a low-pressure discharge tube containing natural mercury. The splitting of both lines into a triplet can be clearly seen. The close structure indicated by Fig. 7.36a cannot be observed, nor is it possible to detect the hyperfine structure or isotope shift effects of these lines.

A quantitative measurement made with a traveling microscope on this plate yields the following average separation of the outer components from the central one

$$\lambda = 5769.6 \text{ Å} \qquad \Delta x = 0.031 \pm 0.002 \text{ cm}^{-1}$$

$$\lambda = 5790.7 \text{ Å} \qquad \Delta x = 0.028 \pm 0.002 \text{ cm}^{-1}$$

Using the measured plate factor (Eq. 7.7) we obtain

$$\lambda = 5769.6 \text{ Å} \qquad \Delta\lambda = 0.28 \text{ Å} \quad \text{or} \quad \Delta\bar{\nu} = 0.835 \text{ cm}^{-1}$$

$$\lambda = 5790.7 \text{ Å} \qquad \Delta\lambda = 0.25 \text{ Å} \quad \text{or} \quad \Delta\bar{\nu} = 0.746 \text{ cm}^{-1}$$

From Fig. 7.36 we see that

$$\Delta\bar{\nu} = \frac{g\mu_0 H}{(hc)}$$

so that we obtain for $H = 15.7$ kilogauss

$$\lambda = 5769.6 \text{ Å} \qquad g = \tfrac{7}{6} \qquad \mu_0/hc = 4.56 \times 10^{-5} \text{ cm}^{-1}/\text{gauss}$$

$$\lambda = 5790.7 \text{ Å} \qquad g = 1 \qquad \mu_0/hc = 4.75 \times 10^{-5} \text{ cm}^{-1}/\text{gauss}$$

both results being in close agreement with the accepted value of μ_0 (see Eq. 1.2)

$$\frac{\mu_0}{hc} = 4.669 \times 10^{-5} \text{ cm}^{-1}/\text{gauss}$$

In concluding we mention the data shown in Fig. 7.35a, b, and c. These are spectrograms of the same lines of mercury taken under similar conditions as in Fig. 7.34. The defect of the slit, however, has been corrected so that the resulting image is a straight line. Also a polarizer has been inserted and rotated so as to eliminate the central component (polarization parallel to the field axis). The exposure of Fig. 35a was taken without a magnetic field, that of Fig. 35b in a field of 11.0 kilogauss and that of Fig. 35c in a field of 15.7 kilogauss.

---

† The same electromagnet as discussed before (see Fig. 7.26) was used but the spacing between polefaces was reduced to $\frac{3}{8}$-in.

7.2 DATA ON THE HYPERFINE STRUCTURE OF NATURAL MERCURY

In Fig. 7.37 and Fig. 7.38 is shown the hyperfine structure of four lines of natural mercury. The exposures were taken† with a 10-in. grating having 3000 lines/cm and blazed at 63°; the 2537 Å line was photographed in the

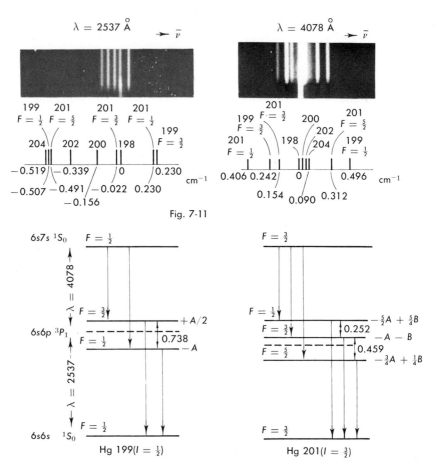

Fig. 7-11

FIG. 7.37 The hyperfine structure of the natural mercury lines at 2537 Å and 4078 Å, as obtained with a high-resolution grating; the longer line indicates the position of Hg198. The structure of each line is indicated and the spacing between components is given in cm⁻¹. In the lower part of the figure the three energy levels involved in the emission of the 2537 Å and 4078 Å lines are shown, separately, for Hg199 and for Hg201. Note that the structure of either line is due to the hyperfine splitting of *only* the $^3P_1$ level.

† By Dr. S. P. Davis at the Spectroscopy Laboratory of the Massachusetts Institute of Technology.

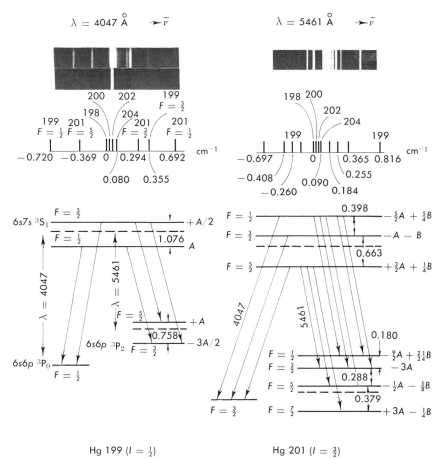

Hg 199 ($I = \frac{1}{2}$)   Hg 201 ($I = \frac{3}{2}$)

FIG. 7.38   The structure is the same as Fig. 7.37 but for the 4047 Å and 5461 Å lines of natural mercury obtained under identical conditions.

24th order and the plate factor was 0.045 Å/mm (at 2537 Å this corresponds to 0.700 cm⁻¹/mm). With each exposure a source containing only Hg¹⁹⁸ was also photographed on the same plate, and the location of this single component is shown beneath the natural mercury structure.

The energy scale increases towards the right-hand side (the wavelengths towards the left). The identified components of each line are also shown and all spacings are given in cm⁻¹ from the position of Hg¹⁹⁸.

Our next task is to reconstruct from these data the hyperfine structure of the energy levels involved in the transition: To facilitate our task, in Fig. 7.39 is reproduced part of the energy-level diagram of mercury including the four lines under examination. We note that the isotope shift is largest in the 2537 Å line and rather small in the other lines, giving rise to a broad unresolved line in the center of the spectrum. This question of the isotope shift has already been discussed in Section 3.3; since, further, for all even

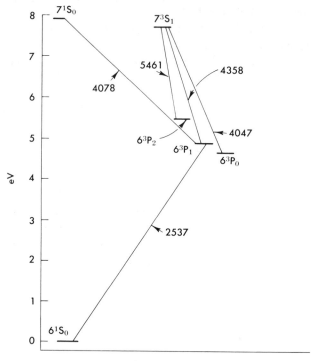

Energy levels of mercury

FIG. 7.39 Some of the energy levels and optical lines of mercury. The hyperfine structure of four of these lines was presented in Figs. 7.35 and 7.36.

isotopes $I = 0$, they do not exhibit hyperfine structure and we need only concern ourselves with the structure of Hg^{199} and Hg^{201} lines.

Let us then consider the 2537 Å and 4078 Å lines. They both connect a $J = 0$ level (where there is no hyperfine structure) with the $^3P_1$ level; thus they must show the same hyperfine structure (but inverted, since the 4078 Å line ends in the $^3P_1$ while the 2537 Å line originates in the $^3P_1$). Indeed, the energy-level diagrams for Hg^{199} and for Hg^{201} shown in the lower part of Fig. 7.37 do completely account for the structures obtained experimentally.†

For the 4047 Å and 5461 Å lines we again note that the first one connects a $J = 0$ level (no hyperfine structure) to a $J = 1$ level, and we can thus establish its structure rather easily. The 5461 Å line now connects the

---

† Some hindsight has been used here in the fact that the components are already correctly identified. Note also that the center of gravity of Hg^{199} is shifted with respect to the center of gravity of Hg^{201} in the 2537 Å line, but not in the 4078 Å line.

$^3S_1$ level with the $^3P_2$; using the previous information we are able to reconstruct the hyperfine structure of that level as shown in the lower part of Fig. 7.38 for both Hg^{199} and Hg^{201}.

Our final task is to extract, from the hyperfine structure of the energy levels, the magnetic dipole interaction constant A, and for Hg^{201} the electric quadrupole constant B as well. To do this the interval rules given by Eqs. 3.6 and 3.11 must be used; these have been evaluated and the displacement of each level from its center of gravity is noted in the diagrams next to the level. By solving the algebraic equations implied by these relations, we find the following values:

$^3P_1$ $A_{199} = 0.492$ cm^{-1} $A_{201} = -0.182$ cm^{-1} $B_{201} = -0.009$ cm^{-1}

$^3P_2$ $A_{199} = 0.303$ $A_{201} = -0.112$ $B_{201} = 0.014$

$^3S_1$ $A_{199} = 0.717$ $A_{201} = -0.265$ $B_{201} = \quad$ —

We could now proceed from the above interaction constants to evaluate the magnetic moments of the Hg^{199} and Hg^{201} nucleus and the electric quadrupole moment of the latter. Although some expressions relating these quantities were given, the actual evaluation is beyond the scope of the present discussion.

8

MAGNETIC RESONANCE EXPERIMENTS

1. Introduction

As we know† when a quantum-mechanical system, for example, an atom (or a nucleus), with angular momentum **L**, (or **I**), different from zero, is placed in a magnetic field **H**, the energy states having different values for the quantum number m acquire an additional energy

$$\Delta E = \frac{\mu}{I} Hm \tag{1.1}$$

where μ is the "magnetic moment" of the system. For systems involving electrons, μ is of the order of the Bohr magneton μ_0, while for nuclei μ is of the order of the nuclear magneton, μ_N. In convenient units

$$\mu_0/h = 1.401 \text{ Mc/gauss}$$
$$\mu_N/h = (\mu_0/h)/1836 = 0.762 \text{ kc/gauss} \tag{1.2}$$

---

† This subject was discussed in detail in Chapter 7; see especially Section 2.

In Fig. 8.1 is shown the splitting of an energy state with $l = 1$ into its three sublevels. As noted in Chapter 7, in atomic systems we do not observe the spontaneous transitions (labeled a in the figure) between sublevels with different m, because they do not satisfy the selection rule $\Delta l = \pm 1$. Instead the splitting of a level is observed through the small change in the frequency of the radiation emitted in the transitions between widely distant levels (with $\Delta l = \pm 1$). It is clear that if we could directly measure the frequency corresponding to a transition between the m sublevels of the same state, a much more precise knowledge of the energy splitting would be obtained.

FIG. 8.1 Splitting of an energy level with $l = 1$ into three components when placed in a magnetic field.

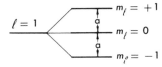

Now, the selection rule $\Delta l = \pm 1$ is applicable to electric dipole radiation; however, transitions with $\Delta l = 0$ $\Delta m = \pm 1$ do occur when *magnetic dipole* radiation is emitted, but the probability for such a transition is reduced by a factor† $(v/c)^2$ from the case of an electric dipole transition. We therefore conclude that spontaneous transitions with $\Delta l = 0$ $\Delta m = \pm 1$ will be very rare, especially if the system can return to its ground state (lowest energy state) by a $\Delta l = \pm 1$ transition. On the other hand, in the presence of an electromagnetic field, *induced* transitions have a probability of occurring if the frequency of the field is equal (or at least fairly close) to the energy difference between the two levels; induced transitions towards higher or lower energy states are equally probable. Further, the transition probability is proportional to the square of the strength of the electromagnetic field (that is, the total number of quanta) so that if a sufficiently strong radiofrequency (rf) magnetic field (of frequency ν_0) is available, magnetic dipole transitions should take place.

By referring to Eq. 1.2, we note that for a field of a few kilogauss the energy splitting corresponds, for nuclei, to a frequency of megacycles, and for electrons, to a few kilomegacycles, that is, to frequencies that can be easily generated in the laboratory. A calculation also shows that a radiofrequency magnetic field of a fraction of a gauss is more than sufficient, in most cases, to cause transitions. A field intensity of 1 oersted corresponds to

$$\bar{S} = \frac{1}{2} \sqrt{\frac{\mu_0}{\epsilon_0}} H^2 = \frac{1}{2} \sqrt{\frac{4 \times 10^{-7}}{8.85 \times 10^{-12}}} \times \left(\frac{10^3}{4\pi}\right)^2 \text{(MKS)} \approx 2.35 \times 10^2 \frac{\text{watts}}{\text{cm}^2}$$

(1.3)

† For atomic systems v is of the order of the velocity in a Bohr orbit, namely, $v/c \approx 5 \times 10^{-6}$.

so that radiofrequency fields of adequate intensity can be generated. Finally, we must be able to detect the fact that a transition took place; this may be done in several ways and is one of the major distinctions between the various types of magnetic resonance experiments.

For example, in the first magnetic resonance experiment, performed by I. Rabi and collaborators in 1939, a beam of atoms having $J = \frac{1}{2}$ was passed in succession through two very inhomogeneous magnets A and B shown in Fig. 8.2. A homogeneous magnetic field existed in the intermediate region C. If a transition took place in region C from a state $m = +\frac{1}{2}$ to $m = -\frac{1}{2}$, that particular atom was deflected in an opposite direction in field B and thus missed the detector. Hence, resonance was detected by a decrease in beam current when the frequency of the radiofrequency was the appropriate one for the magnetic field strength in C.

Another method for detecting the occurrence of resonance is to observe the absorption of energy from the radiofrequency field when transitions toward higher energy levels take place. This is the technique used in most nuclear magnetic resonance (nmr) experiments and in electron magnetic resonance (called "electron paramagnetic resonance," epr) experiments; these will be described in some detail in Sections 4 and 5. In experiments with atomic vapors or transparent materials it is possible to detect the magnetic resonance effect by changes in the polarization of the atomic radiation ($\Delta m \neq 0$) or by selective absorption effects; such an experiment is described in Section 6.

Apart from its intrinsic interest as a way of inducing transitions between the energy sublevels of a quantum-mechanical system (atom, nucleus), magnetic resonance has become an important tool of physics. The atomic beam experiments of Rabi and his coworkers mentioned before led to very precise measurements of the hyperfine structure of atomic systems and thus to accurate values of the nuclear moments. The measurement of the "Lamb shift" (between the $2S_{1/2}$ and $2P_{1/2}$ levels of hydrogen) is an example of such a measurement, as is the very precise determination of the anom-

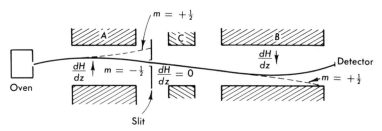

FIG. 8.2   The atomic beam arrangement of I. Rabi and collaborators used to detect magnetic resonance transitions in atomic energy levels.

alous magnetic moment of the electron. (The g factor of the electron is not $g = 2.00$ as stated in Chapter 7 but $g = 2.00232$ (!) as predicted by quantum electrodynamics).

A nuclear magnetic resonance experiment consists in inducing transitions between the sublevels of a nucleus placed in an external magnetic field (in the terminology of Chapter 7, we would call it the Zeeman effect of the nucleus). However, the atom to which the nucleus belongs must have $J = 0$ (diamagnetic material), since otherwise the nuclear spin would be coupled to J and the large electronic magnetic moment would mask the effect. By means of such experiments, nuclear magnetic moments are measured directly and to a high accuracy.

As will be explained in Section 3, the observation of nuclear magnetic resonance in solids and liquids† depends on the relaxation of the nuclear spins through their interaction with the lattice. Thus nuclear magnetic resonance studies have yielded a very large amount of information on the solid-state properties of many materials.

Soon after the first successful nuclear magnetic resonance experiments, it was realized that the width of the observed resonance line for protons was mostly due to inhomogeneities in the constant magnetic field that is used to split the energy sublevels. When a very homogeneous field was applied, the proton resonance line was shown to exhibit a fine structure of the order of 0.01 gauss. This structure depends on the organic compound to which the hydrogens of the sample belong. With even more homogeneous fields a hyperfine structure of the order of 0.001 gauss, is observed. Thus today nuclear magnetic resonance has become a very important tool of analytical chemistry as well.

We speak of paramagnetic resonance when the magnetic moments of electrons are involved in the transitions. This occurs in the study of molecular spectra (where, however, electric dipole transitions are also induced); this field being known as "microwave spectroscopy." Indeed the separation of many molecular energy levels falls conveniently in the microwave bands (millimeter and submillimeter wavelengths), and by using gaseous samples very narrow lines can be obtained.

However, the term paramagnetic resonance is mainly used for transitions between the Zeeman levels of quasi-free electrons in liquids and solids; a more adequate term also frequently used is electron spin resonance. In principle, we should always measure a g factor of 2.00 (if we deal with free electrons), which would not be interesting; instead a great variety of g factors and structure appears in the resonance lines due to the different effective coupling of the electron with the crystalline field. These effects depend on the relative orientation of the magnetic field H_0 and the crystal axis.

---

† Frequently referred to as *nuclear magnetic resonance* in bulk matter.

Thus electron paramagnetic resonance has become a very important tool in the study of crystalline structures as well as in the identification of free radicals in chemistry, medicine, and biophysics.

Section 2 will deal briefly with the conditions for inducing a magnetic dipole transition between two levels of a quantum-mechanical system and we will also discuss the classical analogue to this phenomenon. In Section 3 the mechanisms which are essential for the observation of energy absorption in nuclear magnetic resonance and electron paramagnetic resonance experiments are introduced; namely relaxation and saturation. In Section 4 the techniques and results of nuclear magnetic resonance experiments on protons are presented, and in Section 5 data on electron spin resonance obtained at microwave frequencies are given. Section 6 describes an electron paramagnetic resonance experiment with optical detection of the resonance condition, the so called "double resonance" technique.

REFERENCES

As always, the discussion presented here is limited, and undoubtedly the reader will wish to refer to some of the many excellent monographs and texts written on this subject. The following are suggested:

A. Abragam, *The Principles of Nuclear Magnetism*, Oxford University Press, 1961. The outstanding work on nuclear magnetic resonance. The treatment is theoretical and advanced, but very complete and clear.

E. R. Andrew, *Nuclear Magnetic Resonance*, Cambridge University Press, 1956. A shorter text containing experimental details as well; it is very useful to the students in this course.

C. H. Townes and A. L. Shawlow, *Microwave Spectroscopy*, McGraw-Hill, 1955. An extensive and comprehensive work on the subject, mainly treating the molecular spectra obtained in gases.

G. E. Pake, *Paramagnetic Resonance*, Benjamin, 1962.

D. J. E. Ingram, *Spectroscopy at Radio and Microwave Frequency*, Butterworth, 1955. Very helpful for the study of paramagnetic resonance in solids and crystalline materials.

2. Magnetic Dipole Transitions between the m Sublevels of a Quantum-Mechanical System

Let us consider a system (for example, a nucleus) with angular momentum $\mathbf{I}$ (magnitude $\hbar \sqrt{I(I + 1)}$ and magnetic moment $\boldsymbol{\mu}$ oriented along the spin axis. It is customary (for nuclei) to express the proportionality between the spin $\mathbf{I}$ and magnetic moment $\boldsymbol{\mu}$ by

$$\boldsymbol{\mu} = \gamma \hbar \mathbf{I} \tag{2.1}$$

Here γ is the gyromagnetic ratio, and as can be seen from Eq. 2.3, it has dimensions of radians per sec-gauss.

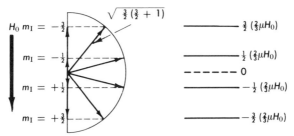

FIG. 8.3 The energy of the four sublevels of a nucleus with spin $I = \frac{3}{2}$ when placed in a magnetic field H_0. Note that the energy depends on the "orientation" of the spin with respect to H_0; also the magnitude of the spin vector is $|\mathbf{I}| = \sqrt{\frac{3}{2}(\frac{3}{2} + 1)}$.

In the presence of an external magnetic field H_0, the nucleus can be in any of the $(2I + 1)$ sublevels labeled by m_I as shown also in Fig. 8.3. We can then write for the energy† of these sublevels (Eq. 1.1)

$$\frac{E}{\hbar} = -\gamma H_0 m \tag{2.2}$$

so that the energy difference between any adjacent sublevels $(\Delta m = \pm 1)$ is simply

$$\frac{\Delta E}{\hbar} = \gamma H_0 = \omega_0 \tag{2.3}$$

Let us then further consider the simplest case, namely, $I = \frac{1}{2}$, in which only two sublevels exist, $m = -\frac{1}{2}$ and $m = +\frac{1}{2}$. In addition to H_0, let a weak field H_1, rotating in a plane normal to H_0 with an angular frequency ω be introduced. Taking as usual the z axis along H_0, we write the two components of H_1 as follows:

$$(H_1)_x = H_x = H_1 \cos \omega t \qquad (H_1)_y = H_y = H_1 \sin \omega t$$

and

$$H_1 \ll H_0$$

The additional energy of the nucleus, due to the field H_1 is

$$\mathfrak{IC}_1 = \mathbf{\mu} \cdot \mathbf{H}_1 = \gamma\hbar(H_x I_x + H_y I_y) = \frac{\gamma\hbar H_1}{2}(I_+ e^{-i\omega t} + I_- e^{+i\omega t}) \tag{2.4}$$

---

† Instead of energy, we use for convenience angular frequency; the transition frequency is $\Delta\nu = (\Delta E/\hbar)/(2\pi) = \omega_0/2\pi$.

where†

$$I_+ = I_x + iI_y \quad \text{and} \quad I_- = I_x - iI_y \tag{2.5}$$

Since the energy specified by Eq. 2.4 is very small as compared to that given by Eq. 2.2, it can be treated as a time-dependent perturbation‡; thus, to first order, the transition probability is proportional to the absolute square of the matrix element

$$M = \frac{\gamma \hbar H_1}{2} \langle f | I_+ e^{-i\omega t} + I_- e^{i\omega t} | i \rangle \tag{2.6}$$

where i and f stand for the initial and final state. As usual the matrix element is evaluated by performing the integral

$$\int \psi_f^* \mathcal{3C}_1 \psi_i \, d^3x \, dt \tag{2.7}$$

and we must include the time dependence of the wave functions

$$\psi_f = u(I, m') \exp\left(-i\frac{E'}{\hbar}t\right)$$
$$\psi_i = u(I, m) \exp\left(-i\frac{E}{\hbar}t\right) \tag{2.8}$$

Here primes refer to the final state, and $u(I, m)$ stands for the space part of the wave function. Evaluating Eq. 2.6 with the help of Eqs. 2.7 and 2.8, we find

$$M = \frac{\gamma \hbar H_1}{2} \left\{ \langle I, m' | I_+ | I, m \rangle \int \exp\left[-i\left(\frac{E - E'}{\hbar} + \omega\right)t\right] dt \right.$$

$$\left. + \langle I, m' | I_- | I, m \rangle \int \exp\left[-i\left(\frac{E - E'}{\hbar} - \omega\right)t\right] dt \right\} \tag{2.9}$$

The matrix elements of the operators I_+ and I_- are§

$$\langle m' | I_+ | m \rangle = \sqrt{I(I + 1) - m(m + 1)} \, \delta_{m',m+1}$$
$$\langle m' | I_- | m \rangle = \sqrt{I(I + 1) - m(m - 1)} \, \delta_{m',m-1} \tag{2.9a}$$

---

† We expand the exponentials and obtain

$(I_x \cos \omega t + iI_y(-i) \sin \omega t) + (I_x \cos \omega t - iI_y(+i) \sin \omega t)$

$$= 2(I_x \cos \omega t + I_y \sin \omega t).$$

‡ See, for example, E. Fermi, *Notes on Quantum Mechanics*, University of Chicago Press, lecture 23.

§ See E. Fermi, *Notes on Quantum Mechanics*, loc. cit. Lecture 28.

and thus I_+ connects only states with $m' - m = 1$ while I_- connects states $m' - m = -1$. For $I = \frac{1}{2}$ the above matrix elements reduce to 1 for either I_+ or I_-. The integrals over the time are essentially δ functions (but see below) expressing the conservation of energy and showing that the transition probability is different from zero only if

$$E' - E = \hbar\omega \qquad \text{for} \qquad m' = m + 1$$

and   (2.10)

$$E - E' = \hbar\omega \qquad \text{for} \qquad m' = m - 1$$

that is, when the angular frequency of the rotating field is equal to the energy difference between adjacent m sublevels. Using Eq. 2.3, the condition of Eqs. 2.10 becomes

$$\hbar\omega = \hbar\gamma H_0 = \hbar\omega_0 \qquad\qquad (2.10a)$$

Indeed, the integrals in Eq. 2.9 would represent δ functions if the integration was from $t = 0$ to $t = \infty$. We know, however, from the treatment of time-dependent perturbation theory† that the integrals are to be taken from $t = 0$ to $t = t$, the time of observation. If we start from an initial state i and integrate over a continuum of final states f, we obtain for the transition rate (transition probability *per unit time*)

$$R_{if} = \frac{2\pi}{\hbar} \mid \mathfrak{M} \mid^2 \rho(E) \qquad\qquad (2.11)$$

Here $\mathfrak{M}$ is the time independent part of the matrix element given by Eq. 2.9 (that is, without the integrals). $\rho(E)$ is the "density of final states" and gives the number of states f per unit energy interval that have energy close to E'. For example, if the final state f has an extremely well-defined energy E_0, then $\rho(E) \to \delta(E - E_0)$; if the final state has a certain width (due for example to a finite lifetime‡ or other broadening effects), then $\rho(E)$ expresses this fact mathematically. We require the function $\rho(E)$ to be normalized and can also express it in terms of frequency

$$\rho(E) = \rho(h\nu) = \frac{1}{h} g(\nu)$$

with

$$\int \rho(E) \, dE = \int g(\nu) \, d\nu = 1 \qquad\qquad (2.12)$$

Combining Eqs. 2.9, 2.11, and 2.12 we finally obtain for the transition rate in the case $I = \frac{1}{2}$

$$R_{-1/2 \to +1/2} = R_{+1/2 \to -1/2} = \frac{\gamma^2 H_1^2}{4} g(\nu) \qquad\qquad (2.13)$$

---

† See E. Fermi, *Notes*, or L. Shiff, *Quantum Mechanics*, McGraw-Hill, Chapter VIII.
‡ Cf. Chapter 6, Section 4 on the Mössbauer effect.

where ν is the frequency of the radiofrequency and $g(\nu)$ gives the shape of the resonance line; clearly $g(\nu)$ will be significantly different from zero only for $\nu \approx \nu_0$.

At this point it is important to note from Eq. 2.9 that the probability that the rotating field H_1 will induce a transition from $m_1 \rightarrow m_2$ (where $m_1 - m_2 = +1$) is always the same as for inducing the transition $m_2 \rightarrow m_1$. Therefore, for the case $I = \frac{1}{2}$ we see that the effect of the radiofrequency field is to *equalize the population* of the two states.†

The reader should keep in mind that Eq. 2.13 was obtained from a perturbation calculation; however, the exact value for the transition probability for $I = \frac{1}{2}$ or for any arbitrary spin can also be calculated.

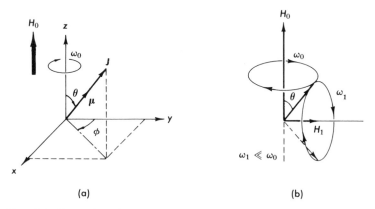

(a)   (b)

FIG. 8.4   Precession of a magnetic moment μ when placed in a magnetic field H_0. (a) The spin precesses with angular frequency $\omega_0 = \gamma H_0$; the angle θ is a constant of the motion. (b) In addition to H_0 a weak magnetic field H_1 is now also applied. H_1 is rotating about the z axis with angular frequency ω_0 and therefore μ precesses about H_1 with angular frequency $\omega_1 = \gamma H_1$; θ is not any more conserved.

Finally, we will briefly show how the effect of a rotating radiofrequency field can be understood also on the basis of a classical model. Consider again a nucleus with spin I and magnetic moment $\mu = \gamma \hbar I$. Let J be the magnitude of the angular momentum, which classically‡ will be just $J = \hbar I$, and let it make an angle θ with the z axis as shown in Fig. 8.4a. If a constant magnetic field H_0 is applied along the z axis, it will exert a torque on the magnetic moment, given by

$$\tau = \mu \times H_0 = \gamma(J \times H_0) \tag{2.14}$$

_____

† This argument can be generalized to $I > \frac{1}{2}$ as well.

‡ As we recall the magnitude of any angular momentum vector I, is given quantum-mechanically by $\hbar\sqrt{I(I + 1)}$.

This must equal the time derivative of the angular momentum

$$\frac{d\mathbf{J}}{dt} = \boldsymbol{\tau} = \gamma(\mathbf{J} \times \mathbf{H}_0) \tag{2.15}$$

As we know from the theory of the gyroscope, the solution† of Eq. 2.15 leads to a precession of the angular momentum vector $\mathbf{J}$ about the z axis, preserving the angle θ, and at an angular frequency ω_0 independent of θ,

$$\boldsymbol{\omega}_0 = -\frac{|\,d\mathbf{J}/dt\,|}{|\,\mathbf{J} \times \mathbf{n}_z\,|}\,\mathbf{n}_z = -\gamma H_0 \mathbf{n}_z \tag{2.16}$$

where $\mathbf{n}_z$ is the unit vector in the z direction.

This phenomenon is called the Larmor precession and the angular frequency given by Eq. 2.16 is the "Larmor" frequency. It is fascinating even though not surprising that the Larmor frequency has the same value as given by Eq. 2.3 for the transition frequency between any adjacent levels ($\Delta m = \pm 1$). Further, since the angle θ is preserved, the energy of the nucleus in the magnetic field remains a constant

$$E = -\boldsymbol{\mu}\cdot\mathbf{H} = -\gamma\hbar I H_0 \cos\theta \tag{2.17}$$

---

† By referring to Fig. 8.4a, we write for the equation of motion (2.15)

$$\frac{dJ_x}{dt} = \frac{d}{dt}\,(J\sin\theta\sin\phi) = \gamma[J_y H_z - 0] = \gamma J \sin\theta\cos\phi H_0$$

$$\frac{dJ_y}{dt} = \frac{d}{dt}\,(J\sin\theta\cos\phi) = \gamma[0 - J_x H_z] = -\gamma J \sin\theta\sin\phi H_0$$

$$\frac{dJ_z}{dt} = \frac{d}{dt}\,(J\cos\theta) = \gamma[0 - 0] = 0$$

J does not change in magnitude because

$$\frac{d}{dt}\,(J^2) = \frac{d}{dt}\,(\mathbf{J}\cdot\mathbf{J}) = 2\mathbf{J}\cdot\frac{d\mathbf{J}}{dt} = 2\mathbf{J}\cdot(\mathbf{J} \times \mathbf{H}_0)\gamma$$

but $\mathbf{J} \times \mathbf{H}_0$ is normal to $\mathbf{J}$ so that the above expression is zero. Given then this fact we note from the third equation that $d\cos\theta/dt = 0$; that is, θ is constant of the motion. For the first two equations we write

$$\frac{d}{dt}\sin\phi = \cos\phi\,\frac{d\phi}{dt} = \gamma H_0 \cos\phi$$

$$\frac{d}{dt}\cos\phi = -\sin\phi\,\frac{d\phi}{dt} = -\gamma H_0 \sin\phi$$

that is,

$$\frac{d\phi}{dt} = \omega_0 = \gamma H_0$$

We now introduce an additional weak magnetic field H_1 oriented in the $x - y$ plane and rotating about the z axis (in the same direction as the "Larmor precessing" spin I) with an angular frequency ω. If the frequency ω is different from ω_0, the angle between the field H_1 and the magnetic moment μ will continuously change so that their interaction will average out to zero. If, however, $\omega \approx \omega_0$, the angle between μ and H_1 is maintained and a net interaction is effective (Fig. 8.4b). If we look at the system in a reference frame that is rotating about the z axis with the angular velocity ω_0, then the spin will appear to make an angle $\psi = 90° - \theta$ with H_1, and according to the previous argument will start to precess (in the rotating frame) about H_1. This corresponds to a "nutation" and a consequent change of the angle θ, which implies a change in the potential energy of the nucleus in the magnetic field (Eq. 2.17). The change in θ is the classical analogy to a *transition* between sublevels with different m. We see that (a) such transitions may take place only if the rotating field has an angular frequency $\omega = \omega_0 = \gamma H_0$, and (b) since the angle θ will continuously change (with an angular frequency $\omega_1 = \gamma H_1$). The effect of the radiofrequency is to populate, on the average, all values of θ, that is *all levels, equally*.

3. Absorption of Energy by the Nuclear Moments

3.1 RELAXATION AND SATURATION

We saw in Section 2 that a radiofrequency magnetic field may indeed induce transitions between the m sublevels of a quantum-mechanical system. In the case of atomic-beam or optical double resonance experiments the atoms are essentially free, while in nuclear magnetic resonance (and electron paramagnetic resonance) experiments the nuclei (electrons) are in constant interaction with their surroundings. There are mainly two types of interactions,† (a) *spin-lattice*, in which the nuclear spin interacts with the entire solid or liquid, transferring energy from the spin system to the lattice; and (b) *spin-spin*, in which the nuclear spin interacts with a neighboring nuclear spin, but the total energy of the spin system remains constant. As a matter of fact, it is the spin-lattice interaction that makes possible the observation of energy absorption from the radiofrequency field when the resonance frequency is reached.

To understand this last statement, consider again the simple case of a nucleus with spin $I = \frac{1}{2}$. In the presence of a magnetic field H_0 it is split into the two energy sublevels with $m = +\frac{1}{2}$ and $m = -\frac{1}{2}$. As remarked before, the rate (Eq. 2.11) for transitions

$$(m = +\tfrac{1}{2}) \rightarrow (m = -\tfrac{1}{2}) \tag{3.1}$$

---

† We will not consider quadrupole interactions or other relaxation mechanisms.

is equal to the rate for transitions

$$(m = -\tfrac{1}{2}) \rightarrow (m = +\tfrac{1}{2}) \qquad (3.2)$$

The number of transitions per unit time is given in either case by

$$R_{if} \times N_i \qquad (3.3)$$

where N_i is the number of nuclei in the initial state. Further, transitions of the type in Eq. 3.1 absorb energy from the radiofrequency field while transitions of the type in Eq. 3.2 give energy to the radiofrequency field.† Thus the net power absorbed from the radiofrequency field is as given in Eq. 3.4. (We also multiply by the energy necessary for one transition.)

$$P = [N_{+1/2} \times R(+\tfrac{1}{2} \rightarrow -\tfrac{1}{2})]\hbar\omega_0$$

$$-[N_{-1/2} \times R(-\tfrac{1}{2} \rightarrow +\tfrac{1}{2})]\hbar\omega_0 = (N_{+1/2} - N_{-1/2})R\hbar\omega_0 \quad (3.4)$$

Thus if $N_{-1/2} = N_{+1/2}$, no net power can be absorbed from the field. However, if we consider a system consisting of a large number of spins in equilibrium with its surroundings, it is known from a very general theorem of statistical mechanics that every state of energy E will be populated according to the Boltzmann distribution

$$N(E) = N_0 \, e^{-E/kT} \qquad (3.5)$$

with k the Boltzmann constant and T the absolute temperature in degrees K. It follows that for a system of N spins **I** in the presence of a magnetic field H_0, each m sublevel will be populated according to

$$N(m) = \frac{N}{2I + 1} \exp\left(+\frac{m\gamma\hbar H_0}{kT}\right) \qquad (3.6)$$

The normalizing factor was approximated by $N/(2I + 1)$, which holds‡ for $\gamma\hbar H_0 \ll kT$; $-m\gamma\hbar H_0$ is the energy of the m sublevel (see Eq. 2.2). Note that T in Eq. 3.6 is the temperature of the spin system and equals the lattice temperature, if no external perturbations (such as the radiofrequency field) are present.

It follows from Eq. 3.6 that the populations $N_{+1/2}$ and $N_{-1/2}$ entering Eq. 3.4 in our previous discussion $(I = \tfrac{1}{2})$ will not be equal. Indeed, there will be a number of excess nuclei N_s, in the lower energy state given by

$$N_s = N_{+1/2} - N_{-1/2} = \frac{N}{2}\left[\exp\left(+\frac{\hbar\omega_0}{2kT}\right) - \exp\left(-\frac{\hbar\omega_0}{2kT}\right)\right] \quad (3.7)$$

---

† Remember Eq. 22.

‡ Indeed, if we expand the exponential through first order, we obtain correctly

$$\sum_{m=-I}^{m=+I} N(m) = N$$

and since $\omega_0 \hbar$ is always much smaller than kT_s, we may write Eq. 3.7 as

$$N_s \approx \frac{N}{2} \frac{\hbar \omega_0}{kT} \qquad (3.7a)$$

It is only these N_s nuclei that can contribute towards a net absorption of energy, and the power (Eq. 3.4) is given by

$$P = N_s \times R \times \hbar \omega_0 = \frac{N}{2} \frac{(\hbar \omega_0)}{kT} \times (\hbar \omega_0) \times R \qquad (3.8)$$

Before proceeding further, we calculate some numerical values: for protons $\gamma = 2.673 \times 10^4$ rad/sec-gauss, so that for $H_0 = 10^4$ gauss and $T = 300°$ K we obtain

$$\frac{N_s}{N} = \frac{\omega_0 \hbar}{2kT} = \frac{(2.67 \times 10^8) \times (6.6 \times 10^{-16}) \text{ eV}}{2(1/40) \text{ eV}} \approx 4 \times 10^{-6}$$

which justifies the approximation used in going from Eq. 3.7 to Eq. 3.7a. If we further consider a sample of 1 cm³ of water, the number of protons contained in it is

$$N = N_0 \times (2/18) = 6 \times 10^{23} \times (2/18) = (2/3) \times 10^{23}$$

If we use for $R = 1$/sec (as can be seen from Eq. 2.13, this is a conservative value; R, however, can be as large as 10^3/sec as discussed below), we obtain for Eq. 3.8

$$P = (\hbar \omega_0) \times \left(N \times \frac{\hbar \omega_0}{2kT_s} \right) \times R \simeq 5 \times 10^{10} \text{ eV/sec} = 8 \times 10^{-9} \text{ watts} \qquad (3.9)$$

This is a very small amount of power, especially since the applied radio-frequency field may be of the order of milliwatts. Therefore, a sensitive null method greatly facilitates the observation of nuclear resonance absorption.

Now in writing down Eq. 3.8, we assume that the power absorbed is proportional to the number of excess nuclei n_s; but as transitions are induced to the upper state, the number n_s will continuously decrease. The decrease will be exponential at a rate R

$$n_s = N_s e^{-Rt} \qquad (3.10)$$

Soon the populations of the two levels will be equalized, $N_{+1/2} = N_{-1/2}$ and no more absorption will be observed.

However, while the radiofrequency field tends to equalize the populations, the "spin-lattice" interaction tends to restore the Boltzmann distribution as given by Eqs. 3.7 and 3.7a at a rate characterized by $1/T_1$. We say that the nuclei are "relaxed" through their interaction with the lattice, and the characteristic time T_1 for this process is called the *spin-lattice relaxation time*. Therefore, in the presence of a radiofrequency field tuned to the resonance frequency, the number of excess nuclei at equilibrium n_s, depends on T_1 and on R; if $R \ll 1/T_1$, then $n_s \to N_s$, while if $R \gg 1/T_1$, $n_s \to 0$. The value of n_s can be easily obtained†

$$n_s = \frac{N_s}{1 + 2RT_1} \tag{3.11}$$

where N_s as given by Eqs. 3.7 and 3.7a is the equilibrium excess of population in the absence of the radiofrequency field.

By using Eq. 2.13 for R, we obtain

$$n_s = \frac{N_s}{1 + \frac{1}{2}\gamma^2 H_1^2 T_1 g(\nu)} \tag{3.11a}$$

From Eq. 3.11a we see that when too much radiofrequency power is used, the number of excess nuclei n_s decreases, and so does the resonance signal. We say that the sample has been saturated, and the ratio n_s/N_s is frequently referred to as the saturation factor Z:

$$\frac{n_s}{N_s} = \frac{1}{1 + \frac{1}{2}\gamma^2 H_1^2 T_1 g(\nu)} \equiv Z \tag{3.14}$$

The maximum useful value of the radiofrequency power therefore depends on the relaxation time T_1. For solids, T_1 is large (it takes a long time

---

† Let $n = n_{+1/2} - n_{-1/2}$ be the instantaneous excess of nuclei in presence of both radiofrequency and relaxation. The effect of the radiofrequency is to make $n \to 0$

$$\left(\frac{dn}{dt}\right)_{\text{rf}} = -2Rn \tag{3.12}$$

(The factor of 2 arises because each transition up decreases $n_{+1/2}$ by one, and also increases $n_{-1/2}$ by one). The effect of relaxation is to return $n \to N_s$

$$\frac{d(N_s - n)}{dt} = -(N_s - n)\frac{1}{T_1} = -\left(\frac{dn}{dt}\right)_{\text{relax}} \tag{3.13}$$

Equilibrium is reached when the sum of the two rates in Eqs. 3.12 and 3.13 is zero; that is,

$$-2Rn + \frac{N_s - n}{T_1} = 0$$

which yields Eq. 3.11

$$n = \frac{N_s}{1 + 2RT_1}$$

for the spins to reorient themselves in the equilibrium position), and therefore only weak radiofrequency fields may be applied. For example, for protons in ice $T_1 = 10^4$ sec. On the contrary, in liquids, especially in solutions containing paramagnetic ions, the relaxation time for protons may be as short as $T_1 = 10^{-4}$ sec.

Finally, a concept often used is the spin temperature T_s; it corresponds to the value of T in the Boltzmann distribution (Eqs. 3.7 and 3.7a) which would produce an excess of nuclei n_s

$$n_s = \frac{N}{2} \frac{\hbar \omega_0}{k T_s} \quad \text{or} \quad T_s = \frac{T N_s}{n_s} = \frac{T}{Z} \tag{3.15}$$

Thus when the spins are saturated, their "temperature" (T_s) becomes very large.

3.2 LINE WIDTH AND THE SPIN–SPIN INTERACTION

Section 3.1 dealt with the interaction of the nuclear spins with the lattice and the way in which this gives rise to a relaxation process which is essential for the observation of energy absorption. It is also possible for a nuclear spin to interact with its neighboring nuclear spins, but without changing the total energy of the spin system.

The effect of the spin-spin interaction, however, is to slightly shift the exact position of the energy level of any individual spin in the external magnetic field. This energy shift does clearly depend on the relative orientation and distance of the spins, and thus is different for each spin, resulting in an apparent broadening[†] of the energy level. Another way of thinking of the spin-spin interaction is that one nuclear spin produces a local magnetic field H_{local} at the position of another spin, which then finds itself in a field

$$\mathbf{H'}_0 = \mathbf{H}_0 + \mathbf{H}_{local}$$

and consequently has a resonance frequency $\omega'_0 = \gamma H'_0$ slightly different from ω_0. To estimate this effect, we calculate the magnetic field produced by a magnetic dipole one nuclear magneton strong, at a typical distance of 1 Å. We use the MKS system.

$$H_{local} \approx \left(\frac{\mu_0}{4\pi}\right) \frac{\mu_N}{r^3} = \left(\frac{\mu_0}{4\pi}\right) \times \frac{e\hbar}{2M_p} \times \frac{1}{r^3}$$

here μ_N is the nuclear magneton and $\mu_0 = 4\pi \times 10^{-7}$ henry/meter is the permeability of free space[‡] and $e\hbar/2M_p$ is the nuclear magneton.

---

† In the classical analogy, we think of the spin–spin interaction as destroying the phase coherence between the precessing spins and the rotating radiofrequency field.

‡ Not the Bohr magneton.

Thus

$$H_{\text{local}} \approx 10^{-7} \frac{1.6 \times 10^{-19}}{10^{-30}} \times \frac{1}{2} \times \frac{6.582 \times 10^{-22} \text{ MeV-sec}}{(938 \text{ MeV})/c^2}$$

$$= 5.6 \times 10^{-21} \times (3 \times 10^8)^2 = 5 \times 10^{-4} \text{ webers/m}^2 = 5 \text{ gauss}$$

$$(3.16)$$

which is a significant broadening of the line, even in a field of 10 kilogauss. In liquids and gases, however, the reorientation of the molecules is so fast that the average local field is very close to zero, and therefore very narrow lines can be obtained.

We have already described the width of the resonance line by the function $g(\nu)$ (Eq. 2.12), and we now see that this width is mainly due to the spin-spin interaction. Since $g(\nu)$ has dimensions of $(\nu)^{-1}$, namely, of time, we *define* one half of its maximum value by T_2

$$\tfrac{1}{2} g(\nu_0) = T_2 \qquad (3.17)$$

where ν_0 is the resonance frequency in the absence of any broadening effects; T_2 is called the *transverse relaxation time*. In view of the normalization condition, Eq. 2.12,

$$\int g(\nu) \, d\nu = 1$$

which also fixes the dimensions of $g(\nu)$, we see that for broad lines T_2 is short, and for narrow lines, T_2 is long.

With the help of the definition in Eq. 3.17, we can then rewrite for the saturation factor Z (Eq. 3.14) *at resonance*

$$Z_0 = Z(\nu_0) = \frac{1}{[1 + \gamma^2 H_1^2 T_1 T_2]} \qquad (3.14a)$$

The width of the resonance line may occasionally also be influenced by very short spin-lattice relaxation, since this limits the "lifetime" of the spins in the upper state; it then follows from the uncertainty principle that the energy level would have a width ΔE. To obtain a width of 5 gauss ($\Delta\nu = \gamma \Delta H/(2\pi)$, which for protons corresponds to 20 kc), T_1 must be[†]

$$\Delta E \Delta t \sim \hbar \qquad T_1 \sim \frac{1}{2\pi(\Delta E/h)} = \frac{1}{6.28 \times 20 \times 10^3} \sim 10^{-5} \text{ sec}$$

Finally, in the experimental observations, spurious broadening effects mainly due to inhomogeneities in the magnetic field must be eliminated first. Clearly, such effects broaden the signal but also decrease its amplitude.

---

† The same argument also holds for T_2.

3.3 THE RADIOFREQUENCY MAGNETIC SUSCEPTIBILITIES†

As is known from electrostatics and magnetostatics, when an electric (or magnetic) field E, (or H) is applied in a region containing matter, the material becomes polarized (or magnetized). We write

$$\mathbf{P} = \chi_e \mathbf{E} \qquad \mathbf{M} = \chi_\mu \mathbf{H} \tag{3.18}$$

where χ_e and χ_μ are the electric and magnetic susceptibilities and measure the strength of this effect. We now know from the microscopic behavior of matter (atomic and molecular physics) that the polarization is due primarily to the alignment of the electric (magnetic) dipole moments of the atoms or molecules in the direction of the applied field. Materials which have such dipole moments and exhibit large polarization should be called *paraelectric* (or for large magnetization, they are indeed called *paramagnetic*).

Now we also know that the refractive index of light (that is, of a high-frequency electromagnetic field) is related to the electric and magnetic susceptibilities, since

$$\epsilon = (1 + \chi_e)\epsilon_0 \qquad \mu = (1 + \chi_\mu)\mu_0$$

and

$$n = \frac{c}{v} = \frac{1/(\sqrt{\epsilon_0\mu_0})}{1/(\sqrt{\epsilon\mu})} = \sqrt{(1 + \chi_e)(1 + \chi_\mu)}$$

Further, the refractive index (and therefore also the susceptibilities) is a function of the frequency, as we know from the familiar phenomenon of the dispersion of light. Thus we have a susceptibility at optical frequencies which differs from the static one and is a function of the frequency‡. As a matter of fact frequently the transmission of the electromagnetic (light) pulse is accompanied by absorption (which may be strongest at a particular resonant frequency of the atomic oscillators); we may account for the absorption by attributing an imaginary part to the susceptibility.

The same formalism can be used for the description of the nuclear magnetic resonance phenomena as well. The static susceptibility arising from the *nuclear* moments in an otherwise diamagnetic material exists, but is very small and difficult to measure. For the radiofrequency susceptibility, we write

$$\chi(\omega) = \chi'(\omega) - i\chi''(\omega)$$

---

† This section may be omitted without a loss of continuity and the reader can proceed directly to the discussion of the experimental technique and results in Section 4. However, the discussion should be quite helpful for understanding the meaning of the "dispersion" curve as well as the observed line shapes for both absorption and dispersion.

‡ For optical frequencies and for almost all materials, χ_μ is 0 and the variation in n arises entirely from χ_e.

where both $\chi'(\omega)$ and $\chi''(\omega)$ exhibit a resonant behavior when ω reaches $\omega_0 = \gamma H_0$. The real part $\chi'(\omega)$ is given by

$$\chi'(\omega) = \tfrac{1}{2}\chi_0\omega_0 T_2\left[\frac{(\omega_0 - \omega)\,T_2}{1 + (\omega_0 - \omega)^2 T_2{}^2 + \gamma^2 H_1{}^2 T_1 T_2}\right] \quad (3.19)$$

while the imaginary part $\chi''(\omega)$ is given by

$$\chi''(\omega) = \tfrac{1}{2}\chi_0\omega_0 T_2\left[\frac{1}{1 + (\omega_0 - \omega)^2 T_2{}^2 + \gamma^2 H_1{}^2 T_1 T_2}\right] \quad (3.20)$$

Here χ_0 is the static susceptibility defined as in Eq. 3.18

$$M_0 = \chi_0 H_0$$

and T_1 and T_2 are the familiar relaxation times introduced before; the term $\gamma^2 H_1{}^2 T_1 T_2$ appearing in the denominator is a measure of the saturation as defined in Eq. 3.14.

Equations 3.19 and 3.20 are shown in Fig. 8.5 under the assumption that $\gamma^2 H_1{}^2 T_1 T_2 \ll 1$; they have the typical behavior of a dispersion and a power resonance curve. We also note that Eq. 3.19 is the derivative, with respect to ω, of Eq. 3.20; by properly adjusting the detection equipment, we may observe experimentally either of those curves (or a combination of both).

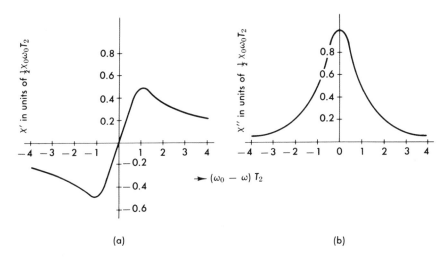

(a)   (b)

Fig. 8.5 The radiofrequency magnetic susceptibilities near resonance. (a) The real part of the susceptibility exhibits a typical dispersion shape (Eq. 3.19 of text). (b) The imaginary part of the susceptibility exhibits a typical absorption shape (Eq. 3.20 of text).

4. Experimental Observation of the Nuclear Magnetic Resonance of Protons

4.1 GENERAL CONSIDERATIONS

To observe nuclear magnetic resonance we clearly will need a sample, a magnet, a source of electromagnetic radiation of the appropriate frequency, and a detection system.

The magnetic field should be quite homogeneous, and therefore it is advisable to choose a good magnet with polefaces at least 4 to 6 in. in diameter. As discussed in Section 3.2, inhomogeneities in the magnetic field broaden the line (and thus, decrease correspondingly the peak amplitude); to obtain reasonable results, the inhomogeneities over the volume of the sample should be less than 1/1000. The choice of the field strength is arbitrary, provided the resulting frequency lies in a convenient radiofrequency band. However, since the signal-to-noise ratio increases (improves) as $\nu_0^{3/2}$, high fields are preferable; commonly magnets of 5000 to 10,000 gauss strength are used, and for protons this corresponds to frequencies of 20 to 40 Mc/sec.

The sample can be any material containing an ample supply of protons: paraffin, water, mineral oil, or any organic substance containing hydrogens will, in general, give a proton nuclear magnetic resonance signal. Some care must be exercised to avoid materials with long spin-lattice relaxation times T_1, since they will saturate at very low levels of radiofrequency power (as explained in Section 3.1) and therefore give weak signals. Similarly it is profitable to have a narrow line, hence materials with long spin-spin relaxation time T_2 are chosen. Liquids will meet this condition, and in most instances the width of the line will be determined by the magnet inhomogeneity ($T_2 = 2 \times 10^{-3}$ sec will give for protons a line width of 0.1 gauss). Plain tap water makes a good sample, or tap water doped with one percent by weight of manganese nitrate [$Mn(NO_3)_2$].

The size of the sample is limited again by the area over which the magnet is homogeneous, but also by practical considerations on the coil which is used to couple the radiofrequency to the sample. In usual practice a 1-cm^3 sample is adequate; it is contained in a small tubular glass (or other) container, around which is wrapped a radiofrequency coil as shown in Fig. 8.6. The whole assembly is then inserted into the magnet gap and should be secured firmly, since vibration is picked up by the coil and appears as noise in the detectors.

In deriving the probability for a transition between the m sublevels, and in all our previous discussion, we have assumed the existence of a *rotating* field at the angular frequency ω close to ω_0. In practice, however, a field oscillating linearly, rather than rotating, is established; such a magnetic field will certainly exist in the interior of the radiofrequency coil shown in Fig. 8.6, and will alternate in direction with a sinusoidal amplitude

$$A \sin \omega t$$

It is well known, however, that linear harmonic motion is equivalent to two rotations in opposite direction and of amplitude $A/2$ as shown in Fig. 8.7; indeed

$$A \cos \omega t \mathbf{n}_x = \frac{A}{2} (\cos \omega t \mathbf{n}_x + \sin \omega t \mathbf{n}_y) + \frac{A}{2} (\cos (-\omega t)\mathbf{n}_x + \sin (-\omega t)\mathbf{n}_y)$$

$$(4.1)$$

where $\mathbf{n}_x$ and $\mathbf{n}_y$ are the unit vectors in the x and y directions. The component rotating in the same direction as the precessing spins will be in resonance and may cause transitions; the other component is completely out of phase and has no effect on the sample.

When the radiofrequency reaches the resonance value ω_0, energy is absorbed from the field of the coil and this fact is sensed by the detector. Because, however, of the low signal levels involved and the difficulty of maintaining a very stable level of radiofrequency power it is advantageous to traverse the whole resonance curve in a relatively short time. This can be achieved either by "sweeping" the frequency of the radiofrequency oscillator while maintaining the magnetic field constant, or by "sweeping" the magnetic field while the frequency remains fixed. The latter condition

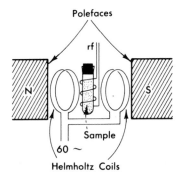

FIG. 8.6   Schematic arrangement of a nuclear magnetic resonance apparatus. The sample is placed in a homogeneous magnetic field and radio frequency is coupled to it by means of the coil. The Helmoltz coils are used to modulate the constant magnetic field.

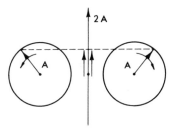

FIG. 8.7   A linearly oscillating field of frequency ω is equivalent to two fields rotating in opposite directions with the same frequency ω.

can be achieved very easily and is therefore widely used. A small pair of Helmholtz coils† is placed across the sample as shown in Fig. 8.6; remember that the constant field H_0, (and thus also the "sweeping field") must be perpendicular to the radiofrequency field H_1, which is determined by the axis of the radiofrequency coil. The sweep coils are fed with a slowly varying current‡, which therefore results in a modulation of the magnetic field H_0. If this sweep covers the value of H_0 which is in resonance with the fixed frequency of the oscillator, a resonance signal modulated at the frequency of the sweep will appear at the detector. The existence of a modulated signal has the advantage of easier amplification and also presents the possibility of using a narrow band-width detector with the corresponding improvement in the signal-to-noise ratio.

The radiofrequency oscillator and detection circuit can be of several types. In the following sections we describe first a system using a radiofrequency bridge for the detection of resonance; in principle this is the most sensitive technique and can be used for point-by-point measurement of a nuclear magnetic resonance line. Next is presented the marginal oscillator system, which is much more compact and is widely used in commercial nuclear magnetic resonance magnetometers; it gives strong signals for protons but cannot be used for detailed studies of line shapes. Finally, a very simple transistor circuit which yields adequate signals for protons is described.

A different experimental technique for the observation of nuclear magnetic resonance consists in the use of two radiofrequency coils placed at 90 degrees with respect to one another, but both lying in the plane normal to H_0. The radiofrequency is supplied in one coil and the second coil acts as a pickup of the radiofrequency voltage induced by the precessing spins. When the radiofrequency is off resonance, the relative phase of the spins is random and no net induction occurs; when, however, resonance is established, all the precessing spins are brought into phase by the action of the H_1 field and an induction signal appears in the other coil. This method is the one used by Bloch and collaborators, whereas the previously described one-coil method was used by Purcell and collaborators. The experiments discussed below were performed with the one-coil method.

Even though nuclear magnetic resonance studies do provide information on relaxation times and mechanisms, we will restrict the discussion of the results obtained in this laboratory to the observation of the resonance conditions for protons.

---

† A pair of coils of diameter d, spaced a distance $d/2$ apart and traversed by current in the same direction, produce a very homogeneous field at the geometrical center of the configuration.

‡ Frequently the 60 cps of the a-c line is used; a transformer and variac are placed in the line for reducing and controlling the voltage applied to the coils. See Fig. 8.9.

4.2 DETECTION OF NUCLEAR MAGNETIC RESONANCE WITH A BRIDGE CIRCUIT

It has been noted that a rotating radiofrequency field can be established in the interior of a small coil, where the sample is placed. We wish now to examine how the small absorption of power from the field (Eq. 3.8) may be detected experimentally: we can think of the absorption as a change in the Q value of the coil when the resonant frequency is reached. The Q value, or quality factor, of a device is defined as 2π times the ratio of the time-averaged energy stored to energy dissipated, in one cycle. For a coil of inductance L and resistance R,

$$Q = \frac{2\pi\omega L}{R} \qquad (4.2)$$

When resonance is reached, the magnetic susceptibility (real part, Eq. 3.19) changes, and thus also the inductance of the coil changes. Alternatively, the absorption of power from the field (imaginary part of the susceptibility Eq. 3.20) corresponds to increased dissipation and therefore increased resistivity of the coil. This small change in the Q value (Eq. 4.2) will result in a change of the voltage across the coil if it is driven by a constant current source (as is the case with the transistor circuit of Section 4.4); if a fixed voltage is applied to the coil, the change in Q is best detected with a bridge circuit.

Such a bridge is shown diagrammatically in Fig. 8.8a. The radiofrequency voltage is applied between points a and g, and therefore radiofrequency current flows through the load L and the dummy branch D; if the bridge

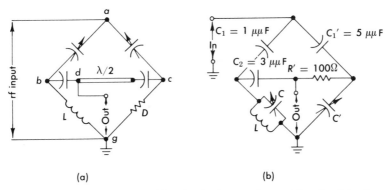

(a)   (b)

FIG. 8.8   A radiofrequency bridge circuit that can be used for the detection of nuclear magnetic resonance. (a) Schematic arrangement; note that L is the load (radiofrequency coil). The $\lambda/2$ line ascertains cancellation at the output of the signals from b and c. (b) A practical radiofrequency bridge circuit. For resonance conditions see Eq. 4.3 in the text.

is balanced, no voltage should appear at the point d (since b and c were in phase and of the same amplitude, and the signal from c to d is shifted by $\lambda/2$). Any slight unbalance of the bridge produces a small voltage at d. In actual practice, the bridge circuit shown in Fig. 8.8b was used where the $R'C'$ elements are effectively generating the $\lambda/2$ phase shift and L is the sample coil. The conditions for balance† are

Resistive balance: $\omega^2 C_1 C_2 \left(1 + \dfrac{C'}{C_1'}\right) R' R_p = 1$

$$(4.3)$$

Reactive balance: $C + C_1 + C_2 \left(1 + \dfrac{C_1}{C_1'}\right) = \dfrac{1}{L\omega^2}$

where R_p is the parallel resistance of the coil. The block diagram of the experimental setup is shown in Fig. 8.9 and consists of the following parts:

(a) The radiofrequency generator, which in this case was the variable frequency oscillator shown in Fig. 8.10. A crystal-controlled oscillator may be used, especially if only a fixed frequency is to be employed.

(b) The standard 93-ohm attenuator, which is very convenient in adjusting the power input into the bridge.

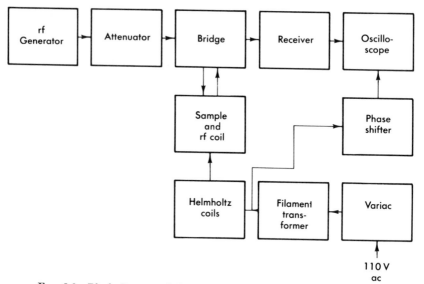

FIG. 8.9 Block diagram of the nuclear resonance measuring apparatus.

---

† See E. R. Andrew, *Nuclear Magnetic Resonance*, Cambridge University Press, p. 47 and references therein.

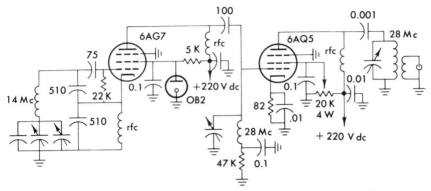

FIG. 8.10 A variable frequency oscillator suitable for providing the radiofrequency for a nuclear magnetic resonance experiment. The frequency is in the range of 28 Mc/sec.

(c) The bridge (Fig. 8.8b)

(d) The radiofrequency coil, which is part of the bridge circuit and consists of 6 turns of number 24 wire wound on a glass tube 8 mm in diameter. Its axis clearly must be perpendicular to the constant field H_0.

(e) The amplifier. For that purpose a National N.C. 400 radio receiver was used and it was tuned to the frequency of the radiofrequency. The bridge output was fed directly to the N.C. 400 input; the audio stages of the receiver were bypassed and an output signal was taken directly after the second detector stage (First I.F. 1721 kc/sec; second I.F. 445 kc/sec).

(f) The oscilloscope used for the display of the resonance. The demodulated signal is fed to the vertical input of the oscilloscope, the horizontal sweep being driven by the sweep voltage.

(g) A phase-shifting network (such as that shown in Fig. 8.11), which is used to allow the positioning of the resonance dip in the center of the oscilloscope sweep.

Frequently, when the voltage for the Helmholtz coils is taken from the line, the oscilloscope may be used with its own linear sweep (1–2 msec/cm on the scope screen is obviously a convenient rate) and triggered from the line voltage.

FIG. 8.11 A simple phase-shifting network that can be used to adjust the position of the resonance signal on the oscilloscope trace.

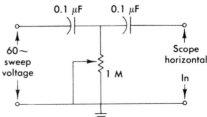

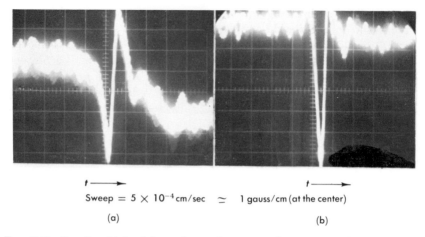

Sweep = 5 × 10⁻⁴cm/sec   ≃   1 gauss/cm (at the center)

(a)   (b)

FIG. 8.12   Results obtained from the nuclear magnetic resonance of protons using a bridge circuit. (a) Dispersion curve. (b) Absorption curve. The oscilloscope sweep was linear at 0.5 msec/cm which corresponds to approximately 1 gauss/cm at the center of the sweep.

The bridge is balanced either in the resistive mode, when the change in the Q of the coil will appear as an absorption curve (Fig. 8.5b); indeed, this is observed experimentally, Fig. 8.12b; or the bridge may be balanced in the reactive mode, when the signal appears as a dispersion curve (Fig. 8.5a) and is so observed experimentally, Fig. 8.12a.

The main problem in obtaining a good null for the bridge is to limit the stray radiofrequency power by appropriate shielding; also the mechanical rigidity of the bridge is important. It is advisable to bypass to ground the Helmholtz coils with a small condenser since they pick up radiofrequency.

The experimental results obtained with this arrangement by a student† are shown in Fig. 8.12. The sample was 1 cm³ of water doped with manganese nitrate $[Mn(NO_3)_2]$ and the extent of the sweep was approximately 6.2 gauss at 60 cps. The horizontal sweep in the figure is 0.5 msec/cm; Fig. 8.12a shows the dispersion curve whereas Fig. 8.12b shows the absorption curve. Since the oscilloscope sweep is linear at 0.5 msec/cm, the corresponding field change is not linear but of the order of 1 gauss/cm at the center of the sweep (see Section 5.3 also).

To measure the resonance frequency exactly, a precise (crystal controlled) frequency meter is required. The BC 221‡ is convenient, and its variable oscillator is adjusted for zero beat against the radiofrequency

---

† T. Walters, class of 1962 with the assistance of F. Reynolds.

‡ U. S. Army Signal Corps, BC 221 frequency meter; this may be purchased as army surplus equipment.

generator either by listening on the earphones or by observing the pattern on the oscilloscope. Since the frequency meter is calibrated at each "checkpoint" against its own crystal, an accuracy of 1 part in 10^5 can be achieved.

If we are interested in finding the gyromagnetic ratio of the protons, we must also know the magnetic field to as good an accuracy as possible. Since the resonance signals appear at some particular value of the sweeping field, we need to know $H_0 + H_{\text{sweep}}$; instead, the frequency may be adjusted so as to resonate when the sweeping field crosses its zero value.† This is achieved by gradually decreasing the amplitude of the sweep field until it is minimal and ascertaining that the absorption curve remains centered on the oscilloscope.

From the experimental curves of Fig. 8.12 it is found that the frequency at resonance is

$$\nu_0 = 28{,}141.48 \pm 0.63 \text{ kc/sec}$$

The magnetic field, unfortunately, cannot be measured to similar accuracy; by using a rotating coil flux meter at the field position previously occupied by the sample, it was obtained

$$H_0 = 6642 \pm 20 \text{ gauss}$$

and hence

$$\gamma = (26.618 \pm 0.08) \times 10^3 \text{ rad/sec-gauss}$$

In good agreement with the accepted value

$$\gamma = 26.73 \times 10^4 \text{ rad/sec-gauss} \tag{4.4}$$

The reader will appreciate at this point that it is much easier to measure ratios of nuclear moments to high accuracy than to obtain their absolute value to the same accuracy.

If it is desired to obtain the g factor of the proton—that is, the connection between magnetic moment and the nuclear magneton—we have

$$\mu = gI\mu_N$$

and from Eq. 1.1 and Eq. 2.3

$$\omega_0 \hbar = 2\mu H_0$$

$$g = \frac{\omega_0 \hbar}{H_0} \frac{1}{2I} \frac{1}{\mu_N} = \frac{1}{2\pi} \frac{\omega_0}{H_0} \frac{1}{2I} \frac{h}{\mu_N} = \frac{1}{2\pi} \gamma \frac{1}{762} \frac{1}{2I} \tag{4.5}$$

---

† Alternatively, the frequency is adjusted so as to resonate at the upper limit of the sweep and it is measured there; next the frequency is adjusted for the lower limit of the sweep and measured again. The difference of the two frequency measurements corresponds to the width of the sweep field (in kilocycles). The central frequency (that is, the frequency corresponding to H_0) is the average value of the two limits.

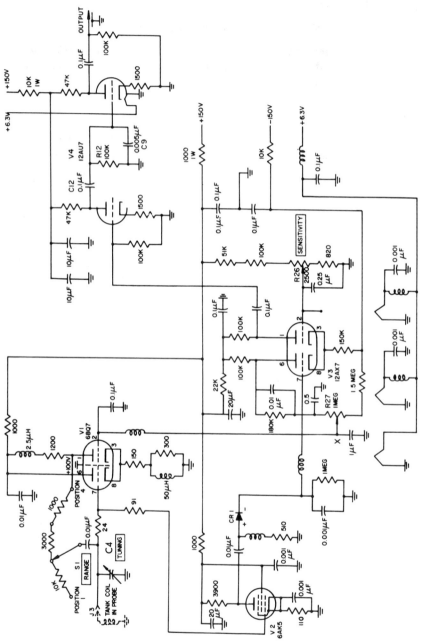

FIG. 8.13   The marginal oscillator circuit for the detection of nuclear magnetic resonance.

using Eq. 1.2 for μ_N/h. Thus

$$g = 5.56 \pm 0.02 \text{ nuclear magnetons}\dagger$$

We have thus demonstrated the phenomenon of nuclear magnetic resonance, and measured to an accuracy of approximately 0.4 percent the magnetic moment of the proton.

4.3 THE MARGINAL OSCILLATOR CIRCUIT

A disadvantage of the system described above is that the bridge must be continuously retuned when the frequency is changed. Thus, in a search for resonance lines of unknown frequency or for a known line in an unknown magnetic field use of the bridge circuit becomes very tedious. The marginal oscillator circuit is compact and is used in most commercial magnetometers; the oscillator frequency can be easily changed by adjusting the variable condenser C_4 (see Fig. 8.13). For larger variations a different coil is used; with four radiofrequency coils the range from 1 to 40 Mc can be covered (switching of the feedback resistor through S_1 is also required). While the marginal oscillator circuit is less sensitive than a bridge, the signal-to-noise ratio is still quite satisfactory, especially for proton resonance.

The sample, the radiofrequency coil, and the modulation and display techniques are all the same as before, but the oscillator and detector are combined in one unit as shown in the circuit diagram,‡ Fig. 8.13, of the magnetometer manufactured by the Laboratory for Electronics Corp. The main idea is to keep the radiofrequency power at a low level so as to allow the direct observation of the absorption and also to avoid saturation of the sample.

Referring to Fig. 8.13, we see that the two sections of V_1 form a cathode-coupled oscillator, the radiofrequency signal being amplified in V_2 and rectified in CR_1. The rectified signal is further amplified in the two sections of V_3. However, through $R27$, negative feedback is applied to the grid of V_{1B}; this has the effect of maintaining the radiofrequency oscillations at a low level since any increase in radiofrequency appears as a drop in the d-c level at point X, which thus tends to suppress the oscillations of V_1. It is possible with this arrangement to sustain a steady radiofrequency at a level as low as 0.1 V across the coil; the level can be adjusted through R_{26} and R_{27}.

---

† The fact that the g factor of the proton is not 2.00, as expected for a *Dirac* particle, tells us that interactions other than electromagnetic are responsible for this large "anomalous" magnetic moment.

‡ Transistorized versions of this circuit are now available; see, for example, the AL 67 nuclear magnetic resonance gaussmeter manufactured by Alpha Scientific Labs Inc.

When resonance is reached, the change in the Q value of the coil results in a small decrease of the radiofrequency level. If the magnetic field is swept at a low frequency, the absorption dip will appear as a small amplitude modulation of the rectified radiofrequency. This audio signal is finally amplified in V_4 whose two sections form a narrow band-width amplifier in order to reduce noise; the amplifier is tuned to 60 cps through the C_{12}, R_{12}, and C_9 network. With this circuit, only the absorption mode is observed; also it is not possible to investigate a line "point by point."

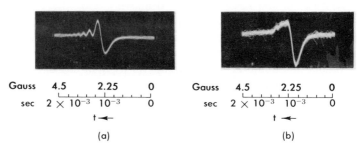

Gauss	4.5	2.25	0
sec	2×10^{-3}	10^{-3}	0

$t \longleftarrow$

(a)

Gauss	4.5	2.25	0
sec	2×10^{-3}	10^{-3}	0

$t \longleftarrow$

(b)

FIG. 8.14   Nuclear magnetic resonance signals of protons obtained with the marginal oscillator circuit. (a) The sample is water doped with manganese nitrate; note the "wiggles" after passage through resonance. (b) The sample is mineral oil.

In Fig. 8.14 are shown oscilloscope traces obtained by students using the marginal oscillator as described above;[†] Fig. 8.14a is obtained with the same sample previously mentioned, water doped with $Mn(NO_3)_2$, and Fig. 8.14b with a sample of mineral oil; for both cases resonance occurs at the same frequency $\nu = 28.189 \pm 0.003$ Mc/sec. The sweeping field was a 60 cps sine wave with maximum amplitude of 6 gauss, and thus a sweeping rate[‡] of

$$\frac{dH}{dt} \approx 2.25 \times 10^3 \text{ gauss/sec} \qquad (4.6)$$

The oscilloscope sweep was linear and set to 0.2 msec/cm, so that the horizontal scale calibration is ~ 0.45 gauss/cm.

We note that Fig. 8.14b looks almost like the predicted absorption line of Fig. 8.5b (inverted), but Fig. 8.14a has a distinctly different shape and is followed, *after* passage through resonance, by decaying oscillations. In both cases the deviation from the theoretical line shape is due to the fast

---

† M. Klein and P. D'Onofrio, class of 1962.

‡ If at $t = 0$ the amplitude of the sweep field is to have its central value, which is 0, then $H_{\text{sweep}} = H_s \sin \omega t$ and $dH/dt = \omega H_s \cos \omega t$, which is almost linear for $t \approx 0$ and $dH/dt \approx \omega H_s$.

passage through resonance, which gives rise to "transient effects"; by fast is meant fast as compared to the characteristic relaxation times of the system.

Indeed, from Fig. 8.14b we estimate the width of the line to be $\Delta H \approx 0.4$ gauss full width at half maximum (fwhm), and using Eq. 3.17† we obtain

$$T_2 = \tfrac{1}{2}g(\nu_0) = \frac{1}{2\Delta\nu} = \frac{\pi}{\Delta\omega} = \frac{\pi}{\gamma\Delta H} = \frac{3.14}{2.67 \times 10^4 \times 0.4} \sim 3 \times 10^{-4} \,\text{sec}$$

(4.7)

which is comparable with the time taken by the field to sweep over the resonance as shown by Eq. 4.6; thus the spin-spin interaction hardly has the time to restore equilibrium conditions during the short time interval‡ that the sample finds itself in resonance with the radiofrequency.

The explanation for the "wiggles" of Fig. 8.14a and Fig. 8.16 is the following. The spins all precess about the z axis at a frequency ω, which slowly varies with the sweeping field as

$$\omega = \gamma(H_0 + H_{\text{sweep}})$$

(4.8)

However, the spins are not in phase; that is, their azimuthal angle ϕ is random. When the field reaches an H value such that the precession frequency ω equals the applied radiofrequency ω_0, resonance occurs and all the spins are brought into phase (that is, they all cluster in the azimuthal plane of the rotating radiofrequency field H_1). As the constant field increases beyond the resonance value, the spins continue to precess at the frequency ω, which again differs from ω_0, but they remain in phase for a time interval of the order of T_2. However, such coherently precessing spins will induce in the radiofrequency coil a voltage at the frequency ω; the two close frequencies ω of the precessing spins and the applied ω_0 which are not exactly equal will interfere (beat) as their relative phase angle

---

† For simplicity we have assumed $g(\nu)$ to have a triangular shape (see Eq. 2.12). Using

$$\int g(\nu)\, d\nu = 1$$

we have

$$1 = \tfrac{1}{2}g(\nu_0)(2\Delta\nu).$$

Similarly, for a Gaussian we obtain

$$T_2 = \tfrac{1}{2}g(\nu_0) = \frac{2}{\gamma\Delta H}$$

‡ The reader may recognize that this is only a case of a generalized uncertainty principle in experimental procedure. With a detector of band width Δf, the signal-to-noise ratio improves with decreasing Δf as $S/N \propto 1/\sqrt{\Delta f}$; however, for a given Δf, we must wait at least a time interval $\Delta t \sim 1/\Delta f$ in order to make even an approximate measurement, and for an accurate value several time constants Δt need to have elapsed.

changes; this results in the "wiggles" we observe. They do decay exponentially as the phase coherence of the precessing spins becomes destroyed with a time constant T_2.

If a linear sweep is assumed, the beat signal has the form

$$e^{-t/T_2} \cos\left[\tfrac{1}{2}\gamma \frac{dH}{dt} t^2 \right] \tag{4.9}$$

where $t = 0$ when resonance is traversed. We also note that the beat frequency increases with time since

$$\omega_b = \tfrac{1}{2}\gamma \frac{dH}{dt} t \tag{4.10}$$

Thus, from a measurement of the wiggle pattern, information about T_2 can be obtained.

When T_2 is long, and sufficient radiofrequency power is applied, it is possible that the phase coherence may not have completely decayed before the magnetic field value H again approaches resonance; in this case wiggles are also observed before resonance as shown in Fig. 8.15a, taken with a linear time sweep of 0.2 msec/cm; Fig. 8.16 is a semilog plot of the amplitude of the wiggles (after resonance) of Fig. 8.15a against time. We see that the data fit an exponential, and yield

$$T_2 \approx 5 \times 10^{-4} \text{ sec}$$

of the same order as given by Eq. 4.7.

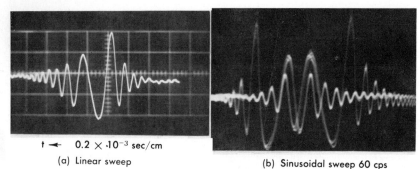

$t \;\leftarrow\;$ 0.2×10^{-3} sec/cm

(a) Linear sweep   (b) Sinusoidal sweep 60 cps

H₂O saturated with LiF

FIG. 8.15 Nuclear magnetic resonance signals for protons when the passage through resonance is very rapid. The sample is water saturated with lithium fluoride. (a) The oscilloscope sweep is linear; note that passage through resonance occurs only once. (b) The oscilloscope sweep is sinusoidal and passage through resonance occurs twice.

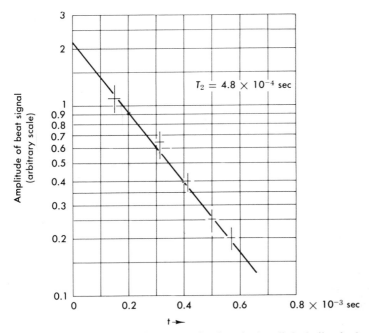

FIG. 8.16 Semilog plot of the amplitude of the "wiggles" of the resonance signal shown in Fig. 8.15 plotted against time. Note that it yields an exponential decay of the amplitude with a characteristic time constant of 4.8×10^{-4} sec.

Just as the inhomogeneity of the magnetic field will broaden the line, however, so it will also destroy the phase coherence (since different parts of the sample precess at different frequencies ω). In fact, a detailed analysis†shows that the pattern of Fig. 8.15a is typical of an inhomogeneous field. Thus both values for the effective T_2 obtained in Eqs. 4.8 and 4.12 are to be interpreted as due to field inhomogeneities; the true T_2 time for these samples is longer.

In concluding, we mention Fig. 8.15b, which is the same as 8.15a but shows the resonance signal when a sinusoidal sweep is applied to the oscilloscope x axis. Therefore resonance appears twice on the oscilloscope, during both increasing and decreasing field; also now the horizontal axis is linear in field rather than in time. The traces shown in Figs. 8.15a and 8.15b were obtained by students‡ using a saturated solution of lithium fluoride (LiF) in water. The sweeping field was at 60 cps and had a maximum value of the order of 10 gauss.

† E. R. Andrew, loc. cit., p. 134.
‡ D. Boyd and P. Nichols, class of 1963.

4.4 TRANSISTORIZED NUCLEAR MAGNETIC RESONANCE DETECTOR

As a final example we present in Fig. 8.17 a simple transistor circuit†
which can be used for the observation of proton nuclear magnetic resonance.
Transistor T_1 (2N502) is the oscillator, feedback being provided by the
10-$\mu\mu$F capacitor from the collector to the emitter. The frequency is determined by the LC circuit formed by the radiofrequency (the sample)
and the variable condenser; it is possible to vary the frequency continuously from 2 to 80 Mc/sec by using coils of the appropriate inductance
for each of four ranges. Diode D_1 (1N56) rectifies the radiofrequency and
T_2 (2N247) amplifies the audio signal. The output is fed directly to an
oscilloscope. The radiofrequency level is adjusted by varying the 10-K
potentiometer in the emitter of T_1.

As before, the sample is 1 cm³ of water that is doped with manganese
nitrate [Mn(NO₃)₂] and is placed inside the radiofrequency coil; the field
is modulated at 60 cps. The important difference between this and the previous circuit is that transistors are *constant current* sources. Thus out of
the collector of T_1 flows a current whose amplitude depends on the base-
emitter bias, and the radiofrequency *voltage* at point A is entirely determined by the impedance of the LC circuit. Any change in the Q of the
coil appears immediately as a change in the radiofrequency level at point
A (or B). This is the fact that accounts for the great simplicity of the circuit
of Fig. 8.17.

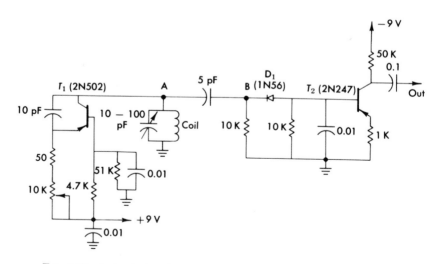

FIG. 8.17   A simple transistorized nuclear magnetic resonance circuit.

† J. R. Singer and S. D. Johnson, *Rev. Sci. Instr.*, **30**, 92 (1959).

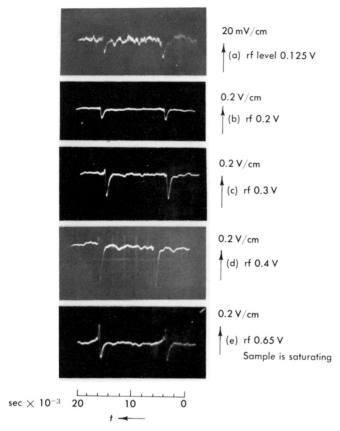

20 mV/cm

(a) rf level 0.125 V

0.2 V/cm

(b) rf 0.2 V

0.2 V/cm

(c) rf 0.3 V

0.2 V/cm

(d) rf 0.4 V

0.2 V/cm

(e) rf 0.65 V

Sample is saturating

sec × 10⁻³   20   10   0

$t \longleftarrow$

FIG. 8.18 Nuclear magnetic resonance signals from protons obtained with the circuit shown in Fig. 8.17 as a function of the amplitude of the radiofrequency. Note that initially the output signal increases with increasing radiofrequency but at a level of approximately 0.5 V the sample is saturated and the signal begins to decrease.

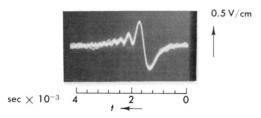

0.5 V/cm

sec × 10⁻³   4   2   0

$t \longleftarrow$

FIG. 8.19 Proton nuclear magnetic resonance signal obtained with the circuit shown in Fig. 8.17 and displayed on an expanded time scale. The magnetic field is of the order of 8 kilogauss.

Data obtained by a student† using this circuit at a frequency of 33,830 Mc/sec are shown in Fig. 8.18 for various levels of radiofrequency voltage across the coil. The magnet was swept at 60 cps, and the oscilloscope sweep is linear at 2×10^{-3} sec/cm; the vertical scale is 200 mV/cm for all but trace (a) where the gain was turned up by a factor of 10. We note that as the level of the radiofrequency is increased, the output at the collector of T_2 increases also; however, for radiofrequency levels beyond ≈ 0.5 V the output level begins to decrease as shown in Fig. 8.18; it almost disappears at ≈ 0.8 V. This is due to the saturation of the sample with increasing H_1 as discussed before and predicted by Eq. 3.14. Finally, in Fig. 8.19 the resonance signal for 0.6 V of radiofrequency and with an expanded horizontal scale of 0.4×10^{-3} sec/cm is shown, indicating clearly the "wiggles" after passage through resonance.

5. Electron Paramagnetic Resonance

5.1 GENERAL CONSIDERATIONS

As noted in Section 1, when radiofrequency resonance is established between electronic rather than nuclear energy levels (always in an external magnetic field), we speak of electron paramagnetic resonance (epr). We know, however, that an energy level is split into sublevels only if its angular momentum J is different from zero; this condition, for example, is met by the *free electron* where the total angular momentum J is just its intrinsic angular momentum, the spin $S = \frac{1}{2}$.

Similarly, from our study of optical spectroscopy in Chapter 7, we know that in many atoms the ground state may have $J \neq 0$ (also many excited states with $J \neq 0$ exist) so that paramagnetic resonance can be established between the Zeeman sublevels of such an atomic state.‡ If we turn to solids, however, it is much more difficult to find electronic states with $J \neq 0$: this is due to the fact that in the chemical binding of atoms into molecules, the valence electrons get paired off, so that each atom appears to have a completely closed shell. For example, in NaCl, the sodium has a $^2S_{1/2}$ electron ($n = 3$, $l = 0$) outside closed shells, and the chlorine has a $^2P_{3/2}$ electron hole ($n = 3$, $l = 1$) inside closed shells. However, in the NaCl molecule, the sodium appears as a Na^+ ion, and hence presents a closed shell configuration, whereas the chlorine appears as a Cl^- ion again with completely closed shells. Consequently, the NaCl molecule is completely diamagnetic.

Nevertheless, it is known from the work on static magnetic susceptibilities (especially at low temperatures) that certain salts show strong

---

† J. S. Weaver, class of 1962.

‡ In that instance, however, it is not usually called electron paramagnetic resonance.

paramagnetism (contain ions with permanent magnetic moments of the order of μ_0). In particular, compounds containing ions of the "transition elements" of the periodic table are frequently found to be paramagnetic. As an example we mention copper sulfate, $[Cu(SO_4)]$, in which compound the double valence results in a Cu^{2+} ion. For copper the $n = 1, 2,$ and 3 shells are completely filled and it has one electron in the $4s$ state, so that Cu^{2+} has a hole in the $3d$ shell; thus the ground state of the Cu^{2+} ion has $l = 2$, $s = \frac{1}{2}$, and, consequently, $J \neq 0$, so that it *does* possess a magnetic dipole moment. In an external magnetic field, the ground state will be split into sublevels and resonance between them can be established and is indeed observed. The actual situation, however, is more complicated due to the electric field of the crystalline lattice; this will be discussed further in Section 5.3.

Electronic moments can also be found in solids when the chemical bond is broken, as in organic free radicals. Especially, the organic salt DPPH, diphenyl-picryl-hydrazil, $(C_6H_5)_2N\text{---}NC_6H_2(NO_2)_3$ shows a very strong and narrow resonance line, with a g factor very close to 2.00 (the free electron value) and is therefore frequently used as a standard. The structure of the molecule is shown in Fig. 8.20, and the "free-electron" behavior comes from the single electron bond in one of the nitrogens.

Paramagnetic resonance has also been observed in other materials where unpaired electrons may exist, as in crystals with lattice defects, in ferromagnetic materials, and in metals and semiconductors.

Since the energy splitting in paramagnetic resonance involves the Bohr magneton rather than the nuclear magneton, the frequencies will be 2000 times larger than the nuclear magnetic resonance frequencies in the same magnetic field. Thus electron paramagnetic resonance can be observed at megacycle frequencies in fields of a few gauss, or otherwise in the microwave region in a magnetic field of a few kilogauss. The latter alternative is almost always chosen, since it presents distinct advantages.

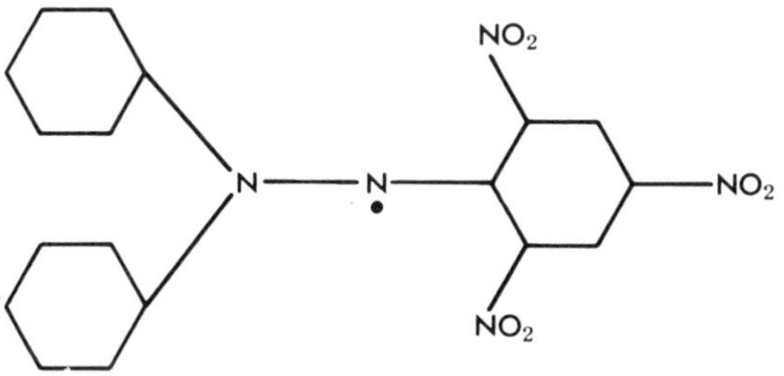

Fig. 8.20 Chemical structure of DPPH (diphenyl-picryl-hydrazil), $(C_6H_5)_2N\text{-}NC_6H_2(NO_2)_3$.

(a) For each transition the absorbed energy is much larger, and thus the signal-to-noise ratio is much improved, as also discussed in Section 4.1.

(b) A high magnetic field is used, thus providing separation between levels that are intrinsically wide and would remain partially overlapped at low fields.

The resonance condition is detected, as in the case of nuclear magnetic resonance by the absorption of energy, and for this reason solids and liquids are much easier to study than gases with their very low densities. Much of our previous discussion on transition probabilities and relaxation mechanisms is equally applicable to electron paramagnetic resonance. Again we depend on the Boltzmann distribution for the creation of differences in the population of the various sublevels; these differences are much larger than in nuclear magnetic resonance, since the energy intervals involved are 2000 times larger (see Eq. 1.2 and Eq. 2.2).

A difficulty with electron paramagnetic resonance, however, is that the width of the resonance lines may be prohibitively large, since both the spin-lattice and spin-spin interactions are stronger than in the nuclear magnetic resonance case (see Eq. 3.16). In order to reduce the line width, the sample may be cooled to low temperatures, which lengthens the spin-lattice relaxation time. To decrease the spin-spin interaction, on the other hand, the paramagnetic ions are diluted in a diamagnetic salt, which effectively increases the distance between the spins (again see Eq. 3.16).

As compared to atomic spectroscopy, where the Zeeman splitting of the sublevels was only a small perturbation on the optical transition, here we measure directly the spacing between these sublevels. However, in optical spectroscopy the atom could be excited to one of any of its higher levels[†]; in electron paramagnetic resonance we are mainly working with the ground state of the ion, and some of the levels closest to it.

When measuring electron paramagnetic resonance lines in solids, a great variety of g factors is obtained. This is due to the differences in the coupling of the unpaired electron's spin with the orbital angular momentum; the strength of this coupling depends very much on the position (in energy) of the adjacent levels of the ion as they are modified by the crystalline field. Further, the electron paramagnetic resonance lines show hyperfine structure characteristic of the interaction of the nucleus with the ionic energy levels; this structure in turn can be used to positively identify small traces of an element contained in some unknown sample.

Similarly, the organic free radicals show characteristic lines (g factors) which can be used to identify them and show hyperfine structure as well.

[†] This is possible because of the strong excitation provided by an arc source, or high-voltage discharge, etc.

In fact, a radical that has no structure (like the DPPH) may exhibit such effects when the sample is prepared in a liquid solution.

It is for these reasons that electron paramagnetic resonance has become of great value in the understanding of crystalline materials as well as of organic structures; the study of these effects has by now so increased that they form independent fields of research, which we will not be able to discuss. In Section 5.2 we will describe the experimental apparatus and show the line observed from DPPH, while in Section 5.3 we will briefly discuss the Cu^{2+} ion and the results obtained from $Cu(SO_4)\cdot 7H_2O$, and $MnCl_2\cdot 4H_2O$.

5.2 ELECTRON PARAMAGNETIC RESONANCE APPARATUS

In this laboratory electron paramagnetic resonance was observed at a microwave frequency of the order of 10,000 Mc ($\lambda_f \approx 3$ cm). A schematic of the interconnections of the equipment is shown in Fig. 8.21, and Figs. 8.23 and 8.25 give some of the details of the microwave plumbing.

The system consists, basically, of the X-band Klystron oscillator and plumbing to transport the power to a microwave cavity in which the sample is placed. The sample, in turn, must be subjected to a permanent field H_0, which is usually continuously varied with a motorized drive. A microwave bridge is used to balance out the power and give a signal only when resonant absorption occurs; this signal is fed to a Keithley electrometer whose output in turn drives the y axis of a recorder. The x axis is driven in proportion to the slow variation of H_0. For very weak signals a phase-sensitive ("lock-in")

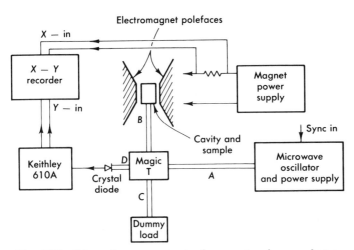

FIG. 8.21 Schematic arrangement of apparatus for an electron paramagnetic resonance experiment in the microwave region.

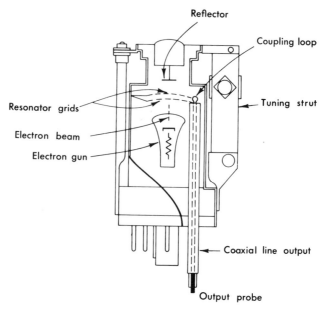

Reflector

Coupling loop

Resonator grids

Tuning strut

Electron beam

Electron gun

Coaxial line output

Output probe

Fig. 8.22   Sketch of the 723A Reflex Klystron.

detector may be used, in which case modulation of the output signal must be provided; either the Klystron may be modulated in an "on-off" mode or the magnetic field may be modulated with sweep coils.

Let us now examine the various elements of the apparatus in slightly more detail†.

(1) *The microwave oscillator.* This is a 723A reflex Klystron‡ mounted in a TS/13-AP power supply; a sketch of the Klystron is given in Fig. 8.22, and in Fig. 8.23 some of the plumbing and auxiliary equipment included in the power supply are shown.

The frequency can be tuned over a small range either by a mechanical adjustment or by varying the reflector voltage. The tube has a few "modes" of operation but most power is achieved at a reflector voltage of -400 V with the grids at ground potential. Figure 8.24 shows the effect of reflector voltage; it can be seen how the superposition of a square-wave voltage on the reflector may modulate either the frequency or amplitude of the tube.

---

† We cannot discuss microwave techniques here, but the reader should consult the well-known *M.I.T. Radiation Laboratory Series*, McGraw-Hill, 1947; also A. Bronwell and R. Beam, *Theory and Application of Microwave Techniques*, McGraw-Hill, 1947.

‡ Both the Klystron and power supply described here are war surplus equipment, but now many improved Klystrons are available commercially.

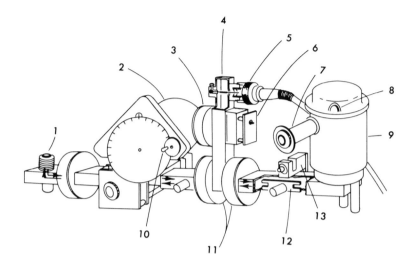

Fig. 8.23 Sketch of the mechanical mounting of the Klystron oscillator, attentuator, power monitor, and wavemeter cavity as arranged in the TS/13AP power supply. (1) Radiofrequency input-output jack, (2) Wavemeter cavity, (3) Wavemeter coupling flange, (4) Power-measuring bead thermistor, (5) Bead thermistor mount, (7) 723A/B cavity tuning strut, (8) 723A/B reflector cap, (9) 723A/B shielded mount, (10) Wavemeter tuning knob, (11) *T* junction and mount, (12) Attenuator strip. Note: Double arrows show direction of radiofrequency power flow.

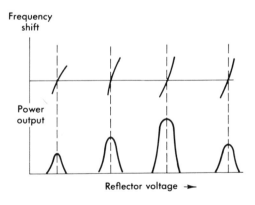

Fig. 8.24 The power output and frequency shift of a Klystron oscillator in its various modes, as a function of reflector voltage.

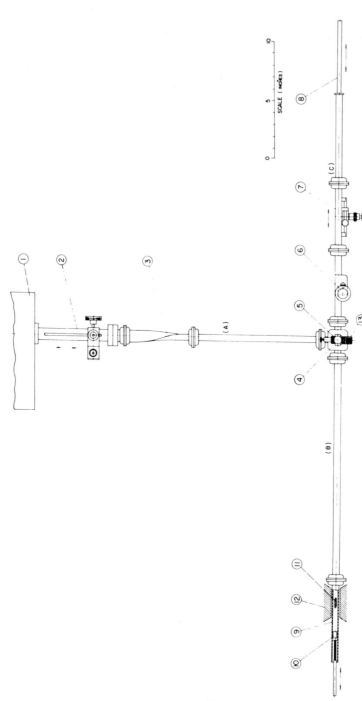

FIG. 8.25 The microwave plumbing used for the electron pramagnetic resonance experiment. The branches (B) and (C) form the two arms of a bridge. (1) TS 13/AP microwave generator, (2) Slotted line, (3) Rigid twist, (4) Magic tee, (5) Detector, (6) Attenuator, (7) Phase shifter, (8) Tuning stub or matched load, (9) Equivalent of tuned cavity, (10) Tuning stub, (11) Sample, (12) Electromagnet pole faces, (13) Bridge output.

SCALE (INCHES)

380

(2) *The wave guide system.* This is shown in Fig. 8.25. It consists of a transport section (A) and two sections (B) and (C) which form the arms of a bridge; (B) contains the sample while (C) contains a dummy load. To balance the bridge we adjust the attenuator (6) (for resistive balance), the phase shifter (7) (for reactive balance), and the shorting plunger (8) †.

At this point we inquire into the configurations of the field inside the rectangular wave guide. We know that only certain modes will propagate‡ (without attenuation) and the wavelength in the guide λ_g is given by

$$\frac{1}{\lambda_g{}^2} = \frac{1}{\lambda_f{}^2} - \frac{1}{\lambda_c{}^2} = \frac{1}{\lambda_f{}^2} - \frac{(m/a)^2 + (n/b)^2}{4} \tag{5.1}$$

where λ_f is the free space wavelength and a and b are the inner dimensions of the guide; m and n are integers. Since

$$a = 2.29 \text{ cm}, \qquad b = 1.02 \text{ cm}, \qquad \text{and} \qquad \lambda_f \approx 3.2 \text{ cm}$$

we find that only the $m = 1$, $n = 0$ mode can propagate, and

$$\lambda_g = 4.5 \text{ cm}$$

In this mode, the electric field is completely transverse to the axis of the guide; this is called the TE_{10} mode. The field lines for the traveling TE_{10} wave are shown in Fig. 8.26, where the density of field lines is proportional to the field strength.

(3) *The microwave cavity and sample.* The cavity is the analogue of the radiofrequency coil in the nuclear resonance apparatus. In the apparatus shown in Fig. 8.25, however, no special cavity was used. Instead a part of the wave guide (9) was ended with a shorting stub (10) and as a result of this, power is reflected and a standing wave is set up. We first adjust the sample to be in the middle of the magnet polefaces, and then adjust the sliding stub (10) so that maximum H field exists at the sample. From the configuration of the *standing wave* pattern (which is different from Fig. 8.26) it is found that maximum H field occurs at a distance x from the short

$$x = \left(\frac{1}{2} + \frac{p}{2}\right)\lambda_g$$

with p an integer. We also note that since the microwave field must be normal to H_0 it is preferable to place the guide in the magnet with its wide side parallel to the polefaces.

---

† Alternatively, a matched load may be used to terminate the dummy arm of the bridge.

‡ See, for example, J. D. Jackson, *Classical Electrodynamics*, John Wiley, Chapter VIII.

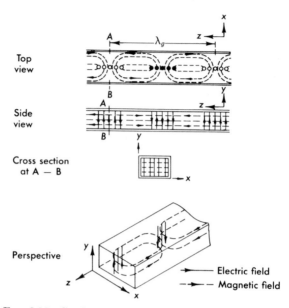

Top view

Side view

Cross section at A — B

Perspective

—————— Electric field
- - -► - - Magnetic field

FIG. 8.26   Configuration of electric and magnetic field lines for a traveling wave in a rectangular waveguide. λ_g is the wavelength in the guide.

(4) *The magic T*. This is the heart of the bridge circuit and is shown in Fig. 8.27. It has the property that microwave power flowing in from (A) can branch into (B) and (C) but not into (D), which is the direction normal to the plane of (A), (B), and (C); on the other hand, power returning (that is, reflected) from (B) or (C) can flow into (D) and is split equally between (A) and (D). This is due to the different polarization of the electromagnetic field that can propagate in the four elements of the magic T, and to the configuration of the slots leading to the four arms. When the T is used in a bridge circuit as in Fig. 8.25, power reflected from (B) and (C) are superimposed at (D) to give null output.

(5) *The detector*. This is connected to arm (D) of the magic T and consists of a microwave diode rectifier. Power is fed in from a small probe protruding into the wave guide, and the output is fed to a Keithley electrometer† which is operated as a millivoltmeter. The Keithley output is then fed to a pen recorder as noted earlier.

(6) *Auxiliary equipment*. This is used for tuning and monitoring the microwave system. For example, (2) in Fig. 8.25 is a slotted line that is

† See Chapter 4, Section 3.4.

used to measure the E field (time average) along the axis of the guide. If a standing wave exists, then the diode connected to the slotted line probe will register maxima and minima (spaced, obviously, $\lambda_g/2$ cm apart). By tuning the system we wish to eliminate such reflections—that is, make the maximum to minimum ratio of the probe readings close to 1.

Also attached to the Klystron output (and incorporated in the TS-13/AP) is an attenuator and a tunable microwave cavity as shown in Fig. 8.23. The cavity affects the frequency and tuning of the Klystron but is mainly intended as a wavemeter. It has a high Q, and when it is tuned onto the Klystron frequency, a dip in the output meter appears; if the cavity has been carefully calibrated against a frequency standard, the reading of the adjusting dial can immediately give the Klystron wavelength to good accuracy. The purpose of the attenuator is to reduce the coupling of the Klystron to the load so that the Klystron operation is not *much* affected by changes in load conditions†. The power supply has a "sync" input which can be used to synchronize the modulation of the reflector voltage with a reference source as is needed when a phase-sensitive detector is used.

(7) *The magnet.* This must be an electromagnet rather than a permanent magnet, since the frequency range is so limited. It is usually swept continuously by a slow motor driving a rheostat. Fast sweep coils for modulation purposes may be added.

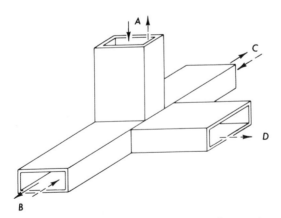

Fig. 8.27   Magic T, for use at microwave frequencies. Power coming in from A divides equally into B and C *but not* into D; however, power coming in from B or C divides equally into A and D.

---

† So-called frequency pulling; a unidirectional device (such as one made out of ferrite) is superior for isolation without loss of transmitted power.

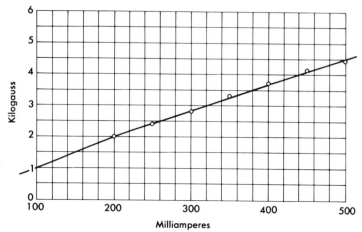

FIG. 8.28   Calibration of the electromagnet used in the electron paramagnetic resonance experiments.

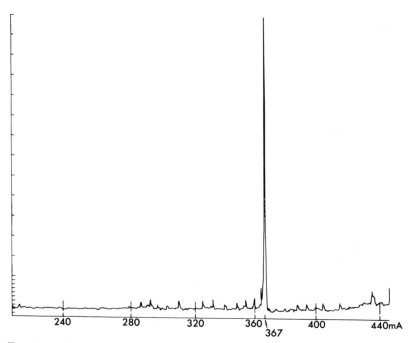

FIG. 8.29   Electron paramagnetic resonance signal from DPPH obtained at a frequency of 10,000 Mc. The abscissa gives the current through the electromagnet in the field of which the sample is located. Note the small width of the line and the good signal-to-noise ratio.

A recommended procedure for setting up is as follows: A sample of DPPH is placed in the cavity, and it is ascertained that the sample is located in the middle of the guide and between the polefaces. The plunger ((10) in Fig. 8.25) is adjusted so that the sample is at maximum H field. Next, the Klystron is tuned for maximum output by adjusting the reflector voltage. At the same time a reasonable standing-wave-voltage ratio (swvr) must exist so that indeed some power flows down the guide. We are then ready to balance out the bridge by adjusting the elements in the dummy arm, mainly, the phase shifter and attenuator. Since the initial unbalance is likely to be quite large, it is convenient to use a scope to observe the output before switching to the Keithley electrometer; as the bridge is balanced, the Klystron may get out of tune, and the procedure should be repeated until a good null is achieved. The magnet is then varied either manually or by the slow sweep until the resonance is found; now the recorder may be connected and the tracing of the line recorded.

The results so obtained by students† with DPPH (also known as DPTH) are shown in Fig. 8.29. The markers on the x axis were introduced when the current of the electromagnet reached specific values as indicated. The magnetic field was measured with a rotating coil gaussmeter, the resulting calibration being shown in Fig. 8.28. To find the g factor we need also to know the frequency of the radiation; by measuring λ_g with the slotted line and using Eq. 5.1, we obtain an *order of magnitude* value for λ_f

$$\lambda_g = 4.5 \pm 0.2 \text{ cm} \rightarrow \lambda_f = 3.2 \pm 0.1 \text{ cm} \qquad (5.2)$$

In convenient units the g factor is given (see Eqs. 2.14 and 1.2 in Chap. 7) by

$$g = \frac{\Delta E}{\mu_0 H} = \frac{hc}{\mu_0} \frac{1}{\lambda H} = \frac{4\pi mc^2}{e} \frac{1}{\lambda H} = \frac{1}{4.669 \times 10^{-5}} \frac{1}{\lambda H} = \frac{21,418}{\lambda H} \qquad (5.3)$$

with λ in centimeters and H in gauss. If we use for λ the value given by Eq. 5.2, and from Fig. 8.28 take $H_0 = 3400 \pm 30$ gauss, we obtain

$$g \text{ (DPPH)} = 1.97 \pm 0.07$$

in good agreement with the accepted value

$$g \text{ (DPPH)} = 2.0036 \qquad (5.4)$$

While microwave frequencies can be determined to great accuracy,‡ this was not done in the present experiment, but instead the DPPH line

---

† E. May and D. Peters, class of 1962.

‡ Usually a higher harmonic of a crystal oscillator is mixed with the microwave frequency, and the resulting intermediate frequency is measured.

was used as a standard to determine the g factors of other samples. In fact, if the magnetic field is linear with current (as it almost is in the present case, see Fig. 8.28), the ratio of the magnet current for DPPH to the current for any other sample, if taken at the same frequency, contains all the information.

We note that the signal-to-noise ratio is quite satisfactory, and from Fig. 8.29 we obtain for the DPPH line a width of less than 8 gauss, the true width being 2.9 gauss.†

5.3 ELECTRON PARAMAGNETIC RESONANCE IN Cu^{2+} AND Mn^{2+}

In Figs. 8.30 and 8.31 are given results obtained with the same electron paramagnetic resonance apparatus when samples of $Cu(SO_4) \cdot 7H_2O$ and $MnCl_2 \cdot 4H_2O$ were used; a small amount of DPPH was also included in the cavity, so that the sharp line shown in the figures was obtained simultaneously and serves as a calibration standard for the product (wavelength $\times$ magnetic field).

We wish now to examine in detail the energy-level structure of the Cu^{2+} ion. Since copper is in a $4s$ state, the Cu^{2+} ion, as explained before, has

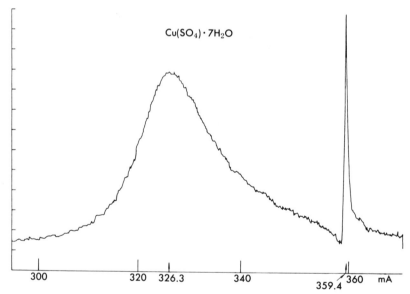

FIG. 8.30   Electron paramagnetic resonance obtained from $Cu(SO_4) \cdot 7H_2O$. The signal from DPPH is also included for calibration purposes. Note the difference in the width and in the g factor of the two lines.

† The apparent width in Fig. 8.29 arises from the response time of the detection equipment, in particular the recorder.

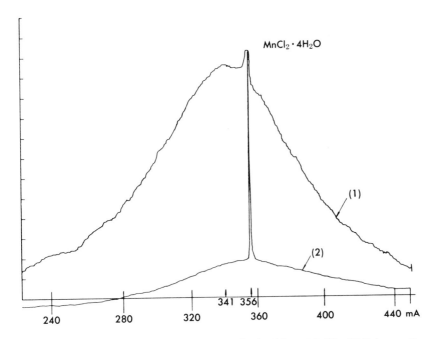

FIG. 8.31   Electron paramagnetic resonance obtained from $MnCl_2 \cdot 4H_2O$ (curve 1). The signal from DPPH is also included for calibration purposes (curve 2).

a hole in the $3d$ shell; however, the orbital angular momentum l and the spin s do not couple into a resultant j because the electric field of the crystal breaks down such (spin-orbit) coupling. The ground state is 10-fold degenerate, having five $(2l + 1)$ levels with $m_l = -2, -1, 0, +1, +2$; and each of these levels is doubly $(2s + 1)$ degenerate, having a sublevel with $m_s - +\frac{1}{2}$ and one with $m_s = -\frac{1}{2}$.

However, the electric field of the crystal interacts differently with each of these m_l states,† so that they have different energies. This is shown in Fig. 8.32 where we note that in a cubic field the ground state is split in two levels while in a tetragonal field (lattice) it is split into four levels; finally, the spin-orbit interaction removes the remaining (orbital) degeneracy between levels E_4 and E_5. Each of the levels E_1 through E_5 contains the two sublevels with $m_s = \pm\frac{1}{2}$. However, the spacing between the levels $E_1 - E_5$ is very large, of the order of 10^4 cm^{-1}, so that even at room temperature all the ions are in the $m_l = 0$ state‡, and the ion appears as a system with

---

† Obviously the quantization is with respect to the z axis defined by the crystalline field.

‡ The Boltzmann factor is $\exp(-1.2 \text{ eV}/kT) \approx \exp(-50)$.

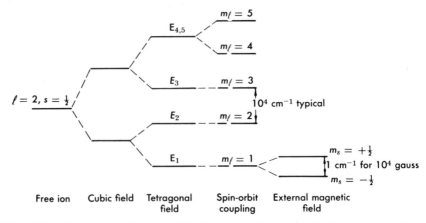

FIG. 8.32   Energy level diagram of the Cu^{2+} ion in the electric field of a crystal. If the field is cubic, only two sublevels arise. If the field has tetragonal symmetry, four sublevels arise. The effect of the spin-orbit coupling is to further completely remove the degeneracy; finally in an external magnetic field each m_l sublevel is split into two components.

$l = 0$ and $s = \frac{1}{2}$ which therefore should have a g factor very close to that of the free electron (this is sometimes called "quenching" of the angular momentum).

When an external magnetic field is applied, the $m_l = 0$ level will split into its two sublevels $m_s = \pm\frac{1}{2}$; however, since the electron (hole) is coupled (spin-orbit) to the other levels of the ion, the g factor is not exactly 2.00. Equation 5.5 gives a theoretical estimate for the g factor, indicating that it should depend on the relative orientation of the external field and the crystal axis

$$g_{||} = 2\left(1 - \frac{4\lambda}{E_3 - E_1}\right)$$

(5.5)

$$g_{\perp} = 2\left(1 - \frac{\lambda}{E_{4,5} - E_1}\right)$$

Here $g_{||}$ and $g_{\perp}$ refer, respectively, to parallel and normal relative orientation†; λ is the spin-orbit coupling parameter for the Cu^{2+} ion ($\lambda = -852$ cm^{-1}) and E_i the energy levels shown in Fig. 8.32. Indeed, single crystals

---

† For intermediate orientation at an angle θ, we have the convenient expression $g = \sqrt{g_{||}^2 \sin^2 \theta + g_{\perp}^2 \cos^2 \theta}$.

of copper sulphate [$Cu(SO_4)$] follow the predictions of Eq. 5.5, the g factor varying from 2.10 to 2.40.

In the present experiment, the results of which are shown in Fig. 8.30, we note that the center of the Cu^{2+} line appears at 326.3 mA, and the DPPH line appears at 359.4 mA.

Referring to the magnetic field calibration (Fig. 8.28), and using the g factor of DPPH as given by Eq. 5.4, we obtain†

$$g\ (Cu^{2+}) = \frac{3350}{3005} \times 2.0036 = 2.25 \pm 0.1 \qquad (5.6)$$

which lies between the accepted values of $g_{||}$ and $g_{\perp}$. Since this sample was not a single crystal but a crystalline powder, the value given by Eq. 5.6 is obviously independent of angle, and it is not possible to observe $g_{||}$ and $g_{\perp}$. The width of the observed line is $\Delta H \approx 170$ gauss.

In the case of $MnCl_2 \cdot 4H_2O$, Mn has five electrons in the $3d$ shell and two electrons in the $4s$ shell. Thus the Mn^{2+} ion has just five electrons in the $3d$ shell (it is half full), which therefore add up to an *orbital* angular momentum $L = 0$; however, the spins add up to a resultant $S = \frac{5}{2}$. The ground state configuration of the ion is $^6S_{5/2}$ and under the influence of the external magnetic field will split into six equally spaced levels‡ with a g factor very close to 2.00; thus the resonance between any adjacent pair of energy levels will always occur at the same frequency.

The results obtained for Mn^{2+} (together with the calibration line from DPPH) are shown in Fig. 8.31, where curve (2) was taken with the gain of the Keithley decreased by a factor ≈ 10. We obtain

Mn^{2+} line at 341 mA

DPPH line at 356 mA

so that

$$g\ (Mn^{2+}) = \frac{356}{341} \times 2.0036 = 2.09 \pm 0.02 \qquad (5.7)$$

in reasonable agreement with the expected value. The width of the line in this case is of the order of $\Delta H \approx 750$ gauss.

Paramagnetic resonance can also be easily observed at room temperature in Cr^{2+} ions; however, a great variety of substances exhibit electron paramagnetic resonance at low temperatures.

---

† We can also directly take the ratios of the currents, $g = (359.4/326.3) \times 2.0036 = 2.21$.

‡ Higher-order effects introduce slight differences in these spacings.

6. The "Double Resonance" Experiment

6.1 THE METHOD

In the experiment we are about to describe, magnetic resonance is established between atomic (electronic) sublevels. It differs, however, from the nuclear magnetic resonance and electron paramagnetic resonance experiments in the following way: (a) The difference in population between the two levels is not due to the Boltzmann factor, but is created artificially by "pumping" atoms into one state; (b) the resonance condition is not detected by power absorption, but by the change in the polarization of the optical radiation emitted after the transition. This technique has had applications in the precise measurement of hyperfine structure in the excited states of atoms, but it is included here mainly because it offers an insight into what can be done by the interconnection of atomic levels.

We will consider the case of atomic mercury already discussed in Chapter 7, and especially the $^3P_1$ excited level. Let us also restrict our discussion to a single isotope Hg^{198},† which as we know has no nuclear spin, $I = 0$. Consider then the apparatus shown in Fig. 8.33: a Hg^{198} source (1) is excited and its radiation is incident on a cell (4) containing Hg^{198} vapor; as is well known, the vapor in the cell will emit the 1849 Å and 2537 Å

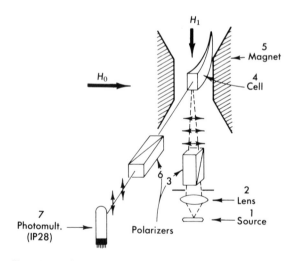

FIG. 8.33 Schematic arrangement of the apparatus used for the observation of optical "double resonance." A radio-frequency coil or microwave cavity should be surrounding the absorption cell.

† The reader may verify by referring also to Section 6.2 that the same conclusions are reached with a natural mercury source and absorption cell.

Fig. 8.34 The energy sublevels of Hg^{198} contributing to the emission of the 2537 Å line. The double arrow indicates the exciting radiation that is assumed to be polarized in a direction parallel to the magnetic field. (a) Conventional energy level diagram. (b) The energy sublevels are now indicated by dots so arranged that $\Delta m = 0$ transitions appear as vertical lines; $\Delta m = +1$ transitions appear as lines slanted to the right and $\Delta m = -1$ transitions appear as lines slanted to the left. The radiofrequency transitions ($\Delta \ell = 0$) appear as horizontal lines (heavy).

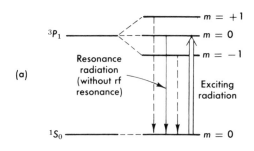

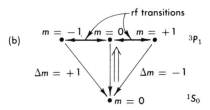

(resonance) lines of mercury. This is because† the mercury atoms in the cell (which are in their ground state) will be excited by the incident radiation to the 6 $^1P_1$ and 6 $^3P_1$ states as shown in Fig. 2.13; when the atoms become de-excited by decaying to the ground state (this can be considered as a relaxation process), they emit the respective optical lines. Since the 1849 Å line lies in the vacuum ultraviolet, it is not observed under usual conditions and we need not include it in our considerations.

Let the cell (4) be located in a magnetic field H_0 (5). The ground state $^1S_0$ is not split, but the excited $^3P_1$ state is split into three sublevels, which would give rise to a triplet line as shown in Fig. 8.34a. In Fig. 8.34b we show the same levels but attempt to emphasize the polarization of the various components of the emitted radiation: the vertical line represents a $\Delta m = 0$ transition—that is, light polarized parallel to the axis of the magnetic field—while the slanted lines represent the $\Delta m = \pm 1$ transitions, that is, light polarized normal to the axis of the field (or circularly, if viewed along the axis). If a polarizer (3) is placed between the source and the absorption cell, allowing only light parallel to the field to be incident, it is clear from Fig. 8.34b that *only* the $^3P_1$, $m = 0$ sublevel may become populated. Indeed, the optical radiation emitted by the cell (observed at right angles to the field axis and to the direction of the incoming radiation) is polarized *parallel* to the axis of the magnetic field.‡

---

† See also the discussion of the Mössbauer effect in Chapter 6, Section 4.

‡ This in turn signifies that only $\Delta m = 0$ transitions back to the ground state take place; since the ground state is uniquely $m = 0$, it means that only the $m = 0$ sublevel of the excited state has been populated.

Thus we find in the excited state a situation where the $m = 0$ sublevel is populated while the $m = +1$ and $m = -1$ sublevels are empty. If now a radiofrequency magnetic field H_1, normal to H_0 and of the appropriate frequency $\omega = gH\mu_0/\hbar$, is applied, it should transfer atoms from the $m = 0$ sublevel to $m = \pm1$; if H_1 is strong enough (as compared to the lifetime of th e^3P_1 level), it will even equalize the populations between the three sublevels. But, as soon as atoms populate the $m = \pm1$ states, it is clear that in returning to the ground state they will emit optical radiation polarized normally to the field (since the transitions are $\Delta m = \pm1$).

To observe this effect an additional polarizer (6) is introduced between the absorption cell and the detecting photomultiplier (7). The polarizer is oriented so as to allow transmission only of light polarized normally to the field; thus, under normal conditions no light is received at the detector. When magnetic resonance is established, optical resonance radiation reaches the detector. This method of detection has also the advantage of being, at least in principle, a null method, since off resonance the signal level is at zero.

The g factor for the $^3P_1$ state can be easily calculated from Eq. 2.13 of Chapter 7, and we obtain $g \, (^3P_1) = \frac{3}{2}$. Further, the energy splitting in frequency units is given by the following (see Eq. 5.3):

$$\nu = \frac{c}{\lambda} = gH\frac{c}{21{,}418} = g \times 1.401 \times H \qquad \text{Mc/sec} \qquad (6.1)$$

Thus

$$\nu = \tfrac{3}{2} \times 1.401 = 2.1 \quad \text{Mc/gauss}$$

as expected, the resonance frequency is typical of electron paramagnetic resonance spacings. The following section describes the experimental procedure and the results obtained.

6.2 PROCEDURE AND RESULTS

The data presented here were obtained at S-band microwave frequencies ($\lambda \approx 10$ cm) even though excellent results can also be obtained in the 100 Mc/sec region which requires only a field H_0 of 50 gauss; the width of the line is only a few gauss (see Fig. 8.35).

The apparatus used was the one shown in Fig. 8.33 and discussed before. Since the 2537 A line is absorbed by ordinary glass, all optical elements must be made of quartz, vycor, or other ultraviolet transmitting material. A natural mercury source and absorption cell may be used, and a low-pressure arc or discharge tube is quite adequate.

The polarizers were quartz Glazebrook prisms cemented with glycerine, and the absorption cell must have flat windows for the entrance of the

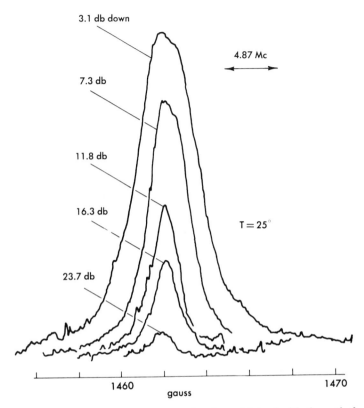

FIG. 8.35 Paramagnetic resonance signals as obtained with the optical double resonance method. The frequency was 3045 Mc/sec and the magnetic field was swept as indicated on the abscissa. Several curves are shown as a function of the microwave power. Note that large levels of microwave power artificially broaden the line.

exciting radiation, we well as for the exit of resonance radiation. The quality of the resonance signals depends very sensitively on the density of mercury vapor. To control the vapor pressure, a small amount of condensed mercury inside the cell is kept in a constant temperature bath.

The microwave power was obtained from a QK61 magnetron oscillator[†] and coupled by means of a coaxial wave guide to a rectangular cavity. The cell was located in the cavity at a position of maximum H field, which was normal to H_0. A standing wave meter, a calibrated tunable cavity, and a tuning stub were part of the microwave plumbing.

---

† This was war surplus material.

The H_0 field was provided by an electromagnet, which could also be modulated with a set of sweep coils. The output signal from the 1P28 RCA photomultiplier was fed to a Keithley electrometer[†], which in turn was connected to the y axis of the recorder; the x axis was driven, as before, in proportion to the modulation field. By using a slow sweep obtained from a motor-driven continuous-card potentiometer, we were able to use long-time constants in the detection system, and thus filter out much of the high-frequency noise.[‡] This can be achieved simply by placing a condenser, C, across the output of the Keithley; since the output impedance is of the order of $R_{out} \approx 20,000\ \Omega$, the time constant is

$$\tau = RC = \frac{C'}{50}\ \mathrm{sec}$$

with C' in microfarads.

The results obtained with this apparatus are shown in Fig. 8.35; the microwave frequency was 3045 ± 1 Mc/sec, and the magnetic field at resonance as measured with a nuclear magnetic resonance probe was 1462 ± 0.5 gauss, so that using Eq. 6.1

$$g = \frac{\nu}{H}\frac{1}{1.401} = 1.486 \pm 0.002 \tag{6.2}$$

The g factor obtained in Eq. 6.2 deviates slightly from the theoretical value of $\frac{3}{2}$; this is due to the interaction of the electrons in the $^3P_1$ state with other levels of the mercury atom. We saw in Section 5.2 how such interactions can greatly modify the g factor; in the present case, however, the effect is much smaller.

Finally, we note that in Fig. 8.35 a series of resonance curves is presented. They were obtained with different levels of microwave power fed into the cavity; the power ratios are indicated in decibels (db)[§] for each curve. It is seen that high power levels artificially broaden the line. Indeed, it can be shown that the width of the line in frequency units is given by

$$(\Delta\omega)^2 = \frac{4}{T^2}[1 + 5.8(\gamma H_1 T)^2] \tag{6.3}$$

where T is the lifetime of the $^3P_1$ state, H_1 the microwave magnetic field, and γ as before the gyromagnetic ratio. For this state $T \approx 1.6 \times 10^{-7}$ sec, and in order to have an appreciable effect it must be $\gamma H_1 = \omega_1 \approx 1/T$; since $\gamma \approx 10^7$ (cf. Eq. 4.4, $2.67 \times 10^4 \times 18,361$ for $g \approx 5.6$), we find that H_1 must be approximately 1 gauss.

[†] See Chapter 4, Section 3.4.

[‡] See also the discussion on p. 369 (footnote).

[§] $db = 10\ \log_{10}(P_1/P_2)$.

9

TIME COINCIDENCE TECHNIQUES

1. Introduction

Some of the preceding chapters have explained how improved resolution in the measurement of physical quantities has led to new results and deeper understanding of the laws of nature. These improvements lie in the measurement of length, electric current, angular deflection, and similar quantities. Measurement of time intervals has not been considered, however, except insofar as frequency is basically a counting of vibrations per unit time.

This chapter is concerned with the measurement of short time intervals, and especially with methods of judging whether or not two events occurred at the same time. We qualify this last statement as follows: (a) at the same time with respect to some clock, and (b) at the same time t_0, within an interval Δt; Δt is the resolution of our measuring equipment. If indeed the two events occurred at the same time, we say that they are in *time coincidence*†. Also in speaking of "short" time intervals we mean short

---

† As long as the measurements are made in a frame of reference with respect to which both the events and the observer are at rest, we can safely talk of time coincidence.

with respect to the time resolution of a human being, which is of the order of 1/1000 sec to 1 sec. Modern techniques permit the measurement of time intervals in the order of 10^{-9} to 10^{-11} sec. If we consider the dimensions of the typical apparatus to be 1 ft, it takes light (and therefore also information) 10^{-9} sec to traverse it; this can be assumed as another natural limit in the time scale.

While the precise measurement of time can be traced several centuries back to the use of chronometers for astronomical observations and navigation, the technique of time coincidences is recent. Coincidence circuits were first built by W. Bothe and by B. Rossi in 1930, using vacuum tubes. A schematic drawing of Rossi's apparatus, used to detect cosmic rays, is shown in Fig. 9.1. The outputs from two similar Geiger counters placed one above the other are fed to the coincidence circuit; if both counters fire within one millisecond of one another, a signal appears at the output of the circuit. From the frequency of the single counts in either of the Geiger tubes, one could conclude that the coincident counts were due to one and the same particle traversing both counters. Such an arrangement, called a particle telescope, can be used to measure the direction (angular distribution) of the flux of cosmic radiation.

Particle telescopes have become a basic instrument in experiments with high-energy particle accelerators. In this case the particles under examination have enough energy to traverse several counters, and a multiple coincidence is used to identify them. As will be explained further, individual particles can be distinguished even if the flux rate through the telescope is as high as 10^{7}/sec.

Another application of the coincidence technique is in correlating the simultaneous emission of two or more particles. An example is positron-electron annihilation, which results in the emission of two gamma rays. These can both be simultaneously detected if two counters are placed at a

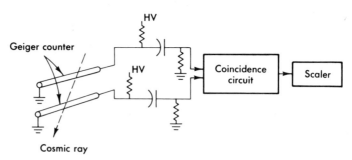

FIG. 9.1   Schematic arrangement of a "counter telescope" using Geiger counters. Such an arrangement can be used for measurement of the cosmic ray flux.

relative angle of 180 degrees with respect to the region of annihilation. The occurrence and the energy of a cosmic ray shower is measured by detecting simultaneously the arrival of a large number of particles.

The lifetime of nuclei or excited states, even as short as 10^{-10} sec, can be determined with these techniques. Similarly, velocities of particles very close to $\beta = 1$ can be measured with relative ease. Also, in the study of optical coherence, the same coincidence techniques are used to measure correlations between photons. These are but a few examples of applications of the time coincidence technique. It has found great use in both basic research and industrial and control systems.

There have also been developed several variations of the basic coincidence circuit, as, for example, the anticoincidence, delayed coincidence, linear gate circuit, etc., which can perform different logic decisions or analogue operations. Implicit use of them has been made on several occasions in other parts of this text.

In Section 2 a basic up-to-date high-resolution coincidence circuit will be described. Also the required shaping, time stability, and other aspects of the pulses that are fed to a coincidence unit will be discussed. In Section 3, data on angular correlations are presented; first the simple case of positron annihilation (from Na^{22}) and next the correlation between the two gamma rays emitted (in the de-excitation of Ni^{60}) following the decay of Co^{60}. In Section 4 a simple discussion of the theory of γ–γ correlation is given. Finally in Section 5 a particle telescope is discussed and the velocity of high-energy π mesons, K mesons, and protons is obtained; note that in a beam of fixed momentum, the velocity measurement uniquely identifies the particle.

REFERENCES

As a general reference to the techniques of high-speed circuits, we recommend the following:

I. A. Lewis and F. H. Wells, *Millimicrosecond Pulse Techniques*, Pergamon Press, 1959.
S. De Benedetti and R. W. Findley, "The Coincidence Method," *Handbuch der Physik* (*Encyclopedia of Physics*), Springer Verlag, 1958.

Both of these references, however, treat vacuum tube circuits.

2. Coincidence Circuits

2.1 BASIC CIRCUITS

We can think of the basic two-element circuit as two switches in parallel which must be simultaneously open in order to give an output. The simple Rossi circuit can be placed in this category; it is shown in Fig. 9.2. The two triodes are connected through the common anode resistor R_L, and their

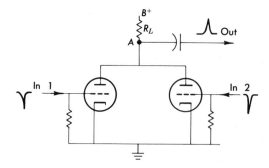

FIG. 9.2 The simple coincidence circuit devised by B. Rossi. When a negative signal appears simultaneously at both inputs, a positive signal appears at the output.

grids are kept at cathode potential; the input signals must be negative and of sufficient amplitude to turn the tubes off.

The description of the *static* behavior of the circuit can be followed in Fig. 9.3. Under quiescent conditions, both tubes conduct, *each* tube drawing a current i_p as determined by the $E_g = 0$ curve and a load line $2R_L$. When an input signal turns one of the tubes off, the other tube conducts a current i_p' determined by a load line R_L; thus the voltage at the point A rises from V_A to V_S. If both tubes are simultaneously turned off, the voltage at A rises to the full B^+ voltage, which can be taken as the signature for a coincidence.

The basic Rossi circuit is capable of resolution of 10^{-6} sec, and is frequently used with diodes instead of the vacuum tubes. It has the disadvantage, however, that the discrimination between one pulse and two simultaneous pulses is approximately $1:2$, and it is not very reliable for multiple-channel coincidences. The reader must always keep in mind that in actual practice the two input pulses may not be of equal amplitude and width, and that they do have a finite rise and fall time.

An improvement in the basic circuit of Rossi is the circuit of Garwin, shown diagrammatically in Fig. 9.4. The diode D_A is biased into conduc-

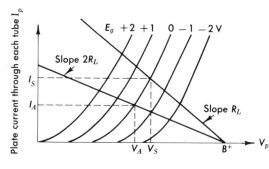

FIG. 9.3 The static behavior of the Rossi coincidence circuit shown in Fig. 9.2. When both tubes are on, the current in *each* tube is determined by the load line with slope $2R_L$ and the plate voltage is V_A; when only one tube is on, the load line has slope R_L and the plate voltage is V_S. When both tubes are cut off, the place voltage rises to the full B^+ value.

tion when one or both of the tubes are on, since the voltage across R_1 lowers the voltage of point Q. Thus, when only one of the tubes is turned off, the surge at point Q is bypassed to ground through the diode and the large condenser C_1, and only a very small signal appears through the output condenser C_2. If both tubes are simultaneously cut off, the diode is turned off, since the voltage at point P is held by C_1, while point Q rises towards B^+; therefore the condenser C_1 is isolated from point Q and the voltage surge appears across C_2. Diode D_B is a simple series discriminator in order to respond only to positive signals and of sufficient amplitude. The resulting discrimination between single and simultaneous pulses is as good as $1:10$; the circuit is also well suited for multiple coincidences.

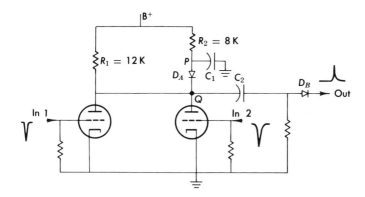

Fig. 9.4 The coincidence circuit of Garwin. Normally the diode D_A is conducting and any voltage changes at the point Q are decoupled through C_1 to ground; even when one tube is cut off the diode still remains open. When both tubes are cut off, D_A becomes reversed biased and the voltage at point Q rises. Diode D_B acts as a series discriminator on the output signal.

Circuits have also been used in which the two inputs are fed to the grids of the same vacuum tube, especially the 6BN6 "gated-beam" tube. These circuits can be thought of as two switches in series; however, they do not have an analogue in transistorized coincidence circuits.

2.2 The Sugarman Transistorized Coincidence Circuit

A great variety of transistor coincidences has been developed, for slow logic or gating as well as for high-speed work. The circuit used in this laboratory and the one which we will discuss was developed by R. Sugarman

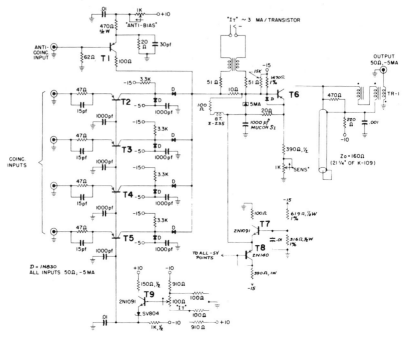

FIG. 9.5 Complete schematic of the Sugarman transistorized coincidence circuit. All transistors are 2N700.

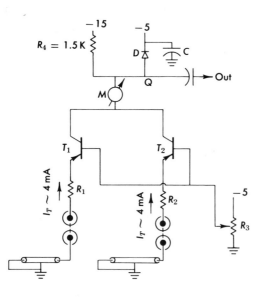

FIG. 9.6 Simplified diagram of the Sugarman coincidence circuit shown in Fig. 9.5; note that the circuit is equivalent to the arrangement of Garwin shown in Fig. 9.4. Under normal operating conditions both transistors T_1 and T_2 are conducting, the d-c return path being provided by the clip lines attached to the photomultiplier output. Note also that in the actual circuit, diode D, is a 5 mA tunnel diode.

as part of an entire system of high-speed logic† for use at Brookhaven National Laboratory; improved versions of these circuits are now available commercially from several manufacturers.‡

The Sugarman circuit is of the Garwin type and is capable of resolving times of the order of 2×10^{-9} sec (10^{-9} sec $\equiv$ nanosecond $\equiv$ nsec). It is designed for inputs and output of 5 mA negative signals and 50 Ω impedance level. The actual circuit is shown in Fig. 9.5; it provides four "yes" (coincidence) inputs and one "no" (anticoincidence) input. The circuit may be better understood, however, if the simplified schematic shown in Fig. 9.6 is considered first.

The two input transistors T_1 and T_2 are *p-n-p* and are d-c biased with their emitters to ground through the 47 Ω resistors R_1 and R_2. The path to ground is completed through the "shorting stubs" required to clip the input signals to the desired width; thus the input is at a ground d-c level.§ A fixed current, approximately 3 to 5 mA, flows through each transistor, depending on the setting of the base bias voltage determined by potentiometer R_3; the collector current can be monitored with the meter M.

The d-c current flows from ground through the emitter resistors R_1 and R_2, through the transistors T_1 and T_2, and from the collectors towards B^-, in part through resistor R_4, and in part through the diode D. As long as both or one of the transistors is on, the (Garwin) diode will conduct, and any surge at point Q is decoupled by the condenser C. If both transistors are turned off, however, the voltage at point Q drops towards -15 V, and the diode is turned off, giving rise to an output signal.

We now return to the actual circuit as shown in Fig. 9.5. At first the diode D is a 5-mA "tunnel diode," and it will adjust itself to supply the collector current for as many of the input transistors as are on. The equivalent circuit and the characteristic of the tunnel diode are shown in Fig. 9.7. The diode operates with a load line determined by R_4, but as long as current is flowing from the transistors the diode is biased in the low voltage direction; thus point Q is at -4.9 V and a current of $(15 - 4.9)/1.5 \approx 6.7$ mA flows through R_4. As the transistors are turned off, the current through the tunnel diode decreases, even reverses direction (when only one transistor is on), but the voltage excursion at point Q is very small (< 100 mV) even in the static limit; dynamically condenser C will decouple any pulse as long

---

† R. Sugarman, F. Merritt, and W. Higinbotham, *Nanosecond Counter Circuit Manual*, Report BNL 711 (T-248), available from the Office of Technical Services, Department of Commerce, Washington 25, D. C. We will frequently refer in the text to this report, as the "*Nanocard Manual*."

‡ For example, from Chronetics Inc., 500 Nuber Ave., Mount Vernon, New York.

§ The reader may have noted that the entire circuit of Fig. 9.5 is d-c coupled in order to prevent shifts in d-c levels due to charging of capacitors at high counting rates.

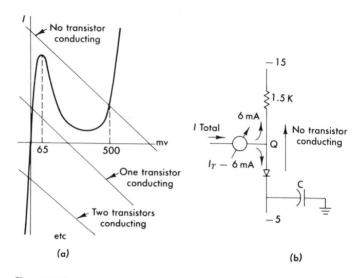

Fig. 9.7(a) Characteristic curve of a 5 mA tunnel diode. The straight lines indicate the operating load line with two transistors conducting, one transistor conducting and so on. (b) Simplified diagram indicating the operation of the tunnel diode and the flow of current from the collectors of the transistors. Note that a fixed bias current of 6 mA flows always through the 1.5 K resistor; the balance of the transistor-current flows through the tunnel diode. When all the transistors are turned off simultaneously, the bias current has to flow through the tunnel diode but in the reverse direction raising the voltage at point Q.

as the tunnel diode is open. When, however, all transistors are turned off, the tunnel diode through its regenerative action will jump to its high-voltage state, dropping point Q to -5.5 V and cutting off C. Thus D acts also as a discriminator, since the total *transistor* current must be reduced to less than 1 mA before the tunnel diode "goes over the hump."

The voltage changes at Q appear at the base of T_6, which is part of the output stage required to shape the pulse, which is delivered at -5 mA and 0 d-c level. Transistors T_7 and T_8 form a temperature-compensated stabilized divider for the -5 V, and similarly, T_9 is used to set the bias level of the input transistors. The series diodes in the collectors of $T_2 - T_5$ are used to standardize the input pulses, while the shunt diodes serve to maintain (restore) the d-c level. The anticoincidence transistor is normally off, and only when an input is applied is it turned on.

The performance of this circuit is discussed in Section 2.3; for more details the reader should consult the *Nanocard Manual*.

2.3 OPERATION OF A COINCIDENCE SYSTEM

Let us consider a system appropriate for detecting time coincidences between pulses originating from radiation detectors; such a system† is shown in Fig. 9.8. It should be emphasized that the performance of the system is usually determined by the quality of the pulses provided by the detector; these pulses are not all of the same amplitude, width, and so on, and do have a time jitter with respect to the origin of the event—that is, the arrival of the particle in the detector. It is therefore advantageous to standardize the input pulses, either with a discriminator as in Fig. 9.8, or with a limiter.

If it is desired to have short resolving times, clearly the "rise time" of the signals must also be short; thus a plastic scintillator is much preferable to inorganic crystals for fast coincidence work. Also it is clear that the resolving time of the system can never be shorter than $\Delta t = \Delta t_1 + \Delta t_2$ where Δt_1 and Δt_2 are the widths of the two signals as shown in Fig. 9.9; here t_1 and t_2 are the times corresponding to the leading edge of the input signals. A convenient method for shortening the length of fast pulses is by "clipping"

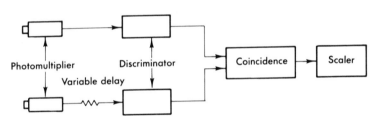

FIG. 9.8   Schematic block diagram of an arrangement for the measurement of coincident counts between two scintillation counters.

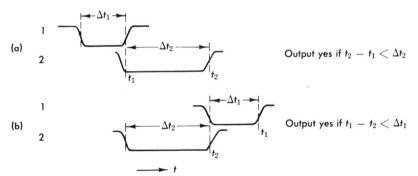

FIG. 9.9   The resolving time of a coincidence circuit Δt is the sum of the width (duration) of the signals in channel (1) and in channel (2). In (a) the signal in channel (2) arrives early, whereas in (b) the signal in channel (1) arrives early.

---

† See *Nanocard Manual*.

with a shorted stub as shown in Fig. 9.10. A short pulse propagating across
the coupling condenser divides equally into the output branch and the
shorting stub since both have the same impedance Z_0: when the pulse
reaches the short, it will be reflected with a voltage inversion and after a
time interval $t = 2L/\beta c$ will return to the junction point, where it will
interfere destructively with the pulse propagating into the output branch.
Here L is the length of the shorting stub, and $\beta = 1/\sqrt{\kappa_e \kappa_\mu}$ is the propagation velocity along the cable. In this fashion, pulses a few nanoseconds
wide can be formed, as shown in Fig. 9.11, where the horizontal scale is
2×10^{-9} sec/cm. It is clear that the shortest pulse that can be usefully
formed by "clipping" is determined by its rise time.

In evaluating a coincidence circuit, the following considerations are
taken into account (a) how efficient the circuit is for detecting true co-

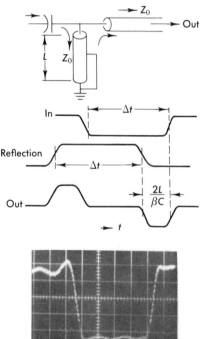

FIG. 9.10   Clipping of a pulse to a
desired width (duration) by the use
of a shorted coaxial cable (stub). Note
that at the shorted end the pulse is
reflected and also inverted; the width
of the pulse is $2L/\beta c$ where L is the
length of the shorting stub and β the
velocity of propagation along the coaxial cable.

FIG. 9.11   (*Below*) A shaped pulse of
10 nsec width, and of 5 mA amplitude.
Note that the rise time (and the fall
time) is of the order of 1 nsec.

incidences, (b) what is the resolving time, and (c) what is the feed-through
rate—that is, the rate of output counts when only *one* signal is fed in. These
points should be checked first with a pulser, and next with the actual signals
provided by the system.

There are many versatile and fast pulsers now available commercially,
but a mercury-switch pulser (as shown in Fig. 9.12) can be constructed

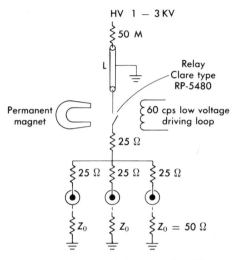

HV 1 — 3 KV

50 M

L

Relay
Clare type
RP-5480

Permanent
magnet

60 cps low voltage
driving loop

25 Ω

25 Ω   25 Ω   25 Ω

Z_0 Z_0 $Z_0 = 50$ Ω

FIG. 9.12 Schematic circuit diagram for a mercury switch pulser capable of providing output pulses with very fast rise time. The width (duration) of the pulse is determined by the length L of the charge cable.

inexpensively and provides very fast pulses; the repetition rate is 120 cps, and the pulse width may be determined by the length of the charge line L. To test the coincidence circuit, we need also to vary the arrival time of the two pulses at the circuit input. This is done by using a variable delay in one of the branches (Fig. 9.8); for microsecond circuits this delay may be provided by a network of capacitors and resistors, but in the nanosecond region, delay is simply provided by inserting additional cable between one of the input signals and the coincidence circuit. (For the coaxial cables commonly used, 1 nsec of delay corresponds to approximately 0.7 ft of cable.)

The efficiency of the coincidence circuits must be close to 100 percent; this can be measured in various ways, but in general a flat top in a delay curve (Fig. 9.13) is an indication of good efficiency. Similarly the feed-through rate must be zero; this is tested by connecting only one input to the coincidence (for the Sugarman circuit as shown in Fig. 9.5, however, the other input must be grounded). To measure the resolving time, we measure the rate of coincidences† as a function of the relative delay between the two inputs to the circuit. Such a curve, obtained in this laboratory, using the two gamma rays from positron annihilation, is shown in Fig. 9.13. The discriminators were clipped with 2 ft of 50 Ω cable (RG 58/U) so that the width of the input signals was approximately 6 nsec;‡ this is in agreement with the observed full width at half maximum (fwhm) of 13 nsec. The full width of the delay curve is then defined as the resolving time Δt of the circuit.§

---

† Coincidences per unit time (or per some other appropriate standard).
‡ For this cable $\beta = 0.659$; hence $2L/\beta C = 4/0.659$ nsec $= 6$ nsec.
§ Frequently the half width of the delay curve is called the resolving time, and the symbol τ is used.

If in a coincidence circuit one input is delayed by a very large time interval with respect to the other, still some output counts can be expected. These counts originate from the accidental overlap of two completely unrelated pulses. We can calculate this "accidental" rate as follows: let R_1 be the rate (counts/sec) of input pulses in channel 1, R_2 the rate in channel 2,

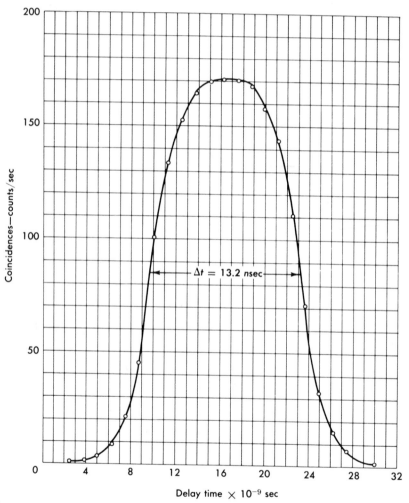

FIG. 9.13 Coincidence counts plotted against delay in one of the input channels. The source of the radiation was Na^{22} and it was viewed by two scintillation counters placed at 180 degrees with respect to the source. Note that the resolving time Δt is 13 nsec and that in the central region (flat top) the coincidence rate is independent of delay; the accidental rate (that is, when a very large delay is inserted between the two counters) is extremely small. See also Fig. 9.23 for a plot of the same data on a semilogarithmic scale.

and the resolving time Δt. Clearly, due to the counts in channel 1 the circuit will be open for a *fraction f*, of the total counting time

$$f = R_1 \times \Delta t$$

therefore the fraction of the R_2 pulses/sec that will accidentally arrive when the circuit is open (due to unrelated pulses in channel 1), is simply

$$R_{\text{acc}} = R_2 \times f = R_1 R_2 \, \Delta t \qquad \text{counts/sec} \qquad (2.1)$$

The accidental rate in a coincidence system is the equivalent of background in counting, or noise in a detector; the accuracy with which we may measure a coincidence rate R_c obviously depends on the ratio

$$R_c / R_{\text{acc}}$$

which should be at least of the order of unity and preferably much larger.

2.4 TIME-TO-PULSE-HEIGHT CONVERTERS

The coincidence circuits described until now must be considered as digital logic blocks since they provide either a *yes* or *no* output. We may also construct analogue coincidence circuits, which measure the time interval between the two pulses, provided that the pulses are coincident to first order—that is, their separation falls within the *time range* of the circuit. Usually, the amplitude of the output signal is made proportional to the difference in time between the two incoming pulses; by feeding this output to a pulse-height analyzer, we obtain a display of the distribution of the time intervals between the pulses (see Fig. 9.30). With such a system it is possible to measure time differences of the order of 10^{-10} sec, and with special precautions, even 10^{-11} sec.

To understand the operation of such a circuit, consider Fig. 9.14. The two input pulses A and B have been appropriately shaped and have a standard length (duration) L_0; pulse A arrived at a time Δt earlier than pulse B (Fig. 9.14a). If we superimposed the two pulses, for example, by linear addition in a crystal diode, a pulse shape as shown in Fig. 9.14b will be generated; next a discriminator with level x_0 is used to select only the part of the pulse with amplitude greater than x_0 as shown in Fig. 9.14c. Finally, if we amplify and integrate this truncated pulse, we obtain a signal (Fig. 9.14d), whose amplitude is proportional to the time overlap P between A and B. Since the input signals have *standard* length, it follows that the difference in time of arrival

$$\Delta t = L_0 - P$$

We note that (1) we may not distinguish whether A arrived before B or vice versa, (2) we may not measure time intervals longer than L_0, and (3)

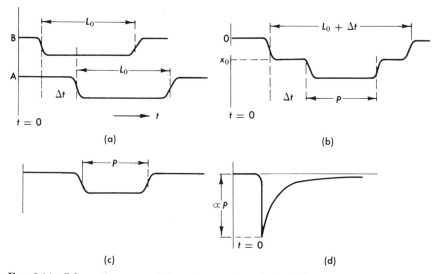

FIG. 9.14   Schematic representation of conversion of time difference to pulse height by the superposition of the two signals. (a) The two signals A and B are of the same duration L_0, but signal A is early by a time interval Δt. (b) Result of the superposition of the two signals A and B. (c) The resulting pulse when the signal shown in (b) is truncated at the level x_0. (d) The signal shown in (c) after integration and amplification.

maximum output is obtained when the two signals overlap exactly; these features are characteristic of most time-to-pulse-height converter circuits.

In practice, the overlap between the two pulses is obtained by a different technique, shown schematically in Fig. 9.15. The idea is to charge condenser C through the resistor R only during the time interval that pulses A and B are overlapped; if then the capacitor is not allowed to saturate, the voltage to which it becomes charged, V_c, is proportional to the time overlap P,

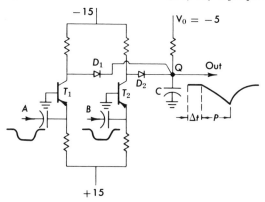

FIG. 9.15   Simplified schematic diagram of a circuit that can be used for time-to-pulse-height conversion. Condenser C is being charged only when an input signal exists simultaneously on both inputs A and B.

since

$$V_c = V_0(1 - e^{-t/RC}) \approx V_0 \frac{t}{RC} \propto P$$

Normally both transistors T_1 and T_2 are conducting, the collectors being held close to ground; therefore point Q is also held at ground due to the clamping action of diodes D_1 and D_2. When input A arrives, T_1 is turned off, D_1 is reverse biased, but D_2 still keeps Q at ground; when now B also arrives, T_2 is turned off and D_2 becomes reverse biased, allowing C to charge towards -15 V. Finally, when pulse A has disappeared, D_1 again clamps Q at ground, stopping the charging process.

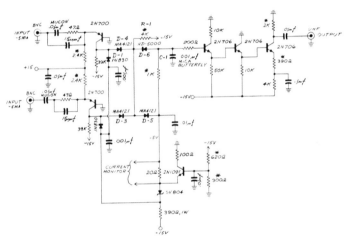

FIG. 9.16 Complete diagram of the time-to-pulse-height conversion circuit designed by R. Sugarman. The last three *n-p-n* transistors serve to amplify and stretch the output signal. Components marked with an asterisk should be kept to close tolerances.

The actual circuit is more complicated and is shown in Fig. 9.16. The last three 2N706 transistors are used to integrate, amplify, and stretch the voltage that appeared on C.

When such small time intervals are to be measured, the shaping of the pulses becomes very important. In particular the "time jitter" between the actual event† and the appearance of the pulse must be kept to a minimum. This time jitter may occasionally arise from faulty equipment, but it is mainly due to physical reasons, as explained below.

In most cases the signals produced in the two detectors are not of the same amplitude: for example, when gamma rays are detected, we know that the output pulse-heights are distributed over a spectrum (see Figs. 5.27

---

† That is the time at which the light quantum reaches the photocathode.

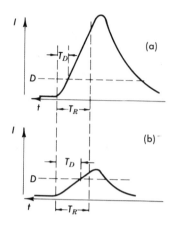

FIG. 9.17 Origin of the time jitter that may affect the operation of a coincidence circuit. Two signals that have the same rise time T_R but different amplitudes are shown. The dotted line marked D represents the discriminator level. Note that the delay T_D between the appearance of the signal and the time at which the discriminator is triggered is longer for signal (b).

to 5.30); when a charged particle has crossed the two counters of a telescope, it will have deposited the same amount of energy only if both counters are of the same thickness. But the light collection efficiency of the counters, or the gain of the photomultipliers, or other factors, may be different. Now, the light pulses reaching the photocathode are usually very weak, so that a very small number of photoelectrons† is emitted; thus the statistical fluctuations in the number of emitted photoelectrons will vary the output pulse heights even when the light input is the same.

Figure 9.17 shows two photomultiplier output pulses, the amplitude of (a) being three times that of (b). The shape of the pulses is determined mainly by the "rise time," T_R of the photomultiplier output network; we may define T_R as the time necessary for the signal to rise from 10 to 90 percent of its peak value, and it is a constant of the system, independent of pulse height. The dotted line D in Fig. 9.17 is the amplitude level at which the discriminator will trigger. It is now clear that even though the two pulses (a) and (b) originated at the same time, the shaped output pulses of the discriminators will differ in time by $T_D{}^{(a)} - T_D{}^{(b)} \neq 0$; this jitter, however, does not exceed the output rise time T_R. Therefore, for good timing, signals with as fast a rise time as possible are desirable. Schemes such as the zero-crossing technique have been devised to compensate for differences in T_D, arising from variations in the amplitude of the pulses.

When detectors with large physical dimensions are used, the time required by the light pulse to reach the photomultiplier is substantial; of the order of 2 nsec/ft.‡ Thus, differences in light path must be compensated or accounted for.

---

† It is frequently as low as a single photoelectron.

‡ The refractive index of scintillon is approximately $n = 1.5$, and the light does not travel directly, but propagates by multiple reflection ($1/\cos 45° \approx 1.4$).

2.5 OSCILLOSCOPE DISPLAY

A different technique for coincidence measurement is to display the pulses on an oscilloscope and record them photographically. This method provides the possibility of measuring the delay between two pulses over a large time scale; also many pulses can be simultaneously displayed, especially with multitrace oscilloscopes. Although the method has the disadvantage that the photographic record must be examined, pulse amplitude and other information can be permanently registered.

Such a technique is very well suited for measuring the lifetime of stopped cosmic ray μ mesons, as shown in Fig. 9.18. The μ mesons are brought to rest in a large NaI crystal, as signaled by a coincidence in the telescope made of scintillators S_1 and S_2, the crystal C, and an anticoincidence in scintillator S_3; we designate this signature by $S_1 S_2 C \overline{S_3}$. Whenever a stopping particle† signal occurs, the oscilloscope trace is triggered at a rate of approximately 0.5 μsec/cm. The output of the NaI crystal is connected to the vertical input of the scope, so that if a pulse appears within 10 cm $\times$ 0.5 μsec/cm = 5 μsec of the triggering time, it will also appear on the trace of the scope. A camera advance circuit is required in order to advance the film by one frame after each pulse; the camera shutter is left open all the time.

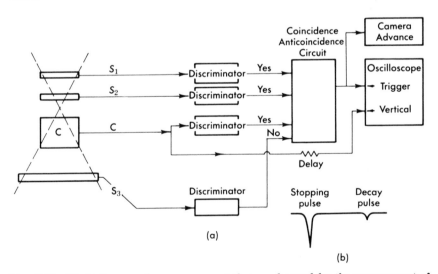

FIG. 9.18   Block diagram of an arrangement that can be used for the measurement of the lifetime of stopping cosmic-ray μ mesons. The signature $S_1 S_2 C \overline{S_3}$ is used to signal the stopping of a μ meson in the counter C, and also starts the oscilloscope trace; if a second signal originates in C in close succession to the stopping signal, it is displayed on the oscilloscope and indicates the decay of the μ meson.

† At sea level these will be mainly μ mesons and some soft electrons.

Since the lifetime of the μ meson is 2.2×10^{-6} sec, our time scale covers two lifetimes, so that most μ mesons will be observed to decay. From a semilog plot of the number of decays at each point of the scope trace (time), the lifetime of the μ meson is immediately deduced. Since the cosmic ray flux is only about 10^{-2} counts/cm²-ster-sec, if our detector has an area of 50 cm² (assuming approximately 2 steradian acceptance), the average time interval between pulses is of the order of 1 sec. Thus, the probability of the second pulse on the scope trace (Fig. 9.18b) being accidentally produced by a new (different) particle is extremely small.

3. Angular Correlation Measurements

3.1 GENERAL CONSIDERATIONS

We will now discuss the measurement of the correlation in angle between two gamma rays that were emitted simultaneously from the same source. The origin of these gamma rays is frequently the cascaded decay of a nucleus, as in the case of Ni^{60} (Co^{60}) already discussed in Chapter 5. (See Fig. 5.27.) We reproduce in Fig. 9.19 the decay scheme of this nucleus and note that the 1.333 MeV gamma ray follows the 1.172 MeV gamma ray, the lifetime of the intermediate state being only about 10^{-12} sec, so that for all practical purposes the two gamma rays are coincident.

The fact that these two gamma rays are correlated in angle can be understood from the following general argument: the first gamma ray will have an angular distribution with respect to the spin axis of the nucleus; thus its observation at a fixed angle $\theta = 0$ conveys information about the probability of finding the spin at some angle ψ with respect to the direction $\theta = 0$. Now the second gamma ray also has some angular distribution about the spin axis which now is known to be at ψ. Thus the probability that the second gamma ray will be emitted at an angle θ can be found; this is called the angular correlation function $C(\theta)$. The time coincidence signal assures us that the two gamma rays have indeed come from the same nucleus and,

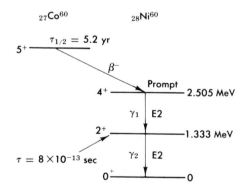

FIG. 9.19   Nuclear decay scheme of Co^{60} by beta decay to Ni^{60} and subsequent deexcitation of the Ni^{60} nucleus to its ground state by the emission of two cascaded gamma rays.

therefore, are the two gamma rays of interest. A more detailed discussion of this correlation between cascaded gamma rays is presented in Section 4. In Na^{22}, the angular correlation arises from a much simpler mechanism. Na^{22} is a positron emitter as is shown from the decay scheme of Fig. 5.29. The positrons are slowed down in a thin copper sheet with which we surround the source; the slow positrons are captured by the electrons of the copper to form positronium, which decays by the annihilation of the positron and electron into two gamma rays. The energy of these gamma rays is precisely 0.511 MeV, and since the center of mass of positronium was at rest, the two gamma rays must be directed in exactly opposite directions in order to conserve momentum as shown in Fig. 9.20.

FIG. 9.20 Capture of a positron by an electron to form positronium, and the subsequent annihilation of the positron-electron pair into two gamma rays.

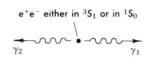

Thus the angular correlation theoretically is given by

$$C(\theta) = \delta(\pi - \theta)$$

and the observed finite width is due to the resolution of the apparatus; obviously the two gamma rays are simultaneous. Since the Na^{22} angular correlation is so sharp, it is frequently used for calibration purposes.

Angular correlations may also be observed between beta and gamma rays, or alpha and gamma rays, etc. This technique has proved very fruitful for the analysis of nuclear decay schemes and the assignment of spin and parity to excited nuclear levels.

In this laboratory the correlation of Na^{22} and of Co^{60} was measured. The apparatus shown in Fig. 9.21 was used; it consists of two similar gamma-ray detectors placed at equal distances from the source; one detector is fixed and the other is free to rotate around the source, varying the angle θ between the detected gamma rays. The detector outputs are fed to a coincidence circuit and the rate of coincident counts $C(\theta)$ is measured and compared with the theoretical correlation function.

It is important to measure $C(\theta)$ with the best possible resolution if the data are to be fitted with a polynomial in $\cos\theta$ of high order. It can be shown that $C(\theta)$ must be a polynomial in *even* powers of $\cos\theta$, the highest power being $2k$; where $k \leq I_b$, l_1, l_2 in the notation of Section 4.2, below. Frequently the experimental measurement may be restricted to the measurement of the *anisotropy* of the coincident gamma rays, that is,

$$\alpha = \frac{C(180°) - C(90°)}{C(90°)}$$

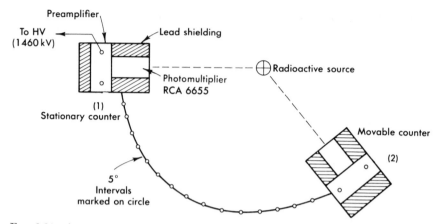

FIG. 9.21   Apparatus that can be used for angular correlation measurements. Two scintillation crystals mounted on photomultipliers are protected by appropriate lead shielding. One counter assembly is fixed, whereas the other one can be rotated about the
position of the source.

The limiting factors in these experiments are two: (a) The coincidence
rate must be high enough to allow statistically significant data to be accumulated in a reasonable time interval. To increase the coincidence rate
a stronger source may be used, or the solid angle may be increased, or the
efficiency of the detector improved (if it has not been maximized already).
(b) The accidental rate must be kept well below the coincidence rate;
again it depends on source strength and solid angle, but also on the resolving time. Let ω_1 and ω_2 be the solid angles subtended at the source by
detectors (1) and (2), and let ϵ_1 and ϵ_2 be their respective efficiencies. Then
the "singles" counting rates are

$$R_1 = N\omega_1\epsilon_1$$

$$R_2 = N\omega_2\epsilon_2$$

(3.1)

where N is the number of disintegrations per unit time of the source. If
the two gamma rays are *uncorrelated* (or if the correlation is small, as
happens mostly in nuclear decay), the coincidence rate† is

$$R_c = N\omega_1\omega_2\epsilon_1\epsilon_2 \tag{3.2}$$

For most experimental arrangements $\omega_1 = \omega_2$ and $\epsilon_1 = \epsilon_2$, so that we find
for the accidental rate R_A (Eq. 2.1),

$$R_A = R_1R_2\Delta t = N^2\omega^2\epsilon^2 \, \Delta t \tag{3.3}$$

---

† The efficiency of the coincidence *circuit* has been set $\epsilon_c = 1$ as it ought to be.

and for the ratio of the accidentals to true coincidences,

$$\frac{R_A}{R_C} = N \, \Delta t \qquad (3.4)$$

We wish to keep this ratio small, of the order of or smaller than 0.1. From Eq. 3.4 we see how important it is for correlation experiments to have a short resolving time; with $\Delta t = 10$ nsec, a source as strong as 0.5 millicurie may be used. We also note that the detector efficiency should be high, since it enters Eq. 3.2 quadratically; however, the solid angle cannot be increased arbitrarily because this will destroy the angular resolution and wash out the correlation $C(\theta)$.

3.2 THE APPARATUS

The apparatus has been shown in Fig. 9.21 and we give here some additional details. The reader should, however, refer to Chapter 5, Section 4.2, in connection with the instrumentation and techniques of gamma-ray detection. The detectors were NaI crystals 1 in. in diameter and 1 in. thick, mounted on RCA 6655 photomultipliers; they were located 8 in. from the source. Both crystals are protected from scattered radiation with some lead shielding, and the movable detector can be rotated about the center in 5-degree intervals.

The block diagram of the electronics is shown in Fig. 9.22, where the individual units are the circuits described in the Nanocard Manual. If it is desired to build or to operate this particular system, the above publication should be consulted. The units are interconnected with 50-ohm coaxial cable, RG 58/U, except for the scaler driver output, which preferalby should be a 100-ohm cable. In the ensuing discussion we will assume that the circuits have been properly adjusted.

The light output of a NaI crystal is a long pulse about 0.2 μsec in duration and (therefore) of low peak amplitude. The discriminators† (Nanocard Manual unit 1H–51) trigger at a level that can be adjusted from 2 to 5 mA, so that the photomultipliers must be operated at a high voltage of approximately 1.5 kV in order to successfully drive the discriminators. The output signals are clipped with a 2-ft shorting stub to a width of about 6 nsec, and are of the standard amplitude (-5 mA).

One of the outputs from each discriminator is fed to the coincidence circuit (Nanocard Manual unit IH–56) and a second output to a scaler driver (Nanocard Manual unit IH–71) and then to a CMC 303B scaler

---

† In the present case the output voltage levels are irrelevant, since the photomultiplier is a current source and the discriminator responds to current pulses.

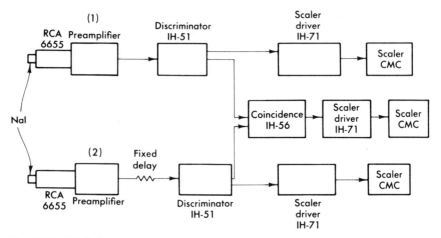

FIG. 9.22   Block diagram of the electronics used for angular correlation measurements. This is basically the same arrangement as shown in Fig. 9.8 with the addition of discriminators, and of scalers for monitoring the "singles" rates in the counters.

capable of a peak rate of 10^5/sec. The coincidence output is also fed to a scaler driver and scaler. In this way the "singles" in each channel, and the "doubles" are counted. The delay between the two inputs to the coincidence circuit may be easily adjusted by inserting appropriate cable lengths between the discriminator and coincidence in one *or* the other of the channels.

Some care is required in order to properly set the discriminator bias levels and photomultiplier high voltage. First the system is checked out with a pulser, to adjust the setting and functioning of the scaler drivers and scalers. Next the actual signals are fed into the circuits and the discriminator outputs "looked at" on a fast oscilloscope† to ascertain that the pulses are "clean" and uniform. The high voltage is set by taking a plateau curve, which will not be completely flat but nevertheless should show a clear knee. If the system is working properly, the "singles" rates R_1 and R_2 in the two channels should be (almost) equal.

It is possible to measure the resolving time of the coincidence circuit either by taking a delay curve (Fig. 9.13) or by making use of Eq. 2.1. When the latter method is used, the two counters are separated by a very large distance and a separate source is placed in front of each. In view of the geometrical arrangement and the fact that an additional delay of 60 ft is placed in one of the channels, all the coincidence counts are accidentals. By varying the distance between the source and the respective counter, the results given in Table 9.1 were obtained; the counting time

† Tektronix 517 or 585.

TABLE 9.1

Determination of Resolving Time from Accidental Coincidences

Counts/sec			Δt (sec)
Channel (1)	Channel (2)	Coincidence	($\Delta t = C/R_1 R_2$)
2151	2056	0.061	13.8×10^{-9}
5920	6262	0.528	14.2×10^{-9}
14,662	13,481	2.912	14.7×10^{-9}
31,207	35,443	14.217	12.8×10^{-9}

was of the order of 10 min at each point. We note that the resolving time so obtained (column 4) is quite consistent in spite of the fact that the accidental coincidence rate increased by a factor of about 2000 between measurements; this resolving time is also consistent with the width of the two input signals (which were about 6 nsec wide) and the data of Fig. 9.13.

The above results as well as those to be presented in the following two sections were obtained by students.†

3.3 The γ–γ Correlation of Na^{22}

A 100 μCi Na^{22} source, wrapped with a 0.001-in. brass foil is placed at the center of the apparatus; the dimensions of the source are kept at a minimum and it is positioned as accurately as possible. Since the solid angle is

$$\Delta\Omega = ((0.5)^2 \times \pi)/(8)^2 \approx 4\pi \times 10^{-3}$$

and assuming a detector efficiency $\epsilon_1 \sim \epsilon_2 \sim 0.3$, the expected rate for "singles" is

$$R_1 \sim R_2 = \frac{3.7 \times 10^{10} \times 10^{-4}}{4\pi} \times (4\pi \times 10^{-3}) \times 0.3 \approx 1000 \text{ counts/sec}$$

Since the two gamma rays are completely correlated when the two counters are at $\theta = 180°$, the expected coincidence rate at this angle is

$$C(\theta) = N\Delta\omega\epsilon^2 = R\epsilon \approx 300 \text{ counts/sec} \tag{3.5}$$

The observed rates are of this order of magnitude. However, the 1.277 MeV gamma ray also contributes to the single rate; on the other hand, the finite size of the source, as well as errors in geometrical alignment, reduce the coincidence rate from the calculated value.

---

† T. Londergan and S. Pieper, class of 1965.

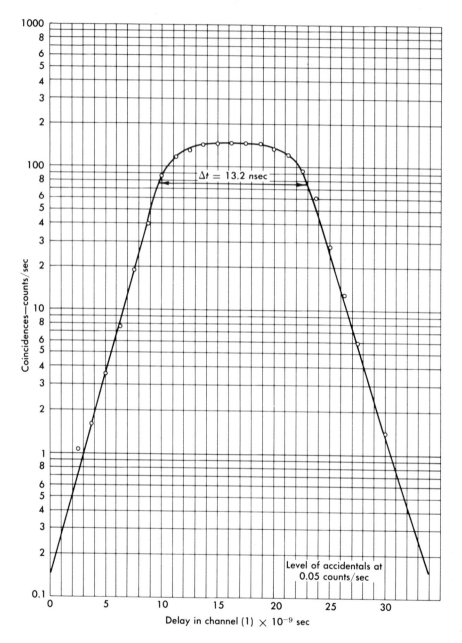

FIG. 9.23 A delay curve for coincidences from a Na²² source; the same curve was shown in Fig. 9.13. Note that the resolving time is of the order of 13 nsec and that the accidental rate is lower than the true coincidence rate by a factor of at least 1000.

We first wish to check if the coincidence is correctly "timed"—that is, if the appropriate delay has been inserted so as to make truly coincident signals arrive to the circuit at the same time. To this effect the movable counter is rotated to 180 degrees and the counting rate is obtained as a function of the variable delay which is introduced in channel (1); for convenience, a fixed delay of 12 ft of cable† has been introduced in channel (2). The data so obtained have already been given in Fig. 9.13 on a linear plot; they are repeated in Fig. 9.23 on a semilog plot, which is the more appropriate representation for a delay curve.

We note that (a) indeed, the peak counting rate occurs when 16 nsec delay are inserted in channel (1) as expected; (b) in the peak region, the delay curve is flat over at least 6 nsec; this indicates good efficiency and consequently that small time jitters will not result in changes in the counting rate (provided the delay is set at the center of the curve); (c) the width of the curve at half maximum, which gives the resolving time of the circuit, is 13.2 nsec, in excellent agreement with the values found in Section 3.2, and what is expected from the width of the input signals; (d) the accidental rate is very low; by inserting 40 ft of delay it is found to be 0.048 ± 0.005 counts/sec, yielding a ratio

$$\frac{\text{signal}}{\text{noise}} = \frac{150}{0.05} \sim 3 \times 10^3 \tag{3.6}$$

which is more than adequate.

The considerable *slope* of the ascending and descending parts of the delay curve, Fig. 9.23 (which appears in the linear plot, Fig. 9.13, as "wings"), is due to the time jitter of the input signals associated with their low peak amplitude, as discussed in Section 2.4. The stability of the system can be judged from the fluctuations of the coincidence rate in the flat region as well as from the fluctuation of the singles rates given in Table 9.2 (page 421).

We are now ready to obtain data on the angular correlation of Na^{22}. The movable counter is rotated in appropriate steps to either side of 180 degrees, and the doubles and singles rates are recorded. The resulting data are shown in Fig. 9.24, and in Table 9.2 some representative points are listed. Columns 2 and 3 of Table 9.2 give the singles rates for the stationary and the movable counter, respectively; the coincidence rate‡ is given in column 4. The counting time at each point was of the order of 1 min, which provides good statistics (about 1 percent in the peak region).

---

† For RG 58/U coaxial cable, $\beta = 0.659$; hence 12 ft = 18 nsec.

‡ The rate for accidentals should have been subtracted from the results of column 4; however, it is so small (see Eq. 3.6) that we neglect it.

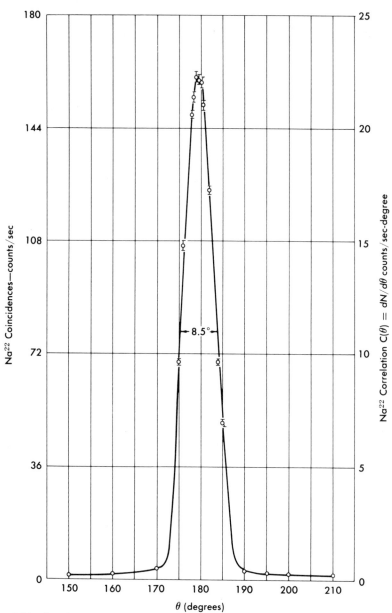

FIG. 9.24 Angular correlation of the gamma rays from a Na²² source. In the figure the coincidence rate is plotted as a function of the angle between the two counters. Note that the full width of the correlation curve is 8.5°, which is entirely due to the angular resolution of the two counters; the isotropic background outside the peak is very small. On the right-hand side scale is given the number of coincident counts per second per degree subtended by the counter.

TABLE 9.2

<small>REPRESENTATIVE DATA ON THE $\gamma-\gamma$ CORRELATION OF Na^{22}</small>

θ (Degrees)	Counts/sec			Coincidences (Counts/sec-degree)
	Stationary Counter	Movable Counter	Coincidence	
90	3011	3086	1.5 ± 0.1	0.21
150	2996	3071	1.7 ± 0.2	0.23
160	3013	3090	1.7 ± 0.2	0.23
170	2994	3064	3.5 ± 0.2	0.49
175	3011	3114	66.8 ± 1.0	9.2
178	2992	3189	148.0 ± 1.5	20.6
180	2995	3035	159.0 ± 1.6	22.1
182	3014	3178	124.0 ± 1.2	17.2
185	2991	3069	50.2 ± 1.0	7.0
190	3039	3127	3.2 ± 0.2	0.42
200	3005	3102	2.0 ± 0.1	0.26
210	3007	3136	1.8 ± 0.1	0.25

Indeed we do notice a very pronounced correlation at $\theta = 180°$, with an angular width of $\pm 4°.25$. This width is of the order of the angular resolution of our system, which might be taken as the angle subtended at the position of the source by one of the counters.

$$\tan\left(\frac{\Delta\theta}{2}\right) = \frac{0.5}{8} \rightarrow \Delta\theta = 7°.2 \qquad (3.7)$$

We therefore conclude that this correlation is compatible with

$$C(\theta) = \delta(\pi - \theta) \qquad (3.8)$$

The anisotropy as defined by Eq. 3.1 is

$$\alpha = \frac{C(180°) - C(90°)}{C(90°)} = \frac{150 - 1.5}{1.5} \approx 100 \qquad (3.9)$$

which is extremely large and compatible with $\alpha \rightarrow \infty$ as predicted by Eq. 3.8.

The counts observed at large angles are still real coincidences, but due mainly to the *isotropic* correlation of the 1.277-MeV gamma ray with one of the annihilation gamma rays; it should be of the order of the correlated counts multiplied by the solid angle for one detector $\Delta\Omega \sim 10^{-2}$, as is indeed the fact. Also, a small fraction of the background originates from annihilation gamma rays that have scattered through a large angle in the source or the converter.

In column 5 of Table 9.2, the coincidence rate has been divided by the angular acceptance $\Delta\theta$ of the movable counter as given by Eq. 3.7. These data can also be read off Fig. 9.24 if the right-hand vertical scale is used. Indeed, since the correlation is a function of θ, it is obvious that our system measures $C(\theta)$ at θ within the differential range $\pm\Delta\theta$.

From the results presented we conclude that Na^{22} provides a very good technique for aligning and adjusting the equipment; especially since the strong correlation from the annihilation gamma rays is quite easy to detect. Also, the obtained correlation provides strong evidence for the annihilation of the positron-electron pair into two gamma rays; if a differential discriminator is used after the detector, it is also possible to measure the energy of the coincident gamma rays. The angular resolution of the equipment may be easily improved by simply increasing the distance between the source and the counters. In fact, precise data on positron annihilation are quite sensitive to the momentum of the positronium just before it annihilates; this in turn provides information on the structure of the Fermi surface of the converter material.

3.4 THE γ–γ CORRELATION OF Co^{60}

Once the equipment has been adjusted and aligned (for example, with Na^{22}) as described before, any correlation may be measured. A Co^{60} source of the same strength as the Na^{22} source ($100\ \mu Ci$) was placed at the center of the apparatus and data were taken every 15 degrees. The discriminator levels could be readjusted, but it is usually preferable to leave everything as is.

TABLE 9.3

REPRESENTATIVE DATA ON THE γ–γ CORRELATION OF Co^{60}

θ Degrees	Counts/sec				$\dfrac{C(\theta)}{C(90°)}$
	Stationary Counter	Movable Counter	Coincidences	Corrected Coincidences	
60	2203	2129	0.880	0.810	1.080
90	2132	2157	0.820	0.750	1.000
105	2152	2127	0.857	0.787	1.049
120	2144	2130	0.864	0.794	1.059
135	2109	2125	0.886	0.816	1.088
150	2132	2136	0.933	0.863	1.151
165	2121	2123	0.931	0.861	1.148
180	2116	2124	0.944	0.874	1.165
210	2086	2134	0.889	0.819	1.087

Since the Co^{60} γ–γ correlation has a small anisotropy (as compared to Na^{22}, Eq. 3.9) the expected coincidence rate is

$$C(\theta) \approx N(\Delta\omega)^2\epsilon^2 \approx 4 \text{ counts/sec} \qquad (3.10)$$

which is much smaller than that given by Eq. 3.5 for the same source strength. Consequently, also, the signal-to-noise ratio (Eq. 3.6) will be only about 30, and the "accidental" rate, which was 0.070 counts/sec, must be subtracted. Furthermore in view of the smaller correlation, better statistical accuracy is required.

Representative data taken in one run are presented in Table 9.3 and plotted in Fig. 9.25. In column 5 the coincidence rate after the subtraction of accidentals is given, while in column 6 the rate at each angle is nor-

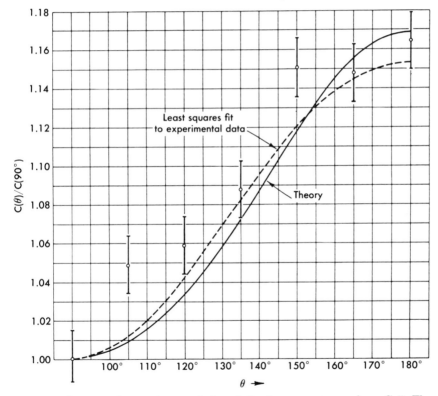

Fɪɢ. 9.25 Data on the angular correlation of the two gamma rays from Co^{60}. The correlation function $C(\theta)/C(90°)$ is plotted against the angle between the two counters. Note, however, that the ordinates begin at the value 1.00. The experimental points are shown and the dotted curve is a least squares fit to the data. The solid line shows the theoretical curve which is given by the function $1 + 0.125 \cos^2 \theta + 0.042 \cos^4 \theta$.

malized to the rate at 90 degrees. At each point sufficient coincidence counts were taken to give 1 percent statistical accuracy (10,000 sec $\approx$ 3 hr); these errors are indicated by the flags shown in Fig. 9.25, where we plot $\alpha(\theta) = C(\theta)/C(90°)$ against angle. We see that the *fractional* errors on $[\alpha(\theta) - 1]$ are now much larger, and of the order of 10 percent.

It is known from theoretical considerations (see Section 4) that the Co[60] correlation function is of the form

$$\alpha(\theta) = \frac{C(\theta)}{C(90°)} = 1 + a_1 \cos^2 \theta + a_2 \cos^4 \theta \qquad (3.11)$$

A least-squares fit† to Eq. 3.11 was made, using the entire set of experimental data‡ and the following values were obtained for the coefficients a_1 and a_2

$$a_1 = 0.190 \pm 0.08 \qquad a_2 = -0.04 \pm 0.08$$

to be compared to the theoretical values

$$a_1 = 0.125 \qquad a_2 = 0.042$$

The correlation function that results from the above coefficients is included in Fig. 9.25; the dotted line represents the least-squares fit, and the solid line the theoretical curve.

From Fig. 9.25 we see clearly that an anisotropy in the angular distribution of the γ–γ coincidences from Co[60] exists; we obtain

$$\alpha = \alpha(180°) - 1 = 0.165 \pm 0.016 \qquad (3.12)$$

The error flags in Fig. 9.25 were set at 1.5 percent, but the data points scatter even more. This is not due to the "statistics," but to random fluctuations and drifts of the equipment over the long counting intervals.

An elementary analysis of these data is presented in the following section.

4. Elementary Discussion of the γ–γ Angular Correlation in Nuclear Decay

4.1 MULTIPOLE EXPANSION OF THE ELECTROMAGNETIC FIELD

From the study of atomic spectroscopy and the Zeeman effect, we are familiar with the fact that emitted (or absorbed) electromagnetic radiation can connect states with $\Delta l = 1$ or 0, and that the transition may involve changes in the m number; only changes $\Delta m = \pm 1$ or 0 are allowed. Nuclear transitions involving gamma rays do, however, in general, connect states with $\Delta l > 1$; the electromagnetic field (equivalently, the emitted quantum

---

† This was performed on an IBM 7074 digital computer.

‡ This included 21 more measurements in addition to those presented in Table 9.3.

of the field) then provides the balance of angular momentum. If the initial state has $\mathbf{J}_a$ and the final state $\mathbf{J}_b$, the emitted quantum must have angular momentum l, such that

$$\mathbf{J}_a = \mathbf{J}_b + l \tag{4.1}$$

where the quantum mechanical law for angular momentum addition must be used.

Indeed, we can ascribe discrete angular momentum to the electromagnetic field even in a classical treatment. When the dimensions of the source are small as compared to the wavelength of the radiation,† the field can be expanded in vector spherical harmonics (multipole expansion),

$$\mathbf{E}(\mathbf{r}) = \sum_{l=1}^{\infty} \sum_{m=-l}^{l} a_E(l, m)\mathbf{E}_E(l, m; \mathbf{r}) + a_M(l, m)\mathbf{E}_M(l, m; \mathbf{r}) \tag{4.2a}$$

and similarly

$$\mathbf{H}(\mathbf{r}) = \frac{-i}{k\mu_0} \nabla \times \mathbf{E}(\mathbf{r}) \tag{4.2b}$$

The functions $\mathbf{E}_E$ and $\mathbf{E}_M$ must satisfy the free wave equations‡

$$\nabla^2 \mathbf{E} - \frac{1}{c^2} \frac{\partial^2 \mathbf{E}}{\partial t^2} = 0 \tag{4.3}$$

and thus Eqs. 4.2 will also be solutions of Eq. 4.3 for any arbitrary values of the expansion coefficients $a_E(l, m)$ and $a_M(l, m)$. For the same order (l, m), the functions $\mathbf{E}_E$ and $\mathbf{E}_M$ correspond to solutions of different parity. That is,

$$\mathbf{E}_E(l, m; \mathbf{r}) = (-1)^l \mathbf{E}_E(l, m; -\mathbf{r}) \quad \text{electric multipole}$$

$$\mathbf{E}_M(l, m; \mathbf{r}) = -(-1)^l \mathbf{E}_M(l, m; -\mathbf{r}) \quad \text{magnetic multipole}$$

---

† This is a condition which is always met in nuclear decay; $r_0 \approx 10^{-13}$ cm, and even for a 10-MeV gamma ray $\lambda \approx 10^{-12}$ cm.

‡ For the exact representation of the expansion functions $\mathbf{E}_E$, and $\mathbf{E}_M$, see the detailed treatment in Chapter XVI of J. D. Jackson's *Classical Electrodynamics*, John Wiley and Sons, 1962. This expansion of a *vector* field is in complete analogy with the more familiar expansion of a scalar field (the potential ϕ) in eigenfunctions of the Laplace equation in spherical coordinates. We then had

$$\phi(\mathbf{r}) = \sum_{l=0}^{\infty} \sum_{m=-l}^{l} a(l, m)\Phi(l, m; \mathbf{r}) \quad r > \text{dimensions of source}$$

where

$$\Phi(l, m; \mathbf{r}) = Y_{lm}(\theta, \phi) r^{-(l+1)}$$

and

$$\nabla^2 \Phi(l, m; \mathbf{r}) = 0$$

It is interesting that in most nuclear transitions the emitted radiation is a pure multipole; that is, from the expansion coefficients $a_E(l, m)$ and $a_M(l, m)$ appearing in Eq. 4.2, only those of a *specific* l value are different from zero. Further, since the indices l and m characterize the angular momentum of the field, we see from Eq. 4.1 that the following restriction is imposed on the possible values of l and m:

$$| J_a - J_b | < l < J_a + J_b$$

and

$$M_a - M_b = m \tag{4.4}$$

where M_a and M_b are the projections of $\mathbf{J}_a$ and $\mathbf{J}_b$ on the z axis with respect to which the expansions of Eq. 4.2 have been made.

The energy radiated by an electromagnetic source is given by the Poynting vector. Let then a pure multipole of order l, m be emitted; since the coefficients for the electric and magnetic field are related (see Eq. 4.2b), we obtain

$$\frac{dU(l, m; \theta,\varphi)}{d\Omega} = \frac{1}{2\pi\omega^2 c} F_l^m(\theta, \varphi) \, | \, a(l, m) \, |^2 \tag{4.5}$$

where $dU/d\Omega$ is the power emitted into the solid angle $d\Omega$ at the angle θ, φ, and $F_l^m(\theta, \varphi)$ is the angular distribution[†] and is given by

$$F_l^m(\theta, \phi) = \frac{1}{2}\left[1 - \frac{m(m + 1)}{l(l + 1)}\right]| \, Y_l^{m+1} \, |^2$$

$$+ \frac{1}{2}\left[1 - \frac{m(m - 1)}{l(l + 1)}\right]| \, Y_l^{m-1} \, |^2 + \left[\frac{m^2}{l(l + 1)}\right]| \, Y_l^m \, |^2 \tag{4.6}$$

with Y_l^m the spherical harmonics.

4.2 CORRELATION BETWEEN MULTIPOLES

It is clear from Eq. 4.6, and under the assumption of emission of a pure multipole, that the gamma rays arising from nuclear decay will have a special angular distribution with respect to some axis; we may choose this axis to be along the spin $\mathbf{J}_a$ of the initial state. If all nuclei (that is, their spins) in a sample were oriented in the same direction, it would be possible to measure this angular distribution experimentally; if, however, the nuclei are randomly oriented, the distribution of the gamma rays will be, by necessity, isotropic.

[†] See J. D. Jackson, *loc. cit.*

We can obtain an oriented sample of nuclei, for example, by applying a very strong magnetic field† at low temperatures, or by any other means that selectively populates certain m sublevels more than others. However, instead of attempting to orient the sample dynamically, we may select (tag) out of the sample those nuclei whose spin axis is oriented in a particular direction. This is in essence what happens in the γ–γ correlation measurement: the first gamma ray establishes, in a probability sense, the spin axis of the nucleus; then the angular distribution of the second gamma ray (obviously from the same nucleus) is measured with respect to the spin axis, which was so established by the first measurement. To clarify these arguments, the following simple quantitative example is helpful.

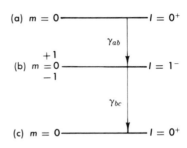

Fɪɢ. 9.26   A possible scheme for the energy levels of a nucleus which is deexcited to the ground state by the cascaded emission of two gamma rays. The spin and parity of each level is indicated to the right, whereas the possible values of the projection of the spin onto the z axis are shown to the left.

Consider a nucleus with the energy level scheme shown in Fig. 9.26. The spin and parity of each state is indicated, and clearly states (a) and (c) can have only $m = 0$, although in the intermediate state the nucleus may have $m = 0, \pm 1$. For the moment we have not determined a z axis, and in general the transition from a to b is a multipole of order

$$l_1 = 1 \text{ (electric)}, \qquad \text{projections } m_1 = -1, 0, +1$$

while the transition from b to c is

$$l_2 = 1 \text{ (electric)}, \qquad \text{projections } m_2 = +1, 0, -1$$

where m_2 and m_1 are correlated, since it must be $m_1 + m_2 = 0$.

Using Eq. 4.6 we find

$$F_1^{-1}(\theta, \varphi) = \tfrac{1}{2} \mid Y_1^0 \mid^2 + \tfrac{1}{2} \mid Y_1^{-1} \mid^2 = \tfrac{3}{4} \cos^2 \theta + \tfrac{3}{8} \sin^2 \theta$$

$$F_1^0(\theta, \varphi) = \tfrac{1}{2} \mid Y_1^1 \mid^2 + \tfrac{1}{2} \mid Y_1^{-1} \mid^2 = \tfrac{3}{4} \sin^2 \theta \qquad (4.7)$$

$$F_1^{+1}(\theta, \varphi) = \tfrac{1}{2} \mid Y_1^0 \mid^2 + \tfrac{1}{2} \mid Y_1^1 \mid^2 = \tfrac{3}{4} \cos^2 \theta + \tfrac{3}{8} \sin^2 \theta$$

Since all three transitions to the intermediate state b have equal probability [that is, the coefficient $a(l, m)$ is independent of m], the angular

† See Chapter 8.

distribution of the first gamma ray is given by the sum of the three distributions of Eq. 4.7. The result is

$$F(\theta, \varphi) = \sum_{m=-1}^{m=+1} F_1^m(\theta, \varphi) = \tfrac{3}{2}$$

an isotropic distribution as it should be.

Let us now fix the z axis along the direction of this first gamma ray[†]. Then immediately we know that the intermediate state b is in an $m = \pm 1$ (and never an $m = 0$) state with *respect* to this axis. This is evident from Eqs. 4.7 if we set $\theta = 0$, because then

$$F_1^1 = F_1^{-1} = \tfrac{3}{4} \qquad \text{and} \qquad F_1^0 = 0$$

This is simply the familiar fact that in a $\Delta m = 0$ transition a photon can never be emitted along the z axis.[‡] Thus, by observing the first gamma ray, we have "tagged" a nucleus which has $m = +1$ or $m = -1$ with respect to the direction of this gamma ray. If this nucleus undergoes now another transition,[§] the emitted radiation will no longer be isotropic; it will contain only the two multipoles

$$a(1, -1), \qquad \text{and} \qquad a(1, +1)$$

and from Eqs. 4.7 we obtain

$$C(\theta) = F_1^{-1}(\theta, \varphi) + F_1^1(\theta, \varphi) = \tfrac{3}{2} \cos^2 \theta + \tfrac{3}{4} \sin^2 \theta = \tfrac{3}{4}(1 + \cos^2 \theta)$$

$$(4.8)$$

This method of calculating angular correlations can be extended to higher spins. In general we combine $F_{l_1}^{m_1}(\theta = 0)$ and $F_{l_2}^{m_2}(\theta, \varphi)$ with the appropriate coefficients, where l_1 and l_2 are the orders of the multipoles emitted in the first and second transition, respectively. Next, we must sum over all substates m_a, m_b, and m_c (of the initial, intermediate, and final state) in such a fashion that $m_a - m_1 = m_b = m_c + m_2$. This topic cannot be pursued further here, but for a very clear presentation see H. Fraunfelder's article on "Nuclear Angular Correlation" in *Beta and Gamma-ray Spectroscopy* by K. Siegbhan, North Holland Co., 1955.

[†] Clearly this introduces no loss of generality since the distribution of this gamma ray is isotropic.

[‡] See, for example, Chapter 7 on the polarization of radiation emitted in the Zeeman effect.

[§] That is, provided that this second transition takes place in a short enough time interval so that the probability of random reorientation of the nucleus (while in the intermediate state) is small.

By the more powerful technique of group theory a general expression for the angular correlation function can be obtained

$$C(\theta) = 1 + \sum_{k=1}^{k_{max}} A_{2k}P_{2k}(\cos\theta) \qquad (4.9)$$

where k_{max} must be the smallest of I_b, l_1 or l_2. Here $P_{2k}(\cos\theta)$ is the Legendre polynomial of order $2k$. The coefficients A_{2k}, as before, can be predicted theoretically.

For Co^{60}, the decay scheme has been given in Fig. 9.19, and we note that

$$I_a = 4^+, \qquad I_b = 2^+, \qquad I_c = 0^+$$

The multipoles are $l_1 = 2^+$ and $l_2 = 2^+$—that is, electric quadrupole in both cases. Thus the highest order will be $\cos^4\theta$ and we can write

$$C(\theta) = 1 + a_1\cos^2\theta + a_2\cos^4\theta$$

The coefficients are found to be $a_1 = \frac{1}{8}$ and $a_2 = \frac{1}{24}$.

Conversely, if the spins of the nuclear states are not known, an experimentally observed angular correlation yields information on l_1 and l_2, which in turn are helpful in inferring or determining, I_a, I_b, and I_c.

5. Velocity Measurements and Particle Identification by Time of Flight Techniques

5.1 EXPERIMENTAL SETUP

The measurement of the velocity of fast particles traversing a counter telescope will be described in this section. Although such an experiment can be performed with cosmic rays, it is much more instructive when performed with a beam of monoenergetic particles. Such beams, of sufficient energy to traverse a counter telescope, are produced by particle accelerators, and therefore are not available in a teaching laboratory. Here, however, we will deviate from the practice followed in this text, and will consider results obtained with beams produced by the *Cosmotron*, the 3-BeV proton-synchrotron at Brookhaven National Laboratory.

In Fig. 9.27 is shown a section of this accelerator including the location of the internal target. The 3-BeV protons circulating in the Cosmotron are made to strike the target, from which are now emitted scattered protons, target protons, and neutrons, and unstable particles produced in the collision such as π mesons, and few K mesons. By providing an appropriate hole in the shielding and by the use of collimators, the particles emitted at 32 degrees were formed into a beam.† The collimated beam is then

---

† These beams are made as parallel as possible. By using focusing "magnetic quadrupoles" which act like lenses, it is possible to improve greatly the intensity of the beam.

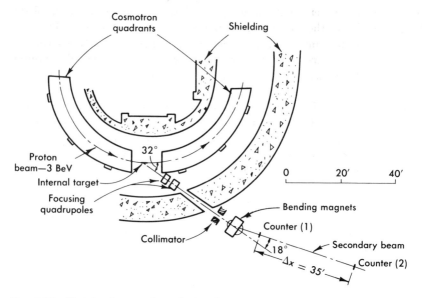

FIG. 9.27 Sketch of a portion of the *Cosmotron*, a 3-BeV proton synchrotron (accelerator). Particles produced in the interaction of the proton beam with an internal target are formed into a beam at 32 degrees with respect to the primary beam direction. A pair of quadrupoles is used to focus the beam and a bending magnet provides momentum selection. Counters (1) and (2) form a telescope which is also used for measurement of the time of flight between the two counters.

allowed to traverse a large magnet, where each momentum is bent differently according to the well-known expression

$$2 \sin \frac{\theta}{2} = \frac{0.3\, Bl}{p} \tag{5.1}$$

where

B is the field in kilogauss,

l is the effective length of the magnet in centimeters, and

p is the momentum of the particle in MeV/c.

We can select a particular angle of bend by using either a collimator or a counter telescope; the selection of the bending angle defines the momentum of the beam. Further, for any particle, the momentum uniquely determines its velocity, since

$$\beta = \frac{p}{E} = \frac{p}{\sqrt{p^2 + m^2}} = \left[1 + \frac{m^2}{p^2}\right]^{-1/2} \tag{5.2}$$

so that a measurement of the velocity (with known momentum) establishes the mass and thus the identity of the particles. The simplest way to measure velocity is to determine the time (of flight) necessary for the particle to traverse the distance between two counters. The counters determining the flight path in this case are shown in Fig. 9.27 as (1) and (2); they are separated by 35 ft.

To measure the time of flight a fast coincidence system as described in Section 3.2 is used. In order to calibrate the system, the two counters are moved physically so that one is next to the other, and a delay curve is taken. Since the passage of all particles through both counters is simultaneous (transit time less than 10^{-9} sec), the coincidence circuit is "timed" at the center of the delay curve and this is called *time zero*. Now, *without* changing any cable lengths the counters are moved to their appropriate positions (1) and (2) separated by 35 ft; clearly no coincidences will be registered,† since even for particles moving with $\beta = 1$ the signal in counter 2 will be generated at a time

$$t_2 = \frac{35 \text{ ft}}{c} = 35 \times 10^{-9} \text{ sec}$$

which is *later* than the time $t_1 = 0$ at which the signal was generated in counter 1. If we now begin to insert delay in channel 1 (the *upstream* counter), we will eventually bring the two signals into coincidence and the circuit will register counts. However, the delay necessary for obtaining coincidences directly measures $t_2 - t_1$, that is, the time of flight.

5.2 Velocity Measurement with a Conventional Coincidence Circuit

A beam of positive particles,‡ such as described above, emerging from the target of an accelerator of sufficiently high energy, contains in varying proportions the known stable and long lived particles, namely protons, π^+ mesons, K^+ mesons, μ^+ mesons, and positrons; the μ mesons and positrons are not produced directly in the target but the μ mesons originate from the decay of π^+ mesons while the positrons are due mainly to materialization (pair production) of high-energy gamma rays resulting from π^0-meson decay.

We will examine two momenta, namely 600 MeV/c and 1100 MeV/c; the corresponding velocities for the various particles are summarized in Table 9.4, which also includes the flight time for a distance of 35 ft.

---

† Provided that the resolving time of the circuit is shorter than $t_2 - t_1$, and excluding accidentals.

‡ Positive particles are selected by using the appropriate polarity of the bending magnet.

TABLE 9.4

VELOCITIES AND FLIGHT TIMES (FOR $\Delta x = 35$ FT) FOR DIFFERENT PARTICLES

Particle	Mass (MeV/c²)	600 MeV/c		1100 MeV/c	
		$\beta = v/c$	Δt (nsec)	$\beta = v/c$	Δt (nsec)
Proton	938	0.539	65.0	0.762	46.0
K^+ meson	494	0.773	45.3	0.912	38.4
π^+ meson	140	0.973	36.0	0.993	35.3
μ^+ meson	105	0.985	35.5	0.995	35.0
Positron	0.5	≈ 1.00	35.0	≈ 1.00	35.0

We note that the difference in flight times between protons and π mesons is 29 nsec for 600 MeV/c and 10.7 nsec at the higher momentum of 1100 MeV/c. These intervals are longer than, or of the order of, the resolving time of the coincidence system and it is therefore possible to separate these two components of the beam. It is clear, however, that the separation of positrons and μ mesons from π mesons cannot be achieved with this technique at such high momenta†. The K mesons are an intermediate case, but their relative abundance in the beam is so small ($K/\pi \approx 10^{-4}$) that a special high-resolution technique is required (this is described in Section 5.3).

The delay curves obtained at the two momenta are shown on a semilog plot in Figs. 9.28 and 9.29.‡ The horizontal scale gives the relative delay in feet of cable (RG 58/U), and we note that indeed the peaks for the π mesons and protons appear at the correct positions. What is more important is the separation between the two peaks, which is very clear at 600 MeV/c and still distinct at 1100 MeV/c; at the high momentum, however, it approaches the limit of this system.

We note that the resolving time (full width at half maximum of the delay curves) of the system is of the order of 6 nsec while the efficiency is still good (flat top of delay curve). From the Figs. 9.28 and 9.29 we can appreciate the usefulness of a short resolving time and of steep slopes in this application of the time coincidence technique.

5.3 VELOCITY MEASUREMENT WITH A TIME-TO-PULSE-HEIGHT CONVERTER

In order to improve the resolution in time beyond that shown in Figs. 9.28 and 9.29, an "analogue" coincidence circuit is used. In the procedure

---

† See Chapter 5, Section 6.2, on Cherenkov counters, for measurement of velocities $\beta \approx 1$.

‡ Courtesy of Drs. R. Ellsworth and M. Tannenbaum.

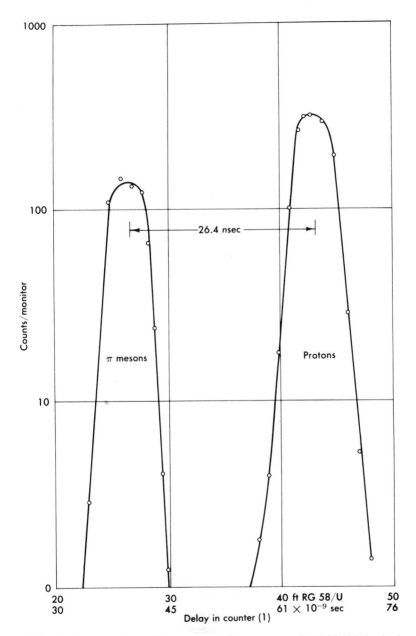

Fɪɢ. 9.28 Delay curve for positive particles of momentum of 600 MeV/c in the beam shown in Fig. 9.27. Note that the π mesons and the protons are clearly separated into two groups differing in flight time by 26.4 nsec as expected when the distance of flight is 35 ft.

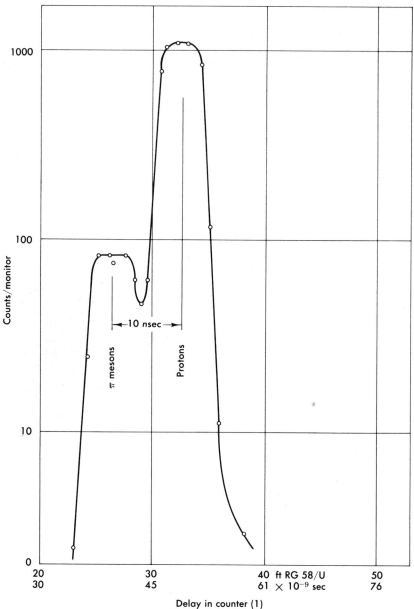

FIG. 9.29 Delay curve obtained under identical conditions as in Fig. 9.28 but for a momentum of 1100 MeV/c. Note that now the separation between π mesons and protons is only 10 nsec, again in agreement with the expected value for a flight path of 35 ft. Note, however, that the relative number of π mesons to protons in the beam is quite different at this momentum from the value it had at the lower momentum of 600 MeV/c.

considered here, the output from counters 1 and 2 of Fig. 9.27, after having been shaped in discriminators (Nanocard Manual unit IH–51), was fed to the time-to-pulse-height converter (Nanocard Manual unit IH–134) described in Section 2.4. The distance between the two counters was now reduced to 16 ft.

The output of the converter circuit was fed to a 400-channel pulse-height analyzer,† where it was sorted and cumulatively stored in the memory. After sufficient data had been collected, the contents of the memory were printed out and also displayed on an oscilloscope. Figure 9.30 is such a photograph of the pulse-height distribution obtained at a beam momentum of 1100 MeV/c. The scale is linear, and the ordinate represents 1000 counts per channel full scale while the abscissa uses only 200 channels full scale; each division of the grid shown in Fig. 9.30 is equivalent to 10 channels.

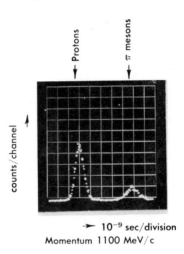

Fig. 9.30   200-channel display of the output of the time-to-pulse-height converter. The abscissa gives pulse height (which is equivalent to time difference between the two counters) and the ordinate gives the number of pulses in a particular channel. We note again two clearly separated groups corresponding to π mesons and protons, with a difference in flight time of approximately 5 nsec. This is indeed the expected difference for the momentum of 1100 MeV/c at which these data were obtained and for a separation of 16 ft between counters.

10⁻⁹ sec/division

Momentum 1100 MeV/c

In order to calibrate the abscissa in time units, we insert a delay (cable of length L_0) into one of the input channels to the converter; the whole pattern of Fig. 9.30 is then shifted by a time interval $\Delta t = L_0/\beta c$. The time scale is, in general, linear except near maximum output (which corresponds to perfect overlap). In this way the horizontal displacement is found to be equivalent to 1 nsec/division of the grid.

In Fig. 9.30 we note two peaks which indicate that the beam contains at least two types of particles with distinct transit times (from counter 1 to counter 2). The difference in transit times is given by the separation of the peaks, which is 4.8 nsec; by consulting Table 9.5 we see that at 1100

† R.I.D.L. Model 34-12b, manufactured by Nuclear Chicago Co., 4501 West North Ave., Melrose Park, Illinois.

TABLE 9.5

VELOCITIES AND FLIGHT TIMES (FOR $\Delta x = 16$ FT) FOR DIFFERENT PARTICLES

Particle	Mass (MeV/c^2)	760 MeV/c		1100 MeV/c	
		$\beta = v/c$	Δt (nsec)	$\beta = v/c$	Δt (nsec)
Protons	938	0.630	25.4	0.762	21.0
K^+ mesons	494	0.838	19.3	0.912	17.5
π^+ mesons	140	0.983	16.3	0.993	16.1
μ^+ mesons	105	0.991	16.1	0.995	16.0
Positrons	0.5	≈ 1.00	16	≈ 1.00	16

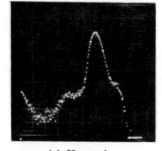

(a) Ungated

FIG. 9.31 Display of the time-to-pulse-height converter output obtained under similar conditions as in Fig. 7.30 but for a momentum of 760 MeV/c. (a) No special selection is made of the signals that are fed to the converter. Note that the proton peak has now moved out of the display because the time-of-flight difference exceeds the range of the converter. The shoulder to the left of the π-meson peak is due to K^+ mesons. (b) The same display as in (a) above but the converter is now "gated." Only signals that have been identified, in a Cherenkov counter, as corresponding to particles with the velocity appropriate to a K meson are fed to the converter. The peak corresponding to the K mesons is now very clear whereas only the few π mesons that were not rejected by the Cherenkov counter appear in a separate group.

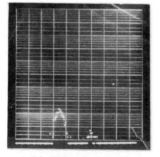

(b) Cherenkov gated
10^{-9} sec/division
momentum 760 MeV/C

MeV/c and for 16 ft flight path this corresponds closely to the difference in the time of flight between protons and π mesons†. The left-hand peak is due to protons and the right-hand peak to π mesons as indicated in the figure‡.

The width of the peaks in Fig. 9.30 is a measure of the time jitter of the signals and of the ultimate resolution of the system; in the present example the full width at half maximum is approximately§ 1 nsec. Finally, in order to compare the relative number of protons and π mesons, we must *integrate* the area under the peaks rather than take the peak values.

Figures 9.31a and 9.31b are similar displays of converter output, but on a semilogarithmic scale. They were taken at a momentum of 760 MeV/c and the time scale has not been changed.¶ The shift of the π meson peak in Fig. 9.31a is irrelevant (due to instrumentation), but the proton peak is shifted (beyond the range of the converter) since now the difference in flight time between π mesons and protons (see Table 9.5) is 9 nsec. Between the two peaks we recognize a shoulder at a position corresponding to particles with a flight time 2.5 nsec longer than that for π mesons. We identify this shoulder as due to K^+ mesons, whose expected flight-time difference is 3 nsec.

That this shoulder is indeed due to K^+ mesons can be seen from Fig. 9.31b. In this instance a differential Cherenkov counter* (with a liquid radiator) was included in the telescope. The Cherenkov counter selected only particles with $\beta = 0.84$, that is, K^+ mesons at this momentum of 760 MeV/c; its output was then used to "gate" the time-to-pulse-height converter. Indeed, in Fig. 9.31b most of the counts have the correct timing (that is, pulse height) for K^+ mesons, and there are also a few counts due to π^+ mesons that have "fed through" the Cherenkov counter.

---

† More precisely, the correspondence is between protons and all particles with $\beta \approx 1$, that is, π^+, μ^+, and e^+.

‡ This is not always so; the relative position of the "early" to the "late" peak depends on the adjustment of the delay of the input signals.

§ When testing the system with a pulser, the width of the peaks is usually only one or two channels.

¶ Since maximum output represents complete overlap, and zero output no overlap at all, in order to expand the time scale (for example, to 0.5 nsec/division) we shorten the width of the input signals (halve it) and raise the output gain (by a factor of two).

* See Chapter 5, Section 6.2.

10

ELEMENTS FROM THE THEORY
OF STATISTICS

1. Definitions

Statistics is the science that tries to draw inferences from a finite number of observations constituting only a sample, so as to postulate rules that apply to the entire population from which the sample was drawn.

In the field of physics, statistics is needed (a) to fit data—that is, to estimate the parameters of assumed frequency functions; (b) to treat random errors; and (c) to treat events that are inherently of a statistical nature.

1.1 Definition of Probability

The probability of occurrence of an event can be axiomatically defined as equal to one (=1) if the event occurred, or equal to zero (=0) if the event did not take place. An alternative definition of probability is based on the *frequency* of occurrence of an event. Suppose that several trials of the same experiment have been made; then the probability of occurrence of an event A, that is, $P(A)$, is given by the number of times event A was obtained divided by the total number of trials (in the limit that the total number of trials → ∞). This definition of probability retains its full value even in the case of nonrepetitive experiments, since the one trial can be considered as the first of a series of trials.

1.2 SAMPLE SPACE

Any *set of points* that represents all possible outcomes of an experiment is a sample space. For example, if a coin is tossed twice, the sample space consists of the 4 points indicated in Fig. 10.1. (Sample spaces can be finite or infinite and discrete or continuous.)

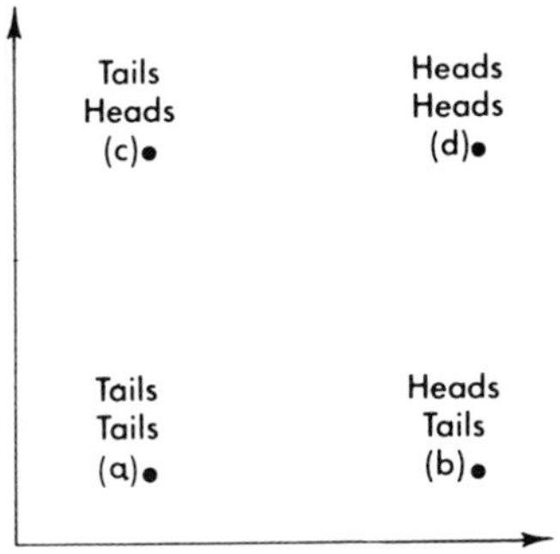

FIG. 10.1 Simple example of a discrete and finite sample space. Here the sample-space points correspond to all possible outcomes of "tossing a coin" twice.

```
Tails        Heads
Heads        Heads
(c)•         (d)•


Tails        Heads
Tails        Tails
(a)•         (b)•
```

Once the sample space for a particular experiment is constructed, we may assign in the sense of Definition 1.1 a probability p_i to each point i of the space. From the definition, we have

$$p_i > 0 \qquad \sum_{\substack{\text{all sample} \\ \text{space points}}} p_i = 1$$

thus

$$p_i \leq 1$$

and the probability of occurrence of an event A is

$$P(A) = \frac{\sum_A p_i(A)}{\sum p_i} = \sum_A p_i(A)$$

where $\sum_A$ indicates summation over all points that include event A.

In most situations treated by statistics, equal probability is assigned to each sample-space point; a condition we will maintain throughout this discussion. Then

$$p_i = \frac{1}{n}$$

n being the total number of sample-space points, and

$$P(A) = \frac{n(A)}{n}$$

where $n(A)$ is the number of sample-space points containing event A.

For example, in the case of the sample space of Fig. 10.1, the probability of obtaining heads at least once is

$$P(\text{heads}) = \frac{n(\text{heads at least once})}{n} = \frac{3}{4}$$

while the probability of obtaining heads once and tails once (irrespective of order) can again be found by counting the appropriate points in the sample space of Fig. 10.1. We obtain

$$P(\text{heads, tails}) = \frac{n(\text{heads, tails})}{n} = \frac{2}{4}$$

1.3 PROBABILITY FOR THE OCCURRENCE OF A COMPLEX EVENT

The probability that both events A and B will occur is called the joint probability

$$P[AB] = \frac{n(A \text{ and } B)}{n}$$

where n = total number of sample-space points. The probability that either A or B will occur is called the either probability

$$P[A + B] = \frac{n(A \text{ or } B)}{n}$$

and the probability that A will occur when it is certain that B occurred is called the conditional probability

$$P[A \mid B] = \frac{n(A \text{ and } B)}{n(B)}$$

All these probabilities are defined in the sense of 1.1 as the number of sample-space points that contain the stated condition divided by the total number of sample-space points *allowed for*, by the statement.

Figures 10.2a and 10.2b illustrate two sample spaces. All points within domain Ⓐ include event A while all points within domain Ⓑ include events B. The points contained in any intersection of the two domains Ⓐ and Ⓑ include both events A and B.

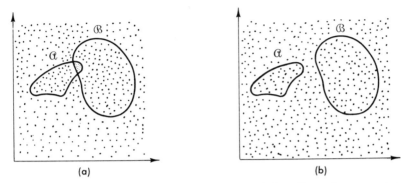

Fɪɢ. 10.2   In the sample spaces shown it is assumed that all sample-space points in domain α contain event A, whereas all points in domain $\mathcal{B}$ contain event B. (a) There exists a region where both event A and event B can occur simultaneously. (b) No such region exists; events A and B are mutually exclusive.

If such a common intersection does not exist in sample space, the two events are *mutually exclusive*, and

$$P[AB] = 0$$

It follows from the consideration of Figs. 10.2 that

$$P[A + B] = P[A] + P[B] - P[AB]$$

For the conditional probability

$$P[A \mid B] = \frac{n(\alpha \text{ intersection } \mathcal{B})}{n(\mathcal{B})}$$

since the condition that event B occurred restricts our sample within domain $\mathcal{B}$. However,

$$P[B] = \frac{n(\mathcal{B})}{n}$$

and

$$P[AB] = \frac{n(\alpha \text{ intersection } \mathcal{B})}{n} \begin{cases} = P[A \mid B] \cdot P[B] \\ \\ = P[B \mid A] \cdot P[A] \end{cases} \tag{1.1}$$

If $P[A \mid B] = P[A]$, it means that the occurrence of B does not affect the probability of occurrence of A. We say that the two events A and B are *independent*. It then follows from Eq. 1.1 that

$$P[AB] = P[A] \cdot P[B] \tag{1.2}$$

Equation 1.2 in turn implies (when combined with Eq. 1.1) that for independent events

$$P[B \mid A] = P[B]$$

To illustrate some of the ideas expressed above, consider the following. For the sample space of Fig. 10.1 we may define: event A = heads in

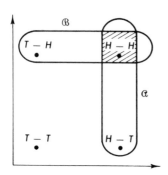

FIG. 10.3   The sample space of Fig. 10.1 including the domain α (heads in the first throw) and the domain $\mathcal{B}$ (heads in the second throw).

first throw; event B = heads in second throw. The domains are shown in Fig. 10.3, and it follows (assigning $p = \frac{1}{4}$ to each point)

$$P[A] = \tfrac{1}{2}; \qquad P[B] = \tfrac{1}{2}$$

$$P[AB] = \tfrac{1}{4}$$

$$P[A + B] = P[A] + P[B] - P[AB] = \tfrac{1}{2} + \tfrac{1}{2} - \tfrac{1}{4} = \tfrac{3}{4}$$

$$P[A \mid B] = \tfrac{1}{2} \qquad P[B \mid A] = \tfrac{1}{2}$$

$$P[AB] = P[A \mid B] \cdot P[B] = \tfrac{1}{2} \cdot \tfrac{1}{2} = \tfrac{1}{4} = P[A] \cdot P[B]$$

Thus events A and B are *not* mutually exclusive but are independent.

1.4 RANDOM VARIABLE

To study a sample space analytically (instead of geometrically), it is convenient to use a numerical variable which takes a definite value for each and *every* point of the sample space; however, the same value may be assigned to several points. Thus, a random variable used for the representation of a finite and discrete sample space will have a definite range and will take only discrete values. As an example, for the sample space of Fig. 10.1, we can make the following assignment if we are *not* interested in the *order* of occurrence of events.

Sample-space point	x (random variable)
(a)	-1
(b)	0
(c)	0
(d)	$+1$

1.5 Frequency Function

Frequency function (of a random variable) is a function $f(x)$ such that $f(x_0)$ is the probability that the random variable x may take the specific value x_0. By Definition 1.1, $f(x)$ gives the number of points in the sample space that have been assigned the value x of the random variable, divided by the total number of sample-space points. The function $f(x)$ is defined only within the range of x and need not have a definite analytic form. For the example considered above (the sample space of Fig. 10.1), $f(x)$ is just a table, as shown in Table 10.1 (see also Fig. 10.4).

TABLE 10.1

Example of a Frequency Function $f(x)$ of the Random Variable x

Sample-space point	x	$f(x)$
(a)	-1	$\frac{1}{4}$
(b)	0	$\left.\begin{array}{c} \\ \\ \end{array}\right\} \frac{1}{2}$
(c)	0	
(d)	$+1$	$\frac{1}{4}$

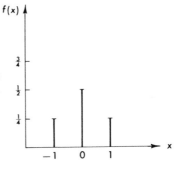

Fig. 10.4   The distribution function of the discrete random variable x defined in Table 10.1 of the text.

The summation of $f(x)$ over the entire range of x must give 1.

$$\sum_{\text{all } x} f(x) = 1$$

The probability that the random variable may take any value smaller or equal to x is given by

$$F(x) = \sum_{t < x} f(t)$$

and is called the *distribution function* of x (or integral distribution function).

It is sometimes convenient to describe a sample space in terms of two or more random variables, a frequency function existing for each of them. If these random variables are independently distributed in the sense of Eq. 1.2, the joint frequency function is

$$f(x_1, x_2 \cdots) = f(x_1) f(x_2) \cdots f(x_n)$$

If the random variable is continuously varying (for example, describes the height of individuals), the probability of occurrence of the specific value x when a measurement is performed defines the frequency function $f(x)\, dx$ of the random variable x. The random variable may now take any value within the range of its definition. Note, however, that the probability of occurrence of the exact value x is zero, while it is the probability of occurrence of some value in the infinitesimal interval dx about x that exists. For a continuously varying random variable, we have

$$f(x) \geq 0 \quad \text{and} \quad \int_{-\infty}^{+\infty} f(x)\, dx = 1$$

Similarly

$$\int_{a}^{b} f(x)\, dx = P[a < x < b] \quad a < b$$

and

$$F(x) = \int_{-\infty}^{x} f(t)\, dt$$

1.6 SOME DEFINITIONS FROM COMBINATORIAL ANALYSIS

(a) *Permutations.* A permutation of n objects in groups of r objects is defined as follows: Consider n objects; any group of r of these objects, when ordered, forms a permutation; the same group of r objects, when ordered in a different fashion, forms a *new* permutation. As an example consider the three objects:

$$\square, \triangle, \bigcirc$$

There are only six possible permutations of three objects in groups of two. They are shown below.

$$\square\triangle, \quad \triangle\square; \quad \square\bigcirc, \quad \bigcirc\square; \quad \triangle\bigcirc, \quad \bigcirc\triangle$$

We state without proof that the number of possible permutations of n objects in groups of r, $_nP_r$, is

$$_nP_r = n(n-1)\cdots(n-r+1) = \frac{n!}{(n-r)!}$$

Then

$$_nP_n = n!$$

as it must be.

(b) *Combinations.* A combination of n objects in groups of r objects is defined as any grouping of r objects out of the original n. The ordering within the grouping is not relevant. Thus for the previous example there are only three possible combinations

$$\square\triangle; \quad \bigcirc\square; \quad \bigcirc\triangle$$

The number of possible combinations of n objects in groups of r, $\begin{bmatrix} n \\ r \end{bmatrix}$, is

$$\begin{bmatrix} n \\ r \end{bmatrix} = \frac{_nP_r}{_rP_r} = \frac{n!}{r!(n-r)!}$$

(c) Note

$$n! = n\cdot(n-1)! \qquad 1! = 0! = 1$$

2. Frequency Functions of One Variable

2.1 DEFINITIONS

Let us assume that a population (for example, all the possible outcomes of an experiment) can be described by a frequency function; we may attempt to find this function in two ways:

(a) By the use of a mathematical model based on the definitions of the previous section, thus obtaining a *"theoretical frequency function."*

(b) By observing a sample of the population and determining its "*empirical frequency function.*"

The advantage of obtaining a frequency function for a population is that the few parameters involved in the frequency function suffice to describe completely the *entire* population and thus provide as much information as the most extensive data.

We will now deal only with populations that can be described by a frequency function depending on a single variable. To obtain the empirical frequency function it is best to divide the members of the sample in classes (defined by the random variable) and then make a graphical plot or *histogram* of the sample. If we try to describe the histogram, the first obvious features are its location and its spread.

A very useful set of measures are the *moments* of a histogram, defined in the usual way (moments of forces, electric moments, etc.). Thus, if x_i is the value of the random variable for the class i and if f_i is the number of events in this class, the kth moment of the *empirical* frequency function about the origin is

$$m'_k = \frac{1}{n} \sum_{\text{all } i} x_i^k f_i$$

where n is the size of the sample. Similarly, the kth moment about any other point x_0 is

$$m_k(x_0) = \frac{1}{n} \sum_{\text{all } i} (x_i - x_0)^k f_i$$

2.2 MEAN AND STANDARD DEVIATION

The first moment about the origin, m'_1 is called the mean, and will be denoted by m:

$$m = m'_1 = \frac{1}{n} \sum_{\text{all } i} x_i f_i \tag{2.1}$$

(commonly called the "average" of x). The second moment about the mean, m_2, is called the *variance*; its square root is called the *standard deviation* and is denoted by s; s has the same dimensions as the random variable x.

$$s = \sqrt{m_2} = \sqrt{\frac{1}{n} \sum_{\text{all } i} (x_i - m)^2 f_i} \tag{2.2}$$

An often used relation pertaining to s is

$$s^2 = \frac{1}{n} \sum_{\text{all } i} (x_i - m)^2 f_i = \frac{1}{n} \sum (x_i^2 - 2mx_i + m^2) f_i$$

$$s^2 = \frac{1}{n} \sum_{\text{all } i} x_i^2 f_i - \frac{2m}{n} \sum (x_i f_i) + m^2$$

$$s^2 = \frac{1}{n} \sum_{\text{all } i} (x_i^2 f_i) - m^2$$

usually written as

$$\overline{\Delta x^2} = \overline{x^2} - (\bar{x})^2 \tag{2.3}$$

In most cases the mean and the standard deviation are the best measures (contain most information) of an empirical frequency function; there are, nevertheless, cases where they are very poor measures and instead it is much better to give other location measures, such as the median or the geometric mean, and so on, and other variation measures such as the range or the mean variation $[(1/n) \sum |x_i - m| f_i]$, and so on.

2.3 THEORETICAL FREQUENCY FUNCTIONS

As mentioned before, a theoretical frequency function $f(x)$ might be of the *discrete* type—that is, the random variable x takes only integer values or of the "*continuous*" type. Most of the discrete random variables usually represent the number of successes, or of counts obtained, etc. In going from discrete frequency functions to continuous ones, obviously all summations are replaced by integrals.

Moments are defined as in Eq. 2.1, but instead of the empirical frequencies f_i, the theoretical frequency function $f(x)$ is used; the theoretical moments are designated by Greek letters, Latin letters being reserved for the empirical moments.

Thus, the kth moment about the origin is

$$\mu'_k = \sum_{x=-\infty}^{+\infty} x^k f(x)$$

The first moment about the origin gives the mean, and is denoted by $\mu = \mu'_1$. The kth moment of a theoretical frequency function about its mean is

$$\mu_k = \sum_{x=-\infty}^{+\infty} (x - \mu)^k f(x)$$

The square root of the second moment about the mean gives the standard deviation and is denoted by $\sigma = \sqrt{\mu_2}$

$$\mu_2 = \sum_{x=-\infty}^{+\infty} (x - \mu)^2 f(x)$$

2.4 THE BERNOULLI OR BINOMIAL FREQUENCY FUNCTION

This basic frequency function is applicable when there are only *two* possible outcomes of an experiment, as for example the occurrence of an event A, or its nonoccurrence (we designate this by B). If the experiment is repeated n times, the random variable x describes the number of times event A occurred. The frequency function—that is, the probability of obtaining a certain x—is given by

$$f(x) = \frac{n!}{x!(n-x)!} p^x q^{n-x} \tag{2.4}$$

where p is the probability that even A will occur in this experiment (defined in the sense of Section 1.1); and $q = 1 - p$ is the probability that B will happen, namely, that event A will not occur.

To prove Eq. 2.4, consider the probability of obtaining event A, x times in a definite sequence

$$\underbrace{A A \cdots A}_{x} \quad \underbrace{B B \cdots B}_{n-x}$$

this joint probability of order n is according to Definition 1.3,

$$\underbrace{p p \cdots p}_{x} \underbrace{q q \cdots q}_{n-x} = p^x q^{n-x}$$

since the outcome of consecutive experiments is independent. However, any other sequence, containing the same number x of occurrences, is also a satisfactory answer, since we are not interested in the order of occurrence of event A. Thus we must sum over all sample space points that give x occurrences; the number of all such sample-space points is given by the permutations of n objects in groups of n when x of them are alike (have probability p), which is

$$\frac{n!}{x!(n-x)!}$$

completing the proof of Eq. 2.4.

The frequency function fulfills the normalization requirement as it should, since

$$\sum_{x=0}^{n} f(x) = \sum_{x=0}^{n} \frac{n!}{x!(n-x)!} p^x q^{n-x} = (p + q)^n = [p + (1 - p)]^n = 1$$

$$(2.5)$$

2.5 MOMENTS OF THE BINOMIAL FREQUENCY FUNCTION

From the definitions of Section 2.3, and since the range of x is from 0 to n, we have

$$\mu = \mu'_1 = \sum_{x=0}^{n} xf(x) = \sum_{x=0}^{n} x \frac{n!}{x!(n-x)!} p^x q^{n-x}$$

$$= \sum_{x=1}^{n} x \frac{n!}{x!(n-x)!} p^x q^{n-x}$$

$$= np \sum_{x=1}^{n} \frac{(n-1)!}{(x-1)!(n-x)!} p^{x-1} q^{n-x}$$

if we let $y = x - 1$, it follows

$$\mu = np \sum_{y=0}^{n-1} \frac{(n-1)!}{y![(n-1)-y]!} p^y q^{[(n-1)-y]}$$

where now the sum is equal to $(p + q)^{n-1} \equiv 1$. Thus

$$\mu = np \tag{2.6}$$

Next we wish to obtain the second moment about the mean, $\mu_2 = \sigma^2$. We first calculate μ'_2, given by

$$\mu'_2 = \sum_{x=0}^{n} x^2 \frac{n!}{x!(n-x)!} p^x q^{n-x}$$

We use

$$x^2 = x(x - 1) + x,$$

so that

$$\mu'_2 = \sum_{x=0}^{n} x(x-1)\, \frac{n!}{x!(n-x)!}\, p^x q^{n-x} + \mu$$

$$= \sum_{x=2}^{n} x(x-1)\, \frac{n!}{x!(n-x)!}\, p^x q^{n-x} + \mu$$

$$= n(n-1)p^2 \sum_{x=2}^{n} \frac{(n-2)!}{(x-2)!(n-x)!}\, p^{x-2} q^{(n-x)} + \mu$$

and letting $y = x - 2$, as before, the sum is equal to $(p + q)^{n-2} \equiv 1$, and we obtain

$$\mu'_2 = n(n-1)p^2 + \mu = n^2 p^2 - np^2 + np$$

Next we use Eq. 2.3 to obtain

$$\mu_2 = \sigma^2 = \mu'_2 - \mu^2 = -np^2 + np = np(1-p) = npq$$

Thus

$$\sigma = \sqrt{npq} \qquad (2.7)$$

The binomial frequency function is applicable to many physical situations, but it is cumbersome to calculate with. When n becomes large, however, the binomial frequency function approaches either the Poisson or the Gaussian frequency function which will be discussed in Sections 2.6 and 2.7. In order for the binomial frequency function[†] to approach the:

Poisson distribution n must be large, for example, $n > 100$,
                           but $\mu = np$ finite and small, for example, $p < 0.05$.

Gaussian distribution n must be large, for example, $n > 30$,
                           and also p large, for example, $p > 0.05$.

2.6 THE POISSON FREQUENCY FUNCTION

This is still a frequency function for the discrete random variable x which describes, as in Section 2.4, the number of times event A will be obtained if the experiment is repeated n times when $n \to \infty$ (for large n). Contrary to Eq. 2.4, however, neither n nor p appear explicitly in the analytic expression of the frequency function, but instead only their product

$$y = np \qquad (2.8)$$

---

† See, however, the detailed discussion in Section 2.9.

which remains finite in spite of $n \to \infty$, since $p \to 0$. The Poisson frequency function is given by

$$f(x) = \frac{y^x e^{-y}}{x!} \tag{2.9}$$

and it is shown in the next section that y is the mean of Eq. 2.9.

To prove Eq. 2.9, let us first note that since n is large, it (but not x) may be treated as a continuous variable; second, we will assume that for a small (differential) number of trials dn, the probability of obtaining event A once is proportional to this number of trials: that is,

$$P\{1, dn\} = \lambda \, dn \tag{2.10}$$

where λ is a constant. Note that Eq. 2.4 fulfills this requirement for $x = 1$ in the limit that $p \to 0$ or $q \to 1$. In terms of sample space our assumption means that the density of sample-space points containing event A is uniform in the limit of a differential element of sample-space area. The Poisson frequency function then follows for all populations for which assumption 2.10 is valid.

Let $P\{x, n\}$ be the probability of obtaining event A, x times in n trials, so that $P\{0, n\}$ is the probability of obtaining no events A in n trials. Then the probability of obtaining no events in $n + dn$ trials is

$$P\{0, n + dn\} = P\{0, n\} \cdot [1 - P\{1, dn\}]$$

since the events are independent.† Using Eq. 2.10 we obtain

$$\frac{P\{0, n + dn\} - P\{0, n\}}{dn} = -P\{0, n\} \cdot \lambda$$

or

$$-\frac{dP\{0, n\}}{dn} = P\{0, n\} \cdot \lambda$$

which has the solution

$$\ln P\{0, n\} = -n\lambda$$

$$P\{0, n\} = e^{-n\lambda} \tag{2.11}$$

---

† Since the increase in the number of trials dn is differential, the possibility of obtaining more than one event in dn is excluded.

and use has been made of the initial condition that for $n = 0$

$$P\{0, 0\} = 1$$

In a similar manner we obtain

$$P\{1, n + dn\} = P\{1, n\} \, P\{0, dn\} + P\{0, n\} \, P\{1, dn\}$$

where the two possible *either* probabilities are summed. Making use again of Eq. 2.10, we may write the above result as

$$P\{1, n + dn\} = P\{1, n\} \cdot [1 - \lambda \, dn] + P\{0, n\} \cdot \lambda \, dn$$

by further transforming and using Eq. 2.11 as well,

$$\frac{dP\{1, n\}}{dn} + \lambda P\{1, n\} - \lambda e^{-n\lambda} = 0$$

The solution of this linear first-order equation is straightforward, leading to

$$P\{1, n\} = e^{-n\lambda} \left[\int e^{n\lambda} \lambda e^{-n\lambda} \, dn + C \right] = (n\lambda) e^{-n\lambda} \qquad (2.12)$$

making use of the initial condition $P\{1, 0\} = 0$.

In general the following recursion formula holds:

$$\frac{dP\{x, n\}}{dn} + \lambda P\{x, n\} - \lambda P\{(x - 1), n\} = 0$$

which is satisfied by

$$f(x) = P\{x, n\} = \frac{(\lambda n)^x e^{-n\lambda}}{x!} \qquad (2.13)$$

as can be verified by substitution.

Thus Eq. 2.9 has been proven, and we can identify the proportionality constant λ as the probability that even A will occur in one trial.[†] As pointed out before, however, it is only the product $y = \lambda n = pn$ that may be properly defined: it is the theoretical mean of the discrete random variable x when the same (large) number of n trials are repeated many times.

Equation 2.9 correctly fulfills the normalization requirement

$$\sum_{x=0}^{n=\infty} f(x) = e^{-y} \sum_{x=0}^{\infty} \frac{y^x}{x!} = e^{-y} e^{y} = 1$$

It is shown in Section 2.9 that Eq. 2.9 is the limiting form of Eq. 2.4 when $p \to 0$ and $n \to \infty$.

[†] $P\{1, 1\} = \lambda e^{-\lambda} \to \lambda$ when $\lambda \ll 1$.

2.7 Moments of the Poisson Frequency Function

Following the approach used in Section 2.5, the moments of the Poisson frequency function will be obtained by direct evaluation of the defining equations; note that as $n \to \infty$ the upper limit of x is also ∞.

$$\mu = \mu'_1 = \sum_{x=0}^{x=n \to \infty} x \frac{y^x e^{-y}}{x!} = \sum_{x=1}^{\infty} \frac{y^x e^{-y}}{(x-1)!}$$

$$= e^{-y} y \sum_{x=1}^{\infty} \frac{y^{(x-1)}}{(x-1)!} = e^{-y} y e^y = y$$

Thus

$$\mu = y \tag{2.14}$$

as expected from our previous discussion. We see that through Eq. 2.14 we obtain the physical significance for the parameter y. Further,

$$\mu'_2 = \sum_{x=0}^{\infty} x^2 \frac{y^x e^{-y}}{x!} = \sum_{x=0}^{\infty} \left(x(x-1) \frac{y^x e^{-y}}{x!} \right) + y$$

$$= e^{-y} \sum_{x=2}^{\infty} \left(\frac{y^x}{(x-2)!} \right) + y = e^{-y} y^2 \sum_{x=2}^{\infty} \frac{y^{(x-2)}}{(x-2)!} + y = y^2 + y$$

and using Eq. 2.3 we obtain

$$\mu_2 = \sigma^2 = \mu'_2 - \mu^2 = y^2 + y - y^2 = y$$

Thus

$$\sigma = \sqrt{y} \tag{2.15}$$

The close analogy of Eq. 2.14 to Eq. 2.6 and of Eq. 2.15 to Eq. 2.7 should be clear; also the derivation of these equations is completely similar.

2.8 The Gaussian or Normal Frequency Function and Its Moments

This is indeed a most important frequency function because (a) it is a limiting case which many frequency functions approach; (b) the distribution of most physical observables is satisfactorily described by it; and (c) measurements containing *random* errors are distributed normally about the true value of the measured quantity.

The Gaussian distribution gives the frequency of the continuous random variable x in terms of two parameters a and b which are the first and second moment of the frequency function. In its normalized form, the Gaussian distribution is given by

$$f(x) \, dx = \frac{1}{b\sqrt{2\pi}} \exp\left[-\frac{1}{2}\left(\frac{x-a}{b}\right)^2\right] dx \qquad (2.16)$$

and is shown in Fig. 10.5. The range of the variable x is from $-\infty$ to $+\infty$. In order to show the normalization of Eq. 2.16, as well as to find the mo-

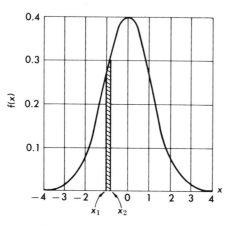

FIG. 10.5 The Gaussian frequency function normalized to zero mean and unit variance

$$f(x) \, dx = (1/\sqrt{2\pi})e^{-x^2/2} \, dx.$$

Note that the probability of finding a value of x between x_1 and x_2 is proportional to the corresponding area under the Gaussian.

ments, it is useful to know the values of the integral of $x^n e^{-ax^2}$ which are summarized in Table 10.2. To obtain the moments we proceed as before

$$\mu = \mu'_1 = \frac{1}{b\sqrt{2\pi}} \int_{-\infty}^{+\infty} x \exp\left[-\frac{1}{2}\left(\frac{x-a}{b}\right)^2\right] dx$$

TABLE 10.2

VALUE OF THE INTEGRAL OF $f(n) = \displaystyle\int_0^\infty x^n \exp(-ax^2) \, dx$

n	$f(n)$	n	$f(n)$
0	$\frac{1}{2}\sqrt{\pi/a}$	1	$1/2a$
2	$\frac{1}{4}\sqrt{\pi/a^3}$	3	$1/2a^2$
4	$\frac{3}{8}\sqrt{\pi/a^5}$	5	$1/a^3$

$$F(n) = \int_{-\infty}^{+\infty} x^n \exp(-ax^2) \, dx = \begin{cases} 2f(n) & \text{when } n \text{ even} \\ 0 & \text{when } n \text{ odd} \end{cases}$$

We let $x = tb + a$, $dx = b\, dt$; thus

$$\mu = \frac{1}{\sqrt{2\pi}} \left[\int_{-\infty}^{+\infty} bte^{-(t^2/2)}\, dt + \int_{-\infty}^{+\infty} ae^{-(t^2/2)}\, dt \right]$$

according to Table 10.2, integrals with odd powers of t vanish, thus

$$\mu = a \tag{2.17}$$

Similarly

$$\mu'_2 = \frac{1}{b\sqrt{2\pi}} \int_{-\infty}^{+\infty} x^2 \exp\left[-\frac{1}{2}\left(\frac{x-a}{b}\right)^2 \right] dx$$

with the same substitution

$$\mu'_2 = \frac{1}{\sqrt{2\pi}} \left[\int_{-\infty}^{+\infty} b^2 t^2 e^{-(t^2/2)}\, dt + \int_{-\infty}^{+\infty} 2\, abte^{-(t^2/2)}\, dt + \int_{-\infty}^{+\infty} a^2 e^{-(t^2/2)}\, dt \right]$$

so that by using Table 10.2 we obtain

$$\mu'_2 = \frac{1}{\sqrt{2\pi}} \left[b^2 \tfrac{1}{2}\sqrt{8\pi} + a^2\sqrt{2\pi} \right] = a^2 + b^2$$

and using Eq. 2.3,

$$\mu_2 = \sigma^2 = \mu'_2 - \mu^2 = b^2$$

Thus

$$\sigma = b \tag{2.18}$$

We see that through Eqs. 2.17 and 2.18, we obtain the physical significance of the parameters a and b of Eq. 2.16. It is sometimes useful to transform the random variable linearly so as to obtain a frequency function with zero mean and unit standard deviation; the transformation is

$$y = \frac{x-\mu}{\sigma}; \qquad dy = \frac{dx}{\sigma}$$

and Eq. 2.16 becomes

$$f(y)\, dy = \frac{1}{\sqrt{2\pi}} e^{-(y^2/2)}\, dy \tag{2.19}$$

2.9 The Gaussian Frequency Function As a Limiting Case

In the previous section we gave Eq. 2.16 without proof. We will now show that it can be obtained from the binomial frequency function, Eq. 2.4, in the limit of $n \to$ large and $|\, np - x\,| \ll np$.

Consider Eq. 2.4:

$$f(x) = \frac{n!}{x!(n-x)!} p^x q^{n-x}$$

If $n \to \infty$ but $np \to \mu$ remains finite, we may write

$$f(x) = \frac{n(n-1)\cdots(n-x+1)}{n^x} \cdot \frac{(np)^x}{x!} \cdot (1-p)^{n-x}$$

$$f(x) = \frac{1[1-(1/n)]\cdots[1-(x-1)/n]}{(1-p)^x} \cdot \frac{(np)^x}{x!} \cdot (1-p)^n$$

$$(2.20)$$

But

$$(1-p)^n = \left[(1-p)^{-(1/p)}\right]^{-np} \to [e]^{-\mu}$$

since from the definition of e,

$$\lim_{z \to 0} (1+z)^{1/z} = e$$

and in the present case we have $p \to 0$. Further

$$\lim_{n \to \infty} \frac{1[1-(1/n)]\cdots[1-(x-1)/n]}{(1-p)^x} = 1$$

because $p \to 0$ and x is finite; by substituting the last two expressions into Eq. 2.20 we obtain the Poisson frequency function, Eq. 2.9:

$$f(x) = \frac{\mu^x e^{-\mu}}{x!} \qquad (2.9)$$

We now use the further condition that x be a continuous variable and $|np - x| \ll np$, namely, its deviations from the mean μ be small; then the following approximate expression is valid

$$\ln \frac{\mu}{x} = \ln\left(1 + \frac{\mu - x}{x}\right) = \left(\frac{\mu - x}{x}\right) - \frac{1}{2}\left(\frac{\mu - x}{x}\right)^2 + \cdots$$

hence

$$\frac{\mu}{x} \approx \exp\left(\frac{\mu - x}{x}\right) \exp\left[-\frac{1}{2}\left(\frac{\mu - x}{x}\right)^2\right]$$

and

$$(\mu)^x \approx x^x \exp(\mu - x) \exp\left[-\frac{1}{2}\left(\frac{\mu - x}{x}\right)^2\right]$$

From Stirling's formula we have

$$x! \simeq \sqrt{2\pi x}\, x^x e^{-x}$$

and by substituting $(\mu)^x$ and $x!$ into Eq. 2.9 we obtain

$$f(x) = \frac{\mu^x e^{-\mu}}{x!} = \frac{e^{-\mu} x^x e^{(\mu - x)} \exp\left\{-\tfrac{1}{2}\left[(\mu - x)/x\right]^2\right\}}{\sqrt{2\pi x}\, x^x e^{-x}}$$

$$= \frac{1}{\sqrt{2\pi x}} \exp\left[-\frac{1}{2}\left(\frac{\mu - x}{x}\right)^2\right] \quad (2.21)$$

Thus the binomial frequency function in its limit approaches a Gaussian frequency function with

$$\text{mean} \qquad \mu = np$$

$$\text{standard deviation} \qquad \sigma = \sqrt{x} \approx \sqrt{npq} \quad (2.22)$$

where $x \approx npq$ follows from $|\mu - x| \ll \mu$ and $p \to 0$. From Eq. 2.22 we see that the moments of the limiting Gaussian frequency function are the limits of the moments of the original binomial frequency function.

2.10 PROPERTIES OF THE GAUSSIAN FREQUENCY FUNCTION

Let us now interpret this frequency function as given by Eq. 2.16. We could refer to our original example of obtaining event A, x times when a choice between A or B is made n times; x then can vary from 0 to n in integer values. It is easier, however, to consider the measurement with a ruler of the length of a rod; we let the continuous random variable x represent the result of *one* measurement. If the true length of the rod is x_0, Eq. 2.16 specifies that a result between x and $x + dx$ will be obtained with a frequency

$$f(x)\, dx = \frac{1}{\sigma\sqrt{2\pi}} \exp\left[-\frac{1}{2}\left(\frac{x_0 - x}{\sigma}\right)^2\right] dx \quad (2.23)$$

One may also say that the probability that the measurement will yield a result x between x and $x + dx$ is given by Eq. 2.23. In simpler words, if N measurements are performed, a result between x_1 and x_2 is likely to be obtained in $n(x_1, x_2)$ of these measurements, where

$$n(x_1, x_2) = N \cdot F(x_1, x_2) = \frac{N}{\sigma\sqrt{2\pi}} \int_{x_1}^{x_2} \exp\left[-\frac{1}{2}\left(\frac{x_0 - x}{\sigma}\right)^2\right] dx \quad (2.24)$$

as shown in Fig. 10.5.

TABLE 10.3
SOME NUMERICAL VALUES OF THE NORMALIZED GAUSSIAN FUNCTION

$$f(x) = \frac{1}{\sqrt{2\pi}} \exp\ (-x^2/2) \qquad F(-c, c) = \int_{-c}^{+c} f(x)\ dx$$

$f(0)$	$= 0.3989$	$F(-1, 1)$ = 0.6826
$f(1) = f(-1)$	$= 0.2420$	$F(-2, 2)$ = 0.9554
$f(2) = f(-2)$	$= 0.0540$	$F(-3, 3)$ = 0.9974
		$F(-0.69, 0.69)$ = 0.5000

Note that in Eqs. 2.23 and 2.24 the standard deviation σ is determined by the conditions of the measurement. The applicability of the Gaussian distribution to the results obtained from such measurements lies in the fact that (a) n, the number of (least) divisions of the ruler is large, and (b) the errors in measurement $|\ x_0 - x\ |$ are small as compared to x.

In Table 1.03 are given the values of $f(x)$ and its integral, $F(c)$ for the normalized Gaussian function (Eq. 2.19).

From Table 10.3, for example, we see that half of the measurements do yield a result x, between

$$x_0 - 0.69\sigma < x < x_0 + 0.69\sigma$$

or that only 2.23 percent of the results may yield x, such that

$$x > x_0 + 2\sigma$$

Also we conclude that a result x in the small interval Δx about x_0 will be obtained $(0.3989)/(0.0540) = 7.4$ times more frequently than a result in the same small interval Δx about $x_0 + 2\sigma$.

3. Estimation of Parameters and Fitting of Data

In Section 1 the basic definitions have been given; in Section 2, analytic expressions for some frequency functions have been obtained. We will now see how statistics can be applied to the interpretation of a measurement or an experiment.

We can consider one or more measurements to form a sample of a population which obeys a certain frequency function; we are then faced with one of two estimation problems:

(a) Given the frequency function and its parameters, what is the probability of obtaining from a measurement the result x?

(b) Given the result x of a measurement, what are the parameters of the frequency function (or the frequency function itself)?

In physics we are usually faced with estimation of type (b), since a set of experimental data are obtained and it is then desired to reduce them to a few parameters which should describe the whole population; and, therefore, also any new measurement that may be performed.

There are several methods for obtaining "estimators" to an unknown parameter. Some of these methods are almost subconsciously applied, but most of them can be derived from the principle of "maximum likelihood" introduced by R. A. Fisher in 1920.

3.1 MAXIMUM LIKELIHOOD

To apply this principle we must have knowledge of the normalized frequency functions of the variables x_i that form the data,

$$f(x_i, \theta)$$

where θ is the parameter to be estimated and upon which obviously the frequency function depends. We may then form the product of the frequency functions for all observed variables.

$$\mathcal{L}(x_1, x_2 \cdots x_n, \theta) = f(x_1, \theta) f(x_2, \theta) \cdots f(x_n, \theta) \qquad (3.1)$$

which is called the likelihood function for the parameter θ (note that $\mathcal{L}$ is *not* a frequency function for the parameter θ). The theorem of maximum likelihood then states that the value of θ, θ^*, that maximizes $\mathcal{L}$ (for the set of observed data) is the best estimator of θ.

$$\frac{\partial \mathcal{L}(x_1, x_2, \cdots; \theta)}{\partial \theta} \bigg]_{\theta = \theta^*} = 0$$

It is almost always convenient in practice to work with the logarithm of $\mathcal{L}$, since when $W = \log \mathcal{L}$ is maximum, so will also be $\mathcal{L}$.

As an example, we consider a set of n data x_i that obey a normal frequency function about a, with a standard deviation σ; let us seek the best value for the parameter a.

$$f(x_i, a) = \frac{1}{\sigma\sqrt{2\pi}} \exp\left[-\frac{1}{2}\left(\frac{a - x_i}{\sigma}\right)^2 \right] \qquad (3.2)$$

then

$$\mathcal{L} = \prod_{i=1}^{n} f(x_i, a)$$

and

$$W = \log \mathcal{L} = -n \log (\sigma\sqrt{2\pi}) - \frac{1}{2} \sum_{i=1}^{n} \left(\frac{a - x_i}{\sigma}\right)^2 \tag{3.3}$$

$$\frac{\partial W}{\partial a} = -\sum_{i}^{n} \frac{a - x_i}{\sigma^2}$$

setting then $(\partial W)/(\partial a) = 0$ leads to the estimator a^*;

$$\sum_{i}^{n} \frac{a^* - x_i}{\sigma^2} = 0 \qquad \frac{na^*}{\sigma^2} - \sum_{i}^{n} \frac{x_i}{\sigma^2} = 0$$

or

$$a^* = \frac{1}{n} \sum_{i}^{n} x_i \tag{3.4}$$

Thus if a set of measurements is distributed normally, the best estimator for the true value of the parameter is the mean of the measurements (first moment).

Similarly we may obtain the estimator, σ^*, for σ, by differentiating Eq. 3.3 with respect to σ

$$\frac{\partial W}{\partial \sigma} = -\frac{n}{\sigma} + \sum_{i}^{n} \left[\left(\frac{a - x_i}{\sigma}\right)\left(\frac{a - x_i}{\sigma^2}\right)\right] = 0$$

$$(\sigma^*)^2 = \frac{1}{n} \sum_{i}^{n} (a - x_i)^2 \tag{3.5}$$

where, in Eq. 3.5, a should be replaced by its estimator a^* given by Eq. 3.4. Again we obtain the familiar result that the best estimator for the standard deviation of the theoretical frequency function is given by the second moment (about the mean) of the observed measurements.

The principle of maximum likelihood can be further extended to give the variance S^2 of the estimator θ^*; that is, if the determination of estimators θ^* is repeated, the values so obtained will have a standard deviation S, where

$$\frac{1}{S^2} = -\frac{\partial^2 W}{\partial \theta^2} \tag{3.6}$$

We may apply Eq. 3.6 to our sample of measurements that obeys a normal frequency function, where W was given by Eq. 3.3. We obtain

$$\frac{1}{S^2} = -\frac{\partial^2 W}{\partial a^2} = \sum_{i}^{n} \frac{1}{\sigma^2} = \frac{n}{\sigma^2}$$

Thus the standard deviation of the estimators will be

$$S = \frac{\sigma}{\sqrt{n}} \tag{3.7}$$

where n is the number of measurements used for obtaining each estimator. Equation 3.7 is a well-known result which we will obtain again when we discuss the combination of errors in Section 4.

3.2 THE LEAST-SQUARES METHOD

Until now we have discussed the case where all n measurements are made on the same physical quantity whose true value is a, for example, the data of Eq. 3.2. But consider now a set of measurements yielding values y_1, y_2, $\cdots$, y_n depending on another variable x; the corresponding true values of y, which we designate by $\bar{y}$, are assumed to be a function of x, and of one or more parameters a_ν common to the whole sample. Thus we write

$$\bar{y}_i = y(x_i; a_\alpha, \cdots, a_\nu) \tag{3.8}$$

Further, each measurement y_i has associated with it a standard deviation σ_i, which is not the same for each point. This situation is shown in Fig. 10.6.

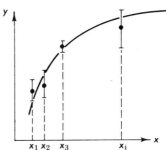

FIG. 10.6 Least-squares fit of a two-dimensional curve to a set of data points obtained for different values of x. Note that each data point has associated with it a different error as indicated by the flags; this is taken into account when forming the least-squares sum.

It is possible that the form of Eq. 3.8 is known or may be correctly inferred from the physics of the process under investigation, in which case the estimation is reduced to finding the best estimators for the parameters a_ν. If, however, the form of Eq. 3.8 is not known, various functional relationships have to be assumed, for example, a polynomial of order k. We then speak of fitting a curve to the data. Even though special techniques are developed in Section 3.4 to ascertain which curve fits best, the following discussion is applicable in either case.

The method of least squares follows directly from the assumption that each individual measurement y_i is a member of a *Gaussian* population with a mean given by the true value of y_i, $\bar{y}(x_i; a_\lambda)$; for the standard deviation of this Gaussian we use the experimental error σ_i of each measurement. Then in analogy to Eq. 3.2 we write for the frequency function of the y_i

$$f(y_i; x_i; a_\lambda) = \frac{1}{\sigma_i\sqrt{2\pi}} \exp\left\{-\frac{1}{2}\left[\frac{y_i - \bar{y}(x_i; a_\lambda)}{\sigma_i}\right]^2\right\} \tag{3.9}$$

And in analogy with Eq. 3.1 we form the likelihood function

$$\mathcal{L}(y_1\cdots y_n; x_1\cdots x_n; a_\lambda) = \prod_{i=1}^{n} f(y_i; x_i; a_\lambda)$$

We seek the estimators $a_\lambda{}^*$ that maximize this function, or its logarithm W

$$W = \log \mathcal{L} = -\sum_{i}^{n} \log (\sigma_i\sqrt{2\pi}) - \frac{1}{2}\sum_{i}^{n}\left[\frac{y_i - \bar{y}(x_i; a_\lambda)}{\sigma_i}\right]^2 \tag{3.10}$$

Since the σ_i are fixed by the measurements, the estimators $a_\lambda{}^*$ are those values of a_λ that *minimize* the sum

$$\mathfrak{M} = \sum_{i=1}^{n}\frac{[y_i - \bar{y}(x_i; a_\lambda)]^2}{\sigma_i{}^2} \tag{3.11}$$

that is, those that give the "least-squares sum." They are obtained by solving the simultaneous equations

$$\frac{\partial \mathfrak{M}}{\partial a_\lambda} = 0 \qquad \lambda = 1 \text{ to } \nu$$

3.3 APPLICATION OF THE LEAST-SQUARES METHOD TO A LINEAR FUNCTIONAL DEPENDENCE

The simplest case of functional dependence $y(x)$ is the linear one

$$y = a + bx$$

If we assume that every measurement y_i has the *same standard deviation* (statistical weight), we may obtain the estimators a^* and b^* that minimize Eq. 3.11 in closed form.

Since $\sigma_1 = \sigma_2 = \cdots = \sigma_n = \sigma$, instead of Eq. 3.11 we need only minimize

$$\mathfrak{R} = \sum_{i=1}^{n}[y_i - (a + bx_i)]^2 \tag{3.12}$$

Hence

$$\frac{\partial \mathcal{R}}{\partial a} = -2 \sum_{i=1}^{n} [y_i - (a + bx_i)] = 0$$

$$\frac{\partial \mathcal{R}}{\partial b} = -2 \sum_{i=1}^{n} \{[y_i - (a + bx_i)]x_i\} = 0$$

(3.13)

which after some manipulation† leads to

$$a^* = \frac{\sum x_i^2 \sum y_i - \sum x_i \sum (x_i y_i)}{n \sum x_i^2 - \sum x_i \sum x_i}$$

$$b^* = \frac{n \sum (x_i y_i) - \sum y_i \sum x_i}{n \sum x_i^2 - \sum x_i \sum x_i}$$

(3.14)

The standard deviations for the above estimators may be obtained by an extension of Eq. 3.6 which now yields a symmetric square matrix

$$\mathbf{H}_{\lambda\mu} = -\frac{\partial^2 W}{\partial a_\lambda \partial a_\mu} = \frac{1}{2\sigma^2} \frac{\partial^2 \mathfrak{M}}{\partial a_\lambda \partial a_\mu}$$

(3.15)

The elements of the *inverse* matrix give the variance of the estimators a^*. A complete discussion of this error matrix is given in Section 4; suffice it to say here that the usually given expressions (Eqs. 3.16) for the standard deviation of the estimators (Eqs. 3.14) are the square roots of the diagonal elements of $\mathbf{H}^{-1}$ (see Eq. 4.18). We then obtain

$$\sigma_{a^*} = \sqrt{(\mathbf{H}^{-1})_{aa}} = \sigma \sqrt{\frac{\sum x_i^2}{n \sum x_i^2 - \sum x_i \sum x_i}}$$

$$\sigma_{b^*} = \sqrt{(\mathbf{H}^{-1})_{bb}} = \sigma \sqrt{\frac{n}{n \sum x_i^2 - \sum x_i \sum x_i}}$$

(3.16)

In case $\sigma_1 \neq \sigma_2 \neq \cdots \neq \sigma_n$, it is $\mathfrak{M}$ and not $\mathcal{R}$ that must be minimized (Eqs. 3.12 and 3.13). If, however, one wishes to maintain the formalism of Eqs. 3.14 and 3.16, the set of data can be rearranged to an equivalent set with some added (fictitious) data but all of them having the same standard deviation. This is based on Eq. 3.7 which we interpret as meaning that one measurement with variance σ^2 is equivalent to n measurements with variance $n\sigma^2$. Thus we can follow the procedure described below. Find the

† Note that the second of the above equations is by no means equal to the first one multiplied by x_i.

least common denominator C of the (σ_i^2); then form the equivalent set of data by including each data point y_i, x_i as many times† as the ratio $C/(\sigma_i^2)$.

Equations 3.14 and 3.16 have already been used repeatedly in the text; for example, see Chapter 2.

3.4 GOODNESS OF FIT; THE χ^2 DISTRIBUTION

We have seen how the least-squares method, as a consequence of the principle of maximum likelihood, may be used to fit a curve to a set of data. Once the curve has been found, however, the necessity arises to ascertain quantitatively how good the fit is. This is important especially if the functional dependence is not known; a poor fit might indicate the necessity for fitting with a curve of higher order; or a poor fit might indicate inconsistencies in the data.

Similarly, we may wish to test whether a certain hypothesis is supported from the data, in which case the goodness of the fit may establish the level of confidence with which the hypothesis should be accepted.

Let us first suppose that we know the true functional relationship of y to x, that is, $\bar{y}(x) = f(x)$; we may then form the least-squares sum

$$\mathfrak{M} = \sum_{i=1}^{n} \frac{[y_i - \bar{y}(x_i)]^2}{\sigma_i^2} \tag{3.11}$$

The range of $\mathfrak{M}$ is $0 < \mathfrak{M} < +\infty$ but we would be surprised if $\mathfrak{M} = 0$ and would be equally surprised if $\mathfrak{M}$ was extremely large. Thus we have already a quantitative indication as to how well the data fit the known (or assumed) curve $y = f(x)$.

If a new set of data are obtained pertaining to the same experimental situation, and Eq. 3.11 is again formed, a new value $\mathfrak{M}_q$ will result. Clearly, if enough such measurements are repeated, each time yielding a value for $\mathfrak{M}$, we will obtain the frequency function for $\mathfrak{M}$. Once the frequency function is known, it is then easy to tell what the probability of obtaining a specific $\mathfrak{M}$ is. We may, for example, calculate that in 95 percent of the cases $\mathfrak{M} < \mathfrak{M}_0$; if then a specific set of data yield $\mathfrak{M}_s \geq \mathfrak{M}_0$, we know that such data should be obtained only in 5 percent of the experiments and can therefore be rejected.

Obtaining the frequency function for the least-squares sum in this way is obviously impractical. Nevertheless, it is true that the *distribution* of $\mathfrak{M}$ is *independent* of the curve $y = f(x)$ and of σ_i, and can therefore be calculated theoretically; it depends only on the number n of points that

---

† In essence each data point is "weighed" by a factor proportional to $1/(\sigma^2)$.

are compared, and is called the χ^2 distribution

$$f(\mathfrak{M}) \, d\mathfrak{M} = \frac{\mathfrak{M}^{(\nu/2)-1} \exp \, (-\mathfrak{M}/2)}{2^{\nu/2} \, \Gamma(\nu/2)} \, d\mathfrak{M} \equiv f(\chi^2) \, d\chi^2 \qquad (3.17)$$

here ν is the number of "degrees of freedom" of $\mathfrak{M}$. In the present case we set

$$\nu = n$$

because this is the number of truly independent points being compared. In Eq. 3.17 $\Gamma(x)$ is the "gamma function" which for positive integer arguments† is simply

$$\Gamma(n) = (n - 1)!$$

Consider next that $y = f(x)$ is not known, but that a two-parameter curve is fitted to n data points, yielding estimators a^* and b^*. Then one forms again the least-squares sum $\mathfrak{M}$ using $\bar{y} = f(x; a^*, b^*)$ but now the frequency function for the $\mathfrak{M}$'s is Eq. 3.17 with the n degrees of freedom reduced by the number of estimators obtained from the data, that is,

$$\nu = n - 2$$

The χ^2 distribution may also be used for comparing the frequency of occurrence of a class of events with the theoretical frequency (function). Let us consider, for example, the 100 measurements of the radioactive sample given in Table 5.2 and divide the sample in seven classes, with approximately equal expected frequencies; the resulting frequency of the experimental observations o_i in each class is given in the following Table 10.4. Next we obtain from the data the estimators for the parameters of a

TABLE 10.4

OBSERVED AND EXPECTED FREQUENCY OF THE RESULTS OF 100 MEASUREMENTS
OF A RADIOACTIVE SAMPLE

Class	0–75	75–79	79–83	83–87	87–91	91–95	95–∞
o_i	15	11	15	15	18	12	14
e_i	13	12	15	16	16	13	15
$(e_i - o_i)^2/e_i$	0.307	0.083	0	0.062	0.25	0.077	0.067

---

† The general definition of the gamma function is:

$$\Gamma(z) = \int_0^\infty t^{z-1} \exp \, (-t) \, dt$$

for more details see any text on advanced calculus.

Gaussian (1) $\mu^* = \bar{N}$, (2) $\sigma^* = \sqrt{\bar{N}}$ and (3) the overall normalization, namely, $\sum o_i = \sum e_i$; thus the degrees of freedom of χ^2 are four, corresponding to seven classes less three estimators. From the Gaussian distribution we calculate the expected frequencies e_i for each class; they are also given in Table 10.4.

In complete analogy with the least-squares sum, Eq. 3.11, we form the χ^2 sum

$$\chi^2 = \sum_{i=1}^{n} \frac{(e_i - o_i)^2}{e_i}$$

Note that χ^2 is now a discrete variable, since frequencies of classes are compared; however, Eq. 3.17 which holds for a continuously variable χ^2 is valid provided the number of classes $n \gtrsim 5$ and the expected frequencies $e_i \geq 5$.

For this experiment (see also Fig. 5.23a) we obtain

$$\chi^2 = 0.846$$

and we explained before that $\nu = 4$. From a table of the χ^2 distribution we find that in 93 percent of the cases the χ^2 distribution would be larger than the result obtained here. Thus one may suspect that the data is "too good" a fit to the estimated Gaussian.

The χ^2 distribution of Eq. 3.17 for different degrees of freedom is shown in Fig. 10.7; tables of this distribution may be found in most reference

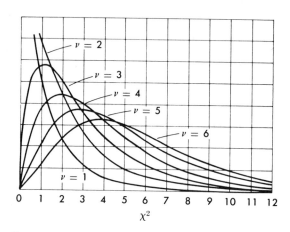

FIG. 10.7   The frequency function for the distribution of χ^2, for different degrees of freedom. All curves are normalized to the same unit area. Note that for large ν the χ^2 distribution approaches a Gaussian.

manuals.† It should not be surprising that when the number of degrees of freedom increases $\nu > 30$, the χ^2 distribution approaches a Gaussian‡ with mean $\mu = \nu - \frac{1}{2}$.

3.5 Additional Remarks on Data Fitting

In the previous paragraphs we have discussed how the basic problem of estimation, namely the fit of a curve to a set of data may be solved. In addition to the χ^2 distribution there are also other tests which may be applied to evaluate the goodness of the fit, such as the F test, the t test, and so on, which offer specific advantages in certain cases.

In fitting a curve to data, unless there exist special reasons not to do so, we normalize the area under the theoretical curve to be the same as the area under the experimental points; this corresponds to the requirement of Section 3.4 that the $\sum o_i = \sum e_i$.

Let us next assume that the functional dependence $y = f(x)$ of the theoretical curve is not known, and that we wish to fit the data with a polynomial. We may start with a polynomial of order n, find the estimators for its coefficients, and then test the goodness of fit. If the χ^2 is too large, we should try a polynomial of order $n + 1$, etc. This leads to considerable labor, because all the $n + 1$ coefficients have to be re-estimated; if, however, the fitting is attempted with a set of orthogonal polynomials (for example, the Legendre polynomials), it turns out that the n coefficients calculated from the polynomial of order n remain unchanged and we need only evaluate the new coefficient of the $(n + 1)$ power of x. In addition, advantage can be taken of any symmetry properties that the data may possess.

Finally, it should be clear that the above estimation procedures are not necessarily limited to curve fitting. Need may arise to fit a surface to a set of data points that depend on two random variables $y = f(x, z; a_\lambda)$, and so on, for cases of higher dimensionality.

4. Errors and Their Propagation

4.1 Introduction

When we perform a measurement of a physical quantity x, it can be expected that the result obtained, x_1, will differ from x; this difference is the *error* of the measurement and consists of a *systematic* and a *random* contribution. Suppose, now, that the measurement is repeated under the same conditions n times; then the results x_n will be distributed (in most

---

† For example, *Handbook of Chemistry and Physics*, 43rd Ed., p. 221.

‡ It is really the distribution of $\sqrt{2\chi^2}$ that approaches the Gaussian with mean $\mu = \sqrt{(2\nu - 1)}$ and unit standard deviation (R. A. Fisher's approximation).

cases) normally about a mean $\bar{x}$ with a standard deviation σ. The difference between $\bar{x}$ and the true value x is then the systematic error, and the standard deviation σ of the Gaussian is a measure of the *dispersion* of the results due to the random error.

The object of the measurement, however, is the determination of the unknown true value x; since this is not possible, we seek to find whether x lies between certain limits, or whether the true value x is distributed about some mean x^* with a standard deviation σ^*. Note that in a rigorous sense, this statement is incorrect, since the unknown true value x is not distributed, but is fixed; what we mean is that the probability $x = x^*$, or $x > x^*$, etc., is given by the normal frequency function with mean $\bar{x}$ and $\sigma = \mu'_2$ the second moment of the measured data about their mean $\bar{x}$.

Thus, by repeating the measurement several times, it is possible in principle to circumvent the random errors because (a) a knowledge of $\bar{x}$ and σ contains all possible information about the unknown true value x, and (b) as n increases, the second moment should decrease as $1/\sqrt{n}$ and may be made arbitrarily small. On the other hand, the systematic errors cannot be extracted from a set of identical measurements. They can either be estimated by the observer, or be judged from a performance of the same measurement with a different technique. Therefore, it is unadvisable to reduce the random errors much below the expected limits of the systematic errors. In what follows we will discuss only the treatment of random errors and work under the assumption that the results of the measurements follow a normal distribution.

Until now we have considered the simple case where the unknown value x is directly measured and an error σ_x can be associated with the measurement; that is, the frequency function of x depends only on one variable.

$$f(x) = \frac{1}{\sqrt{2\pi}\sigma_x} \exp\left[-\frac{1}{2}\left(\frac{\bar{x} - x}{\sigma}\right)^2\right]$$

Most frequently, however, the unknown value x is not directly measured, and we distinguish two cases: (a) x is an explicit function of the quantities $y_1, y_2, \cdots, y_n$ that are measured and have with them associated errors $\sigma_1, \sigma_2, \cdots, \sigma_n$. Namely,

$$x = \phi(y_1, y_2, \cdots, y_n) \tag{4.1}$$

and it is desired to find the estimator x^* and its standard deviation σ_x; (b) x is an implicit function of other unknown variables $u_1, u_2, \cdots, u_m$ and of the quantities $y_1, y_2, \cdots, y_n$ that are measured and have with them associated errors $\sigma_1, \sigma_2, \cdots, \sigma_n$. Namely,

$$\phi(x; u_1, u_2, \cdots, u_m; y_1, y_2, \cdots, y_n) = 0 \tag{4.2}$$

and it is desired to find the estimators x^*; u_1^*, u_2^*, $\cdots$, u_m^*, and the symmetric error matrix $\sigma_{ij}(i, j = 1, \cdots, m + 1)$. Such an example was treated in Section 3.3, and we know that at least $m + 1$ sets of measurements are required to obtain the $m + 1$ estimators.

The techniques for obtaining the best estimators were discussed in Section 3. In this section we will discuss how the random error of x may be determined from knowledge of the errors of the independent variables y_n; this procedure is frequently referred to as the combination or the propagation of the errors of the y_n's.

4.2 Propagation of Errors

Let us first assume x to be an explicit function of the measured y_n as discussed before (Eq. 4.1):

$$x = \phi(y_1, y_2, \cdots, y_n) \tag{4.1}$$

By applying the maximum likelihood method, it can be shown that the estimator x^* is obtained by using the mean values μ_n, of the measured y_n (provided† the y_n are distributed normally). Here the mean values μ_n are obtained from r different measurements

$$\mu_n = \sum_{i=1}^{r} y_n^{\,i}$$

Thus

$$x^* = \phi(\bar{y}_1, \bar{y}_2, \cdots, \bar{y}_n) = \phi(\mu_1, \mu_2, \cdots, \mu_n) \tag{4.3}$$

Next we make a Taylor expansion of Eq. 4.1 about x^*, through first order

$$x = \phi(\mu_1, \mu_2, \cdots, \mu_n) + \left[\frac{\partial\phi}{\partial y_1}\right]_\mu (\mu_1 - y_1) + \left[\frac{\partial\phi}{\partial y_2}\right]_\mu (\mu_2 - y_2)$$

$$+ \cdots + \left[\frac{\partial\phi}{\partial y_n}\right]_\mu (\mu_n - y_n)$$

where $[(\partial\phi)/(\partial y_n)]_\mu$ means evaluation of the derivative at the point about which we expand—that is, $(\mu_1, \mu_2, \cdots, \mu_n)$. We can now form the second moment of the distribution of the x^i's as they result from the ob-

---

† Clearly if x is variable, all measurements $y_n^{\,i}$ are made so as to correspond to the same point x.

served $y_n{}^i$'s. The superscript i here refers to the r different sets of measurements.

$$\sigma_x{}^2 = \frac{1}{r}\sum_{i=1}^{r}(\bar{x} - x^i)^2 = \frac{1}{r}\sum_{i=1}^{r}\left[\left(\frac{\partial\phi}{\partial y_1}\right)_\mu (\mu_1 - y_1{}^i) + \cdots + \left(\frac{\partial\phi}{\partial y_n}\right)_\mu (\mu_n - y_n{}^i)\right]^2$$

$$= \left(\frac{\partial\phi}{\partial y_1}\right)_\mu^2 \frac{1}{r}\sum_{i=1}^{r}(\mu_1 - y_1{}^i)^2 + \left(\frac{\partial\phi}{\partial y_2}\right)_\mu^2 \frac{1}{r}\sum_{i=1}^{r}(\mu_2 - y_2{}^i)^2 + \cdots$$

$$+ 2\left(\frac{\partial\phi}{\partial y_1}\right)_\mu \left(\frac{\partial\phi}{\partial y_2}\right)_\mu \frac{1}{r}\sum_{i=1}^{r}(\mu_1 - y_1{}^i)(\mu_2 - y_2{}^i) + \cdots$$

$$\sigma_x{}^2 = \left(\frac{\partial\phi}{\partial y_1}\right)_\mu^2 \sigma_1{}^2 + \left(\frac{\partial\phi}{\partial y_2}\right)_\mu^2 \sigma_2{}^2 + \cdots + 2\left(\frac{\partial\phi}{\partial y_1}\right)_\mu \left(\frac{\partial\phi}{\partial y_2}\right)_\mu \sigma_{12}{}^2 + \cdots \qquad (4.4)$$

Equation 4.4 is the most general expression for the propagation of errors. If we assume that the errors are uncorrelated, namely, $\sigma_{ij} = 0$ when $i \neq j$, we can obtain the results for the simplest functional relationships.

(a) Addition

$$x = y_1 + y_2 + \cdots + y_n$$

$$\sigma_x = \sqrt{\sigma_1{}^2 + \sigma_2{}^2 + \cdots + \sigma_n{}^2} \qquad (4.5)$$

(b) Subtraction

$$x = y_1 - y_2$$

$$\sigma_x = \sqrt{\sigma_1{}^2 + \sigma_2{}^2} \qquad (4.6)$$

(c) Multiplication

$$x = y_1 \times y_2 \times \cdots \times y_n$$

$$\left(\frac{\partial\phi}{\partial y_1}\right)_\mu = \mu_2 \times \cdots \times \mu_n$$

$$\sigma_x = \sqrt{\sigma_1{}^2 \times (\mu_2 \cdots \mu_n)^2 + \cdots + \sigma_n{}^2 \times (\mu_1\mu_2\cdots)^2} \qquad (4.7)$$

$$= x^*\sqrt{\left(\frac{\sigma_1}{\mu_1}\right)^2 + \left(\frac{\sigma_2}{\mu_2}\right)^2 + \cdots + \left(\frac{\sigma_n}{\mu_n}\right)^2}$$

(d) Division

$$x = \frac{y_1}{y_2}$$

$$\left(\frac{\partial \phi}{\partial y_1}\right)_\mu = \frac{1}{\mu_2} \qquad \left(\frac{\partial \phi}{\partial y_2}\right)_\mu = \frac{-\mu_1}{(\mu_2)^2} \tag{4.8}$$

$$\sigma_x = \sqrt{\frac{\sigma_1^2}{(\mu_2)^2} + \frac{\sigma_2^2(\mu_1)^2}{(\mu_2)^4}} = x^* \sqrt{\left(\frac{\sigma_1}{\mu_1}\right)^2 + \left(\frac{\sigma_2}{\mu_2}\right)^2}$$

From the above examples we see that in general the errors are combined in quadrature—that is, it is their squares that are added. Consequently, if the error in one of the variables σ_i is large, it will dominate all other terms and the error of x, σ_x, will be almost equal to σ_i, in spite of good measurements made on the other independent variables.

Our simple rule for the case of addition, Eq. 4.5, may be used to obtain in a different way the result derived in Eq. 3.7. Let a variable x be measured and let the mean of a set of measurements be $\bar{x}_i$, with a standard deviation σ_i; if this set of measurements is repeated under identical conditions, a new mean result $\bar{x}_j \neq \bar{x}_i$ will be obtained, but let the standard deviations be equal; that is, $\sigma_j = \sigma_i$. If n such sets of measurements are performed, the new estimator for x will be

$$x^* = \frac{1}{n}(\bar{x}_1 + \bar{x}_2 + \cdots \bar{x}_n)$$

and thus

$$\left(\frac{\partial \phi}{\partial \bar{x}_i}\right)_\mu = \frac{1}{n}$$

hence, from Eq. 4.4 or 4.5

$$\sigma_x^* = \sqrt{\left(\frac{\sigma_1}{n}\right)^2 + \left(\frac{\sigma_2}{n}\right)^2 + \cdots + \left(\frac{\sigma_n}{n}\right)^2} = \sqrt{n\frac{\sigma^2}{n^2}} = \frac{\sigma}{\sqrt{n}} \tag{4.9}$$

Namely, the standard deviation of the mean of a Gaussian distribution is $\sigma/\sqrt{n}$, where σ is the standard deviation of the individual measurements.

4.3 EXAMPLE OF CALCULATION OF ERROR PROPAGATION

As an example, let us consider an experiment to determine Stefan's constant b, from the relation

$$E = bT^4$$

where the following values of E and T were obtained with the indicated standard deviations:

T (degrees Kelvin)	E (watts/m^2)
800 (1 ± 0.02)	$(\ 3.0 \pm 0.3) \times 10^4$
1000 (1 ± 0.02)	$(\ 8.0 \pm 0.8) \times 10^4$
1200 (1 ± 0.01)	$(15.6 \pm 0.6) \times 10^4$

We wish to calculate *to slide rule accuracy*, the estimator b^* and its standard deviation σ_b.

There are two ways to proceed in this case. We may either calculate b_j^* from each of the three sets of measurements and then combine these values to obtain $b^* = \bar{b}_j^*$, but weighing each b_j^* according to its standard deviation; or one may use least squares in the observed variables E and T^4. Note that a mean of T or E for the three listed measurements makes *no sense* whatsoever since each measurement is made for a *different* T.

We will follow the first procedure, and first we wish to obtain the error on T^4 from the known error on T. For this we should use the general expression, Eq. 4.4, but since $\phi = T^4$ is a function of only one variable,† simple differentiation gives the desired result directly

$$\frac{d\phi}{dT} = 4T^3 \qquad \frac{\Delta\phi}{\phi} = 4\frac{\Delta T}{T} \qquad\qquad (4.10)$$

We note from Eq. 4.9 that it is easier to work with relative errors, and we thus form Table 10.5.

TABLE 10.5

AN EXAMPLE OF A CALCULATION OF PROPAGATION OF ERRORS

Set of data	T^4	$E/T^4 = b_j^*$	$\sigma(T^4)/T^4$	$\sigma(b_j)/b_j^*$
1	0.41×10^{12}	7.3×10^{-8}	0.08	0.13
2	1.0×10^{12}	8.0×10^{-8}	0.08	0.13
3	2.0×10^{12}	7.8×10^{-8}	0.04	0.06

where

$$\frac{\sigma(b)}{b} = \sqrt{\left[\frac{\sigma(T^4)}{T^4}\right]^2 + \left[\frac{\sigma(E)}{E}\right]^2}$$

since the errors in T and E are uncorrelated.

---

† If we choose to write $\phi = T \times T \times T \times T$, we may *not* apply Eq. 4.7, since these variables are correlated; use of Eq. 4.4 and $\sigma_{TT} = \sigma_T$ gives back the result of Eq. 4.10.

For the best estimator of b, we will use the mean of the three measurements but weighed in inverse proportion to the square of their standard deviation (see Section 3.3). Thus

$$\bar{b} = \tfrac{1}{6}(7.3 + 8.0 + 4 \times 7.8) \times 10^{-8} = 7.75 \times 10^{-8}$$

for $\sigma(\bar{b})$ we use Eq. 4.5,

$$\sigma(\bar{b}) = \tfrac{1}{6}\sqrt{\sigma^2(b_1) + \sigma^2(b_2) + 4\sigma^2(b_3)}$$

or the convenient approximation

$$\frac{\sigma(\bar{b})}{\bar{b}} = \frac{1}{6}\sqrt{\left[\frac{\sigma(b_1)}{b_1}\right]^2 + \left[\frac{\sigma(b_2)}{b_2}\right]^2 + 4\left[\frac{\sigma(b_3)}{b_3}\right]^2} = 0.043$$

so that the final result is

$$b^* = 7.75 \ (1 \pm 0.043) \times 10^{-8} \ \text{joules}/(\text{degree})^4\text{-sec-m}^2$$

4.4 Evaluation of the Error Matrix

In the two previous sections we have discussed the case where only one unknown variable x was sought. We will now consider the random errors when several unknown variables are simultaneously estimated or measured.

When only one variable is measured, we know how to obtain from the data the second moment about the mean

$$\sigma^2 = \frac{1}{n} \sum_{i=1}^{n} (\bar{x} - x_i)^2$$

If now p variables are simultaneously measured in an experiment, we must form the $p(p + 1)/2$ second moments about the mean; for example, if we measure x, y, and z, we must calculate the six expressions

$$\sigma^2_{xx} = \frac{1}{n} \sum_{i=1}^{n} (\bar{x} - x_i)(\bar{x} - x_i); \quad \sigma^2_{yy} = \cdots; \quad \sigma^2_{zz} = \cdots$$

$$(4.11)$$

$$\sigma^2_{xy} = \frac{1}{n} \sum_{i=1}^{n} (\bar{x} - x_i)(\bar{y} - y_i) = \sigma^2_{yx};$$

$$\sigma^2_{xz} = \cdots = \sigma^2_{zx}; \quad \sigma^2_{yz} = \cdots = \sigma^2_{zy}$$

If the distribution of the variables x, y, and z is normal, then these six moments form the symmetric error matrix; if the variables are uncorrelated, the matrix is diagonal.

Clearly, the error matrix must be known if it is desired to apply Eq. 4.4. Consider, for example, that from the measured variables x, y, and z we wish to obtain a new unknown u and its standard deviations $\sigma(u)$, where

$$u = \phi(x, y, z) \qquad (4.12)$$

Then the values of σ_{ij}^2 that were obtained from the data with the help of Eq. 4.11, are substituted in Eq. 4.4 along with the partial derivatives of u, which are obtained from Eq. 4.12.

Conversely, if the frequency function of the three variables x, y, and z, and thus of u, is known,

$$f(u) = f[\phi(x, y, z)]$$

it is possible to calculate theoretically the elements of the error matrix through the usual expression

$$\mu'_2(x, y) = \iiint f(x, y, z) xy \, dx \, dy \, dz \qquad (4.13)$$

or

$$\mu_2(x, y) = \iiint f(x, y, z) (\mu_x - x)(\mu_y - y) \, dx \, dy \, dz$$

where

$$\sigma_{xy}^2 = \mu_2(x, y) \qquad \text{etc.}$$

In most practical applications, however, it is difficult to use Eq. 4.11 or Eq. 4.13. Equation 4.11 may not be usable because the unknown variables may not be measured directly (although they are measured implicitly); also, extensive data are required to yield meaningful results, and the calculation is cumbersome. Equation 4.13 may not be usable because the multidimensional integrals are frequently too difficult to calculate. Instead, the method of maximum likelihood provides an easy way for obtaining the error matrix.

As already discussed in Section 3, if the set of data x_i, y_i, $\cdots$, z_i have been measured, and the estimators for the m unknown variables θ_a, θ_b, $\cdots$, θ_m are sought, we may form the likelihood function

$$\mathcal{L}(x_1, x_2, \cdots, x_n, y_1, y_2, \cdots, y_n, \cdots, z_1, z_2, \cdots, z_n; \theta_a, \theta_b, \cdots, \theta_m)$$

$$= f(x_1, y_1, \cdots, z_1; \theta_a, \theta_b, \cdots, \theta_m) f(x_2, y_2, \cdots, z_2; \theta_a, \theta_b, \cdots, \theta_m) \cdots$$

$$\times f(x_n, y_n, \cdots, z_n; \theta_a, \theta_b, \cdots, \theta_m)$$

where f is the frequency function of the measured variable and is usually assumed to be a product of Gaussians. Then the estimators $\theta_a{}^*$, $\theta_b{}^*$, $\cdots$, $\theta_m{}^*$

are given by the values that *simultaneously* maximize $\mathcal{L}$, namely,

$$\frac{\partial \mathcal{L}}{\partial \theta_a}\bigg]_{\theta_a^*, \theta_b^*, \cdots, \theta_m^*} = \cdots = \frac{\partial \mathcal{L}}{\partial \theta_m}\bigg]_{\theta_a^*, \theta_b^*, \cdots, \theta_m^*} = 0 \qquad (4.14)$$

requiring the solution of m coupled equations. Equation 3.14 is a simple example of such a solution of Eq. 4.14. We note that the number of independent data points taken, n, must be larger than or equal to m.

The elements of the error matrix can be obtained from the inverse of the matrix

$$\mathbf{H}_{kl} = \frac{\partial^2 W}{\partial \theta_k \partial \theta_l}\bigg]_{\theta_a^*, \theta_b^*, \cdots, \theta_m^*} \qquad (4.15)$$

where the second-order partial derivatives must be calculated at the values of the estimators, and $W = \log \mathcal{L}$. We have

$$\sigma_{kl}^2 = (\mathbf{H})_{kl}^{-1}$$

where the rule for matrix inversion is

$$(\mathbf{H}^{-1})_{ij} = (-1)^{i+j} \frac{\text{Det } (ji \text{ minor of } \mathbf{H})}{\text{Det } \mathbf{H}} \qquad (4.16)$$

and the minor is the matrix resulting from $\mathbf{H}$ when the jth row and ith column are removed; obviously, the inverse matrix does not exist unless Det $\mathbf{H} \neq 0$.

We will now apply this method of obtaining the error matrix to the simple example treated in Section 3.3. The measured variables are x and y, and estimators are sought for the variables a and b; we assume that x is known exactly and that y is distributed normally for each measurement, and connected to x through

$$y = a + bx$$

using Eq. 3.10, we have

$$\mathcal{L} = \prod_{i=1}^{n} \left[\frac{1}{\sigma_i \sqrt{2\pi}} \exp\left\{ -\frac{1}{2\sigma_i^2} [y_i - \bar{y}(x_i; a, b)]^2 \right\} \right]$$

and

$$W = \log \mathcal{L} = -\frac{n}{2}\log (2\pi) - \sum_{i=1}^{n} \log \sigma_i - \frac{1}{2}\sum_{i=1}^{n}\left[\frac{y_i - (a + bx_i)}{\sigma_i} \right]^2$$

to simplify the calculations we assume $\sigma_1 = \sigma_2 = \cdots = \sigma_n$, so that

$$-\frac{\partial^2 W}{\partial a^2} = \frac{n}{\sigma^2}; \qquad -\frac{\partial^2 W}{\partial a \partial b} = \frac{\sum x_i}{\sigma^2}; \qquad -\frac{\partial W}{\partial b^2} = \frac{\sum x_i^2}{\sigma^2}$$

Hence

$$H = \frac{1}{\sigma^2} \begin{bmatrix} n & \sum x_i \\ \sum x_i & \sum (x_i^2) \end{bmatrix} \tag{4.17}$$

and

$$\text{Det } H = \frac{1}{\sigma^2} [n \sum (x_i^2) - (\sum x_i)^2]$$

Thus

$$\sigma_{\nu\mu}^2 = H^{-1} = \frac{\sigma^2}{n \sum (x_i^2) - (\sum x_i)^2} \begin{bmatrix} \sum (x_i^2) & -\sum x_i \\ -\sum x_i & n \end{bmatrix} \tag{4.18}$$

which gives the results stated in Eq. 3.16.

4.5 THE MONTE CARLO METHOD

It is clear that the calculation of the propagation of errors may become extremely involved, especially when the frequency functions of the variables cannot be expressed analytically and when intermediate processes of statistical nature take place. It is then preferable to use, with the help of a high-speed digital computer, the so-called "Monte Carlo" method.

By this technique, we follow a particular event through the sequence of processes it may undergo. For each process, all possible outcomes are weighed according to the frequency function and divided into x classes of equal probability. Then, from a table of these classes, one class is selected at random: for example, by looking up a table of x random numbers. The outcome of this process is incorporated in the progress of the event until a new decision point is reached, when again random selection is made. Thus, at the end of the sequence of all processes, certain final conditions will be reached from the initial conditions with which we started and through the intermediary of the random choices made at each decision point.

We follow in this fashion several events, always starting with the same initial conditions; but because of the random choices, the final conditions will be spread over some range. If enough events have been followed through, we are able to find the frequency function of the combined process and its parameters, namely, the mean final conditions and the standard deviation that result from a given set of initial conditions.

As a simple example, consider the following sequence of processes. A beam of N,π mesons of sufficiently high energy E is incident on an absorber of thickness l; we wish to know how many counts C will be obtained behind the absorber. Let the mean free path for the absorption of a π meson be λ_0; then to a first approximation,

$$C = Ne^{-l/\lambda_0} \tag{4.19}$$

When a primary π meson interacts, however, it may produce n secondary charged π mesons of average energy $E/2n$; each value of n has a definite probability. This intermediate process now makes the evaluation of C much more complicated, especially if we wish to include the direction of emission of the secondary mesons and their interactions in the absorber with a mean free path λ_s.

We may obtain a result more accurate than Eq. 4.19 by making a Monte Carlo calculation with the following simplifying assumptions, completely neglecting energy loss and angular spread:

(a) All primary π mesons interact with a mean free path λ_0.

(b) At the interaction point, $n = 0$, 1, or 2 secondary π mesons are produced, any of the three multiplicities having equal probability.

(c) Secondary π mesons interact with a mean free path λ_s, but may not produce tertiary π mesons.

We start with a primary π meson and first find the interaction point x_0. To this effect we divide the area under the curve e^{-x/λ_0} into 1000 intervals of equal area, as shown in Fig. 10.8; we select one such interval at random

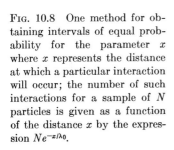

FIG. 10.8 One method for obtaining intervals of equal probability for the parameter x where x represents the distance at which a particular interaction will occur; the number of such interactions for a sample of N particles is given as a function of the distance x by the expression Ne^{-x/λ_0}.

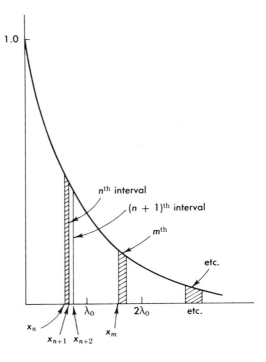

and this specifies x. If $x_0 > l$, clearly this primary will give a count C; if $x_0 < l$, we proceed to choose n. If $n = 0$, the primary does not give a count, but if $n = 1$ or $n = 2$, we find the distance of interaction of the secondary x_s if $n = 1$; or secondaries x_s^1 and x_s^2 if $n = 2$. If $x_s + x_0 > l$, we have a C count; if $x_s + x_0 < l$, we have no C count. If both $x_s^1 + x_0 > l$ and $x_s^2 + x_0 > l$, we have two C counts, and so on.

By following in this way several primaries, say N of them, we can find the transmission ratio C/N as a function of l. By repeating the calculation, we can find the distribution of C/N and its error σ about the true value

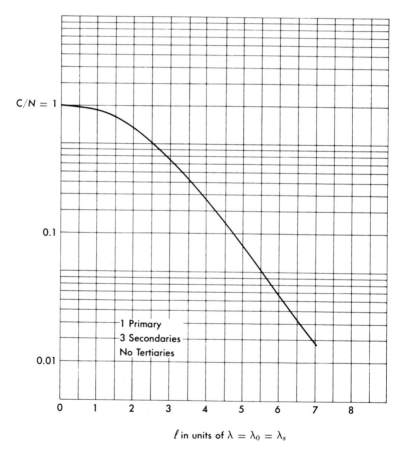

l in units of $\lambda = \lambda_0 = \lambda_s$

FIG. 10.9 Results of a Monte Carlo calculation for the attenuation of π mesons in an absorber. The abscissa gives the distance in the absorber in units of the mean-free path. It is assumed that each primary π meson produces three secondary π mesons, whereas each secondary may not produce tertiary particles. The mean-free path λ_0 for both primary and secondary π mesons has been taken as the same.

C_0/N. Consequently, when an experiment is performed and C/N is measured, we know that the true value may be found in a Gaussian centered at C/N and of standard deviation σ.

Results of such a calculation when all primary π mesons are assumed to produce at the interaction point three secondaries, are shown in Fig. 10.9. The interaction mean free path for both primaries and secondaries has been set equal $\lambda_0 = \lambda_s = \lambda$ and the abscissa gives l in units of λ, while the ordinate is a logarithmic scale of C/N.

5. The Statistics of Nuclear Counting

In many experiments related to nuclear physics, we count the particles or quanta emitted in the decay (or transition) of a nucleus. Usually only a very small fraction of the total sample undergoes such decay. The decay of *one* nucleus is a completely random phenomenon, yet from the number of counts in a given time interval, we may determine the decay probability of this species of nuclei or unstable particles.

5.1 The Frequency Function for the Number of Nuclear Decays

We start with the assumption that the decay of *one* nucleus is purely random and the probability (unnormalized) for decay in a time interval Δt is proportional to Δt and some constant λ with dimensions of inverse time:†

$$p_d = \lambda \, \Delta t \tag{5.1}$$

If we have a sample of N nuclei, since the presence of one nucleus does not affect the decay of another, the probability that *one* nucleus out of the *sample* of N nuclei will decay, in time Δt, is

$$P(1, \Delta t) = \lambda N \, \Delta t \tag{5.2}$$

Equation 5.2 is completely analogous to Eq. 2.10 of Section 2.6, which leads to the Poisson distribution; the only difference is that the constant *product* λN of Eq. 5.2 is the equivalent of the constant λ of Eq. 2.10. Consequently the probability (frequency function) for obtaining n decays in a time interval t is

$$P(n, t) = \frac{e^{-N\lambda t}(N\lambda t)^n}{n!} \tag{5.3}$$

The first moment of Eq. 5.3 (in the discrete unknown variable n), as we know from Eq. 2.14, is

$$\bar{n} = \lambda N t \tag{5.4}$$

---

† E. Schweidler, 1905; this assumption has been proved absolutely correct from the agreement of experiment with the deductions developed in the following paragraphs.

Since $\bar{n}/t$ is the average number of decays per unit time (average rate), we find the physical significance of the constant parameter λ. That is, $N\lambda$ gives the average decay rate of the sample; N is the total number of nuclei in the sample.

Similarly, the second moment about the mean of Eq. 5.3, as we know from Eq. 2.15, is

$$\sigma^2 = \lambda N t = \bar{n}$$

Hence the very frequently used expression,

$$\sigma = \sqrt{\bar{n}} \tag{5.5}$$

Note however, that $\bar{n}/t = N\lambda$ is the theoretical average rate, which is usually unknown (unless λ and N are precisely known for the sample under consideration). The average rate that we measure, $R = n/t$ (counts per unit time), will, in general, differ from the true rate $N\lambda = \bar{n}/t$, but if n is large, R will be distributed normally about $N\lambda$. (See Eq. 5.3a below.)

Note: It is interesting to consider the inverse problem; that is, if the observed number of counts is n, what is the probability that the true average count is given by $N\lambda t$? The frequency function for this distribution is still given by Eq. 5.3, but now in the continuous unknown variable $N\lambda t$,

$$p(N\lambda t, n) = \frac{\exp(-N\lambda t)(N\lambda t)^n}{n!} \tag{5.6}$$

Next we find the moments of this distribution in the variable $N\lambda t$. We let $N\lambda t = x$ so that

$$\mu_1'(x) = \int_0^\infty x P(x, n) \, dx = \frac{1}{n!} \int_0^\infty x \exp(-x) x^n \, dx = \frac{(n+1)!}{n!} = n + 1$$

hence the surprising result that if the observed counts are n, the mean value for the true number of decays is $n + 1$

$$\overline{N\lambda t} = n + 1 \tag{5.7}$$

Similarly

$$\mu_2'(x) = \int_0^\infty x^2 P(x, n) \, dx = \frac{(n+2)!}{n!} = (n+2)(n+1)$$

and

$$\sigma^2 = \mu_2' - \mu^2 = (n^2 + 3n + 2) - (n+1)^2 = n + 1$$

$$\sigma(N\lambda t) = \sqrt{n+1} = \sqrt{\overline{N\lambda t}} \tag{5.8}$$

From the considerations of Section 2.9, it should be clear that when the total number of observed counts n is large, Eq. 5.3 (and for that matter also Eq. 5.6) are well approximated by a Gaussian with mean $\mu = N\lambda t$ and standard deviation $\sigma = N\lambda t$.

$$P(n, t) = \frac{1}{\sqrt{2\pi N\lambda t}} \exp\left[-\frac{(N\lambda t - n)^2}{2N\lambda t}\right] \tag{5.3a}$$

$$P(N\lambda t, n) = \frac{1}{\sqrt{2\pi n}} \exp\left[-\frac{(n - N\lambda t)^2}{2n}\right] \tag{5.6b}$$

Thus, unless we are dealing with very few counts, Gaussian statistics may be safely applied.

Finally, we summarize here some simple consequences of Eq. 5.1 for a *single* nucleus:

(a) If the probability for decay in dt is

$$p_d(dt) = \lambda \, dt$$

(b) Then the probability for not decaying (survival) in the time interval from $t = 0$ to $t = t$ is

$$p_s(t) = e^{-\lambda t}$$

(for proof see Eq. 2.11.)

(c) The probability for decay in dt at time t is

$$p_d(t, dt) = e^{-\lambda t}\lambda \, dt$$

(d) The probability for decay in the time interval from $t = 0$ to $t = t$ is

$$p_d(t) = 1 - p_s(t) = 1 - e^{-\lambda t}$$

Note that only (c) is properly normalized, so that

$$\int_0^\infty p_d(t, dt) = \int_0^\infty e^{-\lambda t}\lambda \, dt = 1$$

expressions (b) and (d) are, correctly, always <1 and reduce to 0 and 1 respectively as t approaches infinity. As to expression (a), we must keep in mind that it holds only for Δt such that $\lambda \, \Delta t \ll 1$.

5.2 Behavior of Large Samples

Having obtained the frequency functions, we may now examine the behavior of the total sample. From Eq. 5.4 we see that given a sample of N nuclei, on the average, in a time interval Δt there will be

$$n = \lambda N \, \Delta t$$

decays; that is, the total sample will be decreased by an amount

$$-\Delta N = N\lambda \, \Delta t \qquad (5.9)$$

Equation 5.9 then leads to the differential equation for the number of nuclei in the sample

$$\frac{dN}{N} = -\lambda \, dt$$

with solution

$$N(t) = N_0 e^{-\lambda t} \qquad (5.10)$$

where N_0 is the number of nuclei at time $t = 0$. Frequently $\tau = 1/\lambda$ is used for the exponent in Eq. 5.10; τ is called the *lifetime* of that particular species of nuclei and is the time in which the population of the sample is reduced to 37 percent $(1/e)$ of its original value. The *half-life*

$$\tau_{1/2} = \tau[\ln_e \tfrac{1}{2}] = 0.693\tau$$

gives the time in which the population of the sample is reduced to half its original value. Using Eq. 5.10 we find, for the *decay rate* as a function of time,

$$\frac{dN}{dt} = R(t) = -\lambda N(t) = -\lambda N_0 e^{-\lambda t} \qquad (5.11)$$

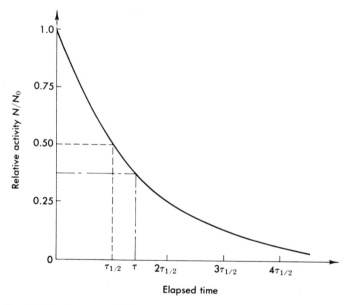

FIG. 10.10 Exponential decay of a sample of radioactive nuclei. The abscissa is calibrated in units of the half-life of the sample; the lifetime is also indicated.

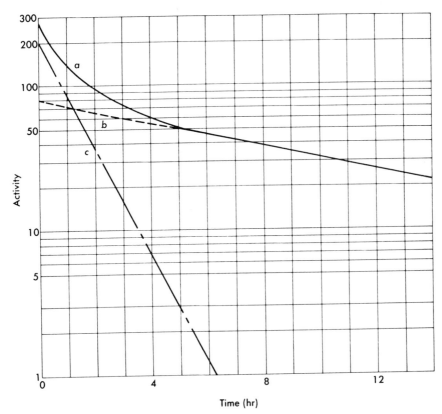

FIG. 10.11 The decay curve for a sample containing two species of radioactive nuclei each decaying with a different lifetime. Note that the composite decay curve a, is the sum of curves b and c.

which obviously has the same time dependence as Eq. 5.10. Experimentally we usually measure $R(t)$ and obtain a curve as shown in Fig. 10.10; from such a plot λ may be obtained.

If the sample contains two or more different species of nuclei with different decay constants λ_1, λ_2, $\cdots$, the time dependence of the decay rate is no longer the simple exponential of Eq. 5.11; instead

$$\frac{dN}{dt} = R(t) = -\lambda_1 N_0^1 e^{-\lambda_1 t} - \lambda_2 N_0^2 e^{-\lambda_2 t} - \cdots$$

If, however, $\lambda_1 \gg \lambda_2$ for small t(that is, $t \sim 1/\lambda_1$), $R(t)$ is dominated by the first term; for large t(for example, $t \approx 1/\lambda_2$), $R(t)$ is dominated by the second term. This is shown in Fig. 10.11, which gives the decay curves on a semilogarithmic plot.

Another situation of interest arises when nuclei of species A decay into species B with a constant λ_A; nuclei B, however, decay in turn into species C with a constant λ_B. Let the original number of nuclei of species A be N_0 and that of species B be 0.

Then the number of nuclei of species A as a function of time is still given by Eq. 5.10, $N_A = N_0 e^{-\lambda_A t}$. However, for the number of nuclei of species B, the following differential equation holds

$$\frac{dN_B}{dt} = +\lambda_A N_A - \lambda_B N_B$$

$$\frac{dN_B}{dt} + \lambda_B N_B = \lambda_A N_0 e^{-\lambda_A t}$$

The solution of this first-order linear differential equation is straight-forward, and with the initial condition $N_B(t = 0) = 0$, we have

$$N_B = N_0 \frac{\lambda_A}{\lambda_B - \lambda_A} \left[e^{-\lambda_A t} - e^{-\lambda_B t} \right] \tag{5.12}$$

Note that Eq. 5.12 always gives $N_B > 0$, as it must be, irrespective of whether $\lambda_A > \lambda_B$ or $\lambda_B > \lambda_A$. Equation 5.12 correctly reduces to $N_B = 0$ for $t = 0$ and for $t = \infty$. The two limiting cases can also be obtained from Eq. 5.12 if we take into account that the decay rate from B to C is given by $R_{BC}(t) = N_B \lambda_B$. Thus

$$\text{for} \quad \lambda_B \gg \lambda_A \qquad R_{BC}(t) \approx N_0 \lambda_A e^{-\lambda_A t}$$

$$\text{for} \quad \lambda_A \gg \lambda_B \qquad R_{BC}(t) \approx N_0 \lambda_B e^{-\lambda_B t}$$

5.3 TESTING OF THE DISTRIBUTION OF RADIOACTIVE DECAY; THE DISTRIBUTION OF THE TIME INTERVALS BETWEEN COUNTS

It is frequently desirable to test whether a sample of counting data does indeed come from the decay of radioactive nuclei, that is, that it follows the frequency function of Eq. 5.3. Such a test can be performed in accordance with the procedures developed in Section 3; in the example of Section 3.4, the distribution of $R = n/t$ about its mean was compared with a Gaussian, so that what was tested was the applicability of Eq. 5.3(a). If we wish to test Eq. 5.3 directly, we must obviously use data where n is small; a very sensitive test can be devised if we plot the distribution of the time intervals between successive decays, or every second, third (etc.) decay.

First we obtain the distribution of the time intervals between two successive decays. Let $t = 0$ when a decay occurs; we then seek the probability that no decay occurs until $t = t$, but a decay occurs within dt at $t = t$. This probability is clearly given by Eq. 5.3 with $n = 0$, multiplied by Eq. 5.2; namely,

$$P(t, dt) \equiv q_1(t) \, dt = e^{-N\lambda t} N\lambda \, dt \tag{5.13} \dagger$$

Equation 5.13 indicates that the shortest time intervals between two counts are much more frequent than the longer ones; this is true for any *random* events, since they obey Eq. 5.1.

Next we consider the distribution of the time intervals between every second, third, $\cdots$, mth count. In practice this need arises when the counts from the output of a scaling circuit are recorded. Consider, therefore, a circuit giving one output count for every m input counts‡; if the true input rate is r, then the output rate R is related through

$$N\lambda = r = Rm$$

Let $t = 0$ when an output pulse arrives, and let $Q_m(t)$ be the probability that another output pulse arrives *in the time interval* t; $q_m(t) \, dt$ will then be the probability that this other output pulse arrives *at* t (between t and $t + dt$).

Another output pulse will arrive if the input counts $n \geq m$, so that

$$Q_m(t) = \sum_{n \geq m}^{\infty} P(n, t) = \sum_{n \geq m}^{\infty} \frac{(rt)^n e^{-rt}}{n!}$$

$$= 1 - \sum_{n=0}^{n=m-1} \frac{(rt)^n e^{-rt}}{n!} \tag{5.13a}$$

where the last equality follows from the normalization of Eq. 5.3

$$\sum_{n=0}^{\infty} P(n, t) = 1$$

Now by considering the sample space of Fig. 10.12 we see that the set of points $Q_m(t)$ is a subset of $Q_m(t + dt)$, so that any sample-space point belonging to $Q_m(t + dt)$ but not to $Q_m(t)$ represents an output count between t and $t + dt$. Thus

$$q_m(t) \, dt = Q_m(t + dt) - Q_m(t)$$

or

$$q_m(t) = \frac{dQ_m(t)}{dt}$$

---

† Compare this equation with the probability for the decay of a single nucleus, as given in Section 5.1(c).

‡ See J. L. Rainwater and C. S. Wu, *Nucleonics*, Vol. 1, 62 (1947).

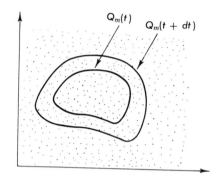

FIG. 10.12 Sample space indicating the domain $Q_m(t)$ which contains all points corresponding to the arrival of an output count in the time interval from 0 to t after the previous count. This domain forms a subset of $Q_m(t + dt)$ which contains all points corresponding to the arrival of the output count from 0 to $t + dt$. Clearly the arrival of the output count at t is $q_m(t) = Q_m(t + dt) - Q_m(t)$.

and taking the derivative of Eq. 5.13a

$$q_m(t) = - \sum_{n=0}^{m-1} \left[\frac{rn(rt)^{n-1}e^{-rt}}{n!} - \frac{r(rt)^n e^{-rt}}{n!} \right]$$

$$= r \sum_{n=0}^{m-1} \frac{(rt)^n e^{-rt}}{n!} - r \sum_{n=1}^{m-1} \frac{(rt)^{n-1} e^{-rt}}{(n-1)!}$$

and by replacing in the second sum n by $l = n - 1$, only the last term of the first sum survives. The result is

$$q_m(t) = r \frac{(rt)^{m-1} e^{-rt}}{(m-1)!} \tag{5.14}$$

Equation 5.14 correctly reduces to Eq. 5.13 for $m = 1$ (since $r = N\lambda$). For $m \geq 2$, Eq. 5.14 has a maximum at $dq_m(t)/dt = 0$, or

$$[r^2(m-1)(rt)^{m-2}e^{-rt}] - [r^2(rt)^{m-1}e^{-rt}] = 0$$

hence $t = (m-1)/r$ and for large m, $t \to m/r = 1/R$. Thus we see that the most probable time interval is not the shortest one, but instead approaches the mean time interval between *output* counts $1/R$; that is, the

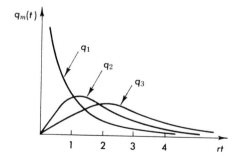

FIG. 10.13 The probability $q_m(t)$ that the mth count will follow any original count at the time interval t. Note that the abscissa is calibrated in units of rt where r is the unscaled rate of events; for m large the curves approach a Gaussian with mean $\langle rt \rangle = m$ or $\langle t \rangle = m/r$.

scaling circuit regularizes the counts. Equation 5.14 is shown in Fig. 10.13 for different values of *m*. Comparison of these curves with experimental data has been presented in Chapter 5, Section 3.5.

REFERENCES

There are many texts, both elementary and advanced, on the subject of statistics, data fitting, and treatment of errors. The references given below were consulted for the preparation of this chapter.

P. G. Hoel, *Introduction to Mathematical Statistics*, John Wiley, 1958. The presentation of Sections 1 and 2 follows Hoel closely.

J. Orear, *Notes on Statistics for Physicists*, University of California report UCRL-8417 (unpublished). Estimation and data fitting are discussed at a fairly mathematical level.

G. Friedlander and J. W. Kennedy, *Nuclear and Radiochemistry*, John Wiley, 1949. A clear presentation of statistics at the intermediate level, especially as applicable to nuclear counting.

OTHER REFERENCES

Y. Beers, *Introduction to the Theory of Error*, Addison-Wesley, 1958.
M. G. Kendall, *The Advanced Theory of Statistics*, Hafner Publishing Co., 1951.
R. A. Fisher, *Statistical Theory of Estimation*, University of Calcutta Press, 1938.

APPENDIX

I. Useful Constants and Relations

In this table all constants and relations are given in the cgs system. They are also given in the rationalized MKSQ system and the MeV, cm, sec system when their value is different in these systems from that in the cgs system. Consult Appendix II also for relations of units between the different systems.

Physical Constants

Constant or relation	cgs	Rationalized MKSQ	MeV, cm, sec
Velocity of light	$c = 2.997925 \times 10^{10}$ cm sec^{-1}	2.997925×10^{8} cm sec^{-1}	—
Charge of the electron	$e = 4.80298 \times 10^{-10}$ esu	1.6021×10^{-19} coulomb	—
Mass of the electron	$m_e = 9.108 \times 10^{-28}$ gm	0.9108×10^{-30} kg	0.511006 MeV/c^2
Ratio, electron charge to electron mass	$e/m = 1.7589 \times 10^{7}$ emu gm^{-1} $= 5.2741 \times 10^{17}$ esu gm^{-1}	1.7589×10^{11} coulomb kg^{-1}	—
Plank's constant	$h = 6.6237 \times 10^{-27}$ erg sec	6.6237×10^{-34} joules sec	6.5850×10^{-22} MeV sec
	$\hbar = h/2\pi = 1.05450 \times 10^{-27}$ erg sec	1.0540×10^{-34} joules sec	
Avogadro's number	$N_0 = 6.02252 \times 10^{23}$ molecules (gm mole)$^{-1}$	6.02252×10^{26} molecules (kg mole)$^{-1}$	—
Mass of the proton	$m_p = 1.67221 \times 10^{-24}$ gm	1.67221×10^{-27} kg	938.256 MeV/c^2
Ratio, proton mass to electron mass	$m_p/m_e = 1836.57$	—	—
Boltzmann's constant	$k = 1.3802 \times 10^{-16}$ erg deg^{-1}	1.3802×10^{-23} joules deg^{-1}	8.6171×10^{-11} MeV deg^{-1}
Universal gas constant	$R = \begin{cases} 8.3143 \times 10^{7} \text{ erg deg}^{-1} \text{ mole}^{-1} \\ 82.06 \text{ cm}^3 \text{ atm deg}^{-1} \text{ mole}^{-1} \end{cases}$	8.3143×10^{3} joules deg^{-1} (kg mole)$^{-1}$	—
Loschmidt's constant	$n_0 = 2.6874 \times 10^{19}$ cm^{-3} atm^{-1}	(molecules per ml of ideal gas at ntp)	—

Constant or relation	cgs	Rationalized MKSQ	MeV, cm, sec

Derived Constants Mainly Related to Atomic Spectroscopy

Constant or relation	cgs	Rationalized MKSQ	MeV, cm, sec
Radius of first Bohr orbit	$a_0 = \hbar^2/m_e e^2 = 0.52967 \times 10^{-8}$ cm	$(4\pi\epsilon_0)\hbar^2/m_e e^2 = 0.52967 \times 10^{-10}$ m	—
Rydberg's wave number	$R_\infty = (e^4 m_e/2\hbar^2)(2\pi c\hbar)^{-1} = 109737.32$ cm^{-1} $R_H = 109677$ cm^{-1}	$[e^4 m_e/(4\pi\epsilon_0)^2 2\hbar^2](2\pi c\hbar)^{-1} = 1.0973732 \times 10^7$ m^{-1} 1.09677×10^7 m^{-1}	13.605 eV
Fine structure constant	$\alpha = e^2/c\hbar = 1/137.0388$ (dimensionless)	$e^2/(4\pi\epsilon_0)\hbar c$	—
Bohr magneton	$\mu_0 = e\hbar/2m_e c = 0.92732 \times 10^{-20}$ erg gauss^{-1}	$e\hbar/2m_e = 0.92732 \times 10^{-23}$ amp m^2	0.578815×10^{-14} MeV gauss^{-1}
(Zeeman displacement per gauss)	$\mu_0/hc = 4.6688 \times 10^{-5}$ cm^{-1} gauss^{-1}	$\mu_0/h = 1.401$ Mc. gauss^{-1}	
Nuclear magneton	$\mu_p = e\hbar/2m_p c = 0.50504 \times 10^{-23}$ erg gauss^{-1}	$e\hbar/2m_p = 0.50504 \times 10^{-26}$ amp m^2	3.1524×10^{-18} MeV gauss^{-1}
Proton nuclear magnetic resonance	—	4.257 kc/gauss	—

Other Derived Constants

Constant or relation	cgs	Rationalized MKSQ	MeV, cm, sec
Classical radius of the electron	$r_e = e^2/m_e c^2 = 2.8177 \times 10^{-13}$ cm	$e^2/(4\pi\epsilon_0)m_e c^2 = 2.8177 \times 10^{-15}$ m	
Thomson cross section	$8/3\pi r_e^2 = 0.66516 \times 10^{-24}$ cm^2	—	
Compton wavelength of the electron	$\lambda_c = h/mc = 2.4262 \times 10^{-10}$ cm $\lambda_c = h/mc = 3.86144 \times 10^{-11}$ cm	2.4262×10^{-12} m	
Stefan's constant	$\sigma = 2\pi^5 k^4/15c^2 h^3 = 5.6687 \times 10^{-5}$ erg cm^{-2} sec^{-1} deg^{-4}	5.6687×10^{-8} joules m^{-2} sec^{-1} deg^{-4}	

I. Useful Contents and Relations (Continued)

Conversion Factors

1 eV $\quad = 1.6021 \times 10^{-12}$ erg

kT at 288°K $\quad = 0.0248 \cong 1/40$ eV

For photons $\lambda = he/E$ $\quad = 12378/E$ (E in eV)
(in Ångstroms)

For electrons $\lambda = h/p = hc/(E^2 - m_e^2 c^4)^{1/2}$; in the x-ray region we may write λ (in Å) $\sim (150/V)^{1/2}$ where V is the accelerating potential in volts

Permittivity of free space $\quad \epsilon_0 = 10^7/4\pi c^2$ farad/meter $\cong 1/4\pi \times 1/(9 \times 10^9) = 10^{-9}/36\pi$ exists only in MKSQ system

Permeability of free space $\quad \mu_0 = 4\pi \times 10^{-7}$ henry/meter

1 calorie = 4.184 joules

1 atm = 1033.2 gm/cm² $\quad = 14.697$ psi

1 amu = 931 MeV

Numerical Constants

1 radian = 57.29578 deg

e = 2.71828

π = 3.14159

ln 2 = 0.69315

$\log_{10} e$ = 0.43429

Miscellaneous

1 year	3.1536×10^7 sec
Density of air	1.205×10^{-3} gm cm^{-3} at 20° C
Acceleration of gravity	$g = 980.67$ cm sec^{-2}
Gravitational constant	$G = 6.664 \times 10^{-8}$ cm^3 kg^{-1} sec^{-2}
Nuclear radius	$R = 1.2 \times 10^{-13} \times A^{1/3}$ cm; A is the mass number of the nucleus
True grating space of calcite (20° C)	$d'_{20} = 3.028 \times 10^{-8}$ cm
Volume of 1 gm mole at 0° C and 1 atm	$v_0 = 22.415 \times 10^3$ (gm mole)$^{-1}$

Radioactivity

1 curie	3.7×10^{10} disintegrations sec^{-1}
1 R (Roentgen)	78 ergs gm^{-1} air
1 curie of Co^{60} produces at a distance of 1 m a radiation of 1.3 R/hour	
Natural background	100 mR/year

II. Dimensions and Units of Some Physical Quantities

In this table are given the units, abbreviations, conversion factor, and dimensions of some of the most frequently encountered physical quantities. Whereas no great difficulty arises in going from the cgs to the MKSQ system and vice versa for the mechanical units, in the case of the electromagnetic units the situation is more confused. Therefore we clarify that a physical quantity which amounts to A units in the MKSQ system, amounts to $A \times C$ units in the cgs system where C is the conversion factor shown in column 4 of the table. When accurate work is done, the factors of 3 in the conversion expressions must be replaced by 2.99793 (the exact value of the velocity of light).

Physical quantity	Common symbol	Rationalized MKSQ	Conversion factor	cgs	Dimensions	Other units
Mechanical units						
Length	l	1 meter (m)	10^2	centimeters (cm)	L	1 millimeter (mm) = 10^{-3} m 1 micron (μ) = 10^{-6} m 1 Ångstrom (Å) = 10^{-10} m 1 Fermi (f) = 10^{-15} m
Mass	m	1 kilogram (kg)	10^3	grams (gm)	M	—
Time	t	1 second	1	second (sec)	T	1 millisecond (msec) = 10^{-3} sec 1 microsecond (μsec) = 10^{-6} sec 1 nanosecond (nsec) = 10^{-9} sec
Force	F	1 newton	10^5	dynes	$\mathrm{MLT^{-2}}$	—
Work Energy	W U	1 joule	10^7	ergs	$\mathrm{ML^2T^{-2}}$	electron volt (eV) = $\quad 1.6 \times 10^{-19}$ joules (keV), (MeV), (BeV)
Power	P	1 watt	10^7	ergs $\mathrm{sec^{-1}}$	$\mathrm{ML^2T^{-3}}$	1 horsepower (HP) = 736 watts
Electromagnetic units						
Charge	q	1 coulomb	3×10^9	stat coulombs (esu)	Q	0.1 ab coulombs (emu)

494

Physical quantity	Common symbol	Rationalized MKSQ	Conversion factor	cgs	Dimensions	Other units
Current	I	1 ampere (coulomb sec^{-1})	3×10^9	stat amperes (esu)	Q/T^{-1}	0.1 ab amperes (emu)
Potential	ϕ, V	1 volt	1/300	stat volt (esu)	$ML^2T^{-2}Q^{-1}$	10^8 ab volts (emu)
Electric field	E	1 volt m^{-1}	$1/3 \times 10^{-4}$	stat volt per cm (esu)	$MLT^{-2}Q^{-1}$	10^6 ab volts per cm (emu)
Permittivity of free space	ϵ_0	8.85×10^{-12}	$10^{-9}/36\pi$	farad meter^{-1}	$T^2Q^2M^{-1}L^3$	—
Displacement	D	1 coulomb m^{-2}	$12\pi \times 10^5$	stat volt cm^{-1} (stat coulomb cm^{-2})	Q/L^2	—
Conductivity	σ	1 mho m^{-1}	9×10^9	sec^{-1}	TQ^2/ML^3	—
Resistance	R	1 ohm (Ω)	$1/9 \times 10^{-11}$	sec cm^{-1}	$ML^2T^{-1}Q^{-2}$	(kΩ), (MΩ)
Capacitance	C	1 farad	9×10^{11}	cm	$T^2Q^2M^{-1}L^{-2}$	(μf)($\mu\mu$f or pf)
Permeability of free space	μ_0	$4\pi \times 10^{-7}$	1.257×10^{-6}	henry meter^{-1}	MLQ^{-2}	—
Magnetic flux	Φ	1 weber	10^8	gauss-cm^2 (Maxwells)	$ML^2T^{-1}Q^{-1}$	—
Magnetic induction	B	1 weber m^{-2}	10^4	gauss	$MT^{-1}Q^{-1}$	(kilogauss)
Magnetic field	H	1 ampere-turn m^{-1}	$4\pi \times 10^3$	öersted	$QT^{-1}L^{-1}$	—
Inductance	L	1 henry	$1/9 \times 10^{-11}$	(cgs esu)	ML^2Q^2	(millihenry), (microhenry)
Magnetic moment	μ	1 ampere-m^2	—	—	QL^2T^{-1}	—

Other units

Physical quantity	Common symbol	Rationalized MKSQ	Conversion factor	cgs	Dimensions	Other units
Cross section	σ	1 barn = 10^{-24}cm^2			L^2	(millibarn)(microbarn)
Frequency	ν, f	cycles sec^{-1} (cps)			T^{-1}	(kc) (Mc) (kMc)
Angular frequency	ω	1 radian sec$^{-1} = 2\pi\nu$			T^{-1}	—
Wave number	$\bar{\nu}$	1 cm^{-1} $1/\lambda = \nu/c$			L^{-1}	—

495

III. Materials Used as Absorbers and Detectors

The following table indicates atomic and nuclear properties $(dE/dx$, collision mean free path, radiation length, etc.) of materials used as absorbers and detectors.†

Material	Z	A	Cross-section[a] σ (barns)	$-dE/dx$ min (MeV/gm/cm²)	Collision[a]		Radiation		Density ρ (gm/cm³)
					Length (gm/cm²)	L_{coll} (cm)	Length (gm/cm²)	L_{rad} (cm)	
H₂	1	1.01	0.063	4.14	26.5	374	58	819.0	0.0708[b]
Li	3	6.94	0.23	1.72	50.4	94.3	77.5	145	0.534
Be	4	9.01	0.28	1.71	55.0	29.9	62.2	33.8	1.84
C	6	12.00	0.33	1.86	60.4	39.0	42.5	27.4	1.55[c]
Al	13	26.97	0.57	1.66	79.2	29.3	23.9	8.86	2.70
Cu	29	63.57	1.00	1.45	105.4	11.8	12.8	1.44	8.9
Sn	50	118.70	1.55	1.27	129.7	17.8	8.54	1.17	7.30
Pb	82	207.21	2.20	1.12	156.2	13.8	5.8	0.51	11.34
U	92	238.07	2.42	1.095	163.6	8.75	5.5	0.29	18.7
Hydrogen (bubble chamber, 27.6° K)				0.243 MeV/cm	26.5	452	58	990	0.0586
Propane (C₃H₈, bubble chamber)				0.935 MeV/cm	48.9	119.3	44.7	109.0	0.41
Freon, CF₃Br				2.3	87.1	58.0	17.25	11.5	1.5
Polystyrene (CH scintillator)				2.14 MeV/cm	54.9	52.3	43.4	41.3	~1.05
Ilford emulsion				5.49 MeV/cm	103	27.0	11.2	2.91	3.815

† From W. Barkas and A. Rosenfeld, University of California Radiation Laboratory report UCRL-8030.

[a] $\sigma_{\text{natural}} \equiv \pi(\hbar/m_\pi c)^2 \times A^{2/3} = 63 \text{ mb} \times A^{2/3}$; $L_{\text{collision}} \equiv (A/N_0\sigma_{\text{natural}}) = [A^{1/3}/N_0\pi(\hbar/m_\pi c)^2] = 26.4\ A^{1/3}$ gm/cm².

[b] Boiling at 1 atm.

[c] Variable.

IV. Some Electronic Properties of Atoms*

This table gives the densities and the electron configuration, spectroscopic terms, and ionization potentials of the ground states of the neutral and singly ionized atoms. In each case the electronic configuration of the singly ionized atom is obtained by removing one electron from the last-named shell in the configuration of the neutral atom, except as indicated.

(Example: Sc ion ground state is $3d\ 4s\ ^3D_1$.)

    g = gas
    l = liquid
    IP = ionization potential.

Z	Element	Density g cm^{-3} at 0°C, 1 atm	Ground state			Ion ground state	
			Elect config	Term	IP, eV	Term	IP, eV
1	H	8.988×10^{-5} (g)	1s	$^2S_{1/2}$	13.595		
2	He	1.785×10^{-4} (g)	$1s^2$	$^1S_0$	24.580	$^2S_{1/2}$	54.503
3	Li	0.534	[He] 2s	$^2S_{1/2}$	5.390	$^1S_0$	75.619
4	Be	1.85	$2s^2$	$^1S_0$	9.320	$^2S_{1/2}$	18.206
5	B	2.535	$2s^2\ 2p$	$^2P_{1/2}$	8.296	$^1S_0$	25.149
6	C	3.52 (dia.) 2.25 (graph.)	$2s^2\ 2p^2$	$^3P_0$	11.264	$^2P_{1/2}$	24.376
7	N	1.257×10^{-3}(g)	$2s^2\ 2p^3$	$^4S_{3/2}$	14.54	$^3P_0$	29.605
8	O	1.428×10^{-3}(g)	$2s^2\ 2p^4$	$^3P_2$	13.614	$^4S_{3/2}$	35.146
9	F	1.695×10^{-3}(g)	$2s^2\ 2p^5$	$^2P_{3/2}$	17.418	$^3P_2$	34.98
10	Ne	0.9005×10^{-3}(g)	$2s^2\ 2p^6$	$^1S_0$	21.559	$^2P_{3/2}$	41.07
11	Na	0.9712	[Ne]3s	$^2S_{1/2}$	5.138	$^1S_0$	47.29
12	Mg	1.741	$3s^2$	$^1S_0$	7.644	$^2S_{1/2}$	15.03
13	Al	2.70	$3s^2\ 3p$	$^2P_{1/2}$	5.984	$^1S_0$	18.823
14	Si	2.42	$3s^2\ 3p^2$	$^3P_0$	8.149	$^2P_{1/2}$	16.34
15	P	wh 1.83, red 2.20 met 2.34, blk 2.69	$3s^2\ 3p^3$	$^4S_{3/2}$	10.55	$^3P_0$	19.65
16	S	2.0–2.1	$3s^2\ 3p^4$	$^3P_2$	10.357	$^4S_{3/2}$	23.4
17	Cl	3.16×10^{-3}(g)	$3s^2\ 3p^5$	$^2P_{3/2}$	13.01	$^3P_2$	23.80
18	A	1.782×10^{-3}(g)	$3s^2\ 3p^6$	$^1S_0$	15.755	$^2P_{3/2}$	27.62
19	K	0.870	[A] 4s	$^2S_{1/2}$	4.339	$^1S_0$	31.81
20	Ca	1.54	$4s^2$	$^1S_0$	6.111	$^2S_{1/2}$	11.87
21	Sc	2.5	$3d\ 4s^2$	$^2D_{3/2}$	6.56	$^3D_1$	12.80
22	Ti	4.5	$3d^2\ 4s^2$	$^3F_2$	6.83	$^2F_{3/2}$	13.57
23	V	5.6	$3d^3\ 4s^2$	$^4F_{3/2}$	6.74	$(3d^4)\ ^5D_0$	14.65

* This table from *Principles of Modern Physics* by R. B. Leighton. Copyright © 1959. McGraw-Hill Book Company. Used by permission.

Z	Element	Density g cm^{-3} at 0°C, 1 atm	Ground state			Ion ground state	
			Elect config	Term	IP, eV	Term	IP, eV
24	Cr	6.93	$3d^5\ 4s$	$^7S_3$	6.764	$^6S_{5/2}$	16.49
25	Mn	7.3	$3d^5\ 4s^2$	$^6S_{5/2}$	7.432	$^7S_3$	15.64
26	Fe	7.86	$3d^6\ 4s^2$	$^5D_4$	7.90	$^6D_{9/2}$	16.18
27	Co	8.71	$3d^7\ 4s^2$	$^4F_{9/2}$	7.86	$(3d^8)\ ^3F_4$	17.05
28	Ni	8.8	$3d^8\ 4s^2$	$^3F_4$	7.633	$(3d^9)\ ^2D_{5/2}$	18.15
29	Cu	8.90	$3d^{10}\ 4s$	$^2S_{1/2}$	7.724	$^1S_0$	20.29
30	Zn	7.0	$3d^{10}\ 4s^2$	$^1S_0$	9.391	$^2S_{1/2}$	17.96
31	Ga	5.93	$3d^{10}\ 4s^2\ 4p$	$^2P_{1/2}$	6.00	$^1S_0$	20.51
32	Ge	5.46	$3d^{10}\ 4s^2\ 4p^2$	$^3P_0$	7.88	$^2P_{1/2}$	15.93
33	As	5.73	$3d^{10}\ 4s^2\ 4p^3$	$^4S_{3/2}$	9.81	$^3P_0$	20.2
34	Se	4.82	$3d^{10}\ 4s^2\ 4p^4$	$^3P_2$	9.75	$^4S_{3/2}$	21.5
35	Br	3.12(l)	$3d^{10}\ 4s^2\ 4p^5$	$^2P_{3/2}$	11.84	$^3P_2$	21.6
36	Kr	3.736×10^{-3}(g)	$3d^{10}\ 4s^2\ 4p^6$	$^1S_0$	13.996	$^2P_{3/2}$	24.56
37	Rb	1.532	[Kr] $5s$	$^2S_{1/2}$	4.176	$^1S_0$	27.5
38	Sr	2.60	$5s^2$	$^1S_0$	5.692	$^2S_{1/2}$	11.027
39	Y	3.8	$4d\ 5s^2$	$^2D_{3/2}$	6.377	$^1S_0$	12.233
40	Zr	6.44	$4d^2\ 5s^2$	$^3F_2$	6.835	$^4F_{3/2}$	12.916
41	Nb	8.4	$4d^4\ 5s$	$^6D_{1/2}$	6.881	$^5D_0$	13.895
42	Mo	9.0	$4d^5\ 5s$	$^7S_3$	7.131	$^6S_{5/2}$	15.72
43	Tc	9.0	$4d^5\ 5s^2$	$^6S_{5/2}$	7.23	$^7S_3$	14.87
44	Ru	12.6	$4d^7\ 5s$	$^5F_5$	7.365	$^4F_{9/2}$	16.597
45	Rh	12.44	$4d^8\ 5s$	$^4F_{9/2}$	7.461	$^3F_4$	15.92
46	Pd	12.16	$4d^{10}$	$^1S_0$	8.33	$^2D_{5/2}$	19.42
47	Ag	10.5	$4d^{10}\ 5s$	$^2S_{1/2}$	7.574	$^1S_0$	21.48
48	Cd	8.67	$4d^{10}\ 5s^2$	$^1S_0$	8.991	$^2S_{1/2}$	16.904
49	In	7.28	$4d^{10}\ 5s^2\ 5p$	$^2P_{1/2}$	5.785	$^1S_0$	18.828
50	Sn	7.29	$4d^{10}\ 5s^2\ 5p^2$	$^3P_0$	7.332	$^2P_{1/2}$	14.63
51	Sb	6.65	$4d^{10}\ 5s^2\ 5p^3$	$^4S_{3/2}$	8.639	$^3P_0$	19
52	Te	6.25	$4d^{10}\ 5s^2\ 5p^4$	$^3P_2$	9.01	$^4S_{1/2}$	21.5
53	I	4.94	$4d^{10}\ 5s^2\ 5p^5$	$^2P_{3/2}$	10.44	$^3P_2$	19.0
54	Xe	5.85×10^{-3}(g)	$4d^{10}\ 5s^2\ 5p^6$	$^1S_0$	12.127	$^2P_{3/2}$	21.21
55	Cs	1.873	[Xe] $6s$	$^2S_{1/2}$	3.893	$^1S_0$	25.1
56	Ba	3.78	$6s^2$	$^1S_0$	5.210	$^2S_{1/2}$	10.001
57	La	6.15	$5d\ 6s^2$	$^2D_{3/2}$	5.61	$(5d^2)\ ^3F_2$	11.43
58	Ce	6.79	$4f\ 5d\ 6s^2$	$^3H_5$	......	$(4f^2\ 6s)\ ^4H_{7/2}$	
59	Pr	6.48	$4f^3\ 6s^2$	$^4I_{9/2}$	......	$^5I_4$	
60	Nd	7.00	$4f^4\ 6s^2$	$^5I_4$	6.3	$(4f^5\ 6s)\ ^6I_{7/2}$	
61	Pm	....	$4f^5\ 6s^2$	$^6H_{5/2}$			
62	Sm	7.7	$4f^6\ 6s^2$	$^7S_3$	5.6	$^8F_{1/2}$	11.2
63	Eu	....	$4f^7\ 6s^2$	$^8S_{7/2}$	5.67	$^9S_4$	11.24
64	Gd	....	$4f^7\ 5d\ 6s^2$	$^9D_2$	6.16	$^{10}D_{5/2}$	12 +
65	Tb	....	$4f^8\ 5d\ 6s^2$				
66	Dy						
67	Ho						
68	Er	4.77					
69	Tm	....	$4f^{13}\ 6s^2$	$^2F_{7/2}$	......	$^3F_4$	
70	Yb	....	$4f^{14}\ 6s^2$	$^1S_0$	6.22	$^2S_{1/2}$	12.10
71	Lu	....	$4f^{14}\ 5d\ 6s^2$	$^2D_{3/2}$	6.15	$^1S_0$	14.7
72	Hf	13.3	$4f^{14}\ 5d^2\ 6s^2$	$^3F_2$	5.5	$^2D_{3/2}$	14.9
73	Ta	16.6	$4f^{14}\ 5d^3\ 6s^2$	$^4F_{3/2}$	7.7	$^5F_1$	
74	W	19.3	$4f^{14}\ 5d^4\ 6s^2$	$^5D_0$	7.98	$^6D_{1/2}$	
75	Re	....	$5f^{14}\ 5d^5\ 6s^2$	$^6S_{5/2}$	7.87	$^7S_3$	
76	Os	22.5	$4f^{14}\ 5d^6\ 6s^2$	$^5D_4$	8.7		

Z	Element	Density g cm⁻³ at 0°C, 1 atm	Ground state			Ion ground state	
			Elect config	Term	IP, eV	Term	IP, eV
77	Ir	22.42	$4f^{14}\,5d^7\,6s^2$	$^4F_{9/2}$	9.2		
78	Pt	21.37	$4f^{14}\,5d^9\,6s$	$^3D_3$	9.0	$^2D_{5/2}$	18.56
79	Au	19.3	$4f^{14}\,5d^{10}\,6s$	$^2S_{1/2}$	9.22	$^1S_0$	20.5
80	Hg	13.596(l)	$4f^{14}\,5d^{10}\,6s^2$	$^1S_0$	10.434	$^2S_{1/2}$	18.751
81	Tl	11.86	$4f^{14}\,5d^{10}\,6s^2\,6p$	$^2P_{1/2}$	6.106	$^1S_0$	20.42
82	Pb	11.342	$4f^{14}\,5s^{10}\,6s^2\,6p^2$	$^3P_0$	7.415	$^2P_{1/2}$	15.028
83	Bi	9.75	$4f^{14}\,5d^{10}\,6s^2\,6p^3$	$^4S_{3/2}$	7.287	$^3P_0$	19.3
84	Po	....	$4f^{14}\,5d^{10}\,6s^2\,6p^4$	$^3P_2$	8.43		
85	At	....	$4f^{14}\,5d^{10}\,6s^2\,6p^5$				
86	Em	....	$4f^{14}\,5d^{10}\,6s^2\,6p^6$	$^1S_0$	10.745		
87	Fr	....	[Em] $7s$	$^2S_{1/2}$			
88	Ra	5(?)	$7s^2$	$^1S_0$			
89	Ac	....	$6d\,7s^2$	$^2D_{3/2}$	......	$^1S_0$	
90	Th	11.00	$6d^2\,7s^2$	$^3F_2$	......	$^4F_{3/2}$	
91	Pa	....	$6d^3\,7s^2$				
92	U	18.7	$5f^3\,6d\,7s^2$	$^5L_6$	4	$^4I_{9/2}$	
93	Np						
94	Pu						
95	Am						
96	Cm						
97	Bk						
98	Cf						
99	E						
100	Fm						
101	Mv						

V. Table of Nuclear Species*

This table gives the following information concerning the known nuclear species. The numbers are those of the table columns.

1. The atomic number Z.
2. The chemical symbol. Below this, the *name* of the element is given.
3. The mass number A. The first entry in this column for each element is the *atomic weight* for the isotopic mixture found in nature.
4. The neutron number N.
5. The ground-state spin quantum number I.
6. The ground-state parity π.
7. The ground-state shell configuration, according to the shell model.
8. The mass excess $M - A$, expressed in MeV. (The mass excess in atomic mass units is equal to the tabulated quantity divided by 931.141.)
9. The abundances of the stable nuclides: (a) the *percentage* contribution of the given nuclide to the naturally occurring mixture and (b) the absolute abundance per 10^6 atoms of elemental silicon.
10. The magnetic moment μ in nuclear magnetons. $\mu = gI$, where g is the nuclear g factor and I is the spin quantum number.
11. The nuclear quadripole moment Q_m, in units of 10^{-24} cm^2.
12. The half-life $T_{1/2}$ of unstable nuclides.
13. The type of decay of unstable nuclides ($\alpha = \alpha$ particle, $\beta = \beta$ particle, IT = isomeric transition, K = K-electron capture, L = L-electron capture, fiss = fission). The symbol indicating the type of decay may be followed by one or more numbers which indicate the approximate energy release in MeV. A number in parentheses in this column signifies the maximum possible energy release (Q value) for the given decay type, in MeV. If more than one type of decay is observed, the Q values for each type may be given in parentheses. Thus $(2.75+)$ means that the Q value for β^+ decay is 2.75 MeV.
14. The binding energy of the most weakly bound neutron, in MeV. If this amount of energy is supplied, a neutron may be evaporated, leaving the residual nucleus in its ground state.
15. The total cross section for "thermal" neutrons.

Numbers in Parentheses. In columns 5, 6, and 7, numbers in parentheses generally indicate either uncertain or theoretically indicated values; in columns 8, 10, 11, they indicate the uncertainty in the last digit of the stated value.

---

* This table from *Principles of Modern Physics* by R. B. Leighton. Copyright © 1959. McGraw-Hill Book Company. Used by permission.

Boldface Type. Boldface type is used for the over-all properties (atomic weight, cosmic abundance, and thermal-neutron cross section) of each *element* and for all stable nuclides.

Isomeric States. The isomeric states of *unstable* nuclides are indicated by multiple entries in columns 12 and 13; of *stable* nuclides, by single or multiple entries in columns 12 and 13.

Example. The element germanium ($Z = 32$) has an atomic weight of 72.60, a cosmic abundance of 50.5 atoms per 10^6 atoms of Si, and a total thermal-neutron-absorption cross section of 2.3×10^{-24} cm² (2.3 barns). Thirteen separate nuclides are listed; of them, five ($A = 70, 72, 73, 74$, and 76) are *stable*. The stable isotope $A = 73$ has 41 neutrons, a ground-state spin quantum number $I = 9/2$, even parity, the corresponding shell state of the odd neutron being $g_{9/2}$. It has a mass *defect* of 49.9 ± 0.1 MeV, or 53.6 mmu, so that its mass is

$$73.0000 - 0.0536 = 72.9464 \text{ amu}$$

It accounts for 7.8 percent of all germanium, or 3.84 atoms per 10^6 atoms of Si. Its magnetic moment is $\mu = -0.8767 \pm 0.0001$ nuclear magneton, so that its g factor is $g = -0.8767 \div 9/2 = -0.1946$; and its electric-quadripole moment is $Q = -0.2 \pm 1.1 \times 10^{-24}$ cm². The neutron-absorption cross section of this isotope is 14 barns. The isotope also can exist in an isomeric state whose half-life is 0.53 s, from which it decays by an isomeric transition γ ray whose quantum energy is 54 keV.

(1) Z	(2) Chem symbol	(3) A	(4) N	(5) I	(6) π	(7) Ground-state config	(8) M-A, MeV	(9) Abundance Per cent	(9) Abundance Cosmic
0	n	1	1	1/2	+	$s_{1/2}$	8.367(2)		
1	H *Hydrogen*	1.0080							
		1	0	1/2	+	$s_{1/2}$	7.584(2)	99.985	4.00×10^{10} 4.00×10^{10}
	D	2	1	1	+	$(1/2, 1/2)_1$	13.725(3)	0.015	5.7×10^6
	T	3	2	1/2			15.835(5)		
2	He *Helium*	4.003							3.08×10^9
		3	1	1/2	(+)	$s_{1/2}$	15.817(5)	0.00013	—
		4	2	0	+		3.607(3)	~100	3.08×10^9
		5	3				12.93(3)		
		6	4				19.40(3)		
3	Li *Lithium*	6.940							100
		5	2				13.0(3)		
		6	3	1	+	$(3/2, 3/2)_1$	15.862(5)	7.4	7.4
		7	4	3/2	(−)	$p_{3/2}$	16.977(6)	92.6	92.6
		8	5				23.310(5)		
		9	6						
4	Be *Beryllium*	9.013							20
		7	3				17.840(5)		
		8	4				7.309(5)		
		9	5	3/2	(−)	$p_{3/2}$	14.010(6)	100	20
		10	6				15.566(7)		
5	B *Boron*	10.82							24
		8	3				24.9(4)		
		9	4				15.081(6)		
		10	5	3	+	$(3/2, 3/2)_3$	15.010(6)	18.8	4.5
		11	6	3/2	(−)	$p_{3/2}$	11.914(6)	81.2	19.5
		12	7				16.917(6)		
6	C *Carbon*	12.011							3.54×10^6
		10	4				18.8(1)		
		11	5				13.895(7)		
		12	6	0	(+)	—	3.541(5)	98.89	3.50×10^6
		13	7	1/2	−	$p_{1/2}$	6.963(5)	1.11	3.92×10^4
		14	8	0			7.157(3)		
		15	9				13.19(5)		
7	N *Nitrogen*	14.008							6.60×10^6
		12	5				21.2(1)		
		13	6				9.185(5)		
		14	7	1	+	$(1/2, 1/2)_1$	7.002(3)	99.63	6.58×10^6
		15	8	1/2	−	$p_{1/2}$	4.528(5)	0.37	2.41×10^4
		16	9				10.40(1)		
		17	10				13.0(2)		
8	O *Oxygen*	16.000							2.14×10^7
		14	6				12.17(2)		
		15	7				7.233(6)		
		16	8	0	+	$d_{5/2}$	0(std)	99.759	2.13×10^7
		17	9	5/2	+		4.222(5)	0.037	8.00×10^3
		18	10	0	+		4.521(8)	0.204	4.36×10^4
		19	11				8.93(1)		
9	F *Fluorine*	19.00							1600
		17	8				6.989(4)		
		18	9				6.19(1)		
		19	10	1/2	+		4.142(7)	100	1600
		20	11				5.90(1)		
		21	12						

(10) μ (nucl mag)	(11) Quad mom, 10^{-24} cm²	(12) Half-life $(T_{1/2})$	(13) Decay — Type and energy	(Q)	(14) Binding energy of last neutron E_n, MeV	(15) σ_{in}, 10^{-24} cm²	
$-1.91315(7)$		13 m	β^- 0.78	(0.78)	—		
$+2.7926(1)$					2.25	0.33 / 0.33	H 1
$+0.85735(1)$	$+0.00274(2)$				2.225	0.00057	2
$+2.9788$		12.26 y	β^- 0.018	(0.018)			3
$-2.127414(3)$					6.255	0.007 / 5400	He 3
	(0)				20.58	0	4
		2×10^{-21} s	n $+ \alpha$				5
		0.82 s	β^- 3.50	(3.50)	1.90		6
		$\sim 10^{-21}$ s	p $+ \alpha$			71	Li 5
$+0.82189(4)$	$(+0.02)$				5.3	950(nα)	6
$+3.2559(1)$					7.24	0.033	7
		0.84 s	β^- 13; 2α 3	(16.0)	2.034		8
		0.17 s	β^-; n $+ 2\alpha$	(14)			9
		53 d	K	(0.86)		0.010	Be 7
		$<4 \times 10^{-15}$ s	2α 0.09		18.88	$\sim 10^4$(np)	8
$-1.177(1)$	(0.02)				1.66	0.010	9
		2.7×10^6 y	β^- 0.56	(0.56)	6.816		10
		0.5 s	β^+ 14; 2α 3	(18)		755	B 8
		$>3 \times 10^{-19}$ s	p $+ 2\alpha$		18.4		9
$+1.800(1)$	$+0.074(5)$				8.5	4020(np)	10
$+2.6886(3)$	$+0.036(2)$				11.46	<0.05	11
		0.025 s	β^- 13.4, 9.0	(13.4)	3.364		12
		19 s	β^+ 1.9	(3.6)		0.0032	C 10
		20.5 m	β^+ 0.96	(1.98)	13.38		11
$+0.7023(2)$					18.77	0.0032	12
					4.95	0.0009	13
		5568 y	β^- 0.158	(0.158)	8.169	$<10^{-6}$	14
		2.3 s	β^- 4.3, 9.8	(9.8)	2.2		15
		0.012 s	β^+ 16.7; $3\alpha \sim 4$	(17.7)		1.88	N 12
		10.0 m	β^+ 1.20	(2.22)	20.45		13
$+0.40365(3)$	$+0.02$				10.55	1.8(np)	14
$-0.28299(3)$					10.83	0.00002	15
		7.4 s	β^- 4.10, . . .	(10.4)	2.6		16
		4.14 s	β^- 3.7; (n 1.0)	(8.8)	5.6		17
		72 s	β^+ 1.83	(5.15)		<0.0002	O 14
		2.1 m	β^+ 1.7	(2.7)	13.2		15
					15.6	<0.00002	16
$-1.8935(2)$	-0.02				4.14	0.5(nα)	17
					8.06	0.00022	18
		29 s	β^- 3.2, 4.4	(4.79)	3.96		19
		66 s	β^+ 1.75	(2.77)		0.009	F 17
		1.87 h	β^+ 0.65	(1.67)	9.13		18
$+2.6273$					10.3	0.009	19
		11 s	β^- 5.42	(7.05)	6.60		20
		5 s	β^-	(5.7)			21

(1) Z	(2) Chem symbol	(3) A	(4) N	(5) I	(6) π	(7) Ground-state config	(8) M-A, MeV	(9) Abundance Per cent	(9) Abundance Cosmic
10	Ne	20.183							8.6 × 10⁶
Neon		18	8				10.4(2)		
		19	9				7.40(1)		
		20	10	(0)	(+)		−1.15(1)	90.8	7.74 × 10⁶
		21	11	3/2	(+)	$(d_{5/2})^3 3/2$	0.46(1)	0.26	2.58 × 10⁴
		22	12	(0)	(+)		−1.53(1)	8.9	8.36 × 10⁵
		23	13				1.64(1)		
		24	14						
11	Na	22.991							4.38 × 10⁴
Sodium		20	9				14.2(2)		
		21	10				3.99(4)		
		22	11	3			1.31(2)		
		23	12	3/2	+	$(d_{5/2})^3 3/2$	−2.74(1)	100	4.38 × 10⁴
		24	13	4			−1.34(2)		
		25	14				−2.1(2)		
12	Mg	24.32							9.12 × 10⁵
Magnesium		23	11				1.35(1)		
		24	12	(0)	(+)		−6.85(2)	78.8	7.21 × 10⁵
		25	13	5/2	(+)	$d_{5/2}$	−5.82(2)	10.1	9.17 × 10⁴
		26	14	(0)	(+)		−8.57(2)	11.1	1.00 × 10⁵
		27	15				−6.64(2)		
		28	16				−6.78(3)		
13	Al	26.98							9.48 × 10⁴
Aluminum		24	11				7.2(3)		
		25	12				−1.57(6)		
		26	13				−4.54(2)		
		27	14	5/2	(+)	$d_{5/2}$	−9.23(2)	100	9.48 × 10⁴
		28	15				−8.59(2)		
		29	16				−9.4(1)		
14	Si	28.09							1.00 × 10⁶
Silicon		26	12						
		27	13				−4.41(2)		
		28	14	(0)	(+)		−13.25(2)	92.17	9.22 × 10⁵
		29	15	1/2	(+)	$s_{1/2}$	−13.35(2)	4.71	4.70 × 10⁴
		30	16	(0)	(+)		−15.60(2)	3.12	3.12 × 10⁴
		31	17				−13.83(2)		
		32	18				−14.77(6)		
15	P	30.975							1.00 × 10⁴
Phosphorus		28	13				+0.5(3)		
		29	14				−8.39(2)		
		30	15				−11.28(5)		
		31	16	1/2	(+)	$s_{1/2}$	−15.31(2)	100	1.00 × 10⁴
		32	17	1			−14.87(3)		
		33	18				−16.62(3)		
		34	19				−14.8(2)		
16	S	32.066							3.75 × 10⁵
Sulfur		31	15				−9.87(8)		
		32	16	0	+		−16.58(3)	95.0	3.56 × 10⁵
		33	17	3/2	+	$d_{3/2}$	−16.86(3)	0.75	2.77 × 10³
		34	18	0	+		−19.89(5)	4.2	1.57 × 10⁴
		35	19	3/2			−18.54(4)		
		36	20	(0)	(+)		−20.1(1)	0.017	51
		37	21				−16.7(3)		
17	Cl	35.457							8850
Chlorine		32	15				−3.5(4)		
		33	16				−11.4(2)		
		34	17				−14.37(6)		

(10) μ (nucl mag)	(11) Quad mom, 10^{-24} cm²	(12) Half-life $(T_{1/2})$	(13) Decay — Type and energy	(Q)	(14) Binding energy of last neutron E_n, MeV	(15) σ_{tn}, 10^{-24} cm²	
						<1	Ne
		1.6 s	β^+ 3.2	(4.2)			18
		18.5 s	β^+ 2.2	(3.2)			19
					16.86		20
−0.6614					6.75		21
					10.36		22
		40 s	β^- 4.2, 3.8, . . .	(4.2)	5.19		23
		3.4 m	β^- 1.95, . . .	(2.42)			24
						0.53	Na
		0.3 s	β^+; $\alpha > 2$	(15)			20
		23 s	β^+ 2.50	(3.52)			21
+1.746		2.6 y	β^+ 0.54, . . . ; K	(2.84)			22
+2.2161					12.25	0.53	23
+1.69		15.0 h	β^- 1.39, . . .	(5.51)	6.95		24
		60 s	β^- 4.0, 3.4, . . . , 2.6	(4.0)	9.2		25
						0.063	Mg
		12 s	β^+ 3.0	(4.0)			23
					16.6	0.03	24
−0.8552					7.33	0.27	25
					11.11	0.03	26
		9.5 m	β^- 1.75, 1.57	(2.59)	6.44		27
		21.3 h	β^- 0.45	(1.83)			28
						0.23	Al
		2.1 s	β^+ 8.5, . . .	(14)			24
		7.3 s	β^+ 3.24	(4.26)	17.0		25
		{ 6.5 s	β^+ 3.21	(4.23) }	11.5		26
+3.6408(4)	+0.149(2)	{ ~10^6 y	β^+ 1.2	(4.0) }	12.99	0.23	27
		2.30 m	β^- 2.87	(4.65)	7.72		28
		6.6 m	β^- 2.5, 1.4	(3.8)			29
						0.13	Si
							26
		1.7 s	β^+				27
		4.4 s	β^+ 3.8	(4.8)	16.8	0.1	28
−0.55492(4)					8.47	0.3	29
					10.60	0.11	30
		2.62 h	β^- 1.48, . . .	(1.48)	6.59		31
		~300 y	β^- 0.1	(0.1)			32
						0.20	P
		0.28 s	β^+ 11, 8, . . .	(14)			28
		4.5 s	β^+ 3.94	(4.96)			29
		2.5 m	β^+ 3.3	(4.3)	11.2		30
+1.1316(2)					12.1	0.20	31
−0.2523		14.5 d	β^- 1.71	(1.71)	7.93		32
		25 d	β^- 0.25	(0.25)	10.09		33
		12.4 s	β^- 5.1, 3.2	(7.2)			34
						0.49	S
		2.6 s	β^+ 4.4	(5.4)			31
					1.47		32
+0.6429(2)	−0.06(1)				8.64	0.002(np)	33
					10.9	0.26	34
	+0.045	87 d	β^- 0.167	(0.167)	7.0		35
					9.2	0.14	36
		5.0 m	β^- 4.7, 1.6	(4.7)			37
						33	Cl
		0.31 s	β^+ 10, 8, . . .	(13)			32
		2.8 s	β^+ 4.2	(5.2)			33
		{ 32.4 m	β^+ 2.5, 1.4, . . . ; IT 0.14		10.8		34
		{ 1.5 s	β^+ 4.5	(5.5) }			

(1) Z	(2) Chem symbol	(3) A	(4) N	(5) I	(6) π	(7) Ground-state config	(8) M-A, MeV	(9) Abundance Per cent	Cosmic
17	Cl	35.457							
		35	18	3/2	+	$d_{3/2}$	−18.71(3)	75.53	6670
		36	19	2			−18.91(4)		
		37	20	3/2	(+)	$d_{3/2}$	−20.91(5)	24.47	2180
		38	21				−18.66(6)		
		39	22				−18.79(8)		
		40	23						
18	A	39.944							1.50 × 10⁵
Argon		35	17				−13.30(6)		
		36	18	(0)	(+)		−19.63(4)	0.337	1.26 × 10⁵
		37	19				−20.10(5)		
		38	20	(0)	(+)		−23.48(5)	0.063	2.4 × 10⁴
		39	21				−21.75(7)		
		40	22	(0)	(+)		−23.23(5)	99.60	—
		41	23				−20.92(6)		
		42	24						
19	K	39.100							3160
Potassium		38	19				−17.60(6)		
		39	20	3/2	(+)	$d_{3/2}$	−22.31(7)	93.2	2940
		40	21	4	(−)	(3/2, 7/2)₄	−21.74(6)	0.0119	0.38
		41	22	3/2	+	$d_{3/2}$	−23.50(6)	6.8	219
		42	23	2			−22.51(7)		
		43	24				−23.90(8)		
		44	25				−22.4(2)		
		45	26						
20	Ca	40.08							4.90 × 10⁴
Calcium		39	19				−15.5(4)		
		40	20	(0)	(+)		−23.07(6)	96.9	4.75 × 10⁴
		41	21				−23.07(6)		
		42	22	(0)	(+)		−26.18(6)	0.64	314
		43	23	7/2	(−)	$f_{7/2}$	−25.75(8)	0.14	64
		44	24	(0)	(+)		−28.55(8)	2.1	1040
		45	25				−27.61(8)		
		46	26	(0)	(+)		−28.4(1)	0.0032	1.6
		47	27				−28.4(1)		
		48	28	(0)	(+)		−30.1(2)	0.18	87.7
		49	29				−26.9(2)		
21	Sc	44.96							2.8
Scandium		40	19				−9.1(5)		
		41	20				−17.11(7)		
		42	21						
		43	22				−23.53(8)		
		44	23				−24.90(8)		
		45	24	7/2	(−)	$f_{7/2}$	−27.87(8)	100	2.8
		46	25				−28.41(8)		
		47	26				−30.5(1)		
		48	27				−30.3(1)		
		49	28				−32.1(1)		
		50	29						
22	Ti	47.90							2440
Titanium		43	21						
		44	22						
		45	23				−25.82(8)		
		46	24	(0)	(+)		−30.77(8)	8.0	194
		47	25	5/2	(−)	$(f_{7/2})^5_{5/2}$	−31.19(8)	7.4	189
		48	26	(0)	(+)		−34.34(8)	73.8	1790
		49	27	7/2	(−)	$f_{7/2}$	−34.09(8)	5.5	134
		50	28	(0)	(+)		−36.71(8)	5.3	130
		51	29				−34.8(1)		

μ (nucl mag) (10)	Quad mom, 10^{-24} cm² (11)	Half-life ($T_{1/2}$) (12)	Decay — Type and energy (13)	(Q) (13)	Binding energy of last neutron E_n, MeV (14)	σ_{in}, 10^{-24} cm² (15)	
							Cl
+0.8219(2)	−0.07894(2)				12.8	{ 44 / 0.30(np)	35
+1.284	−0.017	3.1 × 10⁵y	β⁻ 0.71; K	(0.71⁻)		~90	36
+0.6841(2)	−0.06213(2)				9.9	0.56	37
		{ 1 s / 37.3 m	IT 0.66 / β⁻ 4.8, 2.7, 1.1	(4.8) }	6.3		38
		55 m	β⁻ 3.0, 1.7	(3.3)	8.4		39
		1.4 m	β⁻ ~7				40
						0.6	**A**
							35
		1.8 s	β⁺ 4.95	(5.97)		6	36
					14.7		37
		35 d	K, L	(0.82)	8.82	0.8	38
					11.8		39
		260 y	β⁻ 0.57	(0.57)	6.9	0.53	40
					10.0		41
		1.82 h	β⁻ 1.20, 2.49	(2.49)	6.1	>0.06	42
		>3.5 y	β⁻				
						2.0	**K**
		{ 0.95 s / 7.7 m	β⁺ 5.1 / β⁺ 2.7	(6.1) / (5.8) }			38
+0.39087(1)	+0.1				13.2	1.9	39
−1.2982(4)		1.3 × 10⁹y	β⁻ 1.33; K	(1.33⁻)	7.80	70	40
+0.21453(3)					10.2	1.1	41
−1.137		12.5 h	β⁻ 3.5, 2.0, . . .	(3.5) }	7.34		42
		22 h	β⁻ 0.83, . . .	(1.84)			43
		22 m	β⁻ 4.9, 1.5				44
		34 m	β⁻				45
						0.43	**Ca**
		1.0 s	β⁻ 5.7	(6.7)			39
					15.4	0.2	40
		1.1 × 10⁵y	K	(0.44)	8.37		41
					11.4	40	42
−1.3152(2)					7.93		43
						0.6	44
		160 d	β⁻ 0.25	(0.25)			45
						0.3	46
		4.7 d	β⁻ 0.7, 2.0	(2.0)			47
						1.1	48
		8.7 m	β⁻ 2.0, 1.0	(5.1)	5.0		49
						23	**Sc**
		0.2 s	β⁺ 0.9	(14)			40
		0.87 s	β⁺ 5	(6)			41
		0.66 s	β⁺ ~5	(~6)			42
		3.9 h	β⁺ 1.19, 0.82, 0.39	(2.21)			43
		{ 2.4 d / 4.0 h	IT 0.271 / β⁺ 1.47; K	(3.65) }			44
+4.7563(1)						23	45
		{ 20 s / 85 d	IT 0.14 / β⁻ 0.36, . . .	(2.36) }	8.8	0.25	46
		3.4 d	β⁻ 0.44, 0.60	(0.60)	10.5		47
		44 h	β⁻ 0.64	(4.0)	7.98		48
		57 m	β⁻ 2.0	(2.0)	10.3		49
		1.7 m	β⁻ ~3.5	(~6.3)			50
						6.0	**Ti**
		0.6 s	β⁺				43
		>20 y	K				44
		3.08 h	β⁺ 1.02; K	(2.04)			45
					13.3	0.6	46
−0.7871(1)					8.7	1.6	47
					11.4	7.8	48
−1.1022					8.1	1.8	49
					10.8	0.14	50
		5.80 m	β⁻ 2.1, 1.5	(2.4)	6.3		51

(1)	(2)	(3)	(4)	(5)	(6)	(7)	(8)	(9)	
Z	Chem symbol	A	N	I	π	Ground-state config	$M\text{-}A$, MeV	Abundance	
								Per cent	Cosmic
23 V		**50.95**							220
Vanadium		45	22						
		46	23				$-23.4(4)$		
		47	24				$-28.28(8)$		
		48	25				$-30.31(8)$		
		49	26	7/2			$-33.48(8)$		
		50	27	6	(+)	$(7/2, 7/2)_6$	$-34.3(2)$	0.25	0.55
		51	**28**	**7/2**	**(—)**	**$f_{7/2}$**	**$-37.2(1)$**	**99.75**	**220**
		52	29				$-36.1(1)$		
		53	30						
		54	31						
24 Cr		**52.01**							7800
Chromium		46	22						
		47	23						
		48	24						
		49	25				$-30.92(8)$		
		50	**26**	**(0)**	**(+)**		**$-35.7(2)$**	**4.4**	**344**
		51	27				$-36.5(1)$		
		52	**28**	**(0)**	**(+)**		**$-40.0(1)$**	**83.7**	**6510**
		53	**29**	**3/2**	**(—)**	**$p_{3/2}$**	**$-39.6(1)$**	**9.5**	**744**
		54	**30**	**(0)**	**(+)**		**$-41.0(1)$**	**2.4**	**204**
		55	31				$-38.7(2)$		
25 Mn		**54.94**							6850
Manganese		50	25				$-27.9(5)$		
		51	26				$-33.2(1)$		
		52	27	6?			$-35.3(1)$		
		53	28	7/2			$-39.0(1)$		
		54	29				$-39.8(2)$		
		55	**30**	**5/2**	**(—)**	**$(f_{7/2})^5_{5/2}$**	**$-41.5(2)$**	**100**	**6850**
		56	31				$-40.4(2)$		
		57	32						
26 Fe		**55.85**							6.00×10^5
Iron		52	26				$-33.3(2)$		
		53	27				$-35.2(2)$		
		54	**28**	**(0)**	**(+)**		**$-40.4(1)$**	**5.9**	3.54×10^4
		55	29				$-41.3(2)$		
		56	**30**	**(0)**	**(+)**		**$-44.1(2)$**	**91.6**	5.49×10^5
		57	**31**	**(1/2)**	**(—)**	**$p_{1/2}$**	**$-43.4(2)$**	**2.20**	1.35×10^4
		58	**32**	**(0)**	**(+)**		**$-45.2(2)$**	**0.33**	1980
		59	33				$-43.2(3)$		
		60	34						
		61	35						
27 Co		**58.94**							1800
Cobalt		54	27				$-31.5(7)$		
		55	28				$-37.8(2)$		
		56	29	4			$-39.5(2)$		
		57	30	7/2			$-42.9(3)$		
		58	31				$-42.9(2)$		
		59	**32**	**7/2**	**(—)**	**$f_{7/2}$**	**$-44.8(3)$**	**100**	**1800**
		60	33				$-43.9(3)$		
		61	34				$-45.4(3)$		
		62	35				$-43.8(4)$		
28 Ni		**58.71**							2.74×10^4
Nickel		56	28						
		57	29				$-39.6(3)$		
		58	**30**	**(0)**	**(+)**		**$-43.0(3)$**	**68.0**	1.86×10^4
		59	31				$-43.7(3)$		
		60	**32**	**(0)**	**(+)**		**$-46.7(3)$**	**26.2**	7170

(10) μ (nucl mag)	(11) Quad mom, 10^{-24} cm²	(12) Half-life ($T_{1/2}$)	(13) Decay — Type and energy	(Q)	(14) Binding energy of last neutron E_n, MeV	(15) σ_{tn}, 10^{-24} cm²	
						4.9	**V** 45
		~1 s	β^+				46
		0.4 s	$\beta^+ > 6$				47
		31 m	β^+ 1.89	(2.89)			48
		16.2 d	β^+ 0.69; K	(4.02)	11.5		49
+3.3412(3)		~1 y	K	(0.62)	9.1	~100	50
+5.1478(5)	+0.3(2)	4×10^{14} y			11.0	4.5	51
		3.77 m	β^- 2.6	(4.0)	7.3		52
		2.0 m	β^- 2.50				53
		55 s	β^- 3.3				54
						3.1	**Cr** 46
		1.1 s	β^+				47
		23 h	K				48
		42 m	β^+ 1.54	(2.56)			49
					13.4	16	50
		27 d	K	(0.75)	9.1		51
					11.8	0.8	52
−0.4735(6)					7.93	18	53
					9.72	0.37	54
		3.6 m	β^- 2.8	(2.8)	5.9		55
						13.3	**Mn** 50
		0.28 s	$\beta^+ > 6$				50
		45 m	β^+ 2.2	(3.2)			51
± 5.05		{ 21 m { 5.7 d	β^+ 2.7; IT 0.39 K; β^+ 0.6	(4.7)	10.5		52
		~140 y	K	(0.60)	12.0		53
		{ 2 m { 300 d	IT? K	(1.38)	8.9		54
+3.4681(4)	+0.5				10.1	13.3	55
		2.58 h	β^- 2.8, 1.0, 0.7	(3.7)	7.3		56
		1.7 m	β^- 2.6	(2.7)			57
						2.5	**Fe** 52
		8 h	β^+ 0.80; K	(2.21)	10.5		53
		9 m	β^+ 2.6	(3.6)	13.8	2.2	54
		2.9 y	K	(0.22)	9.3		55
					11.1	2.6	56
+0.05					7.6	2.4	57
					10.1	0.9	58
		45 d	β^- 0.46, 0.27, 1.56	(1.56)	6.4		59
		~3 × 10⁵ y	β^-				60
		5.5 m	β^-				61
						37	**Co** 54
		0.18 s	$\beta^+ > 7$				54
± 3.85		18 h	β^+ 1.50, 1.0, . . . ; K	(3.45)			55
± 4.65		77 d	K; β^+ 1.50, . . .	(4.62)			56
		267 d	K	(0.5)			57
		{ 9 h { 71 d	IT 0.025 K; β^+ 0.48, . . .	(2.31)	9.0		58
+4.648(2)	+0.5(2)				10.2	37	59
		{ 10.5 m { 5.2 y	IT 0.059; β^- 1.5 β^- 0.31, . . .	(2.81)	7.49	{~100} { 6 }	60
		1.65 h	β^- 1.22	(1.29)	9.96		61
		{ 1.6 m { 14 m	β^- β^- 2.8				62
						4.6	**Ni** 56
		6.4 d	K				56
		36 h	K; β^+ 0.84	(3.24)			57
					11.7	4.3	58
		8 × 10⁴ y	K	(1.07)	9.00		59
					11.4	2.6	60

(1) Z	(2) Chem symbol	(3) A	(4) N	(5) I	(6) π	(7) Ground-state config	(8) M-A, MeV	(9) Abundance Per cent	(9) Abundance Cosmic
28	**Ni**	**58.71**							
		61	**33**	(3/2)	(−)	p$_{3/2}$	**−46.8(3)**	1.1	342
		62	**34**	(0)	(+)		**−48.8(3)**	3.7	1000
		63	35				−47.0(2)		
		64	**36**	(0)	(+)		**−48.3(2)**	1.0	318
		65	37				−46.0(2)		
		66	38						
29 *Copper*	**Cu**	**63.54**							212
		58	29				−33(1)		
		59	30						
		60	31	2			−40.4(3)		
		61	32	3/2			−44.6(3)		
		62	33				−44.9(3)		
		63	34	3/2	−	p$_{3/2}$	**−47.1(2)**	69.0	146
		64	35	1			−46.6(2)		
		65	36	3/2	−	p$_{3/2}$	**−48.05(2)**	31.0	66
		66	37				−46.8(2)		
		67	38				−47.4(2)		
		68	39						
30 *Zinc*	**Zn**	**65.38**							486
		60	30						
		61	31						
		62	32				−43.2(3)		
		63	33				−43.7(2)		
		64	**34**	(0)	(+)		**−47.2(2)**	48.9	238
		65	35				−46.7(2)		
		66	**36**	(0)	(+)		**−49.4(2)**	27.8	134
		67	**37**	5/2	−	f$_{5/2}$	**−48.0(2)**	4.1	20.0
		68	**38**	(0)	(+)		**−49.8(2)**	18.6	90.9
		69	39				−47.9(2)		
		70	**40**	(0)	(+)		**−48.9(2)**	0.63	3.35
		71	41				−46.7(3)		
		72	42						
31 *Gallium*	**Ga**	**69.72**							11.4
		64	33				−39.9(6)		
		65	34				−43.6(2)		
		66	35				−44.2(2)		
		67	36	3/2			−46.9(2)		
		68	37				−46.9(2)		
		69	**38**	3/2	−	p$_{3/2}$	**−48.8(2)**	60.1	6.86
		70	39				−48.3(2)		
		71	**40**	3/2	−	p$_{3/2}$	**−49.0(2)**	39.9	4.54
		72	41	3			−47.6(3)		
		73	42				−48.5(2)		
32 *Germanium*	**Ge**	**72.60**							50.5
		66	34						
		67	35				−42.5(4)		
		68	36						
		69	37				−46.6(2)		
		70	**38**	(0)	(+)		**−49.9(2)**	20.5	10.4
		71	39				−48.8(2)		
		72	**40**	(0)	(+)		**−51.6(3)**	27.4	13.8
		73	**41**	9/2	+	g$_{9/2}$	**−49.9(1)**	7.8	3.84
		74	**42**	(0)	(+)		**−51.6(1)**	36.5	18.65
		75	43				−49.7(1)		
		76	**44**	(0)	(+)		**−50.9(2)**	7.8	3.87
		77	45				−48.4(1)		
		78	46						

(10) μ (nucl mag)	(11) Quad mom, 10^{-24} cm²	(12) Half-life ($T_{1/2}$)	(13) Decay — Type and energy	(Q)	(14) Binding energy of last neutron E_n, MeV	(15) σ_{tn}, 10^{-24} cm²	
					8.53	2	**Ni 61**
						15	62
		80 y	β^- 0.063	(0.063)	6.7		**63**
					9.7	2	**64**
		2.56 h	β^- 2.10, 0.6, 1.0	(2.10)	6.0		65
		56 h	β^- 0.3				66
						3.7	**Cu**
		9.5 m	β^+ <0.7	}			58
		3 s	β^+ ~8	}			59
		81 s	β^+ 3.7				60
		24 m	β^+ 2.0, 3.0, 3.9	(6.3)			
		3.3 h	β^+ 1.22, . . . ; K	(2.24)	10.6		61
		9.9 m	β^+ 2.9	(3.9)			**62**
+2.2262(4)	−0.157				10.7	4.4	**63**
±0.40		12.8 h	K; β^- 0.57; β^+ 0.66	[(0.57−) (1.68+)]	7.91		64
+2.3845(4)	−0.145				9.8	2.2	**65**
							6
		5.1 m	β^- 2.63, 1.59	(2.63)	7.1	140	6
		61 h	β^- 0.40, 0.48, 0.58	(0.58)	9.1		67
		32 s	β^- 3.0				68
						1.10	**Zn** 60
		2.1 m					60
		1.5 m	β^+ 5				61
		9 h	K; β^+ 0.66	(1.7)			62
		38 m	β^+ 2.36, 1.40, . . . ; K	(3.38)	9.0		63
					11.8	0.5	**64**
		245 d	K; β^+ 0.33	(1.35)	7.9		65
					11.1		**66**
+0.8738(1)	+0.18				7.0		**67**
					10.1	1.1	**68**
		14 h	IT 0.44	}	6.3		69
		52 m	β^- 0.90	(0.90) }			
					9.2	0.09	**70**
		3 h	β^- 1.5	(3.0) }			71
		2.2 m	β^- 2.4	(2.9) }			
		49 h	β^- 0.3, 1.6				72
						2.9	**Ga** 64
		2.5 m	β^+ ~5	(7)			64
		15 m	IT 0.052; β^+ 2.5	}			65
		8 m	β^+ 2.2	(3.2) }			
+1.84	+0.21	9.4 h	β^+ 4.15, . . . ; K	(5.17)			66
		78 h	K	(1.00)	11.2		67
		68 m	β^+ 1.88, . . . , 0.78; K	(2.90)	8.3		68
+2.16(1)	+0.232(2)				10.1	1.9	**69**
		21 m	β^- 1.65, . . .	(1.65)			**70**
+2.561	+0.146(2)				9.1	4.6	**71**
±0.12		14.1 h	β^- 0.64, . . . , 3.17	(4.00)			72
		5 h	β^- 1.4	(1.5)			73
						2.3	**Ge** 66
		2.5 h	K; β^+				66
		19 m	β^+ 3.4				67
		250 d	K				68
		40 h	K; β^+ 1.21, . . . , 0.6, . . .				69
					3.4		**70**
		12 d	K	(0.23)			71
						1.0	**72**
−0.8767(1)	−0.2(11)	(0.53 s)	IT 0.054	)	14		**73**
						0.7	**74**
		49 s	IT 0.14	}			75
		82 m	β^- 1.18, 0.92, . . .	(1.18) }	6.5		**76**
						0.32	**76**
		52 s	β^- 2.9, . . . ; IT 0.16	}			77
		12 h	β^- 2.20	(2.7) }			
		86 m	β^- 0.9				78

(1) Z	(2) Chem symbol	(3) A	(4) N	(5) I	(6) π	(7) Ground-state config	(8) M-A, MeV	(9) Abundance Per cent	(9) Abundance Cosmic
33 *Arsenic*	**As**	**74.91**							4.0
		68	35						
		69	36						
		70	37						
		71	38				−46.8(2)		
		72	39				−47.2(3)		
		73	40				−49.5(1)		
		74	41				−49.0(1)		
		75	**42**	3/2	−	$p_{3/2}$	−50.8(1)	100	4.0
		76	43	2			−49.8(1)		
		77	44				−51.1(1)		
		78	45				−49.8(2)		
		79	46				−50.2(2)		
		80	47						
34 *Selenium*	**Se**	**78.96**							67.6
		70	36						
		71	37						
		72	38						
		73	39				−46.7(1)		
		74	**40**	0	(+)		−50.4(1)	0.93	0.649
		75	41	5/2			−50.0(1)		
		76	**42**	(0)	(+)		−52.8(1)	9.1	6.16
		77	**43**	1/2	−	$p_{1/2}$	−51.8(1)	7.5	5.07
		78	**44**	0	(+)		−53.9(1)	23.6	16.0
		79	45	7/2			−52.5(2)		
		80	**46**	(0)	(+)		−54.0(2)	49.9	33.8
		81	47				−52.5(2)		
		82	**48**	(0)	(+)		−53.4(2)	9.0	5.98
		83	49						
		84	50						
35 *Bromine*	**Br**	**79.916**							13.4
		74	39						
		75	40				−47.2(1)		
		76	41				−48.2(1)		
		77	42				−50.5(1)		
		78	43				−50.4(2)		
		79	**44**	3/2	−	$p_{3/2}$	−52.7(2)	50.6	6.88
		80	45				−52.1(2)		
		81	**46**	3/2	−	$p_{3/2}$	−53.9(2)	49.4	6.62
		82	47	5			−53.4(2)		
		83	48				−54.6(2)		
		84	49				−52.9(2)		
		85	50				−53.4(2)		
		86	51						
		87	52				−46.7(6)		
		88	53						
		89	54						
36 *Krypton*	**Kr**	**83.80**							51.3
		76	40						
		77	41				−47.6(1)		
		78	**42**	(0)	(+)		−51.3(2)	0.35	0.175
		79	43				−51.1(2)		
		80	**44**	(0)	(+)		−54.1(2)	2.27	1.14
		81	45				−53.8(3)		
		82	**46**	(0)	(+)		−56.4(2)	11.6	5.90
		83	**47**	9/2	(+)	$g_{9/2}$	55.6(2)	11.5	5.89

(10) μ (nucl mag)	(11) Quad mom, 10^-24 cm²	(12) Half-life (T_1/2)	(13) Decay — Type and energy	(13) (Q)	(14) Binding energy of last neutron E_n, MeV	(15) σ_tn, 10^-24 cm²	
		~7 m	β+			4.3	**As** 68
		15 m	β+ 2.9				69
		50 m	β+ 1.4, 2.5	(6.6)			70
		62 h	β+ 0.81, ...				71
		26 h	K; β+ 2.50, 3.34, ...	(4.36)			72
		76 d	K;	(0.37)			73
+1.4347(3)	+0.3(2)	17 d	K; β- 1.36, 0.72; β+ 0.93, 1.53 [(1.36−)(2.55+)]		10.2	4.3	74
−0.906	+1.1	(0.018 s	IT 0.28)				**75**
		26.7 h	β- 2.96, 2.41, ...	(2.96)	7.3		76
		39 h	β- 0.69, ...	(0.69)			77
		90 m	β- 4.1, ...	(4.1)			78
		9 m	β- 2.3	(2.4)			79
		~36 s	β-				80
						13	**Se** 70
							71
		44 m	β+				72
		9.7 d	K				73
		{ 7.1 h	β+ 1.29, 1.65	(2.74)			74
		{ 44 m	β+ 1.7			40	
		127 d	K			85	75 76
+0.53326(5)		(17 s	IT 0.16)		7.4	41	77
					10.5	0.4	78
−1.015	+0.8	{ 3.9 m	IT 0.096;	(0.16)	7.0		79
		{ 7 × 10^4 y	β- 0.16		9.3	0.5	80
		{ 57 m	IT 0.103;	(1.38)	6.8		81
		{ 18 m	β- 1.38		9.8	0.05	82
		{ 69 s	β- 3.4				83
		{ 25 m	β- 1.5				
		~2 m	β-				84
						6.6	**Br** 74
		36 m	β+; K				75
		1.6 h	K; β+ 1.70, ...				76
		17 h	β+ 3.57; K	(4.59)	10.7		77
		57 h	K; β+ 0.34	(1.36)	8.4		
		{ 6.4 m	IT	(3.4)	10.6	11.4	78
+2.1058(4)	+0.26(8)	{ <6 m	β+ 2.4				**79**
		{ 4.6 h	IT 0.05	[(2.0−)(1.9+)]	7.3		80
+2.2696(5)	+0.21(7)	{ 18 m	β- 2.0, 1.4; K; β+ 0.86		10.1	2.6	**81**
±1.6	±0.7	35.9 h	β- 0.46	(3.1)			82
		2.3 h	β- 0.94, ...	(0.98)			83
		32 m	β- 4.68, ...				84
		3.0 m	β- 2.5	(2.8)			85
							86
		56 s	β- 2.6, 8.0; n 0.3;	(8.0)			87
		16 s	β-				88
		4.5 s	β-; n 0.5				89
						28	**Kr** 76
		10 h	K				77
		1.2 h	β+ 1.86, 1.67; K			>2	**78**
		{ 55 s	IT 0.13	(1.62)			79
		{ 34 h	K; L; β+ 0.60, 0.34		11.3	90	**80**
		{ 13 s	IT 0.19				81
		{ 2 × 10^5 y	K			40	**82**
−0.9671	+0.15	(1.86 h	IT 0.032)			200	**83**

(1) Z	(2) Chem symbol	(3) A	(4) N	(5) I	(6) π	(7) Ground-state config	(8) M-A, MeV	(9) Abundance Per cent	(9) Abundance Cosmic
36	**Kr**	**84**	**48**	**(0)**	**(+)**		**−57.7(2)**	**57.0**	**29.3**
		85	49	9/2			−56.2(2)		
		86	**50**	**(0)**	**(+)**		**−57.5(2)**	**17.3**	**8.94**
		87	51				−54.7(2)		
		88	52				−53.6(2)		
		89	53				−51.4(4)		
		90	54						
		91	55						
		92	56						
		93	57						
		94	58						
37	**Rb** *Rubidium*	**85.48**							**6.5**
		81	44	3/2			−51.6(3)		
		82	45				−52.6(5)		
		83	46	5/2					
		84	47	2			−55.0(2)		
		85	**48**	**5/2**	**−**	$f_{5/2}$	**−56.9(2)**	**72.2**	**4.73**
		86	49				−57.2(2)		
		87	50	3/2	−	$p_{3/2}$	−58.8(2)	27.8	1.77
		88	51				−56.5(2)		
		89	52				−55.4(4)		
		90	53				−54.1(4)		
		91	54						
		92	55						
38	**Sr** *Strontium*	**87.63**							**18.9**
		81	43						
		82	44						
		83	45						
		84	**46**	**(0)**	**(+)**		**−56.0(3)**	**0.55**	**0.106**
		85	47				−55.9(5)		
		86	**48**	**(0)**	**(+)**		**−59.0(2)**	**9.8**	**1.86**
		87	**49**	**9/2**	**+**	$g_{9/2}$	**−59.1(2)**	**7.0**	**1.33**
		88	**50**	**(0)**	**(+)**		**−61.7(2)**	**82.7**	**15.6**
		89	51				−59.9(2)		
		90	52				−59.8(2)		
		91	53				−57.1(2)		
		92	54				−56.2(2)		
		93	55						
39	**Y** *Yttrium*	**88.92**							**8.9**
		82	43						
		83	44						
		84	45						
		85	46						
		86	47				−54.8(2)		
		87	48				−57.4(3)		
		88	49				−57.9(2)		
		89	**50**	**1/2**	**−**	$p_{1/2}$	**−61.4(2)**	**100**	**8.9**
		90	51				−60.3(2)		
		91	52				−59.8(2)		
		92	53				−58.1(2)		
		93	54				−57.1(3)		
		94	55				−54.4(5)		
		95	56						

(10) μ (nucl mag)	(11) Quad mom, 10^{-24} cm²	(12) Half-life ($T_{1/2}$)	(13) Decay — Type and energy	(Q)	(14) Binding energy of last neutron E_n, MeV	(15) σ_{in}, 10^{-24} cm²	
					0.16		**Kr 84**
−1.001	+0.25	{4.4 h / 10.4 y	β^- 0.83; IT 0.31 / β^- 0.67, ...		5.95	<15	85
					0.06		86
		78 m	β^- 3.8, 1.3, ...	(4.2)	5.53	<600	87
		2.8 h	β^- 0.52, 2.7, ...	(2.9)	6.8		88
		3.2 m	β^- 4.2,2				89
		33 s	β^- 3.2				90
		10 s	β^- 3.6				91
		3 s	β^-				92
		2 s	β^-				93
		1 s	β^-				94
						0.7	**Rb 81**
+2.05		4.7 h	K; β^+ 1.0	(2.2)			81
		{6.3 h / 75 s	K; β^+ 0.77, ... / β^+ 3.2				82
+1.42		83 d	K				83
+1.32		{21 m / 33 d	IT 0.23, 0.46; K / K; β^+ 1.7, ... ; β^- .4	[(0.4−)] / [(2.7+)]			84
+1.3532(4)	+0.3					0.8	**85**
−1.69		{1.0 m / 18.6 d	IT 0.56 / β^- 1.77, 0.7	(1.77)			86
+2.7501(5)	+0.13	4.3 × 10¹⁰ y	β^- 0.27	(0.27)	10.0	0.14	87
		18 m	β^- 5.2, 3.3, 2	(5.2)	6.0	<200	88
		15 m	β^- 3.9, ...	(3.9)	7.4		89
		2.7 m	β^- 5.7				90
		{1.7 m / 14 m	β^- 4.6 / β^- 3.0				91
		~80 s	β^-				92
						1.3	**Sr 81**
		29 m	β^+				81
		26 d	K; β^+ 3.2				82
		33 h	β^+ 1.2; K				83
						1	**84**
		{70 m / 65 d	IT 0.007; K / K		7.5 / 9.5	>1.3	85 / **86**
−1.089(2)		(2.8 h	IT 0.39	)	8.4		87
					11.1	0.005	88
		{~10 d / 54 d	IT / β^- 1.48, ...	(1.48)	6.6	<130	89
		28 y	β^- 0.54	(0.54)	7.6	1	90
		9.7 h	β^- 0.61, ...	(2.67)	5.7		91
		2.7 h	β^- ~0.55	(1.9)			92
		7 m	β^-				93
						1.3	**Y 82**
		70 m	β^+ 2				82
		3.5 h					83
		3.7 h	β^+ 2.0; K				84
		5 h					85
		15 h	β^+ 1.80, 1.2	(6.01)			86
		{14 h / 80 h	IT 0.38 / K; β^+ 0.7	(2.1)	10.5		87
		105 d	K; β^+ 0.83	(3.7)	9.4		88
−0.14		(16 s	IT 0.91	)	11.7	1.3	**89**
		64.0 h	β^- 2.27;	(2.27)	6.7	6	90
		{50 m / 58 d	IT 0.55; / β^- 1.54	(1.54)	7.8		91
		3.5 h	β^- 3.60, 2.7, 1.3	(3.60)	6.6		92
		10 h	β^- 3.1	(3.1?)	6.8		93
		17 m	β^- 5.4	(5.4?)			94
		10 m	β^-				95

(1)	(2)	(3)	(4)	(5)	(6)	(7)	(8)	(9)	
								Abundance	
Z	Chem symbol	A	N	I	π	Ground-state config	$M\text{-}A$, MeV	Per cent	Cosmic
40	Zr *Zirconium*	91.22							54.5
		86	46						
		87	47				−53.9(3)		
		88	48						
		89	49				−58.5(2)		
		90	50	(0)	(+)		−62.5(2)	51.5	28.0
		91	51	5/2	+	$d_{5/2}$	−61.3(2)	11.2	6.12
		92	52	(0)	(+)		−61.6(2)	17.1	9.32
		93	53				−60.2(2)		
		94	54	(0)	(+)		−59.8(4)	17.4	9.48
		95	55				−57.8(4)		
		96	56	(0)	(+)		−57.2(5)	2.8	1.53
		97	57				−54.5(5)		
41	Nb *Niobium* (also called *Columbium*, Cb)	92.91							1.00
		89	48						
		90	49				−58.1(2)		
		91	50				−59.9(3)		
		92	51				−60.0(3)		
		93	52	9/2	+	$g_{9/2}$	−60.3(2)	100	1.00
		94	53				−59.1(3)		
		95	54				−58.9(4)		
		96	55				−57.5(4)		
		97	56				−57.2(5)		
		98	57						
		99	58						
42	Mo *Molybdenum*	95.95							2.42
		90	48				−55.3(3)		
		91	49				−56.3(3)		
		92	50	(0)	(+)		−60.3(3)	15.7	0.364
		93	51				−59.8(3)		
		94	52	(0)	(+)		−61.1(3)	9.3	0.226
		95	53	5/2	+	$d_{5/2}$	−59.9(4)	15.7	0.382
		96	54	(0)	(+)		−60.6(4)	16.5	0.401
		97	55	5/2	+	$d_{5/2}$	−59.2(5)	9.5	0.232
		98	56	(0)	(+)		−59.1(5)	23.8	0.581
		99	57				−55.9(9)		
		100	58	(0)	(+)		−57.5(5)	9.5	0.234
		101	59						
		102	60						
43	Tc *Technetium*								
		92	49				−53.9(8)		
		93	50				−56.7(3)		
		94	51				−56.8(4)		
		95	52				−58.2(4)		
		96	53				−57.5(6)		
		97	54						
		98	55						
		99	56	9/2			−57.3(9)		
		100	57						
		101	58						
		102	59						

(10) μ (nucl mag)	(11) Quad mom, 10^{-24} cm²	(12) Half-life ($T_{1/2}$)	(13) Decay — Type and energy	(Q)	(14) Binding energy of last neutron E_n, MeV	(15) σ_{tn}, 10^{-24} cm²	
						0.18	**Zr** 86
		~17 h	K				87
		1.6 h	β+ 2.10; K	(3.50)			88
		85 d	K				
		{ 4.4 m	IT 0.59; β+ 0.9, 2.4				89
		} 79 h	K; β+ 0.90	(2.83)			
		(0.8 s	IT 2.30	)	12.0	0.1	90
−1.2980					7.2	1	91
					8.7	0.2	92
		9 × 10⁵ y	β− 0.063, . . .	(0.063)	6.7	<5	93
						0.1	94
		65 d	β− 0.36, 0.39, 0.88	(1.12)	6.4		95
						0.1	96
		17 h	β− 1.91, . . .	(2.66)	3.7		97
						1	**Nb** 89
		{ ~2 h	β+				
		{ 1.9 h	β+ 2.9	(3.9)			90
		{ 24 s	IT 0.12				
		} 14.6 h	β+ 1.50, . .	(3.80)			91
		{ 62 d	IT 0.105				
		} long(?)	K	(>1.1)	10.0		92
		{ 13 h	K				
		{ 10 d	K	(1.9)	8.5		93
+6.1659	−0.3	(3.7 y	IT 0.029	)	8.7	>1	
		} 6.6 m	IT 0.041; β− 1.3		7.2	~15	94
		{ 2 × 10⁴ y	β− 0.5	(2.1)			
		} 84 h	IT 0.23		9.2		95
		{ 35 d	β− 0.16	(0.92)			96
		23 h	β− 0.7, 0.4	(3.1)	6.9		97
		{ 1 m	IT 0.75		8.1		98
		{ 72 m	β− 1.27	(1.93)			99
		30 m	β−				
		2.5 m	β− 3.2	(3.2)			
						2.5	**Mo** 90
		5.7 h	K; β+ 1.2				91
		{ 66 s	IT 0.65; β+ 2.45, 2.78, 3.99				92
		{ 15.6 m	β+ 3.44	(4.46)	13.3		
		{ 6.9 h	IT 0.26		7.9		93
		{ >2 y	K	(0.49)	9.9		94
−0.9140(2)					8.0	14	95
					9.1	1	96
−0.9332(1)					6.9	2	97
					8.3	0.13	98
		67 h	β− 1.23, 0.45, . . .	(1.37)			99
						0.2	100
		15 m	β− 1.2, 2.2	(2.4)			101
		11 m	β− 1				102
							Tc 92
		4.3 m	β+ 4.1				93
		{ 44 m	IT 0.39; K				94
		} 2.7 h	K; β+ 0.80, 0.6	(3.1)			
		53 m	β+ 2.41, . . . ; K	(4.30)	8.7		95
		{ 60 d	K; β+ 0.6; IT 0.039		9.5		
		{ 20 h	K	(1.6)			96
		{ 52 m	IT 0.034; K; β+				
		} 4.3 d	K	(3.0)			97
		{ 91 d	IT 0.099				98
		{ ~10⁵ y	K				
		~10⁴ y	β− 0.3	(1.7)			99
		{ 6.0 h	IT 0.002, 0.142			20	100
+5.657		{ 2.1 × 10⁵ y	β− 0.29	(0.29)			
		16 s	β− 2.8		7.1		101
		15 m	β− 1.2	(1.5)			102
		5 s	β− 4				

(1) Z	(2) Chem symbol	(3) A	(4) N	(5) I	(6) π	(7) Ground-state config	(8) M-A, MeV	(9) Abundance Per cent	(9) Abundance Cosmic
44	Ru	101.1							1.49
Ruthenium		94	50						
		95	51				−56.1(5)		
		96	52	(0)	(+)		−57.8(5)	5.6	0.0846
		97	53						
		98	54	(0)	(+)		−58.5(7)	1.9	0.0331
		99	55	5/2	+	$d_{5/2}$	−57.5(9)	12.7	0.191
		100	56	(0)	(+)			12.7	0.189
		101	57	5/2	+	$d_{5/2}$		17.0	0.253
		102	58	(0)	(+)		−59.2(5)	31.5	0.467
		103	59				−57.1(5)		
		104	60	(0)	(+)		−57.9(7)	18.6	0.272
		105	61				−55.1(5)		
		106	62				−55.7(5)		
		107	63						
		108	64						
45	Rh	102.91							0.214
Rhodium		97	52						
		98	53						
		99	54						
		100	55						
		101	56						
		102	57				−57.0(5)		
		103	58	1/2	−	$p_{1/2}$	−57.8(5)	100	0.214
		104	59				−56.2(5)		
		105	60				−57.1(5)		
		106	61				−55.7(5)		
		107	62				−55.6(4)		
		108	63						
46	Pd	106.4							0.675
Palladium		98	52						
		99	53						
		100	54						
		101	55						
		102	56	(0)	(+)		−58.2(5)	1.0	0.0054
		103	57				−57.3(5)		
		104	58	(0)	(+)		−58.8(5)	11.0	0.0628
		105	59	5/2	+	$d_{5/2}$	−57.7(5)	22.2	0.1536
		106	60	(0)	(+)		−55.2(5)	27.3	0.1839
		107	61				−56.8(4)		
		108	62	(0)	(+)		−57.9(4)	26.7	0.180
		109	63				−55.4(4)		
		110	64	(0)	(+)		−56.2(5)	11.8	0.0911
		111	65				−53.2(5)		
		112	66				−53.3(5)		
		113	67						
47	Ag	107.880							0.26
Silver		102	55						
		103	56						
		104	57	2			−55.1(5)		
		105	58	1/2			−55.4(8)		
		106	59				−56.3(5)		
		107	60	1/2	−	$p_{1/2}$	−56.9(4)	51.4	0.134
		108	61				−55.8(4)		
		109	62	1/2	−	$p_{1/2}$	−56.5(4)	48.6	0.126

(10) μ (nucl mag)	(11) Quad mom, 10^{-24} cm²	(12) Half-life ($T_{1/2}$)	(13) Decay — Type and energy	(Q)	(14) Binding energy of last neutron E_n, MeV	(15) σ_{in}, 10^{-24} cm²	
						2.5	Ru 94
		1 h	K				95
		98 m	K; β^+ 1.2	(2.2)		0.01	**96**
							97
		2.9 d	K				**98**
—0.6					7.1		99
					9.5		**100**
—0.7							101
						1.2	**102**
		40 d	β^- 0.20, 0.13, 0.69, . . .	(0.73)	6.5		103
						0.7	**104**
		4.5 h	β^- 1.15	(2.02)			105
		1.0 y	β^- 0.04, . . .	(0.04)			106
		4.5 m	β^- 4.3				107
		~4 m	β^-				108
						150	Rh 97
		35 m	β^+				98
		9 m	β^+ 3.3				
		$\big\{$ 15 d	β^+	$\big\}$			99
		4.5 h	β^+ 0.74				
		21 h	β^+ 2.62, . . . ; K	(3.64)	8.0		100
		$\big\{$ ~5 y	IT?	$\big\}$			
		4.5 d	K				101
		220 d	K; β^- 1.15; β^+ 1.24, . . .	$\Big[\begin{smallmatrix}(1.15-)\\(2.26+)\end{smallmatrix}\Big]$			102
—0.0879		(54 m	IT 0.040	)	9.4	150	**103**
		$\big\{$ 4.4 m	IT 0.077; β^-	$\big\}$			104
		42 s	β^- 2.5, . . .	(2.5)	6.79		
		$\big\{$ 30 s	IT 0.130	$\big\}$			105
		36 h	β^- 0.25, 0.56	(0.56)	9.1		
		$\big\{$ 2 h	β^- ~1	$\big\}$			106
		30 s	β^- 3.53, . . .	(3.53)			
		22 m	β^- 1.2, ~2				107
		18 s	β^- ~4				108
						8	Pd 98
		17 m	β^+				99
		24 m					100
		4.0 d	K				101
		8 h	K; β^+ 0.5, 2.3(?)	(1.5?)		4.8	**102**
		17 d	K	(0.57)	7.1		103
							104
—0.57(5)		(23 s	IT 0.2	)	7.1		105
							106
		7×10^6 y	β^- 0.04	(0.04)			107
					9.4	(0.17)	**108**
		$\big\{$ 4.8 m	IT 0.17	$\big\}$			109
		13.6 h	β^- 1.0	(1.1)			
						>0.4	**110**
		$\big\{$ 5.5 h	IT; β^-	$\big\}$			111
		22 m	β^- 2.14				
		21 h	β^- 0.28	(0.30)			**112**
		1.5 m	β^-				113
						60	Ag 102
		16 m	β^+				103
		1.1 h	β^+ 1.3; K				104
		27 m	β^+ 2.70				105
		40 d	K				
		$\big\{$ 24 m	β^+ 1.96; K	(2.98) $\big\}$			**106**
—0.11304(1)		8.3 d	K				107
		(44 s	IT 0.093	)		30	
		2.3 m	β^- 1.77, . . . ; K; β^+ 0.8	$\Big[\begin{smallmatrix}(1.77-)\\(1.8+)\end{smallmatrix}\Big]$	7.3		108
—0.12996(1)		(40 s	IT 0.088	)	9.1	84	**109**

(1)	(2)	(3)	(4)	(5)	(6)	(7)	(8)	(9)	
								Abundance	
Z	Chem symbol	A	N	I	π	Ground-state config	M-A, MeV	Per cent	Cosmic
47	Ag	107.880							
		110	63				−54.6(4)		
		111	64	1/2			−55.3(5)		
		112	65				−53.6(5)		
		113	66				−53.7(5)		
		114	67						
		115	68				−51.0(6)		
48	Cd	112.41							0.89
	Cadmium	104	56				−53.1(5)		
		105	57				−52.4(8)		
		106	58	(0)	(+)		−56.3(5)	1.22	0.0109
		107	59				−55.5(4)		
		108	60	(0)	(+)		−55.5(4)	0.88	0.0079
		109	61				−56.3(4)		
		110	62	(0)	(+)		−57.5(4)	12.4	0.111
		111	63	1/2	+	$s_{1/2}$	−56.4(5)	12.8	0.114
		112	64	(0)	(+)		−57.5(5)	24.0	0.212
		113	65	1/2	+	$s_{1/2}$	−55.6(5)	12.3	0.110
		114	66	(0)	(+)		−56.3(5)	28.8	0.256
		115	67				−54.0(5)		
		116	68	(0)	(+)		−54.1(5)	7.6	0.068
		117	69				−51.4(6)		
		118	70						
49	In	114.82							0.11
	Indium	107	58				−52.1(5)		
		108	59						
		109	60				−54.6(4)		
		110	61				−53.5(4)		
		111	62				−55.5(8)		
		112	63				−55.0(5)		
		113	64	9/2	+	$g_{9/2}$	−55.8(5)	4.2	0.0046
		114	65				−54.4(6)		
		115	66	9/2	+	$g_{9/2}$	−55.4(5)	95.8	0.105
		116	67				−53.5(5)		
		117	68				−54.2(5)		
		118	69						
		119	70						
50	Sn	118.70							1.33
	Tin	108	58						
		109	59						
		110	60						
		111	61				−53.0(8)		
		112	62	(0)	(+)		−55.0(5)	1.02	0.0134
		113	63						
		114	64	(0)	(+)		−56.4(6)	0.69	0.0090
		115	65	1/2	+	$s_{1/2}$	−55.9(5)	0.38	0.00465
		116	66	(0)	(+)		−56.8(5)	14.3	0.189
		117	67	(1/2)	(+)	$s_{1/2}$	−55.7(5)	7.6	0.102
		118	68	(0)	(+)		−56.5(5)	24.1	0.316
		119	69	1/2	+	$s_{1/2}$	−54.9(5)	8.5	0.115
		120	70	(0)	(+)		−55.8(5)	32.5	0.433

(10) μ (nucl mag)	(11) Quad mom, 10^{-24} cm²	(12) Half-life ($T_{1/2}$)	(13) Decay — Type and energy	(Q)	(14) Binding energy of last neutron E_n, MeV	(15) σ_{tn}, 10^{-24} cm²	
							Ag
		⎰ 270 d	β⁻ 0.53, 0.1, . . . , IT 0.12				110
		⎱ 24 s	β⁻ 2.22, 2.88, . . .	(2.88)			
−0.145		⎰ 75 s	IT				111
		⎱ 7.5 d	β⁻ 1.04, 0.7, . . .	(1.04)			
		3.12 h	β⁻ 3.5, 4.1, 2.7, . . .	(4.1)			112
		5.3 h	β⁻ 2.0	(2.0)	8.6		113
		2 m	β⁻				114
		21 m	β⁻ 3	(3)			115
						3300	**Cd** 104
		59 m	K				105
		55 m	K; β⁺ 1.69, . . .			1	106
		6.7 h	K; β⁺ 0.32	(1.43)			107
					10.2		108
		1.3 y	K; L	(0.15)		>0.2	109
−0.5949(1)		(49 m)	IT 0.150	)			110 111
						>0.03	112
−0.6224(1)		(5 y)	β⁻ 0.58	)	6.4	2.7 × 10⁴	113
					9.05	1.2	114
		⎰ 43 d	β⁻ 1.63, . . .	(1.63)	5.6		115
		⎱ 54 h	β⁻ 1.11, . . .	(1.45)		1.4	116
		⎰ 3.0 h	IT				117
		⎱ 50 m	β⁻ 1.6, 3.0				
		∼30 m	β⁻ 4				118
						190	**In** 107
		30 m	β⁺ 2				108
		50 m	β⁺ 2.3				109
		4.3 h	K; β⁺ 0.7				
		⎰ 5.0 h	K; IT 0.12				110
		⎱ 66 m	β⁺ 2.25	(3.93)			
		2.8 d	K				111
		⎧ 21 m	IT				
		⎨ 2.5 s	IT 0.15				112
		⎩ 14 m	β⁻ 0.66; K; β⁺ 1.52				
+5.486(3)	+1.144	(1.73 m)	IT 0.39	)	9.2	63	113
		⎰ 49 d	IT 0.190; K	[(1.98−)]			114
		⎱ 72 s	β⁻ 1.98, . . . ; K; β⁺ ∼1	[(2.3+)]			
+5.5095(1)	+1.161	⎰ 4.5 h	IT 0.33 β⁻ 0.83		9.1	197	115
		⎱ 6 × 10¹⁴ y	β⁻ 0.6				
		⎰ 54 m	β⁻ 1.00, . . .	(3.36)	6.6		116
		⎱ 13 s	β⁻ 3.3	(3.3)			
		⎰ 1.9 h	β⁻ 1.77, 1.61; IT 0.31				117
		⎱ 1.1 h	β⁻ 0.74	(1.46)			
		⎰ 4.5 m	β⁻ 1.5				118
		⎱ <1 m	β⁻ 4				
		18 m	β⁻ 2.7	(2.7)			119
						0.6	**Sn** 108
		4 h	K				109
		18 m	K, β⁺				110
		4 h	K				111
		35 m	K, β⁺ 1.51	(2.53)			112
						1.3	113
		112 d	K, L				114
−0.9178(1)					7.7		115
					8.9	>0.006	116
−0.9998(1)		(14 d)	IT 0.159	)			117
					9.3	>0.01	118
−1.0460(1)		(275 d)	IT 0.065	)	6.6		119
						0.1	120

(1)	(2)	(3)	(4)	(5)	(6)	(7)	(8)	(9)	
								Abundance	
Z	Chem symbol	A	N	I	π	Ground-state config	M-A, MeV	Per cent	Cosmic
50	Sn	118.70							
		121	71				−53.6(5)		
		122	**72**	(0)	(+)		**−53.9(5)**	**4.8**	**0.063**
		123	73				−51.6(5)		
		124	**74**	(0)	(+)		**−51.6(5)**	**6.1**	**0.079**
		125	75				−49.0(5)		
		126	76						
		127	77						
51 Antimony	Sb	121.76							0.246
		116	65				−52.1(5)		
		117	66						
		118	67				−52.4(6)		
		119	68						
		120	69				−53.1(5)		
		121	**70**	5/2	+	$d_{5/2}$	**−54.0(5)**	**57**	**0.141**
		122	71				−52.4(5)		
		123	**72**	7/2	+	$g_{7/2}$	**−53.0(5)**	**43**	**0.105**
		124	73				−51.0(5)		
		125	74				−51.3(5)		
		126	75						
		127	76						
		128	77						
		129	78						
		130	79						
		131	80						
		132	81						
52 Tellurium	Te	127.61							4.07
		117	65						
		118	66						
		119	67						
		120	**68**	(0)	(+)		**−53.5(5)**	**0.091**	**0.00420**
		121	69						
		122	**70**	(0)	(+)		**−54.4(5)**	**2.5**	**0.115**
		123	**71**	1/2	+	$s_{1/2}$	**−52.8(9)**	**0.88**	**0.0416**
		124	**72**	(0)	(+)		**−53.9(5)**	**4.6**	**0.221**
		125	**73**	1/2	+	$s_{1/2}$	**−52.1(5)**	**7.0**	**0.328**
		126	**74**	(0)	(+)		**−52.5(5)**	**18.7**	**0.874**
		127	75				−50.6(5)		
		128	**76**	(0)	(+)		**−50.2(5)**	**31.8**	**1.48**
		129	77				−48.8(5)		
		130	**78**	(0)	(+)		**−48.6(5)**	**34.4**	**1.60**
		131	79				−46.5(5)		
		132	80				−46.3(5)		
		133	81				−43.9(7)		
		134	82						
53 Iodine	I	126.91							0.80
		119	66						
		120	67						
		121	68						

(10) μ (nucl mag)	(11) Quad mom, 10^{-24} cm²	(12) Half-life ($T_{1/2}$)	(13) Decay — Type and energy	(Q)	(14) Binding energy of last neutron E_n, MeV	(15) σ_{tn}, 10^{-24} cm²	
							Sn
		>1 y	β⁻ 0.42		6.2		121
		27 h	β⁻ 0.38	(0.38)		0.2	122
		130 d	β⁻ 1.42				123
		40 m	β⁻ 1.26	(1.41)	8.5	0.2	124
		9.5 m	β⁻ 2.1, . . .	(2.4)	5.75		125
		10 d	β⁻ 2.4	(2.4)			125
		50 m	β⁻				126
		1.5 h	β⁻				127
					5.5		**Sb**
		15 m	β⁺ 2.4	(4.7)			116
		60 m	β⁺ 1.4	(4.7)			117
		2.8 h	K				117
		3.5 m	β⁺ 3.1; IT 0.11	(4.1)			118
		5.1 h	K; β⁺ 0.7				118
		38 h	K	(0.59)			119
		5.8 d	K				120
		17 m	β⁺ 1.70	(2.72)			120
+3.360	**−0.7**				9.2	7.0	121
		3.5 m	IT 0.075	(1.98−)	6.8		122
		2.8 d	β⁻ 1.4, 1.98, 0.73; K; β⁺ 0.5	(1.5+)			122
+2.547	**−0.8**				9.3	3.46	123
		21 m	IT 0.018; β⁻				124
		1.3 m	β⁻ 3; IT 0.012				124
		60 d	β⁻ 2.31, . . .	(2.91)	6.6		124
		2.7 y	β⁻ 0.30, 0.12, . . .	(0.76)	8.6		125
		9 h	β⁻ 1				126
		28 d	β⁻ 1.9				126
		93 h	β⁻ 0.86, 1.57, 1.11				127
		~1 h	β⁻				128
		4.6 h	β⁻ 0.92, . . .				129
		10 m	β⁻ 2.9				130
		40 m	β⁻				130
		22 m	β⁻ 1.1				131
		2 m	β⁻				132
						4.6	**Te**
		2.5 h	β⁺ 2.5				117
		6.0 d	K; β⁺ 3.1				118
		16 h	K				119
		4.5 d	K			<140	119
							120
		150 d	IT 0.082				121
		17 d	K				121
						3	122
−0.73188(4)		(104 d	IT 0.088	)		400	123
						7	124
−0.88235(4)		(58 d	IT 0.11	)	6.5	1.5	125
					7.2	0.9	126
		110 d	IT 0.008				127
		9.3 h	β⁻ 0.68	(0.68)			127
						0.16	128
		33 d	IT 0.106				129
		72 m	β⁻ 1.46, 1.01, . . .	(1.49)			129
						0.2	130
		30 h	β⁻ 0.4; IT 0.18				131
		25 m	β⁻ 2.1	(2.3)			131
		77 h	β⁻ 0.22, . . .				132
		63 m	IT 0.4				133
		2 m	β⁻ 1.4, 2.4	(3.0)			133
		44 m	β⁻				134
						6.3	**I**
		18 m	β⁺				119
		>1.3 h	β⁺ 4.0				120
		1.4 h	β⁺ 1.13, . . .				121

(1)	(2)	(3)	(4)	(5)	(6)	(7)	(8)	(9)	
								Abundance	
Z	Chem symbol	A	N	I	π	Ground-state config	$M\text{-}A$, MeV	Per cent	Cosmic
53	**I**	**126.91**							
		122	69				−50.3(5)		
		123	70	5/2					
		124	71	2			−50.7(5)		
		125	72	5/2			−51.9(5)		
		126	73				−50.4(5)		
		127	**74**	**5/2**	**+**	**d$_{5/2}$**	**−51.4(5)**	**100**	**0.80**
		128	75	1			−49.7(5)		
		129	76	7/2			−50.5(5)		
		130	77				−48.6(5)		
		131	78	7/2			−48.7(5)		
		132	79				−46.7(5)		
		133	80				−46.9(7)		
		134	81				−45.4(7)		
		135	82						
		136	83				−40.1(5)		
		137	84						
54 *Xenon*	**Xe**	**131.30**							**4.0**
		121	67						
		122	68						
		123	69						
		124	**70**	**(0)**	**(+)**		**−50.9(5)**	**0.094**	**0.00380**
		125	71						
		126	**72**	**(0)**	**(+)**		**−51.7(5)**	**0.092**	**0.00352**
		127	73				−50.4(9)		
		128	**74**	**(0)**	**(+)**		**−51.7(5)**	**1.92**	**0.0764**
		129	**75**	**1/2**	**+**	**s$_{1/2}$**	**−50.7(5)**	**26.4**	**1.050**
		130	**76**	**(0)**	**(+)**		**−51.6(5)**	**4.1**	**0.162**
		131	**77**	**3/2**	**+**	**d$_{3/2}$**	**−49.7(5)**	**21.2**	**0.850**
		132	**78**	**(0)**	**(+)**		**−50.3(5)**	**26.9**	**1.078**
		133	79				−48.7(7)		
		134	**80**	**(0)**	**(+)**		**−48.8(5)**	**10.4**	**0.420**
		135	81						
		136	**82**	**(0)**	**(+)**		**−46.5(5)**	**8.9**	**0.358**
		137	83				−41.7(4)		
		138	84						
		139	85						
		140	86						
		141	87						
55 *Cesium*	**Cs**	**132.91**							**0.456**
		125	70						
		126	71				−46.9(7)		
		127	72	1/2			−48.3(9)		
		128	73				−47.6(7)		
		129	74	1/2			−49.6(8)		
		130	75				−48.6(5)		
		131	76	5/2			−49.4(5)		
		132	77	2			−48.5(7)		
		133	**78**	**7/2**	**+**	**g$_{7/2}$**	**−49.2(7)**	**100**	**0.456**
		134	79	{ 8 } { 4 }			−47.5(8)		
		135	80	7/2					
		136	81						
		137	82	7/2			−45.7(10)		
		138	83				−42.9(10)		
		139	84						
		140	85						

(10) μ (nucl mag)	(11) Quad mom, 10⁻²⁴ cm²	(12) Half-life ($T_{1/2}$)	(13) Decay — Type and energy	(Q)	(14) Binding energy of last neutron E_n, MeV	(15) σ_{tn}, 10⁻²⁴ cm²	
							I
		3.5 m	β^+ 3.12	(4.14)			122
		13 h	K				123
		4.5 d	K; β^+ 2.20, . . .	(3.22)			124
± 2.6	-0.66	60 d	K; L	(0.15)			125
							126
$+2.8090(4)$	$-0.6(2)$	13.3 d	K; β^- 0.87, . . . ; β^+ 1.11, . . .	$\left[\begin{smallmatrix}(1.25-)\\(2.13+)\end{smallmatrix}\right]$	9.1	6.3	**127**
		25.0 m	β^- 2.12, 1.67, . . . ; K	(2.12−)	6.6		128
$+2.603$	-0.49	1.7 × 10⁷ y	β^- 0.15	(0.19)		30	129
		12.6 h	β^- 1.02, 0.60, . . .	(2.95)			130
	-0.35	8.05 d	β^- 0.61, . . .	(0.97)		~600	131
		2.3 h	β^- 0.9, . . .	(3.57)			132
		21 h	β^- 1.3, 0.4	(1.8)			133
		52 m	β^- 1.5, 2.5	(3.4)			134
		6.7 h	β^- 1.0, 0.5, 1.4				135
		86 s	β^- 6.4, 5.0, 3.6	(6.4)			136
		22 s	β^-; n 0.6				137
						35	**Xe**
							121
		40 m	β^+				122
		19 h	K; β^+ 3.12				123
		1.8 h	K; β^+ 1.7				**124**
		⎰ 55 s	IT 0.075	⎱			125
		⎱ 18 h	K	⎰			**126**
		⎰ 75 s	IT 0.175	⎱			127
		⎱ 36.4 d	K	⎰		<5	**128**
$-0.77255(2)$		(8 d	IT 0.196	)		45	**129**
						<5	**130**
$+0.68680(2)$	-0.12	(12 d	IT 0.164	)		120	**131**
						<5	**132**
		⎰ 2.3 d	IT 0.233	⎱			133
		⎱ 5.27 d	β^- 0.35;	(0.43) ⎰		<5	**134**
		⎰ 15 m	IT 0.52	⎱		3.2 × 10⁶	135
		⎱ 9.2 h	β^- 0.91, 0.54	(1.16) ⎰		0.15	**136**
		3.8 m	β^- 3.5				137
		17 m	β^- 2.4				138
		41 s	β^-				139
		16 s	β^-				140
		3 s	β^-				141
						31	**Cs**
		45 m	K; β^+ 2.05				125
		1.6 m	β^+ 3.8, . . . ; K	(4.8)			126
± 1.41		6.2 h	K; β^+ 1.06, 0.7, . . .				127
		3.8 m	β^+ 3.0, 2.5, 1.5	(4.0)			128
± 1.47		31 h	K				129
		30 m	β^+ 1.97; K; β^- 0.44	$\left[\begin{smallmatrix}(0.44-)\\(2.99+)\end{smallmatrix}\right]$			130
$+3.48$		9.7 d	K; L	(0.35)			131
$+2.20$		6.2 d	K				132
$+2.5771$	-0.003				9.0	31	**133**
$+2.973$		⎰ 3.1 h	IT 0.13; β^- 0.55	⎱	6.7		134
		⎱ 2.3 y	β^- 0.65	(2.05) ⎰			
$+2.7134$		2.0 × 10⁶ y	β^- 0.21	(0.21)		~15	135
		13 d	β^- 0.34, 0.66				136
$+2.8219$		30 y	β^- 0.52, 1.18	(1.18)	7.1	<2	137
		32 m	β^- 3.40, . . .		4.9		138
		9.5 m	β^- ~4				139
		66 s	β^-				140

(1)	(2)	(3)	(4)	(5)	(6)	(7)	(8)	(9) Abundance	
Z	Chem symbol	A	N	I	π	Ground-state config	M-A, MeV	Per cent	Cosmic
56	**Ba**	**137.36**							3.66
Barium		126	70						
		127	71						
		128	72						
		129	73				−47.0(8)		
		130	**74**	(0)	(+)		**−49.0(5)**	0.101	0.00370
		131	75						
		132	**76**	(0)	(+)			0.097	0.00356
		133	77						
		134	**78**	(0)	(+)		−49.6(8)	2.42	0.0886
		135	**79**	3/2	+	$d_{3/2}$		6.6	0.241
		136	**80**	(0)	(+)			7.8	0.286
		137	**81**	3/2	+	$d_{3/2}$	−46.9(10)	11.3	0.414
		138	**82**	(0)	(+)		−47.8(10)	71.7	2.622
		139	83				−44.6(10)		
		140	84				−42.5(10)		
		141	85						
		142	86						
		143	87						
57	**La**	**138.92**							2.00
Lanthanum		131	74						
		132	75						
		133	76						
		134	77				−45.9(8)		
		135	78						
		136	79						
		137	80						
		138	81	5		(7/2, 3/2)	−46.5(10)	0.089	0.0018
		139	**82**	7/2	(+)	$g_{7/2}$	**−47.0(10)**	99.911	2.00
		140	83				−43.8(10)		
		141	84				−42.5(10)		
		142	85						
		143	86						
58	**Ce**	**140.13**							2.26
Cerium		133	75						
		134	76						
		135	77						
		136	**78**	(0)	(+)			0.19	0.0044
		137	79						
		138	**80**	(0)	(+)		**−47.5(10)**	0.26	0.00566
		139	81				−46.9(10)		
		140	**82**	(0)	(+)		**−47.6(10)**	88.47	2.00
		141	83	7/2			−45.0(10)		
		142	**84**	(0)	(+)		**−43.7(10)**	11.08	0.250
		143	85				−40.5(10)		
		144	86				−38.7(10)		
		145	87						
		146	88				−33.2(10)		
59	**Pr**	**140.92**							0.40
Praseodymium		135	76						
		136	77						
		137	78						
		138	79				−44.0(10)		
		139	80						
		140	81				−44.3(10)		
		141	**82**	5/2	+	$d_{5/2}$	**−45.5(10)**	100	0.40
		142	83				−43.0(10)		
		143	84				−41.9(10)		
		144	85				−39.0(10)		
		145	86						
		146	87				−34.3(10)		

(10) μ (nucl mag)	(11) Quad mom, 10^{-24} cm²	(12) Half-life ($T_{1/2}$)	(13) Decay — Type and energy	(Q)	(14) Binding energy of last neutron E_n, MeV	(15) σ_{tn}, 10^{-24} cm²	
						1.2	**Ba**
		97 m	K; β^+ 3.8				126
		~12 m					127
		2.4 d	K; β^+ 3.0, . . .				128
		1.9 h	β^+ 1.6				129
						6	**130**
		11.6 d	K				131
						3	**132**
		{ 39 h	IT 0.276	}			133
		{ 8 y	K	}			**134**
						<4	
+0.835(3)		(29 h	IT 0.268	)		5	**135**
						<1	**136**
+0.935(3)		(2.60 m	IT 0.662	)	6.8	4	**137**
					8.6	0.55	**138**
		85 m	β^- 2.22, 0.8, 2.38	(2.38)	5.2	4	139
		12.8 d	β^- 1.02, 0.48				140
		18 m	β^- 2.8				141
		6 m	β^-				142
		<0.5 m	β^-				143
						8.9	**La**
		58 m	β^+ 1.6				131
		4.5 h	β^+ 3.5				132
		4 h	K; β^+ 1.2				133
		6.5 m	K; β^+ 2.7	(3.7)			134
		19 h	K				135
		9 m	K; β^+ 2.1				136
		>10⁵ y					137
+3.685	± 0.9	2 × 10¹¹ y	K; β^- 1.0				138
+2.776(3)	+0.3(1)				8.8	8.9	**139**
		40.2 h	β^- 1.34, 0.8, . . .	(3.75)	5.1	3	140
		3.8 h	β^- 2.43, 0.9	(2.43)	6.9		141
		77 m	β^- >2.5				142
		~19 m	β^-				143
						0.7	**Ce**
		6.3 h	K; β^+ 1.3				133
		72 h	K				134
		22 h	K; β^+ 0.8				135
						~22	**136**
		{ 35 h	IT 0.26	}			137
		{ 9 h	K	}			**138**
						1	139
		140 d	K	(0.3)	9.0	0.6	**140**
					5.5		141
± 0.89		32 d	β^- 0.43, 0.57	(0.57)	7.1	1	**142**
					5.1	6	143
		33 h	β^- 1.09, 0.3, . . .	(1.44)			144
		285 d	β^- 0.30, . . .	(0.30)			145
		3.0 m	β^- 2.0				146
		14 m	β^- 0.7				
						11	**Pr**
		22 m	β^+ 2.5				135
		70 m	β^+ 2.0				136
							137
		2.0 h	β^+ 1.4				138
		4.5 h	β^+ 1.0				139
		3.4 m	β^+ 2.3; K	(3.3)	7.8		140
+3.8(4)	−0.054				9.4	11	**141**
		19.1 h	β^- 2.16, 0.6	(2.16)		20	142
		13.8 d	β^- 0.92	(0.92)			143
		17 m	β^- 2.98, 0.8, 2.3	(2.98)			144
		5.9 h	β^- 1.7				145
		24 m	β^- 3.7, 2.3	(4.2)			146

(1)	(2)	(3)	(4)	(5)	(6)	(7)	(8)	(9)	
Z	Chem symbol	A	N	I	π	Ground-state config	M-A, MeV	Abundance	
								Per cent	Cosmic
60	Nd	144.27							1.44
Neodymium		138	78						
		139	79						
		140	80				−44.2(10)		
		141	81				−43.8(10)		
		142	**82**	**(0)**	**(+)**		**−45.2(10)**	**27.1**	**0.39**
		143	**83**	**7/2**	**(−)**	f$_{7/2}$	**−42.8(10)**	**12.2**	**0.175**
		144	84	(0)	(+)		−42.0(10)	23.9	0.344
		145	**85**	**7/2**	**(−)**	f$_{7/2}$		**8.3**	**0.119**
		146	**86**	**(0)**	**(+)**		**−38.5(10)**	**17.2**	**0.248**
		147	87	5/2			−35.8(10)		
		148	**88**	**(0)**	**(+)**		**−33.5(10)**	**5.7**	**0.0824**
		149	89				−31.0(10)		
		150	**90**	**(0)**	**(+)**		**−29.9(10)**	**5.6**	**0.0806**
		151	91						
61	Pm								
Promethium		141	80						
		142	81						
		143	82						
		144	83						
		145	84						
		146	85				−38.2(10)		
		147	86				−36.7(10)		
		148	87				−33.3(10)		
		149	88				−32.7(10)		
		150	89				−28.8(10)		
		151	90						
62	Sm	150.35							0.664
Samarium		143	81						
		144	**82**	**(0)**	**(+)**		**−41.0(16)**	**3.1**	**0.0108**
		145	83						
		146	84				−38.9(10)		
		147	85	7/2	(−)	f$_{7/2}$	−37.0(10)	15.0	0.100
		148	**86**	**(0)**	**(+)**		**−36.0(10)**	**11.2**	**0.0748**
		149	**87**	**7/2**	**(−)**	(f$_{7/2}$)	**−34.0(10)**	**13.8**	**0.0920**
		150	**88**	**(0)**	**(+)**		**−34.1(10)**	**7.4**	**0.0492**
		151	89						
		152	**90**	**(0)**	**(+)**		**−30.4(10)**	**26.8**	**0.176**
		153	91						
		154	**92**	**(0)**	**(+)**		**−27.5(10)**	**22.7**	**0.150**
		155	93				−24.7(11)		
		156	94				−23.2(10)		
63	Eu	152.0							0.187
Europium		144	81						
		145	82						
		146	83						
		147	84						
		148	85						
		149	86						
		150	87				−31.5(10)		
		151	**88**	**5/2**	**(+)**	d$_{5/2}$		**47.8**	**0.0892**
		152	89	3					
		153	**90**	**5/2**	**+**	d$_{5/2}$		**52.2**	**0.0976**
		154	91	3			−25.2(10)		
		155	92				−26.9(10)		
		156	93				−24.1(10)		
		157	94				−25.1(10)		
		158	95						
		159	96						

(10) μ (nucl mag)	(11) Quad mom, 10^{-24} cm²	(12) Half-life $(T_{1/2})$	(13) Decay Type and energy	(Q)	(14) Binding energy of last neutron E_n, MeV	(15) σ_{tn}, 10^{-24} cm²	
						48	**Nd**
		22 m	β^+ 2.4				138
		5.5 h	K; β^+ 3.1	(4.1)			139
		3.3 d	K	(0.1)			140
		2.4 h	K; β^+ 0.7	(1.7)			141
						17	**142**
−1.0(2)					5.0	320	**143**
						5	144
−0.6(1)		1.5 × 10¹⁵ y	α 1.8			44	**145**
						2	146
± 0.6							147
		11.6 d	β^- 0.81, 0.37	(0.90)		4	**148**
		1.8 h	β^- 1.5	(1.6)			149
						3	**150**
		15 m	β^- 1.9	(2.0)			151
							Pm
							141
		20 m	β^+ ~2.6				142
		~300 d	K				143
		~300 d	K				144
		25 y	K; L				145
		~2 y	β^- 0.7				146
		2.6 y	β^- 0.23	(0.23)		~60	147
		{ 5.3 d / 42 d	β^- 2.5 / β^- 0.6, 2.4	}			148
		50 h	β^- 1.05	(1.34)			149
		2.7 h	β^- 2.0, 3.0				150
		27 h	β^- 1.1				151
						(10⁴)	**Sm**
		8 m	β^+ 2.3	(3.3)			143
						~0.03	**144**
		1.0 y	K				145
		~5 × 10⁷ y	α 2.5				146
−0.7(1)		1.3 × 10¹¹ y	α 2.18				147
							148
−0.6(1)						6.6 × 10⁴	**149**
							150
		80 y	β^- 0.076	(0.096)		1.2 × 10⁴	151
						140	**152**
		47 h	β^- 0.71, 0.64, 0.81, . . .	(0.81)		5	153
							154
		23 m	β^- 1.8	(2.2)	5.6		155
		~10 h	β^- 0.9				156
						4300	**Eu**
							144
		18 m	β^+ 2.4				145
		5 d	K				146
		38 h	K				147
		24 d	K; α 2.9				148
		58 d	K				149
		120 d	K				150
+3.6	+1.2	14 h	β^- 1.1			8600	151
± 2.0 / +1.6	+2.5	{ 9.3 h / 13 y	β^- 1.88 (1.88) / K; β^- 0.70, . . .	}		5000	152
						400	**153**
± 2.1		16 y	β^- 1.5			1400	154
		1.7 y	β^- 0.15, 0.25	(0.25)		1.3 × 10⁴	155
		15 d	β 0.4, 2.4	(2.4)			156
		15 h	β^- 1.0				157
		60 m	β^- 2.6				158
		20 m	β^-				159

(1) Z	(2) Chem symbol	(3) A	(4) N	(5) I	(6) π	(7) Ground-state config	(8) M-A, MeV	(9) Abundance Per cent	Cosmic
64 *Gadolinium*	**Gd**	**157.26**							**0.684**
		148	84				−34.1(16)		
		149	85						
		150	86				−32.5(10)		
		151	87						
		152	**88**	(0)	(+)			**0.20**	**0.00137**
		153	89						
		154	**90**	(0)	(+)		**−28.2(10)**	**2.15**	**0.0147**
		155	**91**	(3/2)			**−27.1(11)**	**14.7**	**0.101**
		156	**92**	(0)	(+)		**−26.5(10)**	**20.5**	**0.141**
		157	**93**	(3/2)			**−26.8(10)**	**15.7**	**0.107**
		158	**94**	(0)	(+)		**−24.7(10)**	**24.9**	**0.169**
		159	95						
		160	**96**	(0)	(+)		**−20.7(10)**	**21.9**	**0.149**
		161	97						
65 *Terbium*	**Tb**	**158.93**							**0.0956**
		149	84						
		150	85						
		151	86						
		152	87						
		153	88						
		154	89						
		155	90						
		156	91						
		157	92						
		158	93						
		159	**94**	3/2	(+)	$d_{3/2}$		**100**	**0.0956**
		160	95				−20.7(20)		
		161	96						
		162	97						
66 *Dysprosium*	**Dy**	**162.51**							**0.556**
		156	**90**	(0)	(+)			**0.052**	**0.00029**
		157	91						
		158	**92**	(0)	(+)			**0.090**	**0.00050**
		159	93						
		160	**94**	(0)	(+)		**−22.5(20)**	**2.29**	**0.0127**
		161	**95**	(5/2)				**18.9**	**0.105**
		162	**96**	(0)	(+)		**−21.5(20)**	**25.5**	**0.142**
		163	**97**	(5/2)				**25.0**	**0.139**
		164	**98**	(0)	(+)		**−18.2(20)**	**28.2**	**0.157**
		165	99				−16.4(20)		
		166	100						
67 *Holmium*	**Ho**	**164.94**							**0.118**
		160	93						
		161	94						
		162	95						
		163	96						
		164	97				−16.0(10)		
		165	**98**	7/2	+	$g_{7/2}$	**−17.6(20)**	**100**	**0.118**
		166	99						
		167	100						
		168	101						
		169	102						
68 *Erbium*	**Er**	**167.27**							**0.316**
		160	92						
		161	93						
		162	**94**	(0)	(+)			**0.136**	**0.000316**
		163	95						
		164	**96**	(0)	(+)		**−17.0(10)**	**1.56**	**0.00474**

μ (nucl mag)	Quad mom, 10^{-24} cm²	Half-life ($T_{1/2}$)	Decay — Type and energy	(Q)	Binding energy of last neutron E_n, MeV	σ_{in}, 10^{-24} cm²	
						3.8×10^4	**Gd** 148
		>35 y	α 3.2				149
		9 d	K; α 3.0				150
		>10^5 y	α 2.7				151
		~150 d	K			<180	152
		236 d	K				153
							154
± 0.3	$+1.1$					~7×10^4	155
							156
± 0.4	$+1.0$					1.8×10^5	157
						4	158
		18 h	β^- 0.95, 0.60, . . .	(0.95)			159
						0.8	160
		3.7 m	β^- 1.6	(2.0)			161
						45	**Tb** 149
		4.1 h	α 3.95				150
		19 h	α 3.4				151
							152
		5.1 d	K				153
		{ 7 h	K; β^-				154
		{ 17.2 h	K; β^- 2.75, 1.66				155
		{ 5 h	β^- 0.14				156
		{ 5 d	K; β^- 0.6, 0.2				157
							158
						45	159
		72 d	β^- 0.56, 0.85, . . .	(1.81)		~600	160
		7 d	β^- 0.55, . . .	(0.6)			161
		14 m					162
						1100	**Dy** 156
							157
		8.2 h	K				158
							159
		134 d	K; L				160
-0.37	$+1.1$						161
							162
$+0.51$	$+1.3$						163
						2700	164
		{ 1.3 m	IT 0.108; β^- 0.1			4700	165
		{ 2.32 h	β^- 1.25, . . .	(1.25)			
		82 h	β^- 0.3	(0.3)			166
						64	**Ho** 160
		5.0 h	K				160
		2.5 h	K				161
		22 m	K; β^+ 1.3				162
							163
		37 m	K; β^- 0.99, 0.90	(0.99)			164
± 3.3						64	165
		{ >30 y	β^- 0.2, . . .	(2.4)			166
		{ 27.2 h	β^- 1.85	(1.85)			
		3.0 h	β^- 1.0, 0.28				167
							168
		1.6 h	β^-				169
						170	**Er** 160
		29 h	K				161
		3 h	K; β^+ 1.2				162
		75 m	K				163
							164

(1) Z	(2) Chem symbol	(3) A	(4) N	(5) I	(6) π	(7) Ground-state config	(8) M-A, MeV	(9) Abundance Per cent	(9) Abundance Cosmic
68	Er	167.27							
		165	97						
		166	98	(0)	(+)			33.4	0.104
		167	99	7/2	(−)	$(f_{7/2})$		22.9	0.770
		168	100	(0)	(+)		−15.0(10)	27.1	0.0850
		169	101						
		170	102	(0)	(+)		−9.9(20)	14.9	0.0228
		171	103						
69	Tm	168.94							0.0318
	Thulium	165	96						
		166	97						
		167	98						
		168	99						
		169	100	1/2	(+)	$s_{1/2}$		100	0.0318
		170	101						
		171	102						
		172	103						
		173	104						
		174	105						
70	Yb	173.04							0.220
	Ytterbium	166	96						
		167	97						
		168	98	(0)	(+)			0.14	0.00030
		169	99						
		170	100	(0)	(+)			3.03	0.00666
		171	101	1/2	(−)	$p_{1/2}$		14.3	0.0316
		172	102	(0)	(+)		−15.0(30)	21.8	0.0480
		173	103	5/2	(−)	$f_{5/2}$		16.2	0.0356
		174	104	(0)	(+)		−17.9(30)	31.8	0.0678
		175	105						
		176	106	(0)	(+)			12.7	0.0278
		177	107						
71	Lu	174.99							0.050
	Lutecium	170	99						
		171	100						
		172	101						
		173	102						
		174	103						
		175	104	7/2	(+)	$g_{7/2}$		97.40	0.0488
		176	105			(11/2, 13/2)	−2.4(20)	2.60	0.0013
		177	106						
		178	107						
		179	108						
72	Hf	178.50							0.438
	Hafnium	170	98						
		171	99						
		172	100						
		173	101						
		174	102	(0)	(+)			0.18	0.00078
		175	103						
		176	104	(0)	(+)		−3.4(20)	5.2	0.0226
		177	105	7/2				18.5	0.0806
		178	106	(0)	(+)		+0.2(20)	27.1	0.119
		179	107	9/2				13.8	0.0604
		180	108	(0)	(+)		+1.9(20)	35.2	0.155
		181	109				+3.7(20)		

(10) μ (nucl mag)	(11) Quad mom, 10^{-24} cm²	(12) Half-life ($T_{1/2}$)	(13) Decay — Type and energy	(13) (Q)	(14) Binding energy of last neutron E_n, MeV	(15) σ_{tn}, 10^{-24} cm²	
							Er
		10 h	K				165
							166
±0.5	±10						167
						2	**168**
		9.4 d	β^- 0.33	(0.33)			169
						9	**170**
		7.5 h	β^- 1.06, 0.67, 1.48	(1.48)			171
						125	Tm
		25 h	K				165
		7.7 h	K; β^+ 2.1				166
		9.6 d	K				167
		87 d	K; β^- 0.5				168
						125	169
		129 d	β^- 0.97, 0.89	(0.97)		2000	170
		1.9 y	β^- 0.10	(0.10)			171
		19 m	β^-				172
							173
		~2 d	β^-				174
						37	Yb
		54 d	K				166
		18 m	K				167
		32 d	K			1.1 × 10⁴	**168**
							169
+0.45							170
							171
							172
−0.65	+3.9(4)						173
						~60	174
		4.2 d	β^- 0.47, 0.07, 0.36	(4.7)			175
						7	176
		2.0 h	β^- 1.30, . . .	(1.30)			177
						111	Lu
		1.7 d	K				170
		{ 8.5 d	K				171
		{ 1.6 y	K				171
		{ 4.0 h	β^+ 1.2				172
		{ 6.7 d	K				173
		1.4 y	K				174
+2.6	+5.9	165 d	K; β^- 0.6			18	175
+3.8	+7(1)	{ 3.7 h	β^- 1.2	(1.2)		3600	176
		{ 7.5 × 10¹⁰ y	β^+ 0.42; K	(1.02)			176
		6.8 d	β^- 0.50, 0.18, 0.39, . . .	(0.50)			177
		22 m	β^-				178
		~5 h	β^-				179
						105	Hf
		1.8 h	β^+ 2.4				170
		16 h	K				171
		~5 y	K				172
		24 h	K				173
						1000	174
		70 d	K				175
						<30	176
+0.61	+3					370	177
						80	178
−0.47	+3	(19 s	IT 0.16	)		65	179
		(5.5 h	IT	)		10	180
		46 d	β^- 0.41	(1.02)			181

(1)	(2)	(3)	(4)	(5)	(6)	(7)	(8)	(9)	
								Abundance	
Z	Chem symbol	A	N	I	π	Ground-state config	$M\text{-}A$, MeV	Per cent	Cosmic
73 *Tantalum*	**Ta**	**180.95**							**0.065**
		176	103						
		177	104						
		178	105						
		179	106						
		180	**107**				**2.0(20)**	**0.012**	**0.000007**
		181	**108**	7/2	+	$g_{7/2}$	**2.7(20)**	**99.988**	**0.065**
		182	109				5.0(20)		
		183	110						
		184	111						
		185	112						
		186	113						
74 *Tungsten*	**W**	**183.86**							**0.49**
		176	102						
		177	103						
		178	104						
		179	105						
		180	**106**	(0)	(+)		**1.3(20)**	**0.14**	**0.0006**
		181	107						
		182	**108**	(0)	(+)		**3.3(20)**	**26.2**	**0.13**
		183	**109**	1/2	(—)	$p_{1/2}$	**4.9(20)**	**14.3**	**0.070**
		184	**110**	(0)	(+)		**5.9(20)**	**30.7**	**0.15**
		185	111						
		186	**112**	(0)	(+)		**9.7(20)**	**28.7**	**0.14**
		187	113				11.6(22)		
		188	114						
75 *Rhenium*	**Re**	**186.22**							**0.135**
		180	105						
		181	106						
		182	107						
		183	108						
		184	109						
		185	**110**	5/2	+	$d_{5/2}$		**37.1**	**0.0500**
		186	111				10.0(23)		
		187	**112**	5/2	+	$d_{5/2}$	10.3(22)	62.9	0.0850
		188	113				15.2(20)		
		189	114						
		190	115						
		191	116						
76 *Osmium*	**Os**	**190.2**							**1.00**
		182	106						
		183	107						
		184	**108**	(0)	(+)			**0.018**	**0.00018**
		185	109						
		186	**110**	(0)	(+)		**8.9(23)**	**1.59**	**0.0159**
		187	**111**	(1/2)	(—)	$(p_{1/2})$	10.3(22)	1.64	0.0164
		188	**112**	(0)	(+)		**13.1(20)**	**13.3**	**0.133**
		189	**113**	3/2	(—)	$p_{3/2}$	16.9(20)	16.1	0.161
		190	**114**	(0)	(+)		**16.2(20)**	**26.4**	**0.264**
		191	115				20.0(21)		
		192	**116**	(0)	(+)		**21.0(30)**	**41.0**	**0.410**
		193	117				24.6(20)		
		194	118						

(10) μ (nucl mag)	(11) Quad mom, 10^{-24} cm²	(12) Half-life ($T_{1/2}$)	(13) Decay Type and energy	(Q)	(14) Binding energy of last neutron E_n, MeV	(15) σ_{tn}, 10^{-24} cm²	
						22	**Ta** 176
		8.0 h	K				176
		2.2 d	K				177
		{ 9.3 m	K; β^+ 1.1	}			178
		{ 2.1 h	K; β^+ ~1	}			
		~600 d	K				179
		(8.1 h	K; β^- 0.70, . . .	)			180
+1.9	+5.9	(0.33 s	IT	)	7.6	22	**181**
		{ 16 m	IT 0.18	}			182
		{ 112 d	β^- 0.51	(1.7) }	6.0	~2 × 10⁴	
		5.2 d	β^- 0.62, . . .	(1.07)			183
		8.7 h	β^- 1.26, 0.15				184
		49 m	β^- 1.7				185
		10 m	β^- 2.2				186
						18	**W** 176
		1.3 h	K; β^+ ~2				176
		2.2 h	K				177
		21 d	K				178
		30 m	K				179
		(5 ms	IT	)		<20	**180**
		140 d	K; L	(0.9)			181
						20	**182**
		(5.5 s	IT	)	6.2	11	**183**
						2.0	**184**
		{ 1.7 m	IT	}			185
		{ 74 d	β^- 0.43	(0.43) }			
					7.1	36	**186**
		24 h	β^- 0.62, 1.31	(1.31)		~80	187
		65 d	β^-				188
						86	**Re** 180
		2.4 m	β^- 1.1				180
							181
		{ 13 h	K	}			182
		{ 64 h	K	}			
		150 d	K				183
		{ 50 d	K	}			184
		{ 2.2 d	K	}			
+3.1714(6)	(2.8)					105	**185**
		91 h	β^- 1.07, 0.93, . . . ; K	(1.07)			186
+3.2039(6)	+2.6	~5 × 10¹⁰ y	β^- <0.008		7.3	75	187
		} 20 m	IT 0.10	}		<3	188
		{ 17 h	β^- 2.12	}			
		~200 d	β^- 0.2				189
		3 m	β^- 1.7				190
		10 m	β^- 1.8				191
						15	**Os** 182
		24 h	K				182
		} 10 h	K	}			183
		{ 15 h	K	}			
						~20	**184**
		95 d	K; L				185
							186
+0.12		(35 h	IT	)	6.3		**187**
							188
+0.651	+0.6						189
		{ 9 m	IT 0.62	}		8	**190**
		{ 6 h	IT	}			
		} 14 h	IT 0.074	}			191
		{ 16 d	β^- 0.14	(0.31) }			
						1.6	**192**
		31 h	β^- 1.10, . . .	(1.10)		200	193
		~2 y	β^-				194

(1) Z	(2) Chem symbol	(3) A	(4) N	(5) I	(6) π	(7) Ground-state config	(8) M-A, MeV	(9) Abundance Per cent	(9) Abundance Cosmic
77 *Iridium*	Ir	192.2							0.821
		187	110						
		188	111						
		189	112						
		190	113						
		191	**114**	3/2	+	d$_{3/2}$	**19.9(21)**	**38.5**	**0.316**
		192	115				23.0(20)		
		193	**116**	3/2	+	d$_{3/2}$	**23.5(20)**	**61.5**	**0.505**
		194	117				24.6(15)		
		195	118				26.7(15)		
		196	119						
		197	120						
		198	121				31.4(20)		
78 *Platinum*	Pt	195.09							1.625
		187	109						
		188	110						
		189	111						
		190	112					0.012	0.0001
		191	113						
		192	**114**	(0)	(+)		**21.5(20)**	**0.78**	**0.0127**
		193	115				23.6(15)		
		194	**116**	(0)	(+)		**22.3(15)**	**32.8**	**0.533**
		195	**117**	1/2	−	p$_{1/2}$	**24.6(15)**	**33.7**	**0.548**
		196	**118**	(0)	(+)		**25.0(15)**	**25.4**	**0.413**
		197	119				27.3(30)		
		198	**120**	(0)	(+)		**27.0(20)**	**7.2**	**0.117**
		199	121						
79 *Gold*	Au	**197.0**							0.145
		191	112						
		192	113	1					
		193	114	3/2					
		194	115	1					
		195	116						
		196	117				26.2(30)		
		197	**118**	3/2	+	d$_{3/2}$	**26.5(30)**	**100**	**0.145**
		198	119	2			28.4(30)		
		199	120	3/2			28.9(30)		
		200	121				32.0(30)		
		201	122				33.1(30)		
		202	123						
		203	124						
80 *Mercury*	Hg	200.61							0.284
		191	111						
		192	112						
		193	113						
		194	114						
		195	115						
		196	**116**	(0)	(+)		**25.5(30)**	**0.15**	**0.00045**
		197	117	1/2					
		198	**118**	(0)	(+)		**27.0(30)**	**10.0**	**0.0285**
		199	**119**	1/2	−	p$_{1/2}$	28.4(30)	**16.9**	**0.0481**
		200	**120**	(0)	(+)		**29.7(30)**	**23.1**	**0.0656**

(10) μ (nucl mag)	(11) Quad mom, 10^{-24} cm²	(12) Half-life ($T_{1/2}$)	(13) Decay — Type and energy	(Q)	(14) Binding energy of last neutron E_n, MeV	(15) σ_{in}, 10^{-24} cm²	
						460	Ir 187
		12 h	K; β^+ 2.2				188
		41 h	K; β^+ 2				189
		11 d	K				190
		{ 3 h	K; β^+ 2.0				
		{ 11 d	K				
+0.17(3)	+1.2(7)	(5 s	IT 0.042	)		1000	191
		{ 1.4 m	IT 0.057; β^-				192
		{ 74 d	β^- 0.67, . . . ; K	(1.58)			193
+0.18(3)	+1.0(5)				7.8	120	
		19 h	β^- 2.24, 1.191, . . .	(2.24)			194
		2.3 h	β^- 1.2, 2.1, . . .	(2.1)	6.5		195
		9.7 d	β^- 0.08				196
		7 m	β^- 1.6, . . .				197
		50 s	β^- 3.6	(4.4)			198
						10	Pt 187
		3 h	K				187
		10 d	K				188
		11 h	K				189
		$\sim 10^{12}$ y(?)	α 3.3(?)			$\sim$90	190
		3.0 d	K				191
						8	192
		{ 3.4 d	IT 0.135				193
		{ "long"	L				
					9.5	1.2	194
+0.60592(8)		($\sim$6 d)	IT 0.130	)	6.1	27	195
					8.2	1.2	196
		{ 1.4 h	IT				197
		{ 19 h	β^- 0.67, 0.48, 0.47	(0.75)			
						4	198
		30 m	β^- $\sim$1.2	($\leq$1.7)			199
						98	Au 191
		3 h	K				191
		4.8 d	K; β^+ 1.9				192
+0.1		{ 4 s	IT 0.032				193
		{ 17 h	K				
+0.07		39 h	K; β^+ 1.55, 1.22	(2.57)			194
		{ 30 s	IT 0.057, 0.318				195
		{ 180 d	K	(0.27)			
		{ 14 h	K or IT				196
		{ 5.6 d	K; β^- 0.27	(0.70$-$)			
+0.13(1)	+0.56	(7.4 s	IT 0.130, 0.409	)	7.9	98	197
+0.50		2.70 d	β^- 0.96, 0.28, 1.37	(1.37)	6.4	2.6×10^4	198
+0.24		3.15 d	β^- 0.30, 0.25, 0.46	(0.46)		$\sim$30	199
		48 m	β^- 2.2	(2.2)			200
		26 m	β^- 1.5	(1.5)			201
		$\sim$25 s	β^-				202
		55 s	β^- 1.9	(1.9)			203
						350	Hg 191
		57 m	K				191
		6 h	K; β^+ 1.2				192
		{ 12 h	IT 0.101; K				193
		{ 5 h	K				
		{ 0.4 s	IT?				194
		{ $\sim$130 d	K				
		{ 40 h	K IT 0.128				195
		{ 9.5 h	K				
						2500	196
+0.52		{ 25 h	IT 0.165; K				197
		{ 65 h	K				
							198
+0.50413(3)		(43 m	IT 0.368			2000	199
						<50	200

(1) Z	(2) Chem symbol	(3) A	(4) N	(5) I	(6) π	(7) Ground-state config	(8) M-A, MeV	(9) Abundance	
								Per cent	Cosmic
80	**Hg**	**200.61**							
		201	121	3/2	(−)	$p_{3/2}$	**31.7(30)**	13.2	0.0375
							Based on ^{16}O std ↑		
							Based on ^{208}Pb std ↓		
		202	122	(0)	(+)		**32.9(6)**	29.8	0.0844
		203	123				34.0(4)		
		204	124	(0)	(+)		**34.7(2)**	6.8	0.0194
		205	125				37.6(2)		
81	**Tl**	**204.39**							0.108
Thallium		195	114						
		196	115						
		197	116						
		198	117						
		199	118						
		200	119						
		201	120						
		202	121				33.9(5)		
		203	122	1/2	+	$s_{1/2}$	**33.5(4)**	29.5	0.0319
		204	123				35.1(1)		
		205	124	1/2	(+)	$s_{1/2}$	**35.8(1)**	70.5	0.0761
RaE″		206	125				37.67(2)		
AcC″		207	126				39.24(4)		
ThC″		208	127				43.77(2)		
		209	128				47.19(6)		
RaC″		210	129				51.79(3)		
82	**Pb**	**207.21**							0.47
Lead		197	115						
		198	116						
		199	117						
		200	118						
		201	119						
		202	120				34.0(5)		
		203	121				34.8(5)		
		204	122	(0)	(+)		**34.3(1)**	1.3	0.0063
		205	123				35.89(5)		
		206	124	(0)	(+)		**36.15(1)**	26	0.122
		207	125	1/2	−	$p_{1/2}$	**37.79(1)**	21	0.0995
		208	126	(0)	(+)		**38.774(std)**	52	0.243
		209	127				43.27(5)		
RaD		210	128				46.40(2)		
AcB		211	129				51.0(1)		
ThB		212	130				54.17(3)		
		213	131				58.7(4)		
RaB		214	132				62.00(8)		
83	**Bi**	**209.00**							0.144
Bismuth		198	115						
		199	116						
		200	117						
		201	118						
		202	119						
		203	120						
		204	121						
		205	122						
		206	123				39.7(2)		
		207	124				40.19(4)		
		208	125				41.70(7)		
		209	126	9/2	(−)	$h_{9/2}$	**42.64(5)**	100	0.144

(10) μ (nucl mag)	(11) Quad mom, 10^{-24} cm²	(12) Half-life ($T_{1/2}$)	(13) Decay — Type and energy	(13) (Q)	(14) Binding energy of last neutron E_n, MeV	(15) σ_{tn}, 10^{-24} cm²	
−0.5990(1)	+0.6				6.3	<50	**Hg 201**
							202
		48 d	β⁻ 0.21	(0.49)	3		**203**
						0.4	**204**
		5.2 m	β⁻ 1.6, 1.4	(1.6)			**205**
						3.3	**Tl 195**
		1.2 h					**196**
		~4 h					
		⎰ 0.54 s	IT?				**197**
		⎱ 2.8 h	K				
		⎰ 1.9 h	K; IT 0.26				**198**
		⎱ 5 h	K				
		7.4 h	K				**199**
		27 h	K; β⁺				**200**
		3.0 d	K				**201**
		12 d	K; L		8.8	11	**202**
+1.6117(1)							**203**
± 0.07		4.1 y	β⁻ 0.76; K	$\left[\begin{array}{c}(0.76-)\\(0.34+)\end{array}\right]$	6.5		**204**
+1.6275(1)					7.6	0.11	**205**
		4.20 m	β⁻ 1.51	(1.51)	6.2		**206**
		4.78 m	β⁻ 1.45	(1.45)	6.8		**207**
		3.1 m	β⁻ 1.79, 1.28, . . .	(4.99)	3.8		**208**
		2.2 m	β⁻ 1.8, 2.3	(3.9)	4.8		**209**
		1.32 m	β⁻ 1.9	(5.4)	3.9		**210**
						0.17	**Pb 197**
		42 m	K; IT(?)				**198**
		2.3 h	K				**199**
		⎰ 12 m	IT 0.42				
		⎱ 1.5 h	K				**200**
		21 h	K				**201**
		⎰ 1.0 m	IT 0.66				
		⎱ 9 h	K				**202**
		⎰ 3.5 h	IT 0.79, 0.13; K				
		⎱ ~10⁵ y	L				**203**
		⎰ 6 s	IT 0.86				
		⎱ 52 h	K	(1.4)			**204**
		(68 m	IT 0.91)		6.4	0.9	**205**
		>10⁶ y	L		8.1	0.03	**206**
+0.58750(7)		(0.80 s	IT 1.06)		6.7	0.73	**207**
					7.4	0.00045	**208**
		3.3 h	β⁻ 0.62	(0.62)	3.9		**209**
		20 y	β⁻ 0.020	(0.067)	5.2		**210**
		36.1 m	β⁻ 1.4, 0.5	(1.4)	3.8		**211**
		10.64 h	β⁻ 0.34, 0.58	(0.58)	5.2		**212**
							213
		26.8 m	β⁻ 0.7, . . .	(1.0)			**214**
						0.033	**Bi 198**
		7 m	K; α 5.83				
		~25 m	K; α 5.47				**199**
		35 m	K				**200**
		⎰ 1.0 h	K; α 5.15				**201**
		⎱ ~2 h	K				
		1.6 h	K				**202**
		12 h	K; α 4.85				**203**
		12 h	K				**204**
		14 d	K				**205**
		6.4 d	K	(3.6)			**206**
		8.0 y	K; L	(2.4)			**207**
							208
+4.082	−0.4				7.4	0.033	**209**

(1) Z	(2) Chem symbol	(3) A	(4) N	(5) I	(6) π	(7) Ground-state config	(8) M-A, MeV	(9) Abundance Per cent	(9) Abundance Cosmic
83	**Bi**	**209.00**							
RaE		210	127	1			46.34(2)		
AcC		211	128				49.60(5)		
ThC		212	129				53.58(3)		
		213	130				56.77(6)		
RaC		214	131				61.01(4)		
		215	132				64.2(3)		
84	**Po**								
Polonium		201	117						
		202	118						
		203	119						
		204	120						
		205	121						
		206	122						
		207	123						
		208	124				43.1(1)		
		209	125	1/2			44.47(9)		
RaF		210	126				45.17(2)		
AcC'		211	127				48.98(2)		
ThC'		212	128				51.33(1)		
		213	129				55.38(6)		
RaC'		214	130				57.84(3)		
AcA		215	131				62.1(1)		
ThA		216	132				64.68(3)		
		217	133				69.0(3)		
RaA		218	134				71.72(8)		
85	**At**								
Astatine		203	118						
		204	119						
		205	120						
		206	121						
		207	122						
		208	123						
		209	124						
		210	125				49.0(2)		
		211	126				49.77(5)		
		212	127				53.0(6)		
		213	128				55.6(1)		
		214	129				58.90(6)		
		215	130				61.36(6)		
		216	131				65.14(5)		
		217	132				67.52(8)		
		218	133				71.4(2)		
		219	134				74.2(3)		
86	**Em**								
Emanation		206	120						
		207	121						
		208	122						
		209	123						
		210	124						
		211	125						
		212	126				53.1(1)		
		213	127						
		214	128				58.1(3)		
		215	129				61.4(1)		
		216	130				63.11(4)		
		217	131				66.88(8)		
		218	132				68.7(4)		
Actinon: An		219	133				72.6(1)		
Thoron: Tn		220	134				74.69(4)		
		221	135				78.7(3)		
Radon: Rn		222	136				80.92(8)		

(10) μ (nucl mag)	(11) Quad mom, 10^{-24} cm²	(12) Half-life ($T_{1/2}$)	(13) Decay — Type and energy	(Q)	(14) Binding energy of last neutron E_n, MeV	(15) σ_{tn}, 10^{-24} cm²	
							Bi
		5.0 d	β^- 1.17; α	(1.17)	4.7		210
		2.6 × 10⁶ y	α 4.94; β^-	(1.14)	5.1		211
		2.15 m	α 6.62, 6.27; β^-	(0.63−)			212
		60.5 m	β^- 2.25; α 6.05, 6.09, . . .	(2.25−)	4.4		213
		47 m	β^- 1.39, 0.96, α 5.9	(1.39)	5.1		214
		19.7 m	β^- 1.6, 3.17; α 5.5	(3.17)	4.2		215
		8 m	β^-				
							Po
		18 m	K; α 5.67				201
		56 m	K; α 5.59				202
		47 m	K				203
		3.8 h	K; α 5.37				204
		1.5 h	K; α 5.2				205
		9 d	K; α 5.22, . . .				206
		5.7 h	K; α 5.10				207
		2.9 y	K; α 5.11				208
		~100 y	α 4.88; K		6.6		209
		138.40 d	α 5.30, . . .		7.7		210
		{ 25 s ; 0.52 s	α 7.1, 8.7, . . . ; α 7.43, . . .		4.6		211
		0.30 μs	α 8.78		6.0		212
		4 μs	α 8.34		4.3		213
		160 μs	α 7.68		5.9		214
		1.8 ms	α 7.36; β^-		4.1		215
		0.16 s	α 6.77		5.8		216
		<24 m	α 6.5				217
		3.05 m	α 6.00; β^-				218
							At
		7 m	α 6.10				203
		~25 m	K				204
		25 m	K; α 5.90				205
		2.6 h	K				206
		1.8 h	K; α 5.75				207
		{ 1.7 h ; 6 h	K; α 5.65 ; K				208
		5.5 h	K; α 5.64				209
		8.3 h	K; α 5.35				210
		7.5 h	K; L; α 5.86				211
		0.22 s	α				212
		<2 s	α 9.2				213
		<5 s	α 8.78				214
		~100 μs	α 8.00		5.9		215
		~300 μs	α 7.79		4.6		216
		0.018 s	α 7.02		5.9		217
		1.3 s	α 6.63		4.6		218
		0.9 m	α 6.27; β^-				219
							Em
		7 m	α 6.22; K				206
		11 m	K; α 6.12				207
		23 m	K; α 6.14				208
		30 m	K; α 6.04				209
		2.7 h	α 6.04; K				210
		16 h	K; α 5.78, 5.85, 5.61				211
		23 m	α 6.26				212
							213
							214
		<1 m	α 8.6				215
		<9 m	α 8.01		6.7		216
		~1 ms	α 7.74		4.6		217
		19 ms	α 7.13, 6.53		6.6		218
		3.92 s	α 6.82, 6.56, . . .		4.4		219
		52 s	α 6.28, 5.75		6.3		220
		25 m	β^-; α				221
		3.825 d	α 5.48			0.7	222

(1) Z	(2) Chem symbol	(3) A	(4) N	(5) I	(6) π	(7) Ground-state config	(8) M-A, MeV	(9) Abundance Per cent	Cosmic
87 Fr *Francium*		217	130				67.7(2)		
		218	131				70.5(1)		
		219	132				72.41(8)		
		220	133				75.56(6)		
		221	134				77.55(8)		
		222	135				81.1(3)		
	AcK	223	136				83.4(2)		
88 Ra *Radium*		219	131				73.1(2)		
		220	132				74.29(5)		
		221	133				77.32(9)		
		222	134				78.99(5)		
	AcX	223	135				82.2(1)		
	ThX	224	136				84.08(4)		
		225	137				87.4(1)		
	Ra	226	138				89.39(9)		
		227	139				93.4(2)		
	MsTh₁	228	140				95.50(9)		
		229	141				99.1(2)		
		230	142				101.6(3)		
89 Ac *Actinium*		221	132				79.1(2)		
		222	133				81.2(1)		
		223	134				82.78(9)		
		224	135				85.45(9)		
		225	136				87.06(9)		
		226	137				90.2(4)		
	Ac	227	138	3/2			92.1(1)		
	MsTh₂	228	139				95.46(6)		
		229	140				97.3(3)		
		230	141				100.7(5)		
90 Th *Thorium*		223	133				84.4(2)		
		224	134				85.16(6)		
		225	135				87.6(1)		
		226	136				89.06(6)		
	RdAc	227	137				92.0(1)		
	RdTh	228	138				93.22(5)		
		229	139				96.1(1)		
	Io	230	140				97.8(1)		
	UY	231	141				101.1(1)		
	Th	232	142	(0)	(+)		103.17(6)	100	~0.06
		233	143				106.5(1)		
	UX₁	234	144				108.7(1)		
		235	145				112.6(3)		
91 Pa *Protoactinium*		225	134		.		90.0(3)		
		226	135				91.8(2)		
		227	136				92.97(9)		
		228	137				95.26(9)		
		229	138				96.5(1)		
		230	139				99.3(4)		
	Pa	231	140	3/2			100.8(1)		
		232	141				103.53(6)		
		233	142				105.2(1)		
	UX₂ *UZ*	234	143				108.5(4)		
		235	144				110.8(3)		
		236	145				114.4(3)		
		237	146				116.5(5)		
92 U *Uranium*		227	135				94.9(3)		
		228	136				95.56(8)		
		229	137				97.8(1)		
		230	138				98.66(6)		

(10) μ (nucl mag)	(11) Quad mom, 10^{-24} cm²	(12) Half-life ($T_{1/2}$)	(13) Decay — Type and energy	(Q)	(14) Binding energy of last neutron E_n, MeV	(15) σ_{tn}, 10^{-24} cm²	
							Fr
		<2 s	α 8.3				217
		<5 s	α 7.85				218
		0.02 s	α 7.30		6.5		219
		28 s	α 6.69		5.2		220
		4.8 m	α 6.30, 6.07		6.3		221
		15 m	β⁻; α				222
		22 m	β⁻ 1.0, 1.3; α 5.3				223
						20	**Ra**
		<1 m	α 8.0				219
		<9 m	α 7.43		7.2		220
		30 s	α 6.71		5.3		221
		38 s	α 6.55, 6.23		6.7		222
		11.6 d	α 5.70, 5.60, 5.42, . . .		5.2	125	223
		3.64 d	α 5.68, 5.44;		6.4	12	224
		14.8 d	β⁻ 0.32	(0.36)	5.1		225
		1620 y	α 4.78, 4.59		6.3	20	226
		41 m	β⁻ 1.30	(1.30)	4.6		227
		6.7 y	β⁻ <0.02	(<0.05)	6.1	~36	228
		<5 m	β⁻				229
		1 h	β⁻ 1.2				230
						520	**Ac**
		<2 s	α 7.6				221
		5.5 s	α 6.96				222
		2.2 m	α 6.64; K		6.8		223
		2.9 h	K; α 6.17		5.7		224
		10.0 d	α 5.80		6.6		225
		29 h	β⁻ 1.2	(1.2)	5.4		226
+1.1	−1.7	22 y	β⁻ 0.046; α 4.94		6.6	~520	227
		6.13 h	β⁻ 1.11, 0.45, −2.18	(2.18)	4.8		228
		66 m	β⁻				229
		<1 m	β⁻ 2.2				230
						7.5	**Th**
		<1 m	α 7.5				223
		<9 m	α 7.13		7.6		224
		8 m	α 6.57; K		5.9		225
		31 m	α 6.34, 6.23, 6.10		7.0		226
		18.2 d	α 5.97, 5.65–6.03		5.4	1500 (fiss)	227
		1.90 y	α 5.42, 5.34		7.0	120	228
		7300 y	α 4.85, 4.94, 5.02		5.5	45 (fiss)	229
		8 × 10⁴ y	α 4.68, 4.61, . . .		6.7	35	230
		25.6 h	β⁻ 0.09, 0.30, 0.23	(0.32)	5.2		231
		1.39 × 10¹⁰ y	α 3.99, 3.93; spon fiss		6.2	7.5	232
		23.3 m	β⁻ 1.23, . . .	(1.23)	5.2	1400	233
		24.10 d	β⁻ 0.19, 0.10	(0.19)	6.0	1.8	234
		<5 m	β⁻				235
						260	**Pa**
		2.0 s	α				225
		1.8 m	α 6.81				226
		38 m	α 6.46; K		7.1		227
		22 h	K; α 6.09, 5.85		6.1		228
		1.5 d	K; α 5.69		7.0		229
		17 d	K; β⁻ 0.40; β⁺ 0.2, 0.4; α			1500 (fiss)	230
		3.4 × 10⁴ y	α 5.00, 4.63–5.05			200	231
						⎡0.1 (fiss)⎤ ⎣700 ⎦	232
		1.31 d	β⁻ 0.28, 0.4–1.24	(1.24)	5.3	⎡700 (fiss)⎤	
		27.4 d	β⁻ 0.26, 0.14, 0.57	(0.57)	6.9	⎣66 ⎦	233
		{ 1.18 m	β⁻ 2.31, . . . ; IT				234
		{ 6.66 h	β⁻ 0.5, . . .		5.4		
		24 m	β⁻ 1.4	(1.4)	5.9		235
							236
		10 m	β⁻				237
						3.5 + 4.2 (fiss)	**U**
		1.3 m	α 6.8				227
		9.3 m	α 6.67; K		7.73		228
		58 m	K; α 6.42		6.1		229
		21 d	α 5.89, 5.82, 5.66		7.6	~25 (fiss)	230

(1) (2)		(3)	(4)	(5)	(6)	(7)	(8)	(9)	
								Abundance	
Z	Chem symbol	A	N	I	π	Ground-state config	$M\text{-}A$, MeV	Per cent	Cosmic
92	U								
		231	139				101.1(2)		
		232	140				102.24(5)		
		233	141	5/2			104.7(1)		
UII		234	142	(0)	(+)		106.2(1)	0.0055	
AcU		235	143	7/2	(+)		109.4(2)	0.72	
		236	144				111.4(1)		
		237	145				114.3(2)		
UI		238	146	(0)	(+)		116.6(1)	99.27	~0.02
		239	147				120.3(2)		
		240	148				122.7(1)		
93	Np								
Neptunium		231	138				103.0(1)		
		232	139				104.9(2)		
		233	140				105.7(1)		
		234	141				108.3(4)		
		235	142				109.6(1)		
		236	143				112.21(8)		
		237	144	5/2			113.8(2)		
		238	145	2			116.7(1)		
		239	146				119.0(2)		
		240	147				122.4(1)		
94	Pu								
Plutonium		232	138				105.9(1)		
		233	139				107.9(2)		
		234	140				108.6(1)		
		235	141				110.7(2)		
		236	142				111.70(8)		
		237	143				114.0(2)		
		238	144				115.4(1)		
		239	145	1/2			118.2(2)		
		240	146				120.2(1)		
		241	147	5/2			123.0(2)		
		242	148				125.2(1)		
		243	149				128.6(2)		
		244	150				131.0(3)		
		245	151				134.2(5)		
		246	152				137.0(6)		
95	Am								
Americium		237	142				115.4(2)		
		238	143				117.9(5)		
		239	144				119.0(2)		
		240	145				121.6(2)		
		241	146	5/2			123.0(2)		
		242	147				125.9(1)		
		243	148	5/2			128.0(2)		
		244	149				131.2(2)		
		245	150				133.0(2)		
		246	151				136.7(3)		
96	Cm								
Curium		238	142				118.8(1)		
		239	143				120.8(2)		
		240	144				121.7(1)		
		241	145				123.9(2)		
		242	146				125.2(1)		
		243	147				128.0(2)		

(10) μ (nucl mag)	(11) Quad mom, 10^{-24} cm²	(12) Half-life ($T_{1/2}$)	(13) Decay — Type and energy	(Q)	(14) Binding energy of last neutron E_n, MeV	(15) σ_{tn}, 10^{-24} cm²	
							U
		4.3 d	K; α 5.45		5.8	~400 (fiss) ~300	231
		74 y	α 5.32, 5.26, 5.13; spon fiss		7.2	80 (fiss)	232
±0.55	±3.4	1.62×10^5 y	α 4.82, 4.78, 4.73		6.0	$\begin{bmatrix} 60 \\ 520 \text{ (fiss)} \end{bmatrix}$	233
		2.5×10^5 y	α 4.78; spon fiss		6.7	80	234
±0.34	±4.0	7.1×10^8 y	α 4.40, 4.58; spon fiss		5.4	$\begin{bmatrix} 108 \\ 590 \text{ (fiss)} \end{bmatrix}$	235
		2.39×10^7 y	α 4.50, spon fiss		6.3	8	236
		6.75 d	β⁻ 0.24	(0.51)	5.5		237
		4.51×10^9 y	α 4.18, . . . spon fiss		6.0	2.8	238
		23.5 m	β⁻ 1.21	(1.28)	4.9	$\begin{bmatrix} 22 \\ \sim 12 \text{ (fiss)} \end{bmatrix}$	239
		14 h	β⁻ 0.36	(0.36)	5.8		240
						170 + 0.019 (fiss)	**Np**
		50 m	α 6.28				231
		~13 m	K				232
		35 m	K; α 5.53				233
		4.4 d	K; L; β⁺ 0.8			~900 (fiss)	234
		1.1 y	L; K; α 5.06				235
		{ 2.2 h	β⁻ 0.52; K	(0.52)	5.6	2800 (fiss)	236
		{ ≥5000 y	β⁻				
	±6	2.2×10^6 y	α 4.79; 4.52–4.87		6.9	$\begin{bmatrix} 170 \\ 0.019 \text{ (fiss)} \end{bmatrix}$	237
		2.10 d	β⁻ 1.26, 0.27, . . .	(1.30)	5.2	1600 (fiss)	238
		2.33 d	β⁻ 0.33–0.72	(0.72)	6.4	80	239
		{ 7.3 m	β⁻ 2.16	(2.16)	4.8		240
		{ 60 m	β⁻ 0.90	(2.06)			
							Pu
		36 m	K; α 6.58				232
							233
		9 h	K; α 6.19				234
		26 m	L; K; α 5.85		6.2		235
		2.7 y	α 5.75; . . . ; spon fiss		7.3		236
		40 d	K				237
		90 y	α 5.49, 5.45, . . . ; spon fiss			$\begin{bmatrix} 450 \\ 18 \text{ (fiss)} \end{bmatrix}$	238
±0.02	±0.4	2.43×10^4 y	α 5.15, 5.14, 5.10; spon fiss		5.7	$\begin{bmatrix} 300 \\ 730 \text{ (fiss)} \end{bmatrix}$	239
		6600 y	α 5.16, 5.12, . . . ; spon fiss		6.3	~510	240
±1.4		13 y	β⁻ 0.02; α 4.89, 4.85	(0.020)	5.6	$\begin{bmatrix} \sim 380 \\ \sim 1100 \text{ (fiss)} \end{bmatrix}$	241
		3.8×10^5 y	α 4.90, 4.85; spon fiss		6.1	23	242
		5.0 h	β⁻ 0.57, 0.48	(0.57)	5.2	~100	243
		8×10^7 y	α; spon fiss			~1.4	244
		11 h	β⁻			~260	245
		11 d	β⁻ 0.15, . . .				246
							Am
		~1 h	K; α 6.01				237
		1.9 h	K				238
		12 h	K; α 5.78				239
		47 h	K				240
+1.4	+4.9	470 y	α 5.48, . . . ; spon fiss			$\begin{bmatrix} 750 \\ 3.2 \text{ (fiss)} \end{bmatrix}$	241
		{ 16.0 h	β⁻ 0.62; K	(0.62)	5.4	~2500 (fiss) $\begin{bmatrix} \sim 4500 \\ \sim 3500 \text{ (fiss)} \end{bmatrix}$	242
		{ ~100 y	β⁻ 0.59, . . .; K; α				
+1.4	+4.9	~8000 y	α 5.27, 5.22, 5.17–5.34		6.4	82	243
		26 m	β⁻ 1.5; K	(0.15−)			244
		2.0 h	β⁻ 0.90				245
		25 m	β⁻ 1.22	(2.4)			246
							Cm
		2.5 h	K; α 6.50				238
		~3 h	K				239
		27 d	α 6.25; spon fiss				240
		35 d	K; α 5.95				241
		163 d	α 6.11, 6.07, . . . ; spon fiss			~20	242
		35 y	α 5.78, 5.73–5.99		5.7	~300	243

(1)	(2)	(3)	(4)	(5)	(6)	(7)	(8)	(9)
								Abundance
Z	Chem symbol	A	N	I	π	Ground-state config	$M\text{-}A$, MeV	Per cent	Cosmic
96	**Cm**	244	148				129.7(1)		
		245	149				132.3(3)		
		246	150				134.2(1)		
		247	151				137.3(3)		
		248	152				139.8(5)		
		249	153				143.0(7)		
97	**Bk** *Berkelium*	243	146				129.5(2)		
		244	147				131.8(2)		
		245	148				133.1(2)		
		246	149				135.7(3)		
		247	150				137.5(4)		
		248	151				140.3(4)		
		249	152				142.1(2)		
		250	153				145.9(2)		
98	**Cf** *Californium*	244	146				132.6(1)		
		245	147				134.6(3)		
		246	148				135.7(1)		
		247	149				138.2(3)		
		248	150				139.7(1)		
		249	151				142.0(3)		
		250	152				144.0(2)		
		251	153				147.1(4)		
		252	154				149.6(5)		
		253	155				152.6(7)		
		254	156				155.3(5)		
99	**E** *Einsteinium*	246	147				139.8(6)		
		247	148				140.5(2)		
		248	149				142.6(4)		
		249	150				143.6(4)		
		250	151				145.9(5)		
		251	152				147.7(5)		
		252	153				150.6(2)		
		253	154				152.4(2)		
		254	155				156.0(2)		
		255	156				158.1(7)		
		256	157				162.2(7)		
100	**Fm** *Fermium*	250	150				146.8(5)		
		251	151				149.0(6)		
		252	152				150.5(5)		
		253	153				152.9(4)		
		254	154				154.9(2)		
		255	155				158.0(4)		
		256	156				160.2(6)		
101	**Mv** *Mendeleevium*	256	155				161.9(7)		

(10) μ (nucl mag)	(11) Quad mom, 10^{-24} cm²	(12) Half-life ($T_{1/2}$)	(13) Decay — Type and energy	(Q)	(14) Binding energy of last neutron E_n, MeV	(15) σ_{tn}, 10^{-24} cm²	
							Cm
		18 y	α 5.80, 5.76; spon fiss		6.5	~15	244
		1.1×10^4 y	α 5.34			$\left[\begin{array}{c}\text{~200}\\ \text{1800 (fiss)}\end{array}\right]$	245
		4000 y	α 5.36; spon fiss			~15	246
		≫1 y				180	247
		4×10^5 y	α 5.05				248
		"short"	β⁻				249
							Bk
		4.5 h	K; α 6.55, 6.72, 6.20				243
		4.4 h	K; α 6.66				244
		50.0 d	K; α 6.17, 6.33, 5.90				245
		1.8 d	K				246
		7000 y	α 5.50, 5.67, 5.30				247
		~18 h	β⁻ 0.67; K				248
		290 d	β⁻ 0.09; α 5.40, 5.08; spon fiss			~500	249
		3.1 h	β⁻ 0.9, 1.9				250
							Cf
		25 m	α 7.17				244
		44 m	K; α 7.11				245
		36 h	α 6.75, 6.71, . . . ; spon fiss				246
		2.5 h	K				247
		225 d	α 6.26; spon fiss				248
		500 y	α 5.81, 6.00, . . . ; spon fiss			$\left[\begin{array}{c}\text{~270}\\ \text{~600 (fiss)}\end{array}\right]$	249
		10 y	α 6.02, 5.98; spon fiss			~1500	250
		~700 y	α			~3000	251
		2.2 y	α 6.11, 6.07; spon fiss			~30	252
		18 d	β⁻				253
		54 d	spon fiss			≤2	254
							E
		"short"	K				246
		7.3 m	α 7.35; K(?)				247
							248
		2 h	α 6.76				249
							250
		1.5 d	K; α 6.48				251
		~150 d	α 6.64				252
		20 d	α 6.64, 6.60, 6.55, . . . ; spon fiss			~200	253
		{ 37 h	β⁻ 1.1; K	}		≤15	254
		{ ~1 y	α 6.44; spon fiss				
		~30 d	β⁻			~40	255
		"short"	β⁻				256
							Fm
		30 m	α 7.43				250
							251
		30 h	α 7.1				252
		3 d	K				253
		3.4 h	α 7.22, 7.18, . . . ; spon fiss				254
		20 h	α 7.08			<100	255
		3.1 h	spon fiss				256
							Mv
		1 h	K				256

VI. Masses and Mean Lives of Elementary Particles

The following table lists the masses and mean lives of elementary particles.† (The antiparticles are assumed to have the same spins, masses, and mean lives as the particles listed.)

Particle	Spin	Mass (Errors represent standard deviation) (MeV)	Mass difference (MeV)	Mean life (sec)	
Photon					
γ	1	0	γ —	γ Stable	(r)
Leptons					
$\nu_e, \bar{\nu}_e; \nu_\mu, \bar{\nu}_\mu$	$\tfrac{1}{2}$	0	ν	ν Stable	
$e^\mp$	$\tfrac{1}{2}$	0.510976 ± 0.000007	$e^\mp$	$e^\mp$ Stable	
$\mu^\mp$	$\tfrac{1}{2}$	105.655 ± 0.010	$\mu^\mp$ $\big\}$ 33.93 ± 0.05	$\mu^\mp$ $(2.212 \pm 0.001) \times 10^{-6}$	(w)
Mesons					
π^+	0	139.59 ± 0.05	$\pi^\pm$	$\pi^\pm$ $(2.55 \pm 0.03) \times 10^{-8}$	(d)
π^0	0	135.00 ± 0.05	π^0 $\big\}$ 4.59 ± 0.01	π^0 $(2.2 \pm 0.8) \times 10^{-16}$	(h)
$K^\pm$	0	493.9 ± 0.2	K^+ $\big\}$ 3.9 ± 0.6	K^+ $(1.224 \pm 0.013) \times 10^{-8}$	
K^0			K^0	K^0 50% K_1, 50% K_2	
$\left.\begin{array}{c}K_1\\K_2\end{array}\right\}$	0	497.8 ± 0.6	K_1 $\big\}$ $(1.5 \pm 0.5)\hbar/\tau(K_1)$ K_2	K_1 $(1.00 \pm 0.038) \times 10^{-10}$	(e)
				K_2 $6.1(+1.6/-1.1) \times 10^{-8}$	(c)
Baryons					
p	$\tfrac{1}{2}$	938.213 ± 0.01	p $\big\}$ 1.2939 ± 0.0004	p Stable	
n	$\tfrac{1}{2}$	939.507 ± 0.01	n	n $(1.013 \pm 0.029) \times 10^{3}$	(y)
Λ	$\tfrac{1}{2}$	1115.36 ± 0.14	Λ —	Λ $(2.51 \pm 0.09) \times 10^{-10}$	(u)
Σ^+	$\tfrac{1}{2}$	1189.40 ± 0.20	Σ^+	Σ^+ $0.81(+0.06/-0.05) \times 10^{-10}$	(m)
Σ^-	$\tfrac{1}{2}$	1197.4 ± 0.30	Σ^- $\big\}$ 4.45 ± 0.4	Σ^- $1.61(+0.1/-0.09) \times 10^{-10}$	(o)
Σ^0	$\tfrac{1}{2}$	1193.0 ± 0.5	Σ^0	Σ^0 $<0.1 \times 10^{-10}$	(s)
Ξ^-	?	1318.4 ± 1.2	Ξ^- —	Ξ^- $1.28(+0.38/-0.30) \times 10^{-10}$	(f)
Ξ^0	?	1311 ± 8	Ξ^0 —	Ξ^0 1.5×10^{-10} (1 event)	(q)

† From Walter H. Barkas, Arthur H. Rosenfeld, University of California Radiation Laboratory report UCRL-8030. Published courtesy the University of California Lawrence Radiation Laboratory, Berkeley, California.

VII. The Lorentz Transformation and Some Relativistic Kinematics Formulas

Let the system of coordinates S' move along the z-axis with velocity $\bar{v} = \bar{\beta}c$ with respect to the system S. Let p_x, p_y, p_z, and E be the momentum and energy of a particle of rest mass m moving in frame S and let p_x', p_y', p_z', and E' be the momentum and energy of the same particle in frame S'. The Lorentz transformation relates these quantities.

$$\bar{\beta} = \bar{v}/c \qquad \bar{\gamma} = 1/\sqrt{1 - \bar{\beta}^2}$$

$$p = \sqrt{p_x^2 + p_y^2 + p_z^2} \qquad\qquad p' = \sqrt{p_x'^2 + p_y'^2 + p_z'^2}$$

$$\tan \theta = \left(\sqrt{p_x^2 + p_y^2}\right)/p_z \qquad\qquad \tan \vartheta = \left(\sqrt{p_x'^2 + p_y'^2}\right)/p_z'$$

$$E^2 = p^2c^2 + m^2c^4 \qquad\qquad E'^2 = p'^2c^2 + m^2c^4$$

$$\beta = cp/E \qquad \gamma = E/mc^2 \qquad\qquad \beta' = cp'/E' \qquad \gamma' = E/mc^2$$

$$cp_{||} = cp_z = \bar{\gamma}(cp' \cos \vartheta + \bar{\beta}E') \qquad\qquad cp_{||}' = cp_z' = \bar{\gamma}(cp \cos \theta - \bar{\beta}E)$$

$$p_x = p_x' \qquad\qquad p_x' = p_x$$

$$p_y = p_y' \qquad\qquad p_y' = p_y$$

$$E = \bar{\gamma}(E' + \beta cp' \cos \vartheta) \qquad\qquad E' = \bar{\gamma}(E - \bar{\beta}cp \cos \theta)$$

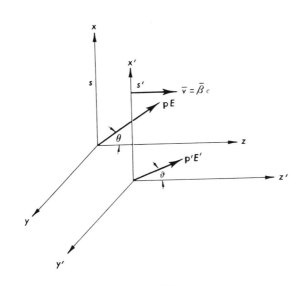

Quantities obtained by forming the scalar product of two 4-vectors, or of a 4-vector with itself are invariant under Lorentz transformations. Such invariants frequently lead to the desired transformation relation in a very simple fashion. As examples we mention:

(1) Rest mass $\{0; imc^2\}^2 = \{\mathbf{cp}; iE\}^2$ leading to $-m^2c^4 = c^2p^2 - E^2$.

(2) Consider two particles 1, and 2 of energy E_1 and E_2 and momentum $\mathbf{p_1}$ and $\mathbf{p_2}$ colliding. Then the *invariant* mass μ of the system is given by

$$c^4\mu^2 = (E_1 + E_2)^2 - c^2(\mathbf{p_1} + \mathbf{p_2})^2$$

The velocity of the system (center of mass velocity) is

$$\beta = c(\mathbf{p_1} + \mathbf{p_2})/(E_1 + E_2) \qquad \gamma = (E_1 + E_2)/\mu c^2$$

The above expressions can be generalized to an arbitrary number of particles.

Another invariant is

$$E_1E_2 - c^2p_1p_2 \cos \theta_{12}$$

(A) Similarly,

$$\frac{d^2\sigma}{d\Omega\, dE} \frac{1}{p}$$

is invariant.

INDEX